W9-ALW-786

# PROBABILITY and STATISTICS

JULIUS R. BLUM
Professor of Mathematics and Statistics
University of New Mexico

JUDAH I. ROSENBLATT
Professor of Information Science, Mathematics and Statistics
Case Western Reserve University

W. B. Saunders Company · PHILADELPHIA · LONDON · TORONTO

W. B. Saunders Company: West Washington Square
Philadelphia, Pa. 19105

12 Dyott Street
London, WC1A 1DB

833 Oxford Street
Toronto 18, Ontario

Probability and Statistics ISBN 0-7216-1763-8

 Made in the United States of America. Press of W. B. Saunders Company. Library of Congress Catalog card number 75-176202. AMS Subject Classification number 60-01; 62-01.

Print No.: 9 8 7 6 5 4 3 2

# Preface

A worthwhile problem requires carefully developed theory and, conversely, good theory acquires value through meaningful applications. That is the theme of this book. Probability and statistics have elegant and worthwhile mathematical foundations, yet they would become quite sterile if they failed to receive the continuing stimulation supplied by modern problems.

Much of today's technology is concerned with observation and prediction of measurements. The practical ability to make satisfactory predictions from measurements already made requires adequate theory. In fact, it is only within the framework of a theory that data have any meaning. We have tried to illustrate this point by means of numerous carefully constructed examples drawn from a great variety of situations. Our hope is that when the reader is later confronted with practical problems he will have the knowledge and experience to develop a suitable theory in which to construct his solutions.

Today there are many fine books in probability and statistics oriented toward mathematics and many applied texts in this area. To the best of our knowledge, no single book at this level is sufficiently oriented toward combining theory with applications. We have attempted to fill this gap and the reader may judge whether we have succeeded.

### *Scope*

This text was written for use at the junior or senior level. The first seven chapters can be covered in a full semester, as can Chapters 8 through 11. Chapters 12–14 can themselves constitute a full third semester. Chapter 14 is written at a somewhat more sophisticated level but requires no further mathematical background.

### *Organization*

The first eight chapters are devoted to probability and the last six to statistics, since a firm background in probability is required for an understanding of statistical inference. However, in Chapter 1 we introduce and illustrate the kinds of problems that are treated in both probability and statistics.

Generally the exercises are included to test understanding of the concepts. The problems are of a more challenging or more peripheral nature. Exercises and problems with a small circle, e.g., 2.°, are those for which a full solution is provided in the student solutions manual.

Sections which are in small print should probably be omitted on first reading. They contain some of the more technical proofs and optional topics.

There are two indices—the normal one for principal subjects, and a special topics index, useful for finding certain specific examples and models of importance.

JUDAH ROSENBLATT
JULIUS R. BLUM

# Acknowledgments

Our greatest debt goes to Lisa Rosenblatt, who not only typed the manuscript but read it critically from the viewpoint of a student new to the subject, and did *all* of the exercises and problems.

Professors Michael Gemignani, David Moore and David Adorno read the manuscript critically and made many worthwhile suggestions.

Many students contributed in no small measure, sometimes by their blank looks, and sometimes, as in the case of Steve Levin, by many searching and penetrating questions.

We are also indebted to our copy editor, Carolyn Buckwalter, who frequently corrected our errant syntax and generally made the text much more pleasant to read.

Finally we thank George Fleming and Carlos Puig, our editors, and all the others at W. B. Saunders Company who accepted so many collect telephone calls.

J. R.
J. R. B.

# Contents

# 1

# Introduction and Motivation

## SECTION 1
## Mathematical Models for Scientific Phenomena and Experiments

Much of modern science, social as well as physical, is concerned with the description and prediction of measurements. The problem of determining the effects of atmospheric pollution, for instance, involves predicting many future measurements—like the population mortality rate—from various other measurements that provide input data—like the carbon monoxide level.

In order to arrive at adequate descriptions, we perform experiments: We make measurements under conditions that we attempt to control as much as possible. Instead of trying to give a precise definition of an experiment (a task that is probably impossible), we shall list what we consider its most relevant features:

(i) the *inputs*,
(ii) the *actions* to be performed, and
(iii) the *possible outcomes*.

By the *inputs* we mean the equipment, materials (including personnel if need be), environment, and "input" data associated with the experiment. The *actions* constitute the actual performance of the experiment, and the *possible outcomes* form an a priori list of all conceivable results of the experiment. The necessity for listing the possible outcomes in advance can be thought of as a measuring instrument limitation: if a meter has a scale from 0 to 100, then the outcomes measured by this instrument must be in the interval [0; 100]. The only additional outcomes possible might be "off-scale" and "blown meter." Once an experiment is described in this way, any suitably trained person should be capable of performing it.

### *1.1 Example: Galileo's Experiment on Falling Bodies*

Galileo is said to have demonstrated that the speed of a falling body does not depend upon its weight. He simultaneously dropped two bodies of the same size but different weights from the Leaning Tower of Pisa and noted their simultaneous arrival on the ground. In describing this experiment we might take the *inputs* to be the two bodies of different weights, the tower, the ground, and Galileo. The *actions* are the simultaneous dropping of the objects and the measuring of impact times. The *possible outcomes* then correspond to ordered pairs of possible impact times, where the first element corresponds to the first body and the second element to the second one. The set of all such outcomes would of course be limited by the accuracy of the available timepieces and any human limitations (such as an observation period not exceeding 12 hours) imposed by Galileo. △

*Ideally*, an experiment should be repeatable under identical conditions as many times as is desirable. We say "ideally" because in practice we can never repeat an experiment exactly. Some conditions, such as location or time, must change. For this reason, we may classify as experiments observations of a kind that permit almost no control of experimental conditions. For example, measurement of sunspot activity, measurement of pollution levels and measurement of the gross national product might constitute experiments.

By a *model* of an experiment we mean any of its many possible mathematical descriptions. More generally, a model of any physical situation is just a mathematical description, or in other words, a theory.

If we have an adequate model or theory of the physical situation, it is not necessary to experiment. For instance, suppose we want an accurate determination of how much gasoline a car will use on a 1000-mile trip. If we know that the gas mileage of the car is about 12 miles per gallon, then we need only do a simple computation to obtain our answer. Suppose, however, we only know that the car's gas mileage is between 7 and 20 miles per gallon. If we want a better estimate of the gasoline usage then we would first want to do some experimenting to narrow down the range of possibilities. In most scientific investigations as in the simple situation we have just described, we are dealing with some *class* of reasonable models, and the purpose of experimentation is to narrow down this class.

We want to emphasize that the model applied to a given physical situation depends on the intended use of this model. For instance, if we are interested in purchasing a timepiece, our description of its error might involve only three possible outcomes; runs accurately, runs too fast, runs too slow. However, if we are interested in carefully calibrating this timepiece, the possible outcomes might consist of all integers (positive and negative) corresponding to the number of seconds lost or gained in one day. Since our descriptions are usually for the purpose of predicting or describing future measurements, we have no interest in "true" theories; our only interest is in adequate ones.

### *1.2 Example: Geometry*

In laying out plans for new streets, we would probably describe positions by means of a two-dimensional rectangular coordinate system. Such a description of positions on the earth's surface is convenient, simple, and usually adequate for small regions. In planning shipping lanes on the ocean, however, we would use the description furnished in spherical trigonometry, because the error in using a two-dimensional rectangular coordinate system would be too great. △

This example illustrates that there is usually no single theory that is best for all experiments to which it might apply. Which model we use depends on the state of our knowledge, the results we desire, convenience, and so forth. All theories are merely simplified, idealized descriptions.

---

### *1.3 PROBLEMS*

In each of the following experiments, decide what you would consider to be the *inputs*, *actions*, and *possible outcomes*. Since the answers you give should correspond to your views of the purposes of the experiments, your answers may not be the same as those furnished. Thus, you might also indicate a purpose for which your answer is reasonable.

1. Toss a coin one time.

2.° Toss a pair of dice one time.

3.° Draw one card from a deck of 52 distinguishable cards.

4. Determine the accumulated error made by a clock over a 24 hour time period.

5.° Record the temperature at noon at the airport.

6. Record the temperature at noon and at midnight at the airport.

7.° Record the temperature over a full 24-hour day at the airport.

8. Record wind speed at a fixed time and location.

9.° Record wind velocity at a fixed time and location.

10. Record wind velocity for a full day at a given location.

11°. Record customer arrivals at a given store over a given one-hour time period.

12. Measure the antibody level for some disease in a given person.

13.° Toss a coin until the first head occurs.

## SECTION 2
## Deterministic and Non-Deterministic Descriptions

In our discussion of models for experiments, no mention was made of how we describe the relations among various parts of an experiment, such as input data and outcomes. These descriptions may be rather simple or quite complicated.

### 2.1 *Example: Constant Speed*

The distance $d$ (output) covered by an auto traveling at a constant speed $s$ for a time interval of duration $t$ is very simply related to the inputs $s$ and $t$ by the equation $d = st$. △

### 2.2 *Example: Earth's Orbit*

The "exact" orbit (output) of the earth around the sun in the presence of the other planets and asteroids is a very complicated function of the masses, positions, shapes, and velocities (inputs) of these bodies. △

Notice that a common feature of these examples is that if we are given precise inputs, then the outputs can be determined precisely. A model with these features—that is, a description in which *precisely determined inputs yield precisely determined outputs*—is called a *deterministic model.*

There are many situations, however, in which a deterministic model is neither feasible nor desirable, for economic or other reasons. A few examples will illustrate this point.

### 2.3 *Example: Tossing a Coin*

Consider a coin toss in which the possible outcomes are heads ($H$) and tails ($T$). For one such toss, it appears that available measurements of the inputs do not allow any degree of certainty in predicting whether $H$ or $T$ will result. △

*Remark.* Despite appearances to the contrary, it is not the habit of most statisticians to spend any appreciable time or effort tossing coins. Coin tossing is simply the statistician's way of describing a non-deterministic experiment with two possible outcomes. These outcomes can be called *success* and *failure* or 1 and 0 rather than heads and tails. Similar remarks hold for dice tossing and drawing from urns.

### 2.4 *Example: Transistor Life*

Suppose we wish to determine the length of life of a given transistor. We assume that we have some criterion for determining proper functioning of such a unit. The only way to determine the lifetime exactly is to run the transistor until it ceases to function properly. Unfortunately, such a test suffers the disadvantage of rendering the transistor useless. On the other hand, it would be reasonable to predict something about the life of this transistor from knowledge of the life of other transistors manufactured under nearly identical conditions. This prediction might rest on data gathered from transistors in actual use or from life tests made on units chosen for this purpose. △

In both of the preceding examples, it is impossible to predict the actual outcome very accurately when only the given inputs and actions are known. A model with these features—that is, one in which even *exact knowledge of the inputs and actions does not allow exact prediction of the outcome*—is called *non-deterministic*.

Under some circumstances, we will prefer a deterministic model. Under other circumstances, we might prefer or at least be satisfied with a non-deterministic description of the same phenomenon. The following example illustrates how our description depends on its intended use.

## 2.5 *Example: Calibration of Wristwatch*

If you own a quality wristwatch, an adequate description of its behavior might be a deterministic model of the type

*watch gains $s$ seconds per hour,*

where $s$ is some rational number (possibly negative). The number $s$ might be determined by "checking" your watch against "correct time" at two successive hours. For purposes of keeping appointments during a 12-hour period, this procedure would probably provide excellent results, even though vibrations and atmospheric changes cause variations that prevent perfect prediction. However, if it is vitally important to avoid being late for the train to your office, then a non-deterministic description involving an average gain per unit time *and an upper bound for error* would seem preferable. △

We remark that frequently an extremely simple deterministic model is adequate for purposes in which great accuracy is not required. When small deviations from predicted values become significant, then we shift to the simplest model (deterministic or non-deterministic) that we feel is adequate.

---

### 2.7 *PROBLEMS*

In the following situations, determine whether a deterministic or a non-deterministic description is more appropriate. As is often the case in realistic problems, there are no clearly "right" or "wrong" answers. Rather, the judgment of whether the model is adequate must be made subjectively and only after comparison has been made between predictions and the actual measurements. More will be said about such judgments when we come to the subject of statistics in Chapter 9.

1.° At an airport, the number of scheduled airline arrivals over one-hour time periods for a full year is examined to determine whether the number of runways is adequate:

(a) in Cleveland, Ohio (to which planes may be diverted from nearby airports that have been closed because of bad weather).

(b) in Tucson, Arizona (a relatively isolated city).

2.° The number of patients arriving at a hospital emergency room is recorded for a period of a month to see whether the number of doctors on duty should be changed.

3. The gas mileage of your car on the highway is measured:
(a) to determine when to stop at gas stations, given the amount of gas in your tank,
(b) to estimate the cost of a trip.

4.° To determine whether the engine overheats under various conditions, the water temperature in a car's cooling system is measured after the car has been idling in 85-degree shade for two hours, with the air conditioning on:
(a) with no people in the car.
(b) with three given people in the car, assuming the air conditioner is controlled by a thermostat.

5. We select 1000 people in a representative way from the population and question them:
(a) for the purpose of determining the potential sales of electric cars.
(b) for the purpose of ascertaining who will win the next election.

6.° Cars are driven into a brick wall at 20 miles per hour to determine how much dollar damage is done to them.

7.° Using radar, we measure the velocity of a space vehicle traveling at constant velocity, with the intention of predicting its position:
(a) in the distant future.
(b) in the near future.

8. The temperature of boiling water is measured at sea level and at various given heights above sea level:
(a) to determine a general relation between altitude and boiling temperature.
(b) to determine a setting on a stove that permits as high a temperature as possible without boiling over.

9. The precipitation at a given location is measured for a full year in an attempt to predict future yearly precipitation in this area.

## EXPERIMENTS

Most people do not have a particularly well-developed intuition for non-deterministic situations, despite the fact that much of our experience is non-deterministic. A few simple experiments may help in this development: Predict the results, perform the experiments, and compare the predictions with reality.

1. Toss a nickel 150 times and made a graph plotting the number of tosses versus the number of heads.

2. Toss two dice 200 times (or 200 pairs of dice once) and record:

the proportion of 2's (sum = 2)

the proportion of 3's (sum = 3)

.
.
.

the proportion of 12's (sum = 12).

3. Pick a stock and try to predict each day whether the stock price will rise or fall. Plot a graph over a 60-day period of the proportion of correct guesses. (That is, on the $k$th day plot the proportion of the first $k$ guesses that are correct.) You may use back issues of a newspaper.

## SECTION 3
## The Phenomenon of Statistical Regularity

Consider a coin tossing experiment. If we perform this experiment only once, then all we can predict with certainty is that the outcome will be either heads or tails (disregarding any edge landings). However, it has been observed that the proportion of heads in a large number of coin tosses tends to be near 1/2.

More generally, it has been observed that as the number of repetitions of an experiment increases, the proportion of times a particular phenomenon appears seems to settle down to some specific number. For instance, the proportion of rainy days in a given location, the proportion of "even sum" outcomes in tosses of a pair of dice, and the proportion of output transistors that last more than a year under a given kind of use all seem to settle down. This "settling down" property as the number of repetitions increases is called *statistical regularity*. A situation displaying statistical regularity falls somewhere between a completely deterministic situation and one in which it is beyond our ability to make any kind of useful prediction. One example of the latter is the long-range behavior of the stock market. Unforeseeable events have an unpredictable and often significant influence on market behavior.

In this book we shall be concerned primarily with non-deterministic experiments that exhibit the phenomenon of statistical regularity. In the next section we begin development of the mathematical models appropriate to such experiments. We conclude this section with a few examples in which statistical regularity seems to be present.

### *3.1 Example: Tire Life*

The proportion of passenger automobile tires used on a given type of automobile that last at least $t$ hours appears to be fairly stable. More precisely, in the experiment of measuring the lifetime of a specific type of tire on a given type of automobile, the phenomenon $L_t$, that the tire lasts at least $t$ hours, seems to exhibit statistical regularity. △

### *3.2 Example: Birth Records*

In the United States, the proportion of males born in any hospital each year seems to be nearly constant at slightly over 0.5. Thus, in the experiment of successively determining the characteristics of each child born in any hospital, the phenomenon "male" seems to exhibit statistical regularity. On the other hand, because of the changing types of drugs used by the population, we might not expect the phenomenon "birth defects" to exhibit statistical regularity. △

### 3.3 *Example: Telephone Calls*

Trunk lines are usually designed so that although overloads might occur, they occur only rarely. It is reasonable to believe that this can be done in view of the apparent statistical regularity of the phenomenon $C_{p,n}$: that for at most $p$ per cent of the business day more than $n$ phone calls are in progress. One experiment in which this phenomenon can be observed consists of keeping a tally of the number of phone calls $N(t)$ in progress at time $t$. △

---

### 3.4 *PROBLEMS*

In each of the following partially described experiments, under what additional assumptions would you feel it reasonable to postulate statistical regularity of the given phenomenon? (Answers to these questions may vary because of different past experiences.) Note that you expect statistical regularity if you feel that conditions are not changing radically from trial to trial.

1.° Examining records of total rainfall in a 24-hour period. The phenomenon of interest is that there is measurable rainfall.

2.° Counting the number of cars passing a given intersection in a 12-hour period of a normal summer day. The phenomena of interest are of the type $L_n$, that at least $n$ cars pass this intersection.

3. Counting the number of people and automobiles in the United States each year. The phenomenon of interest is that the ratio $P/C$ of people to cars lies in an interval $[a, b]$.

4. Measuring the number of atoms that decay in a block of pure uranium-235 in 10 seconds. The phenomenon of interest is that more than $n$ atoms decay.

5.° Measuring the average costs of food and entertainment for a typical family each month. The phenomenon of interest is that the ratio $f/e$ of food cost to entertainment cost lies in some interval $[a, b]$.

6. Determining the weekly gas mileage of a car. The phenomena of interest are of the form $L_m$, that at least $m$ miles per gallon is achieved.

7.° Measuring the amount of time gained each day on a watch. The phenomena concerning us are of the form $G_{[a,b]}$, that the amount gained is in some interval $[a, b]$ (here $a$ or $b$ or both $a$ and $b$ could be negative).

8. What is an expression for the proportion of the business day (8 A.M. to 12 noon, 1 P.M. to 5 P.M.) during which the number of phone calls in progress exceeds $n$? Hint: In Example 3.3, let

$$Q_n(t) = \begin{cases} 1 & \text{if} \quad N(t) > n \\ 0 & \text{otherwise.} \end{cases}$$

Draw a picture of how $Q_n$ might look.

## SECTION 4
## Experiments with Finitely Many Possible Outcomes. The General Finite Model

In practice, the limitations of our measuring instruments allow only a finite number of possible outcomes for any given experiment. Later we shall see that it is often convenient, and in a sense necessary, to investigate models with infinitely many possible outcomes. In this section, however, we shall consider experiments with finitely many possible outcomes. The most general such model is developed as follows: After deciding what phenomena are significant, we choose some positive integer $n$ and distinguish $n$ possible outcomes, which we label $e_1, e_2, \ldots, e_n$. Each of the possible outcomes $e_j$ is called an *elementary event*, and the set $\mathcal{S}$ of elementary events is called the *sample space* for the experiment. An outcome $e_j$ also called a *point* of $\mathcal{S}$.

When we repeat the experiment $k$ times, we can compute the number of times $k_j$ that the outcome $e_j$ has resulted. Note that

$$k_1 + k_2 + \cdots + k_n = k.$$

If the experiment has been performed $k$ times, then precisely one of the $n$ possible outcomes must result on each repetition. We observe that $k_j/k$ is the *proportion* of times, in our $k$ repetitions, that the outcome $e_j$ has resulted. Because we are concerned solely with phenomena that exhibit statistical regularity, we assume that as $k$ grows large, each of the proportions $k_j/k$ approaches a definite number, which we denote by $p_j$ in our mathematical model. We call $p_j$ *the probability of the elementary event* $e_j$. Our interpretation of $p_j$ (the so-called *frequency interpretation*) leads to the conditions that

$$0 \le p_j \le 1 \qquad \text{for each } j \qquad \text{and} \qquad p_1 + p_2 + \cdots + p_n = 1.$$

### 4.1 *Example: Number of Heads in Two Coin Tosses*

Suppose we are interested in the total number of heads obtained in tossing two fair coins. Then a natural choice for the sample space is the set $\mathcal{S} = \{0, 1, 2\}$; 0, 1, and 2 being the elementary events, corresponding to 0, 1, or 2 heads in two tosses. We shall see later that it is reasonable to assign probabilities as follows:

| Elementary event $e_j$ (number of heads) | 0 | 1 | 2 |
|---|---|---|---|
| Assigned probability $p_j$ | 1/4 | 1/2 | 1/4 |

△

### 4.2 *Example: Choosing from a Population with Known Proportions*

In the presidential election of 1948, there were three main candidates: Truman, Dewey, and Henry Wallace. Denote by $p_T$, $p_D$, and $p_W$ the proportions of voters who voted for the respective candidates; let $p_M$ be the proportion who voted for one of the minor candidates. Suppose we select a voter and record his choice for president. In this case we might let the sample space consist of the set $\{T, D, W, M\}$. If care is taken to ensure that each voter is as likely to be chosen as any other ("random sampling"), then the probability of $T$ is $p_T$, and similarly for the probabilities of $D$, $W$, and $M$. △

In most instances we are not especially interested in which particular elementary event has resulted. We are usually concerned with some phenomenon that occurs when the actual outcome is one of a *specified set* of possible outcomes.

### 4.3 *Example: "Reds" at Roulette*

The game of roulette, as played in certain countries, has thirty-eight possible outcomes, 1, 2, . . . , 36, 0, 00. Of the outcomes 1, 2, . . . , 36, eighteen are colored red and eighteen are black; the outcomes 0 and 00 are colored green. We may bet on "reds," and we win if the actual outcome is one of the red ones. If we bet on "reds," we do not care which number actually comes up, so long as it is red. △

### 4.4 *Example: Landslide Election*

In a given election involving two candidates, we might choose the sample space to be the set of integers $\{0, 1, \ldots, 100\}$, where the outcome $j$ means that a given candidate received $j$ per cent of the vote (rounded up to the nearest integer). A candidate may be interested to know whether the outcome was in the set $\{60, 61, \ldots 100\}$ (landslide); he is definitely interested in whether the outcome was in the set $\{51, 52, \ldots, 100\}$ (victory). △

These examples lead us not only to consider the probabilities of elementary events, but also to define the probability of any subset of the sample space. Let $A$ be any subset of the sample space, a collection of some (generally not all) elementary events. We will call $A$ an *event* and define the *probability of* $A$ to be the sum of those $p_j$'s for which $e_j$ is in $A$. We denote the probability of the event $A$ by $P(A)$ and interpret $P(A)$ as the limiting proportion of observed outcomes in $A$ (the limiting proportion of outcomes on which $A$ *occurs*). A bit of reflection should convince you that if our interpretation of $p_j$ is applicable, then so is this interpretation of $P(A)$. Note that for each event $A$

$$0 \leq P(A) \leq 1$$

and

$$P(\mathcal{S}) = 1,$$

where $\mathcal{S}$ represents the entire sample space.

### 4.5 *Example: Dice*

If we toss two dice and are interested in the sum that appears, it seems natural to select as the sample space the set $\mathcal{S} = \{2, 3, 4, 5, 6, 7, 8, 9, 10, 11, 12\}$. In the next chapter we shall see that the following probability assignment is reasonable.

| Elementary event | 2 | 3 | 4 | 5 | 6 | 7 | 8 | 9 | 10 | 11 | 12 |
|---|---|---|---|---|---|---|---|---|---|---|---|
| Probability | 1/36 | 2/36 | 3/36 | 4/36 | 5/36 | 6/36 | 5/36 | 4/36 | 3/36 | 2/36 | 1/36 |

The event "7 or 11," that is, the subset $\{7, 11\}$, has probability $p_7 + p_{11} = 6/36 + 2/36 = 8/36$. The event "even sum," that is, the subset $\{2, 4, 6, 8, 10, 12\}$, has probability $p_2 + p_4 + p_6 + p_8 + p_{10} + p_{12} = 1/36 + 3/36 + 5/36 + 5/36 + 3/36 + 1/36 = 18/36 = 1/2$. If one die is colored green and the other is colored red, then the phenomenon "1 on the green die" does not correspond to a subset of the given sample space $\mathcal{S}$. △

### 4.6 *Example: Human Height*

In a homogeneous human adult population, it appears that an adequate description of the probability $p_j$ that a person's height is between $j$ and $j + 1$ inches is furnished by the formula

$$p_j = \frac{\displaystyle\int_j^{j+1} \frac{1}{\sqrt{2\pi\sigma^2}} e^{-(x-\mu)^2/2\sigma^2}\, dx}{\displaystyle\int_{36}^{96} \frac{1}{\sqrt{2\pi\sigma^2}} e^{-(t-\mu)^2/2\sigma^2}\, dt}, \qquad j = 36, 37, \ldots, 95,$$

where $\sigma^2$ and $\mu$ are determined by the particular population under consideration. (In Chapter 7 we shall investigate certain theorems that make this description theoretically reasonable.) The event "height at least 6 feet" has probability

$$p_{72} + p_{73} + \cdots + p_{95} = \frac{\displaystyle\int_{72}^{96} \frac{1}{\sqrt{2\pi\sigma^2}} e^{-(x-\mu)^2/2\sigma^2}\, dx}{\displaystyle\int_{36}^{96} \frac{1}{\sqrt{2\pi\sigma^2}} e^{-(t-\mu)^2/2\sigma^2}\, dt}. \qquad △$$

We now extract the basic *mathematical content* of the previous discussion.

**4.7 Definition: Finite Probability System.** A *finite probability system* consists of a finite set $\mathcal{S} = \{e_1, e_2, \ldots, e_n\}$, together with numbers $p_1, \ldots, p_n$ such that $0 \le p_j \le 1$ for each $j$ and $p_1 + p_2 + \cdots + p_n = 1$. The set $\mathcal{S}$ is called the *sample space* for the associated experiment, and the $e_j$ are its *elementary events*. For each $j$ the number $p_j$ is called *the probability of the elementary event* $e_j$. If $A$ is any subset of $\mathcal{S}$, we define the *probability*, $P(A)$, of $A$ as the sum of those $p_j$ for which $e_j$ is in $A$. A subset $A$ of $\mathcal{S}$ is called an *event*. (Strictly speaking, for consistency we must define $p_j$ as the probability of the set $\{e_j\}$ whose only element is $e_j$, rather than as the probability of $e_j$. We will disregard this subtle distinction henceforth.)

We interpret $\mathcal{S}$ to include the collection of all possible outcomes of an experiment; in many repetitions of the experiment, $p_j$ will be interpreted as the limiting proportion of occurrences of $e_j$, and $P(A)$ as the limiting proportion of outcomes in $A$. Although this interpretation suggests many interesting results, we should not assume that it proves any mathematical results. Rather, we employ such an interpretation to suggest results that should be provable if our mathematical model is sufficiently complete. If a result is invalid for a given mathematical model when we have good reason to expect it to hold, then we may have to extend or alter the model.

For example, let $A$ and $B$ be two events that cannot occur simultaneously on one performance of the experiment. Let $A \cup B$ denote the union of $A$ and $B$, the collection of all elementary events of $A$ and $B$. The frequency interpretation certainly suggests that the relation

$$(*) \qquad P(A \cup B) = P(A) + P(B)$$

should hold. However, it would be incorrect to consider the frequency interpretation to be a proof. Rather, if

$$A = \{e_1, \ldots, e_j\} \qquad \text{and} \qquad B = \{e_{j+1}, \ldots, e_k\},$$

then we know from our definition of a probability system that (*) holds, because

$$P\{A \cup B\} = p_1 + \cdots + p_{j+k} \qquad \text{and} \qquad P(A) = p_1 + \cdots + p_j,$$
$$P(B) = p_{j+1} + \cdots + p_k.$$

---

### *4.8 EXERCISES*

1. In the dice example (4.5), describe the following phenomena as events (subsets of the sample space) and compute their probabilities:
   (a) odd sum.
   (b)° sum a perfect square.
   (c) sum exceeding 7.

2.° Suppose that clothing for people whose heights are outside the range from 5 feet to 6 feet 6 inches must be custom tailored. If the model for height distribution given in Example 4.6 is used, what is the probability that a person chosen from the population will require custom tailoring?

3. In simulating experiments we often make use of so-called *random numbers*. A particular model for choosing a number at random (from the numbers 0 to 99) would be to let $\mathcal{S} = \{0, 1, \ldots, 99\}$, where each elementary event has probability 1/100. In this

case, compute the probabilities of the following phenomena concerning a randomly chosen number:

(a)° the number is even.
(b) the number is a perfect square.
(c) the number is a perfect cube.
(d)° the number is in the set $\{0, 1, \ldots, 24\}$.

4. Suppose that you had at your disposal an experiment described by the model in the preceding problem. Show how, by performing only the experiment of that problem, you can obtain the same effect as that of an experiment with the following sample space and probabilities:

(a) $\{e_1, e_2, e_3\}$, $p_1 = .1$, $p_2 = .6$, $p_3 = .3$.

Hint: Note that if $A = \{0, 1, \ldots, 9\}$, then $P(A) = .1$. That is, we can think of $e_1$ of our new experiment as occurring when $A$ occurs in the original experiment.

(b)° Sample space $\{e_1, e_2, \ldots, e_n\}$, where $1 \leq n \leq 100$, $e_1, \ldots, e_n$ are just labels and $p_1, \ldots, p_n$ are given two-digit decimals whose sum is 1; for example, $n = 4$, $p_1 = .11$, $p_2 = .45$, $p_3 = .20$, $p_4 = .24$.

5. If three dice are tossed and the outcome is taken to be the sum on the three faces then we may let $\mathcal{S} = \{3, 4, 5, 6, 7, 8, 9, 10, 11, 12, 13, 14, 15, 16, 17, 18\}$. We shall see later that a reasonable probability assignment is the following:

| Elementary event | 3 | 4 | 5 | 6 | 7 | 8 | 9 | 10 | 11 | 12 | 13 | 14 | 15 | 16 | 17 | 18 |
|---|---|---|---|---|---|---|---|---|---|---|---|---|---|---|---|---|
| Probability | 1/216 | 3/216 | 6/216 | 10/216 | 15/216 | 21/216 | 25/216 | 27/216 | 27/216 | 25/216 | 21/216 | 15/216 | 10/216 | 6/216 | 3/216 | 1/216 |

With this assignment, compute the probabilities of the following:

(a)° an odd sum.
(b) the sum is not a perfect square.
(c)° the sum differs from 10.5 by more than 4 (in either direction).
(d) all three faces are 1's.
(e)° all three faces are the same.
(f)° all three faces are different.

## SECTION 5
## Problems of Probability, Problems of Statistics

If you compute the probability of at least 600 heads in 1000 tosses of a fair coin, you are solving a problem in probability. If you attempt to determine whether or not a coin is fair by observing the result of 1000 tosses, then you have posed a statistical problem. What distinguishes these two examples?

As we shall see later, the probability system (see Definition 4.7) involved in the first problem is completely determined by the assumption that the coin is "fair." The problem becomes one of *computing* (with the aid of certain further assumptions) the probability of a somewhat complicated event from the known probabilities of heads on any one toss. Recall that a probability system is specified only when we are given both the sample space and the assigned probabilities. With this in mind, we see that in the second problem the probability system is not completely specified, since the probability $p$ of heads on any one toss is known only to satisfy $0 \leq p \leq 1$. Thus, our problem becomes one of drawing some reasonable conclusions about $p$.

Typically, these are the relevant features that distinguish probability from statistics. In probability theory, we are given the probabilities of all events in some given class, and our task is to *compute* probabilities (or approximations, or bounds for probabilities) of events related to those given. In statistics, we are given a *family* of probability systems, among which there is assumed to be an adequate model for the situation of interest. Here our task is to "narrow down" this family on the basis of the observed data. One ultimate goal of our efforts is to use this refinement for prediction.

Because the theory of statistics rests completely on probability theory, we shall develop probability first. In Chapter 9 we will begin a systematic study of statistical theory. In most actual scientific investigations, problems in both statistics and probability arise.

### 5.1 *Example: Radioactivity*

Radioactivity level (in decays, or atomic splittings, per second) is often described by means of a probability model because the *exact* number of decays in any given time period in a radioactive block of material seems to defy exact prediction, yet appears to exhibit statistical regularity. The initial determination of a reasonable description of the observed decay behavior in a block of a pure radioactive substance is a statistical one. To be specific, the determination of something about the probability of at least $k$ decays in a time interval of length $\ell$ is a statistical problem. On the other hand, if we are given a block of a pure radioactive substance whose properties are well known, such as uranium-235, the computation of the probability of at least $k$ decays in a time interval $[t_0, t_1]$ is a probability problem. Radioactive dating, based on known properties of carbon-14, is a statistical problem, since it is desired to find out which population of the form

"decay without replenishment began in year $x$"

gave rise to the present observations. △

### 5.2 *Example: Speed of Light*

Owing to measurement error, the initial reasonably precise determination of the speed of light was a statistical problem. Now that the speed of light seems well established, the prediction of measurements of the speed of light by an instrument whose error is governed by a known probability model can be considered a probability problem. △

### 5.3 *Example: Lifetimes*

When a new type of transistor is manufactured, the initial problems of determining the probability of survival for at least $t$ hours are statistical. Once these probabilities have been well established, the computation of the probability that at least 50 per cent of some given lot will survive longer than 200 hours is a probability problem. △

### 5.4 *Example: Intelligence*

Fairly precise determination of the distribution of "intelligence" in our population is a statistical problem. In all likelihood, this problem has not been satisfactorily solved because of the complex nature of intelligence. If the initial problem is solved, the computation of the probability of at least one "genius" in a class of size 100 is a probability problem. △

### 5.5 *Example: Epidemic Spread*

Reasonably accurate determination of the present incidence of a communicable disease (such as the probability that a randomly selected person has this disease) by means of sampling the population (blood tests and so forth) is a statistical problem. Prediction of the future spread or incidence of a disease based on a completely specified model of present incidence is a probability problem. △

### 5.6 *Example: Election Problems*

If the number of those eligible to vote who plan to vote Democratic and the probability that each given individual will go to the polls are known, it is a probability problem to predict something about the number of people who actually will vote Democratic in a given election. We might, for instance, want to compute the probability of at least 40,000,000 Democratic votes or determine a useful lower bound for this probability. On the other hand, sampling from the population of eligible voters to learn something about the proportion of Democratic voters in the overall population is a statistical problem. △

### 5.7 *Example: Road Design*

If "rush hour conditions" has a specified meaning, such as "500 cars per hour on the average," then the prediction of the extent of time delay on a given freeway under rush hour conditions is thought of as a probability problem. If, however, "rush hour" is taken to mean the time period from 5 P.M. to 6 P.M., then the determination of a model for rush hour traffic by counting the number of vehicles would be a statistical problem. The design of an "adequate" roadway based on a model obtained earlier from observed data would most likely be considered as a probability problem. △

### 5.8 *Example: Machine-People Interaction*

There are now in use many very complicated control panels, such as those used in high-speed jet aircraft, that require human operation. Much work is involved in the design of these panels so as to reduce the human errors in their operation.

The testing of a given type of panel with people to determine its suitability for actual use (say, to determine accurately the probability that a person makes an error in using the panel once) is a statistical problem of great importance. △

### 5.9 *Example: Oil Exploration*

Test drillings are done at a variety of spots to determine whether or not there is sufficient oil to warrant development of the field. These drillings are part of a statistical investigation. A specified geological model of a given field would permit determination of the probability of at least a 100,000-barrel yield. △

---

## *5.10 PROBLEMS*

In each of the following problems, some area of interest will be described. Try to state some natural statistical and probability problems that might arise for each area. Of course, your answers cannot be expected to be identical to those we will present. You should recognize that we have not yet developed enough theories to solve the problems that will be generated. However, this is almost always the case; we start with the problems and base our theory on the problems we hope to solve.

1.° A manufacturer guarantees for 90 days the television sets he builds.
Hints: (a) If a new type of television set has just been produced, what is the first thing the manufacturer might like to know? (b) If the manufacturer expects to produce 10,000 sets of a given type, what would he like to know, and what would he need to do in order to find out this information?

2. An airline wishes to purchase planes that will help it show a profit. A new type of jet has just been put on the market; suppose we know the amount of profit, $q(n)$ (positive or negative), corresponding to any given number $n$ of passengers.
Hints: (a) For any single plane, what *numbers* would the airline like to know initially? (It must find out something about the number of passengers to expect.) (b) What would the airline want to determine about the airplane performance if it had figures on passenger usage for each trip (for example, cost per mile)?

3.° A supermarket chain wants to know how many cash registers it should install in one of its stores.
Hints: (a) What would the supermarket like to know about the length of time it takes to check out one customer? What would the supermarket like to know about the number of customers at different times during the day and week? What would it like to learn about the relationship between waiting time for service at the register and loss of a customer? (b) If the foregoing elements are known, what would the supermarket like to compute?

4. A supermarket is concerned with how large a supply of certain canned goods it should stock.
Hints: (a) What should be estimated concerning the demand for this product? (b) Assuming the answer to part (a) and the storage cost per unit of this product are known, what would the store like to compute?

5. The telephone company wants to know how large a trunk line to install between New York and Cleveland.

6.° Metal blades installed in a turbine tend to crack after a while, because of vibration. If a blade breaks loose, the cost of complete repair amounts to $75,000. The cost of replacing all blades simultaneously before any break occurs is $2500.

7.° In the hope of avoiding collisions, aircraft make measurements on distance and angle sequentially in time. If their measurements were perfect, then under constant velocity courses the minimum "miss distance" could be computed perfectly. Unfortunately, there is error because of random noise in their measurements.

8. A drug has been developed to treat a contagious disease that is sometimes fatal.

9.° Physicians want to make use of electrocardiograms in diagnosis and treatment (preventive or otherwise).

10.° In determining the relationship between speed and fuel consumption in an automobile, we often postulate an equation such as

$$f = a + bs + cs^2 + ds^3 + \varepsilon,$$

where $s$ is speed and $f$ is fuel consumption per mile; $\varepsilon$ represents random error (error in measurement or variation due to uncontrollable outside factors). Initially, $a$, $b$, $c$, and $d$, as well as some of the characteristics of $\varepsilon$, would be considered unknown. What would we want to compute after completion of the statistical part of the investigation?

Frequently it is not feasible to perform some given experiment, but we can perform an experiment governed by the same model (say in a wind tunnel, or on a computer). For example, on a digital computer one can simulate, one at a time, the various types of target conditions that might be seen on a radar set. The probabilities of some rather complicated events, such as false alarm or missing a target in a device to process data for target detection, are often estimated by using such *simulation*. In Chapter 9, we shall see how many times we must run such experiments in order to obtain the desired information. Simulation is frequently useful when analytic methods of computing probabilities are too difficult.

It is widely thought that a statistical model is necessarily cruder than a deterministic model of the same situation. Sometimes this may be the case, but the opposite may also be true, as we shall see in the following example.

### 5.11 *Example: Radioactive Decay*

The usual deterministic model for radioactive decay postulates that the amount of a pure radioactive material present at time $t$ is

$$a(t) = a(0)e^{-\lambda t},$$

where $\lambda$ is a constant dependent on the material and $a(0)$ is the amount of pure radioactive material present at time $t = 0$. In the usual statistical model, starting out with some reasonable assumptions concerning the probability of decay of a single atom, we are led to a description yielding the probability of exactly $k$ decays in a time interval of any given length. The deviations from the deterministic value $a(0)e^{-\lambda t}$ that are actually observed seem to display a statistical regularity that is quite well described by the computed probabilities. △

Even when the deterministic model for a phenomenon appears more refined than the statistical one, the latter may be preferred, as the next example shows.

### *5.12 Example: Thermodynamics*

In the Newtonian deterministic model for the behavior of a gas, on the order of magnitude of $10^{23}$ (differential) equations are involved, and accurate knowledge of the velocities and positions of about $10^{23}$ particles at some given ("initial") time is required. Such a description is clearly unmanageable. A cruder description, which gives only probabilities of specified velocity and position ranges of the particles, is considerably more manageable and useful. △

# 2

# Discrete Probability Theory

## SECTION 1
## Finite Uniform Models

We begin our study of discrete probability theory by recalling Definition 4.7 (Chap. 1) of a finite probability system: A *finite probability system*† consists of:

(i) a positive integer $n$ interpreted as the number of possible outcomes,

(ii) a set $\mathcal{S} = \{e_1, \ldots, e_n\}$ called the *sample space* and interpreted as being the set of all possible outcomes, that is, the set of *elementary events*, and

(iii) non-negative numbers $p_1, \ldots, p_n$ whose sum is 1, where $p_j$ is the probability of the elementary event $e_j$.

We describe the sample space by describing a typical outcome. For example, you might say "a typical outcome is a sequence of length 5 of 0's and 1's." We try to choose the sample space so that it represents accurately what we want to measure.

The probability $P(A)$ of a set $A$ of possible outcomes is the sum of those $p_j$ for which $e_j$ is in $A$, namely,

$$P(A) = \sum_{j:e_j \in A} p_j .$$

In the particular case that

$$p_1 = p_2 = \cdots = p_n = \frac{1}{n}$$

we shall refer to the model as being *uniform*. In this situation, if a subset $A$ of $\mathcal{S}$ contains exactly $k$ elementary events, we have

$$P(A) = \frac{k}{n} .$$

† A *probability system* is sometimes referred to as a *probability space*.

Thus, it *appears* that in a uniform model the problem of computing probabilities is rather simple: If we specify a set $A$, then its probability may be computed by counting the number of points in the sample space and the number of points in $A$. For this reason, it is often worthwhile to spend some effort in trying to choose a sample space for which a uniform model is reasonable.

### 1.1 *Example: Even on a Die*

For the toss of one "fair" die, we usually let $\mathcal{S} = \{1, 2, 3, 4, 5, 6\}$ with $p_1 = p_2 = \cdots = p_6 = 1/6$. Then the event "even" is $\{2, 4, 6\}$, whose probability is evidently $p_2 + p_4 + p_6 = 3/6 = 1/2$. △

Despite the simplicity of this example, we shall see in the next section that counting problems can be quite formidable. The reason is that it is usually quite easy to decide whether or not a given outcome belongs to a prescribed set $A$, but this fact may give us almost no clue as to the number of possible outcomes in $A$. Try—but not for too long—to count the number of "strings" of 10 coin tosses possessing *at least* two successive heads. Note that if we write down a particular string, such as

$$H\ T\ T\ H\ H\ H\ T\ T\ H\ T,$$

we can tell at a glance whether or not it is in the event "at least two successive heads." Counting difficulties usually arise from the fact that the number of outcomes in a set $A$ is generally available only *implicitly* from the criterion for membership in $A$. Despite the counting difficulties which arise in a uniform model, many problems are still of such a nature that they are best attacked within this framework.

### 1.2 *Example: Sum of Dice*

In Example 4.5 of Chapter 1, we postulated a probability model for the sum on two dice. Here we show that this probability space arises from a conceptually simpler uniform model, whose sample space is not the "natural" one, $\{2, 3, \ldots, 12\}$. For tossing two dice, we let a typical outcome be an ordered pair $(i, j)$ representing the result that the "first" die turned up an $i$ and the "second" die turned up a $j$. Note that we distinguish, say, (2, 5) from (5, 2). We can list the sample space as shown in Table 2.1.

| sum = 2 | (1, 1) | (1, 2) | (1. 3) | ··· | (1, 6) |
|---|---|---|---|---|---|
| sum = 3 | (2, 1) | (2, 2) | | ··· | (2, 6) |
| sum = 4 | (3, 1) | | | ··· | |
| | · | | | | |
| | · | | | | |
| | · | | | | |
| | (6, 1) | (6, 2) | | ··· | (6, 6) |

**TABLE 2–1**

This space has 36 elements. Experimentation indicates that a uniform model is in fact reasonable for the "fair" dice experiment with the given 36-point sample space. Accordingly, we assign

**1.3** $$P(i, j) = 1/36.$$

The event "sum $= k$" is seen in Table 2.1 to consist of all pairs lying along an upward diagonal as indicated for $k = 2, 3, 4$. Thus we see the origin of Table 2.2,

| $k$ | 2 | 3 | 4 | 5 | 6 | 7 | 8 | 9 | 10 | 11 | 12 |
|---|---|---|---|---|---|---|---|---|---|---|---|
| Probability sum $= k$ | 1/36 | 2/36 | 3/36 | 4/36 | 5/36 | 6/36 | 5/36 | 4/36 | 3/36 | 2/36 | 1/36 |

**TABLE 2.2**

which was given in Example 4.5 (Chap. 1). Notice that we are unable to determine the probability of the event "doubles," that is, $\{(1, 1), (2, 2), (3, 3), (4, 4), (5, 5), (6, 6)\}$ from Table 2.2, although we see easily from 1.3 that $P(\text{doubles}) = 1/6$. △

We might think of the flexibility in our choice of the sample space as the flexibility we have in the choice of measuring instruments. In the dice example just given, having only an instrument to measure the sum corresponds to a sample space $S = \{2, \ldots, 12\}$. The 36-point sample space corresponds to a separate instrument to measure the result on each die.

Here is an example of a situation in which we expend a substantial amount of effort to obtain a uniform model with its accompanying advantage of simple analysis by counting.

### *1.4 Example: Acceptance Sampling*

In attempting to assess the quality of a shipment of goods, a subset of the shipment is usually tested and the number of defective items is recorded. In order for the test to be useful, this subset should be a "representative" sample. Hence, great care must be taken to ensure that each subset is as likely to be chosen as any other in this preliminary experiment of choosing a subset of a given size. In the next section we shall see how to count the number of points in the sample space for this experiment, that is, how to count the number of subsets of a given size which can be chosen from some initially given set. △

Here is another example in which the "right" choice of sample space makes a uniform model a reasonable description.

### *1.5 Example: Simultaneous Toss of m Fair Coins*

If you toss a nickel and a dime, then a natural sample space would consist of ordered pairs $(n, d)$ where $n$ is the face on the nickel and $d$ is the face on the dime.

This suggests that the sample space for the simultaneous tossing of $m$ fair coins should consist of all possible $m$-tuples $(n_1, n_2, \ldots, n_m)$, where each $n_j$ stands for either $H$ or $T$ (and is the face on the $j$th coin). It appears that a uniform model on this space seems empirically justified. △

In Chapter 12 we shall investigate how to test the reasonableness of specified models—in particular, uniform ones. Here is another example in which a uniform model has proved useful.

### *1.6 Example: Accident Distribution*

In New York City, five auto accidents are known to have occurred on a weekday between 9:30 and 10:30 A.M. Let us divide this time interval into one-second intervals and assume that no two of these accidents occur in the same one-second interval. If we put a 1 in each interval during which there was an accident, and a 0 in each one in which there was no accident, then we obtain a sequence of 3600 elements, with five 1's and three thousand five hundred and ninety-five 0's. Each such sequence represents a possible outcome (description of when the five accidents occurred) and a uniform model seems to describe the situation well, because 9:30 to 10:30 A.M. is a relatively homogeneous time interval with regard to occurrence of traffic accidents. Such a model is probably not useful for the interval 4:15 to 5:15 P.M., in which the number of accidents might peak at 5:00 P.M. △

---

### *1.7 PROBLEMS*

For each of the following experiments, choose a sample space which you believe will lead to a uniform model. Explain the interpretation of the sample points.

1.° $m$ distinguishable fair dice are tossed simultaneously.

2. A fair coin is tossed $m$ times.

3. A fair die is tossed $m$ times.

4.° A bridge hand is drawn from a deck of cards. (A bridge hand consists of 13 distinguishable cards. A bridge deck consists of 52 distinguishable cards.)

5. A poker hand (five distinguishable cards) is drawn from a deck of 52 distinguishable cards.

6.° Five cards are drawn (one at a time) from a deck of cards numbered 1 to 52. We wish to determine the probability that an even-numbered card is drawn before an odd-numbered one, so this event must correspond to a subset of our sample space.

7. A target consists of three regions, as shown. The area of the outer shell is five times the bullseye area, and the area of the middle shell is three times the bullseye area. The experiment consists of shooting at the target. We "count" only those trials in which the target is hit.

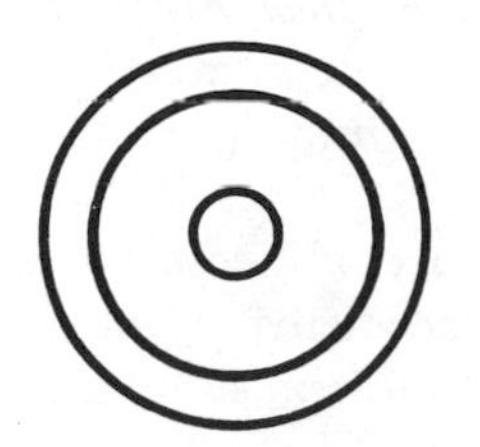

8.° Your wristwatch stands an equal chance of gaining 1, 0, or $-1$ second each half hour. You are interested in its behavior over a 24-hour period.

9. In a telephone directory with 300,000 listings, there are known to be 422 mistakes in phone numbers.

Hint: Each listing here can be thought of as similar to the one-second time intervals of Example 1.6.

## SECTION 2
## Elementary Combinatorial Analysis

In the previous section, we discussed the importance of counting in the determination of probabilities in a finite uniform model; the computation of $P(A)$ reduces the problem to counting the number of elements of $\mathcal{S}$ and $A$. This section is devoted to the development of systematic counting techniques.

For simplicity, let us first note that the elements of any collection of $m$ distinguishable objects can be labeled in an arbitrary fashion; a most convenient labeling consists of the numbers $1, 2, \ldots, m$. Let $M$ denote the set $\{1, 2, \ldots, m\}$ and suppose that $k$ is a fixed integer with $1 \leq k \leq m$. We consider the problem of choosing a sample of size $k$ from the set $M$. In particular, we want to know how many distinct such samples can be drawn. For $k = 1$, it is evident that there are $m$ distinct such samples. However, for $k > 1$, an additional complication arises, namely, whether or not we wish to distinguish between two samples containing the same objects, but drawn in different order. It turns out that in some problems this distinction is needed; in others it is not. For instance, in acceptance sampling where we draw a subset of size $k$ from a set of size $m$, we usually do not need to take account of the order in which the $k$ objects are drawn, whereas in quality control a consecutive number of defective items is significant, so we must take the order into account. We therefore give the following definition.

**2.1 Definition: Samples.** (a) *An ordered sample of size $k$ from the set $M = \{1, 2, \ldots, m\}$ without replacement* is a sequence $(n_1, n_2, \ldots, n_k)$, where all of the $n_j$ are distinct elements of $M$. Note that the word *ordered* implies, for example, that $(1, 2, 5) \neq (2, 5, 1)$. Ordered samples of this type are also called *permutations* of $m$ objects taken $k$ at a time.

(b) *An unordered sample of size $k$ from the set $M = \{1, 2, \ldots, m\}$* is a subset of $M$ consisting of $k$ *distinct* elements. Note that in this case the unordered sample $\{1, 2, 5\}$ is considered the same as the unordered sample $\{2, 5, 1\}$. Unordered samples of this type are called *combinations of $m$ objects taken $k$ at a time.*

(c) *An ordered sample of size $k$ from the set $M = \{1, 2, \ldots, m\}$ with replacement* is a sequence $(n_1, n_2, \ldots, n_k)$ where the $n_j$ are (not necessarily distinct) elements of $M$. In this case only, we permit $k > m$.

We remark that a sequence of $k$ coin tosses yields an ordered sample of size $k$ from the set $\{T, H\}$ with replacement; dipping your hand in a bowl of chips labeled $1, 2, \ldots, m$ and pulling out $k$ of them simultaneously yields an unordered sample of size $k$ from $\{1, 2, \ldots, m\}$, and pulling chips from this bowl one at a time without replacing any chip after it has been drawn, for $k$ drawings, yields an ordered sample of size $k$ from $\{1, 2, \ldots, m\}$ without replacement.

## 2.2 *Theorem: Sample Sizes*

(a) For $1 \leq k \leq m$, the number of distinct ordered samples of size $k$ without replacement from the set $M = \{1, 2, \ldots, m\}$ is denoted by $m_{(k)}$, where

$$
\begin{aligned}
m_{(1)} &= m \\
m_{(2)} &= m(m-1) \\
&\cdot \\
&\cdot \\
&\cdot \\
m_{(k)} &= m(m-1)\cdots(m-k+1).
\end{aligned}
$$

(b) For $1 \leq k \leq m$, the number of distinct unordered samples of size $k$ from the set $M$ is denoted by $\binom{m}{k}$, where

$$
\begin{aligned}
\binom{m}{1} &= \frac{m}{1} = \frac{m_{(1)}}{1!} \\
\binom{m}{2} &= \frac{m(m-1)}{1 \cdot 2} = \frac{m_{(2)}}{2!} \\
&\cdot \\
&\cdot \\
&\cdot \\
\binom{m}{k} &= \frac{m(m-1)\cdots(m-k+1)}{1 \cdot 2 \cdots k} = \frac{m_{(k)}}{k!}
\end{aligned}
$$

where $k! = k_{(k)} = 1 \cdot 2 \cdots k$.

(c) The number of distinct ordered samples of size $k$ from $M$ with replacement is $m^k$ (even when $k > m$).

*Proof:* We shall present only informal arguments, although rigorous proofs using induction are not difficult.

(a) For the case $k = 2$, we note that there are $m$ possible choices for the first element, $n_1$, of the sequence. After we have chosen $n_1$, there are $m - 1$ choices possible for the second element, $n_2$. Hence, we can list the set of all ordered samples of size $k = 2$ without replacement as the following array:

$$m \text{ rows} \begin{cases} (1, 2) & (1, 3) & \cdots & (1, m) \\ (2, 1) & (2, 3) & \cdots & (2, m) \\ \cdot & & & \\ \cdot & & & \\ \cdot & & & \\ (m, 1) & (m, 2) & \cdots & (m, m-1) \end{cases}$$
$$\underbrace{\qquad\qquad\qquad\qquad\qquad\qquad}_{m - 1 \text{ columns}}$$

It is therefore evident for the case $k = 2$ that there are $m(m - 1)$ such samples. Now let us examine the effect of increasing the sample's size by one element. If there are $j$ elements left in $M$ from which to choose this additional element, then each sequence of the original sample yields $j$ sequences with one more element. Since we have $j = m - 2$ elements to choose from in going from sequences of length 2 to sequences of length 3, we go from

$$m(m - 1) \text{ sequences of length 2}$$

to

$$m(m - 1)(m - 2) \text{ sequences of length 3.}$$

A similar argument shows that there are $m(m - 1)(m - 2)(m - 3)$ such sequences of length 4 and generally $m(m - 1) \cdots (m - k + 1)$ distinct ordered samples of size $k$ which can be chosen from $M = \{1, 2, \ldots, m\}$ without replacement.

(b) In order to handle the case of an unordered sample of size $k$, we notice that we can always obtain any ordered sample of size $k$ without replacement from $\{1, 2, \ldots, m\}$ in two stages:

(i) Pick an unordered sample of size $k$ from $\{1, 2, \ldots, m\}$.
(ii) From this unordered sample (set of $k$ elements), form the desired ordered sample which is a sequence of $k$ distinct elements (that is, arrange the set in the desired order).

In order to use this scheme in counting the number of unordered samples of size $k$ from $M$, *we must determine the number of ordered samples of size k*. But this is just the number of ordered samples of size $k$ from a set of $k$ elements without replacement. From the results of part (a), we know this number to be $k_{(k)} = k!$. We are now in a position to use the reasoning described in (i) and (ii), that is:

$$\begin{pmatrix} \text{number of unordered} \\ \text{samples of size } k \\ \text{from } \{1, 2, \ldots, m\} \end{pmatrix} \cdot \begin{pmatrix} \text{number of ordered} \\ \text{samples of size } k \\ \text{without replacement} \\ \text{which can be obtained} \\ \text{from each unordered} \\ \text{sample of size } k \end{pmatrix} = \begin{pmatrix} \text{number of ordered} \\ \text{samples of size } k \\ \text{from } \{1, 2, \ldots, m\} \\ \text{without replacement} \end{pmatrix},$$

that is, denoting the number of unordered samples of size $k$ from $\{1, 2, \ldots, m\}$ by $\binom{m}{k}$, we have

$$\binom{m}{k} \cdot k! = m_{(k)}.$$

Solving for $\binom{m}{k}$, we obtain the formula

$$\binom{m}{k} = \frac{m_{(k)}}{k!}.$$

(c) Here we simply note that each addition of one unit to the sample size multiplies the number of distinct possible sequences by $m$, because then each sequence of length $\ell$ gives rise to $\ell \cdot m$ sequences of length $\ell + 1$. Hence there are

$$\underbrace{m \cdot m \cdots m}_{k \text{ times}} = m^k$$

possible such sequences of length $k$. This concludes the proof.

*2.3 Remarks*

1. Note that we have shown in the course of the preceding discussion that the number of ways to rearrange a sequence of $k$ distinct elements is $k!$.

2. It will be useful to define $0! = 1$. We see that then the formula $(k + 1)! = (k + 1)(k!)$ still holds for the case $k = 0$.

3. We note that $\binom{m}{1} = m$ for all $m \geq 1$, and for $1 \leq k \leq m - 1$ we have $\binom{m}{k} = \binom{m}{m-k}$. This last identity is associated with the fact that in choosing a subset of size $k$ from $\{1, 2, \ldots, m\}$, we are at the same time choosing a subset of size $m - k$, namely those elements *not* chosen for the subset of size $k$. It is for this reason that we let $\binom{m}{0} = \binom{m}{m} = 1$, so that $\binom{m}{k} = \binom{m}{m-k}$ for $0 \leq k \leq m$. We call $\binom{m}{k}$ a *binomial coefficient* (see Problem 2.19(4) for the reason for this terminology).

4. We see that for $0 \leq k \leq m$, we obtain the formula

$$\binom{m}{k} = \frac{m!}{k!(m-k)!}$$

and notice that this formula now agrees with our convention to let $0! = 1$.

5. $\binom{m}{k}$ is not only the number of distinct subsets of size $k$ which can be chosen from a set of size $m$, but also the number of sequences of $H$'s and $T$'s of length $m$ which have exactly $k$ $H$'s. To see this, think of a sequence of $m$ $H$'s and $T$'s as a set

of $m$ boxes labeled $1, 2, \ldots, m$ and each filled with either $H$ or $T$, as illustrated:

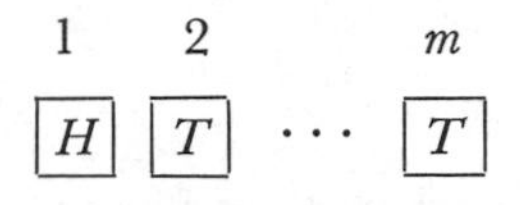

Each sequence of length $m$ having $k$ $H$'s determines a unique subset of size $k$ of the set of labels $\{1, 2, \ldots, m\}$, and conversely. Hence, the number of such sequences is simply the number of subsets of size $k$ of the set $\{1, 2, \ldots, m\}$, that is, $\binom{m}{k}$.

6. When a sample of one of the preceding types is drawn and the model is uniform, we sometimes say that the sample has been *chosen at random.* (This is to be distinguished from a *random sample;* see Definition 4.2 in Chap. 5).

## 2.4 *Example: Number of Elements in Sample Space for Acceptance Sampling*

In order to determine whether or not to accept delivery of a shipment of $m$ items, we may first choose a representative sample of size $k$ from the shipment for examination. If we let the sample space $\mathcal{S}$ for this experiment (of "choosing") consist of all unordered samples of size $k$ which can be chosen from the shipment, then there are $\binom{m}{k}$ elements in $\mathcal{S}$. (Note that $\mathcal{S}$ does *not* consist of the $m$ units of the shipment. $\mathcal{S}$ is the collection of all subsets of size $k$ chosen from the shipment.) In the uniform model we have tried to obtain, each elementary event has probability $\dfrac{1}{\binom{m}{k}}$. For instance, if we test two items from a shipment of five items, then $\mathcal{S}$ has $\dfrac{5 \cdot 4}{1 \cdot 2} = 10$ elements, and each elementary event has probability 1/10 △

## 2.5 *Example: n Coin Tosses*

If an outcome of $n$ coin tosses consists of a sequence of length $n$ of $H$'s and $T$'s, then the sample space consists of $2^n$ elements, since we are obtaining an ordered sample of size $n$ from the two-point set $\{H, T\}$, with replacement. If our model is uniform, then we see from 2.3.5 that the *probability* of exactly $k$ heads is

$$\frac{\text{number of points in } k \text{ heads}}{\text{number of points in } \mathcal{S}} = \frac{\binom{n}{k}}{2^n}. \qquad \triangle$$

## 2.6 EXERCISES

1.° Describe a sample space in which three items of a shipment of six items are chosen to be tested.

2. In Exercise 1, assume the second item of the shipment is defective and all other items are good. Under a uniform model, what is the probability of:

(a)° 0 defectives?

(b) 1 defective?

You might check your result by listing the elements of the sample space and determining which are members of the given event.

3. If in Exercise 1, the second and third items are defective, compute $p(i)$, the probability of $i$ defectives in the sample for $i = 0, 1, 2$.

4.° Describe a sample space $\mathcal{S}$ for the experiment of tossing a die $n$ times that contains events corresponding to the phenomena "$i$ on the $j$th toss." How many elements are in $\mathcal{S}$?

5. Describe an adequate sample space for a preliminary experiment in which five items are sampled from a shipment of size 100. If two successive items in the sample are defective, the shipment will be returned.

6. How many "words" (a word is assumed to be any sequence of letters) of exactly five letters can be formed:

(a)° allowing repeated letters?

(b) not allowing repeated letters?

7. Determine a simple expression for $\binom{n}{0} + \binom{n}{1} + \cdots + \binom{n}{n}$.

Hint: Look at Example 2.5 and use the fact that if $A_k$ is the event, "$k$ heads in $n$ tosses," then $P(A_0) + P(A_1) + \cdots + P(A_n) = 1$.

8.° Without any computing, guess which is more likely: "three heads out of five tosses of a fair coin," or "three heads out of six tosses of a fair coin." Using 2.5, determine whether your guess was correct.

The following result allows us to solve many of the more difficult counting problems that will arise.

## 2.7 *Theorem: Number of Sequences of Length k*

Suppose we have a sequence of length $k$ ($k$ positions) in which the element in the $i$th position can be any one of $\ell_i$ distinct objects. Then the total number of distinct such sequences is $\ell_1\ell_2 \cdots \ell_k$.

*Proof.* For $k = 1$, the number of possible sequences (of length 1) is evidently $\ell_1$. Consider the effect of adding a second coordinate (sequence position) which can be filled by any of $\ell_2$ distinct objects. We see that each distinct sequence of length

1 gives rise to $\ell_2$ distinct sequences of length 2. Hence, there are $\ell_1\ell_2$ distinct sequences of length 2. By considering the effect of adding a third coordinate which can be filled by any of $\ell_3$ distinct objects, we see that there are $\ell_1\ell_2\ell_3$ such distinct sequences of length 3. Repeated application of this argument (which can be rigorously accomplished using mathematical induction) yields the result that there are $\ell_1\ell_2 \cdots \ell_k$ such distinct sequences of length $k$, concluding the proof.

*2.8 Remarks.* The idea behind the proof of this theorem was actually already used in establishing parts (a) and (c) of Theorem 2.2 on sample sizes; it is interesting to note that in part (a) of this theorem (sampling without replacement) the objects which can fill the $i$th position cannot be any of those which have been used to fill previous positions. Nevertheless, the reasoning is applicable because the *number* of available objects for the $i$th position does not depend on which objects filled positions $1, 2, \ldots, i-1$. So, our result is applicable, even when the choice of objects to fill a given position in the sequence may depend on which objects have already been chosen to occupy the previous positions, as long as the *number* of available objects, $l_i$, is not affected.

### 2.9 *Example: Committee Formation*

Suppose that the Senate currently consists of 52 Democrats and 48 Republicans. How many different committees of three Democrats and two Republicans can be formed (where two different committees have different membership)? We may think of such a committee as a sequence of two sets:

$$(\{3 \text{ Democratic members}\}, \{2 \text{ Republican members}\}).$$

Now the number $\ell_1$ of possible first elements is simply the number of subsets of size three of the Democratic membership, that is, $\ell_1 = \binom{52}{3}$ by 2.2(b). Similarly, $\ell_2 = \binom{48}{2}$. Hence, using 2.7, we see that when the Senate consists of 52 Democrats and 48 Republicans, then precisely $\binom{52}{3}\binom{48}{2}$ different committees consisting of three Democrats and two Republicans can be formed. △

This example suggests the following corollary.

### 2.10 *Corollary to 2.7: Size of Unordered Samples under Constraints*

Suppose that the set $M$ consists of $k$ non-overlapping sub-populations:

$A_1$ having $m_1$ elements,

$A_2$ having $m_2$ elements,

.

.

.

$A_k$ having $m_k$ elements.

Note that $M$ has $m_1 + m_2 + \cdots + m_k$ elements. Then the number of distinct unordered samples from $M$ of size $b$ with

$j_1$ elements from $A_1$,

$j_2$ elements from $A_2$,

.

.

.

$j_k$ elements from $A_k$,

where $j_1 + j_2 + \cdots + j_k = b$, is given by

$$\binom{m_1}{j_1}\binom{m_2}{j_2}\cdots\binom{m_k}{j_k}.$$

*Proof.* We can think of the unordered sample as being a sequence of boxes such that the $i$th box contains the subset of size $j_i$ from $A_i$. But by 2.2(b), the number of distinct choices for the contents of the $i$th box is $\ell_i = \binom{m_i}{j_i}$. The result follows from 2.7, concluding the proof.

Another consequence of 2.7 concerns the number of ways of forming sequences from the set $\{1, 2, \ldots, k\}$.

### 2.11 *Corollary to 2.7: Multinomial Coefficients*

The number of distinct sequences of length $m$ having

the value 1 in $j_1$ of its coordinate positions,

the value 2 in $j_2$ of its coordinate positions,

.

.

.

the value $k$ in $j_k$ of its coordinate positions,

where $j_1 + j_2 + \cdots + j_k = m$, is the *multinomial coefficient*

$$\binom{m}{j_1}\binom{m - j_1}{j_2}\cdots\binom{m - j_1 - j_2 - \cdots - j_{k-2}}{j_{k-1}} = \frac{m!}{j_1!j_2!\cdots j_k!}.$$

*Informal Proof.* We may choose $j_1$ positions for the 1's in $\binom{m}{j_1}$ ways. After this, we may choose $j_2$ positions for the 2's from the remaining $m - j_1$ places in $\binom{m - j_1}{j_2}$ ways. Thus, there are $\binom{m}{j_1}\binom{m - j_1}{j_2}$ ways to choose the positions for the 1's and 2's. Continuing in this fashion, we get the desired result.

Those who are not fully convinced should read the optional section.

*More Precise Proof.* Examine any sequence $a = (a_1, \ldots, a_m)$, where each $a_i$ is one of the integers $1, 2, \ldots, k$. Let $j_i$ be the *number* of positions in $a$ occupied by the integer $i$, and let $A_i$ be the *set* of those $j_i$ positions in $a$ occupied by the integer $i$. We claim that to each sequence $(a_1, \ldots, a_m)$ there corresponds precisely one sequence $(A_1, \ldots, A_k)$; conversely, given any sequence $(A_1, \ldots, A_k)$, where each element of $\{1, 2, \ldots, m\}$ is in precisely one of the sets $A_i$, there corresponds precisely one sequence $(a_1, \ldots, a_m)$. For example, the sequence $(a_1, a_2, a_3, a_4, a_5, a_6, a_7, a_8) = (1, 1, 2, 4, 3, 1, 4, 4)$ is associated with the sequence of subsets $(A_1, A_2, A_3, A_4) = (\{1, 2, 6\}, \{3\}, \{5\}, \{4, 7, 8\})$, since positions 1, 2, and 6 are occupied by 1's (i.e., $a_1 = a_2 = a_6 = 1$); position 3 by a $2 (a_3 = 2)$; position 5 by a 3; and positions 4, 7, and 8 by 4's. Hence, in order to determine the number of sequences $(a_1, \ldots, a_m)$ satisfying the given conditions, we can instead count the number of associated sequences $(A_1, \ldots, A_k)$. This latter counting problem is considerably simpler. In fact, there are

$$\ell_1 = \binom{m}{j_1} \text{ choices for } A_1,$$

$$\ell_2 = \binom{m - j_1}{j_2} \text{ choices for } A_2 \text{ once } A_1 \text{ is chosen,}$$

$$\vdots$$

$$\ell_{k-1} = \binom{m - j_1 - j_2 - \cdots - j_{k-2}}{j_{k-1}} \text{ choices for } A_{k-1} \text{ once } A_1, A_2, \ldots, A_{k-2} \text{ are}$$

chosen, and $\ell_k = 1$ choice for $A_k$ once $A_1, A_2, \ldots, A_{k-1}$ are chosen. Thus, by 2.7, the number of possible sequences $(A_1, A_2, \ldots, A_k)$ is

$$\ell_1 \ell_2 \cdots \ell_k = \binom{m}{j_1}\binom{m - j_1}{j_2} \cdots \binom{m - j_1 - j_2 - \cdots - j_{k-2}}{j_{k-1}}$$

and the result is proved.

## 2.12 *Example: Distribution of Sample in Acceptance Sampling*

A shipment of 100 items is received. One of the suggested acceptance sampling procedures is to test 10 items chosen in a representative way from the shipment and accept delivery if fewer than three items in the sample are defective. Assuming a uniform model (that is, equally likely unordered samples of 10 items):

(a) What is the probability of the sample having exactly two defective units if the shipment has exactly four defectives?

(b) What is the probability that the sample has exactly $d$ defectives when the shipment has exactly $D$ defectives?

(c) What is the probability of accepting delivery if the shipment has

(i) exactly 4 defectives?
(ii) exactly 10 defectives?
(iii) exactly 30 defectives?
(iv) exactly 60 defectives?

(d) Plot a graph of probability of acceptance of the shipment versus number of defectives in the shipment. We will want to compute the probability of acceptance for 20, 40, and 50 defectives in the shipment for this part.

*ANSWERS:*

(a) In a uniform model, we know that the probability of an event $A$ is

$$\frac{\text{number of elements of } A}{\text{number of elements of } \mathcal{S}}.$$

In this case, we let $M = \{1, \ldots, 4, 5, \ldots, 100\}$, thinking of $\{1, 2, 3, 4\}$ as the subset of defectives. Let $A_2$ consist of those subsets of $M$ of size 10 having exactly two elements from $\{1, \ldots, 4\}$ and eight elements from $\{5, \ldots, 100\}$. By 2.10, the number of elements of $A_2$ is $\binom{4}{2}\binom{96}{8}$. The number of elements of the sample space $\mathcal{S}$ is $\binom{100}{10}$, that is, the total number of unordered samples of size 10 chosen from the set $\{1, 2, \ldots, 100\}$. Hence

$$P(A) = P(\text{exactly 2 defectives}) = \frac{\binom{4}{2}\binom{96}{8}}{\binom{100}{10}}$$

$$= \frac{4 \cdot 3}{1 \cdot 2} \cdot \frac{96 \cdot 95 \cdot 94 \cdot 93 \cdot 92 \cdot 91 \cdot 90 \cdot 89}{1 \cdot 2 \cdot 3 \cdot 4 \cdot 5 \cdot 6 \cdot 7 \cdot 8}$$

$$\cdot \frac{10 \cdot 9 \cdot 8 \cdot 7 \cdot 6 \cdot 5 \cdot 4 \cdot 3 \cdot 2 \cdot 1}{100 \cdot 99 \cdot 98 \cdot 97 \cdot 96 \cdot 95 \cdot 94 \cdot 93 \cdot 92 \cdot 91}$$

$$= 6 \cdot \frac{10 \cdot 9 \cdot 90 \cdot 89}{100 \cdot 99 \cdot 98 \cdot 97} \cong .046.$$

(The symbol $\cong$ means "is approximately equal to.")

(b) Reasoning as in the answer to (a), we see that the answer is

$$\frac{\binom{D}{d}\binom{100-D}{10-d}}{\binom{100}{10}}.$$

Note that the formula holds even for $d = 0$; the convention $\binom{D}{0} = 1$ is most reasonable.

In general, we say that the assigned probability is *hypergeometric* if the $p_d$ are of the form

$$p_d = \frac{\binom{D}{d}\binom{G}{g}}{\binom{D+G}{d+g}}.$$

(c) (i) As in the previous parts, we compute

$$P_4(A_0) = P_4(0 \text{ defectives}),$$
$$P_4(A_1) = P_4(1 \text{ defective}),$$
$$P_4(A_2) = P_4(2 \text{ defectives}),$$

the subscript 4 indicating that the computation is made for the case of four defectives in the shipment of size 100. We find

**2.13**
$$P_4(A_0) \cong .653$$
$$P_4(A_1) \cong .3$$
$$P_4(A_2) \cong .046 \text{ (from part (a))}.$$

Note that if $B_3$ is the event "fewer than three defectives in an unordered sample of size 10 from $\{\underbrace{1, \ldots, 4}_{\text{defectives}}, 5, \ldots, 100\}$" then

**2.14**
$$P_4(B_3) = P_4(A_0) + P_4(A_1) + P_4(A_2),$$

where all probabilities are computed for populations with four defectives. This follows from the fact that each elementary event $e_j$ in $B_3$ is in precisely one of these $A_i$, and hence the corresponding probability $p_j$ appears exactly once on the right side of equation 2.14, as it should.

Since $B_3$ is equivalent to "acceptance of the shipment," the probability of accepting a shipment having four defectives is, by 2.13 and 2.14,

$$P_4(B_3) \cong .999.$$

(ii) By the same reasoning as in (i), and using similar notation,

$$P_{10}(A_0) = \frac{90 \cdot 89 \cdots 81}{100 \cdot 90 \cdots 91} \cong .331$$

$$P_{10}(A_1) = \frac{100}{81} P_{10}(A_0) \cong .408$$

$$P_{10}(A_2) = \frac{45 \cdot 90}{82 \cdot 81} P_{10}(A_0) \cong .202.$$

Hence

$$P_{10}(\text{accepting shipment}) = P_{10}(B_3) \cong .941.$$

(iii) As in (ii),

$$P_{30}(A_0) = \frac{70 \cdot 69 \cdots 61}{100 \cdot 99 \cdots 91} \cong .023$$

$$P_{30}(A_1) = \frac{300}{61} P_{10}(A_0) \cong .113$$

$$P_{30}(A_2) = \frac{1350 \cdot 29}{62 \cdot 61} P_{10}(A_0) \cong .238.$$

When there are 30 defectives, we find the probability of acceptance to be about .374.

(iv) As in (ii),

$$P_{60}(A_0) = \frac{40 \cdot 39 \cdots 31}{100 \cdot 99 \cdots 91} \cong .0000488$$

$$P_{60}(A_1) = \frac{600}{31} P_{60}(A_0) \cong .000945$$

$$P_{60}(A_2) = \frac{2700 \cdot 59}{32 \cdot 31} P_{60}(A_0) \cong .00783.$$

So, when there are 60 defectives, the probability of acceptance is about .0088.

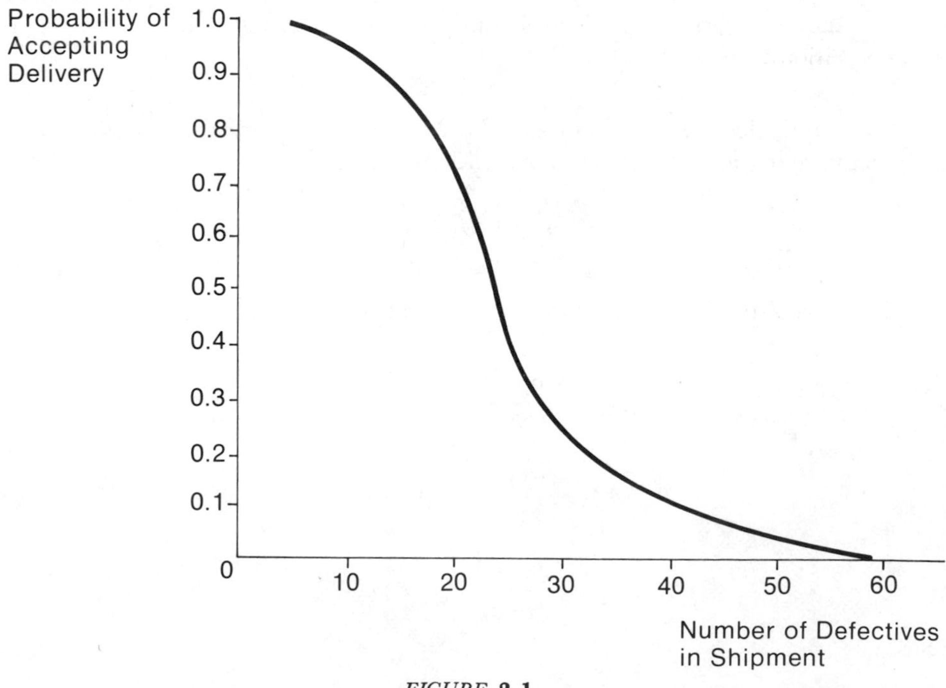


*FIGURE* **2–1**

(d) By similar computations, we find

$$P_{20}(B_3) \cong .685$$
$$P_{40}(B_3) \cong .123$$
$$P_{50}(B_3) \cong .046,$$

and hence the requested graph looks like Figure 2.1.

A few remarks are in order here. First, we see that if misclassification is costly, then this procedure does only a barely satisfactory job of discriminating between lots with 10 and lots with 40 defectives. (This may come as a surprise.) Secondly, assuming destructive testing, no such scheme can ensure even a high probability of obtaining good quality. To see this, imagine any acceptance sampling scheme and ask what will happen if lots with too many defectives are always shipped to you. All that this method ensures is that there will be a low probability of accepting a bad lot and a high probability of accepting a good one. △

### 2.15 *Example: Busy Signal on Weather Report*

The weather report is repeated over the telephone every half minute. Suppose that in a given 30-minute period it is known that 20 calls for the weather report were made. For simplicity, we assume that if two or more phone calls are initiated in any one of the half-minute periods during which the report is being given, all callers receive a busy signal. What is the probability that none of the 20 calls receives a busy signal?

To tackle this problem, we make the assumption that each of the 20 telephone calls must be initiated in precisely one of the 60 half-minute intervals, and therefore we are choosing an ordered sample of size 20 from the set $\{1, 2, \ldots, 60\}$ with replacement. The $j$th element is the time interval during which person $j$ initiates his call. The replacement is allowed in order to take into account the possibility of two or more calls being initiated in the same half-minute period. We assume a uniform model, which has some basis in theory. The event of interest, $A$, corresponds to precisely those ordered samples of size 20 *without* replacement. Thus, the desired probability is

$$P(A) = \frac{\text{number of elements in } A}{\text{number of elements in sample space}} = \frac{60_{(20)}}{60^{20}} \cong .02789. \qquad \triangle$$

In the next example we show how to compute the probability that 1 occurs $k$ times in $n$ tosses of a fair die. This example is a special case of computation of the probability of $k$ successes in $n$ experiments each having probability $p$ of success.

### 2.16 *Example: Number of 1's in n Tosses of a Die*

For $n$ tosses of a fair die, we take a uniform model on the space of sequences of length $n$ of elements from the set $\{1, 2, 3, 4, 5, 6\}$. The probability of $k$ 1's is

$$\frac{\text{number of points of } \mathcal{S} \text{ with } k \text{ 1's}}{6^n}.$$

But the number of sequences with $k$ 1's and $n - k$ 2's through 6's can be computed by noting that there are $\binom{n}{k}$ choices for the positions of the 1's. Each such choice gives rise to $5^{n-k}$ distinct sequences of the type we desire to count, because each of the $n - k$ remaining positions can be occupied by any element of $\{2, 3, \ldots, 6\}$. Thus our desired probability is

$$\frac{\binom{n}{k}5^{n-k}}{6^n}. \qquad \triangle$$

---

## 2.17 *EXERCISES (counting)*

1.° How many three-digit positive integers can be formed:
(a) with all digits even?
(b) with no repeated digits?

2. How many positive integers less than $10^{23}$ are there:
(a) with all digits even?
(b) with no repeated digits?

3. A Chinese menu contains 10 types of seafood, 15 types of vegetables, and 8 types of meat. The blue plate special allows 3 choices of seafood, 5 choices of vegetable, and 2 choices of meat. How many distinct possible blue plate specials are there?

4°. The Cleveland Orchestra has 66 strings, 18 woodwinds, 14 brass, and 5 percussion. How many distinct combos can they form using 3 strings, 2 woodwinds, 2 brass, and 1 percussion?

5. An octave contains 12 distinct notes (five black keys, seven white ones). How many distinct eight-note melodies could Mozart write:
(a) using white keys only?
(b)° allowing no two successive notes to be the same?

6. How many bridge hands are there which contain:
(a)° no hearts?
(b) at least one heart?

7. How many distinct seating arrangements are there of two women and four men around a table in which the two women are not seated next to each other?

---

## 2.18 *EXERCISES*

Please state all needed assumptions.

1.° What is the probability that in a class of 23, no two people will have the same birthday (disregarding the year)?

2. In drawing a 13-card bridge hand, what is the probability of drawing:

(a) at least two aces?

(b)° the ace of hearts followed immediately by the king of hearts?

(c) no hearts (where hearts are cards 1, 2, . . . , 13)?

(d) four hearts, three clubs, three diamonds, and three spades (clubs are 14, . . . , 26, diamonds are 27, . . . , 39, and spades are 40, . . . , 52)?

3. In drawing a 5-card poker hand (from a deck of 52 cards), what is:

(a)° the probability of a flush in spades (that is, that all cards come from the subset of 13 spades)?

(b) the probability of a flush?

4.° Given that in seven consecutive weekdays there were four automobile accidents involving only one car, what is the probability that each occurred on a different day?

5. An urn contains 10 red balls and 30 white balls. Four balls are selected without replacement. What is the probability that:

(a) the first two balls will be white?

(b)° the second ball is red?

(c) at least one ball is red?

6.° An urn contains 10 red and 90 black balls.

(a) If sampling is done with replacement, how large a sample should be taken so as to yield a probability of at least .95 that at least one red ball will be included in the sample? (That is, how many times should you repeat an experiment whose probability of success is .1 in order to be 95 per cent sure of at least one success?)

(b) Determine whether as large a sample is needed for (a) if we sample without replacement.

7. A man has nine keys in his pocket.

(a)° He samples from his pocket one at a time, with replacement (since he has no place to put keys and his other hand is not free). What is the probability that he will need no more than four tries to find his car key?

(b) Suppose he samples without replacement (he puts his keys on a table nearby or drops them on the floor)?

8. Two cards are drawn from a bridge deck $\{1, \ldots, 52\}$ without replacement. What is the probability that:

(a)° the larger card exceeds 26?

(b) the smaller card exceeds 26?

9. Three cards are drawn from a bridge deck without replacement. What is the probability that:

(a) the largest card exceeds 39?

(b)° the smallest card exceeds 39?

10. What is the probability that in a well-shuffled deck of cards the king of hearts will be:

(a)° directly below the king of spades?

(b) adjacent to the king of spades?

(c) directly below a spade?

## *2.19 PROBLEMS*

Please state all assumptions.

1.° A fair die is tossed $n$ times. What is the probability that the total number of 1's and 2's which occur is $k$?

Hint: We can pick the positions for the 1's and 2's in $\binom{n}{k}$ ways. For each choice of positions, there are $2^k \cdot 4^{n-k}$ choices of number (two for each of the $k$ positions, four for each of the $n - k$ positions).

2. Based on your answer to Problem 1, suppose you repeat an experiment $n$ times, with the probability of success on each repetition being $p$. What would you conjecture to be the probability of exactly $k$ successes? Give a reasonable justification of your conjecture.

3. Your closet contains six pairs of shoes. You choose two shoes at random in the dark. What is the probability that you get a pair?

4. Show that

$$(a + b)^m = \binom{m}{0} a^0 b^m + \binom{m}{1} ab^{m-1} + \binom{m}{2} a^2 b^{m-2} + \cdots + \binom{m}{m-1} a^{m-1} b + \binom{m}{m} a^m b^0.$$

(This result is called the *Binomial Theorem.*)
Hint: In the expansion of

$$(a + b)^m = \underbrace{(a + b)(a + b) \cdots (a + b)}_{m \text{ factors}}$$

you must choose one term from each factor. In so doing, you obtain terms $a^k b^{m-k}$ ($m$ factors). The coefficient of $a^k b^{m-k}$ must be the number of distinct choices for the $k$ factors of $a$.

5. Show that in computing probabilities for bridge hands, the uniform model on ordered samples without replacement yields the same probabilities for any event $A$ as does the uniform model on unordered samples, when the latter permits computation of $P(A)$.

6.° Let $j_1, j_2, \ldots, j_k$ be non-negative integers with $j_1 + j_2 + \cdots j_k = m$.
(a) How many distinct partitions are there of a population of $m$ objects into $k$ non-overlapping sub-populations,

the first of which has $j_1$ elements,

the second of which has $j_2$ elements,

.

.

.

the $k$th of which has $j_k$ elements?

Hint: Represent the original population as a sequence of $m$ boxes. Assigning $j_1$ 1's to $j_1$ of the $m$ boxes picks out the first sub-population, assigning $j_2$ 2's to $j_2$ of the remaining $m - j_1$ boxes picks out the second sub-population, and so forth.

(b) How many distinct partitions are there of the original population into non-overlapping sub-populations, one of which has $j_1$ elements, *another* one of which has $j_2$ elements, and so forth?

7. Let us assign $j$ distinguishable balls in a uniform manner to $m$ boxes. What is the probability that exactly:

(a) no balls are assigned to the first box?

(b) one ball is assigned to the first box?

(c)° $k$ balls are assigned to the first box?

Hints: We assume that we are dealing with a uniform model over ordered samples of size $j$ from the set $\{1, 2, \ldots, m\}$ with replacement. The outcome $(n_1, n_2, \ldots, n_j)$ indicates that ball $i$ was assigned to box $n_i$. By allowing replacement, we are permitting more than one ball per box. For part (c), if balls $1, \ldots, k$ were assigned to box 1, then there would be $j - k$ balls to assign to boxes $2, 3, \ldots, m$. The number of such assignments is $(m - 1)^{j-k}$. That is, each choice of $k$ of the $j$ balls to be assigned to box 1 yields $(m - 1)^{j-k}$ assignments of those remaining. But there are $\binom{j}{k}$ choices for the $k$ balls to be assigned to box 1.

8. Suppose $j$ balls are to be assigned to $m$ boxes.

(a)° What is the total number of possible assignments, if we do not distinguish one ball from another, but we do distinguish boxes? For example, for $j = 4$, $m = 2$ we do not consider the assignment

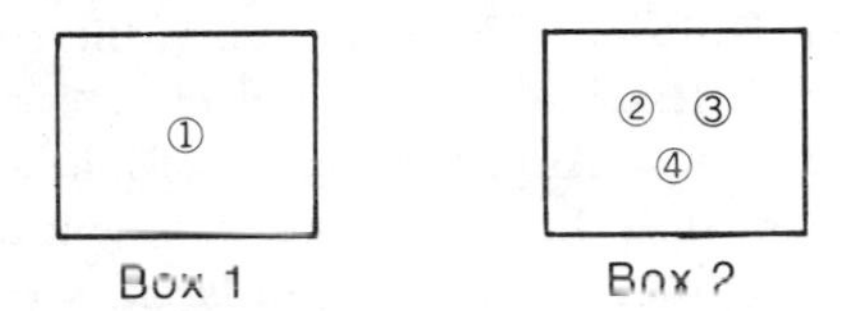


to be different from the assignment

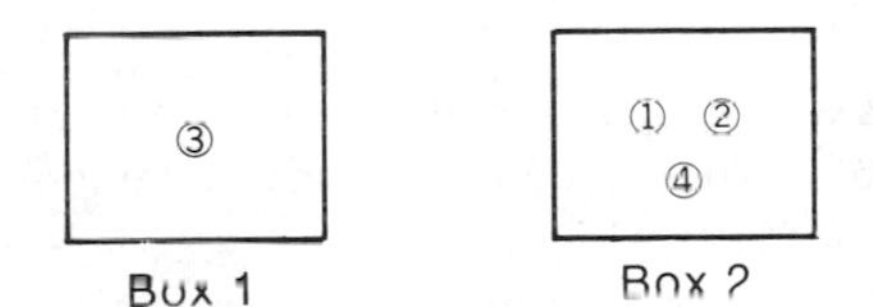


but both are distinguished from

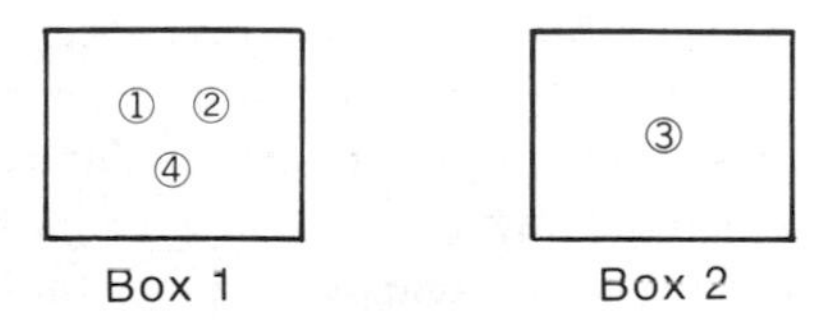


Hint: We represent the $m$ boxes by the spaces between $m + 1$ vertical bars, and the $j$ balls by $j$ asterisks

|*|***|

for the first two cases. Now we consider that each asterisk and bar take up one position of a total of $m + 1 + j$ available positions, where the two end positions must be occupied by bars. Thus, each assignment of balls to boxes corresponds to a choice of the set of positions for the $j$ asterisks from the set of $m + 1 + j - 2$ available positions.

(b) What is the number of distinguishable assignments if each box must contain at least one ball? Hint: Take the first $m$ balls and place one in each box. Then we are left with the problem of part (a) with $m$ boxes and $j - m$ balls.

Problems of assigning balls to boxes are of real importance in certain parts of theoretical physics; for example, balls might correspond to electrons and boxes to different possible orbits.

## SECTION 3
## Discrete Infinite Models

Until now we have only considered experiments with finitely many possible outcomes. We will now present several examples to show that this is too restrictive in many situations. Typically, this is true when we do not know a priori how many trials are required before the experiment is concluded.

### *3.1 Example: Tossing Coin until First Head*

Assume that we toss a coin until the first head appears. Let the outcome be the number of tosses required. In reality, we can only continue tossing the coin until the onset of total exhaustion. Thus, in reality, there can only be finitely many possible outcomes. Nonetheless, it is difficult to decide a priori on an upper bound for the number of tosses required. Also, for reasons of mathematical convenience and simplicity, we prefer to idealize this coin toss experiment and take the sample space to be the set of all positive integers, which is an *infinite* set. △

### *3.2 Example: Gambler's Ruin*

A gambler plays a sequence of games against an adversary in which the probability that the gambler wins \$1.00 on any given game is a known value $p$, and the probability of his losing \$1.00 is $1 - p$ (here $0 \leq p \leq 1$). We assume that initially the gambler possesses $a$ dollars and his adversary possesses $(b - a)$ dollars, so that the total amount of money between the two players is $b$ dollars. The games continue until either of the two players runs out of money, since no credit is allowed. If the gambler runs out of money first, we say that the gambler has been ruined. As in the previous example, physical limitations lead to a finite sample space, and yet it is conceptually easier to allow an infinite sample space. If our only interest is in the duration of the game, the points in the first sample space to suggest itself might correspond to the possible numbers of plays required for one or the other player to be ruined. It can be shown that almost surely one of the players must be ruined (that is, the probability that the game will ultimately terminate is 1). We shall see later that this natural sample space, the set of positive integers, is really not adequate for an analysis of the gambler's ruin problem.

The gambler's ruin problem may seem somewhat frivolous, but the same model is used for some important applications to physics, in particular for describing the diffusion of particles in a gas. △

We shall make free use of the notation and elementary results of the theory of infinite series. A summary of what we use from this subject is found in the Appendix (1.7). In most of our examples convergent infinite series may be manipulated as if they were simply finite sums.

**3.3 Definition: The Infinite Discrete Model.** An *infinite discrete probability system* consists of an infinite set $\mathcal{S} = \{e_1, e_2, \ldots\}$ called the sample space,† together with non-negative real numbers $p_j = P(e_j)$ with $\sum_{k=1}^{\infty} p_k = 1$. The probability $P(A)$ of an event $A$ (subset of $\mathcal{S}$) is given by

$$P(A) = \sum_{e_j \text{ in } A} p_j \cdot$$

(It can be shown that this sum is always meaningful.)

### *3.4 Example: Tossing Coin Until Heads Comes Up*

Let us consider a possibly biased coin whose probability of heads is $p \neq 0$. Using reasoning similar to that in the answer to Problem 2.19(2), we postulate that the probability of $k - 1$ tails followed by a head is $(1 - p)^{k-1}p$. (In the next chapter, this model will be developed from another viewpoint.) But since the event "$k - 1$ tails followed by a head" is the same as "$k$ tosses until the first head," it follows that the probability that $k$ tosses will be required for the first head should be taken as $(1 - p)^{k-1}p$.

Thus, we see why, in the experiment of tossing a (biased) coin until the first head, we take

$$\mathcal{S} = \{1, 2, \ldots\}$$

(Here an outcome of $k$ means that the first head appeared on the $k$th toss.) and

$$p_k = (1 - p)^{k-1}p \qquad k = 1, 2, \ldots .$$

This probability assignment is called the *geometric distribution*. Using the properties of the geometric series summarized in the Appendix, we see that

$$\begin{aligned}
\sum_{k=1}^{\infty} p_k = \sum_{k=1}^{\infty} (1 - p)^{k-1}p &= p\sum_{k=1}^{\infty} (1 - p)^{k-1} \\
&= p\sum_{j=0}^{\infty} (1 - p)^j \qquad (\text{let } j = k - 1) \\
&= p \cdot \frac{1}{1 - (1 - p)} \\
&= 1 \qquad\qquad \text{for} \quad p \neq 0
\end{aligned}$$

as required.

---

† For mathematical convenience we often allow the sample space to include points which we may not believe to be possible outcomes.

Note that the probability that at least 100 throws will be required is

$$\sum_{k=100}^{\infty}(1-p)^{k-1}p = p(1-p)^{99}\sum_{j=0}^{\infty}(1-p)^{j} = (1-p)^{99}$$

(again using the geometric series—see Appendix). For $p = 1/2$, this is a rather small number, approximately $1.6 \times 10^{-30}$. For $p = 1/99$, it is about $1/2.7$, which we see from the well-known result that $\lim_{n\to\infty}\left(1-\frac{1}{n}\right)^{n} = \frac{1}{e}$. △

In many situations, we would like to know in advance a high probability upper bound for the number of trials required to achieve a success. Suppose, in fact, we are given a small positive number $\alpha$, and we desire to find $j$ such that

$$P\{\text{number of trials} \leq j\} \geq 1-\alpha.$$

But for any given $j$,

$$\begin{aligned} P\{\text{number of trials} \leq j\} & \sum_{k=1}^{j}(1-p)^{k-1}p \\ &= p\sum_{r=0}^{j-1}(1-p)^{r} \quad \text{(letting } r = k-1\text{)} \\ &= p\,\frac{1-(1-p)^{j}}{1-(1-p)} \quad \text{(using the geometric series—see Appendix)} \\ &= 1-(1-p)^{j}. \end{aligned}$$

Thus, we want $1-(1-p)^{j} \geq 1-\alpha$ or

$$(*) \qquad (1-p)^{j} \leq \alpha.$$

Taking logarithms, we reduce this to

$$j \geq \frac{\log \alpha}{\log (1-p)}.$$

Another use for such a computation is a statistical one. Suppose we perform the experiment and observe that $j_0$ trials were required. We might want to determine which values of $p$ yield a "poor explanation" for the observed number of trials, that is, for which $p$ is

$$(**) \qquad P\{\text{number of trials} \geq j_0\} \leq \alpha.$$

But

$$P\{\text{number of trials} \geq j_0\} + P\{\text{number of trials} < j_0\} = 1.$$

Hence

$$P\{\text{number of trials} \geq j_0\} = 1 - P\{\text{number of trials} < j_0\}$$

and (**) is equivalent to

$$1 - P\{\text{number of trials} < j_0\} \leq \alpha,$$

that is,

$$P\{\text{number of trials} < j_0\} \geq 1-\alpha$$

or

$$P\{\text{number of trials} \leq j_0 - 1\} \geq 1-\alpha.$$

From reasoning like that which led to (*), we see that (**) holds when $p$ satisfies

$$(1-p)^{j_0-1} \leq \alpha,$$

or

$$1-p \leq \alpha^{1/(j_0-1)},$$

or

$$p \geq 1-\alpha^{1/(j_0-1)}.$$

That is, $p \geq 1-\alpha^{1/(j_0-1)}$ would yield a small probability ($\alpha$) of requiring as many trials as were actually made. Hence we might use the observed number $j_0$ of trials to exclude all $p$ in $[1-\alpha^{1/(j_0-1)}; 1]$ as possible "explanations" of what we observed. For instance, if $\alpha = 1/2^{10}$ and $j_0 = 6$, then all $p$ in $[1-1/4; 1] = [3/4; 1]$ are poor explanations for the six trials required, since for $p$ in $[3/4; 1]$ the probability that six or more trials are needed for at least one success is less than $1/2^{10} = 1/1024$. This idea will be more carefully formulated when we study the theory of confidence intervals in Chapters 9 and 10.

### 3.5 *Example: Repeat Experiment until jth Success*

Suppose we are given an experiment whose probability of success is $p \neq 0$ and we repeat this experiment $k$ times. Using reasoning similar to that in the answer to Problem 2.19(2), we find it reasonable to assign to the event

"$j-1$ successes in the first $k-1$ trials and a success on the $k$th trial"

the probability $\binom{k-1}{j-1} p^{j-1}(1-p)^{(k-1)-(j-1)}p$ for $k = j, j+1, \ldots$. But, similarly to the previous example, the event

"$j-1$ successes in the first $k-1$ trials and a success on the $k$th trial"

is the same as

"$j$th success achieved on the $k$th trial."

It follows that the probability that $k$ tosses will be required for the first achievement of $j$ successes should be taken as

$$\binom{k-1}{j-1} p^j (1-p)^{k-j} \qquad k = j, j+1, \ldots.$$

Therefore, for the composite experiment of repeating a given experiment until $j$ successes are achieved, we now see why we take the sample space $\mathcal{S} = \{j, j+1, \ldots\}$, where an outcome of $k$ means that the $j$th success was achieved on the $k$th trial, and we take

$$p_k = \binom{k-1}{j-1} p^j (1-p)^{k-j}.$$

This probability assignment is called a *Pascal distribution*. △

### 3.6 *Example: Gambler's Ruin*

Suppose that the gambler of Example 3.2 starts out with \$1.00 and his adversary with \$100.00. Further suppose that every sequence of $m$ trials of the form

$$(W, L, \ldots, W)$$

has probability $1/2^m$, where $W$ (or $L$) in the $i$th position means that the gambler wins (or loses) \$1.00 on the $i$th trial. (This uniform model corresponds to taking the parameter $p$ introduced in Example 3.2 to be 1/2.) Thus we are postulating a uniform model for every sequence of $m$ trials. We compute the probability that the game terminates at the $n$th step for a few values of $n$. Certainly the probability of termination in one step is 1/2, since this corresponds only to the sequence $(L)$. Termination in precisely two steps is impossible, for if the game does not terminate in one step, the gambler must win on the first trial, and he then has a fortune of \$2.00. He therefore cannot be ruined in only one more trial. Termination in three steps corresponds to the sequence $(W, L, L)$ and hence has probability 1/8. Termination in four steps is again impossible. Termination in five steps corresponds to the sequences $(W, L, W, L, L)$ and $(W, W, L, L, L)$ and hence has probability $1/2^5 + 1/2^5 = 1/2^4$; that is,

$$p_1 = \tfrac{1}{2},\ p_2 = 0,\ p_3 = \tfrac{1}{8},\ p_4 = 0,\ p_5 = \tfrac{1}{16}\,. \qquad \triangle$$

### 3.7 *Example: Poisson Model*

A model that has proved empirically to be a reasonable description for the number of flat tires due to nails in a time period of length $t$, the number of telephone calls in a time period of length $t$, the number of atomic decays in such an interval, or the number of power failures due to simultaneous activation of too many electrical appliances, may be constructed by taking $\mathcal{S}$ to be the set of non-negative integers and

$$p_k = e^{-\lambda t}\frac{(\lambda t)^k}{k!}, \qquad k = 0, 1, 2, \ldots,$$

where $\lambda$ is some fixed positive number (representing the expected number of occurrences of the specific event in question per unit time). This is called a *Poisson model,* and it is also frequently used as a convenient approximation to certain finite models (in particular, to describe the behavior of the number of successes in $n$ repeated trials when $n$ is large and $np^3$ is small).

In Chapter 12 we shall study some of the statistical procedures used to determine the adequacy of such a description. In Chapter 7 we shall investigate a theoretical justification for the Poisson model. $\triangle$

## *3.8 EXERCISES*

1.° In tossing a possibly biased coin until the first head appears, what is the probability that the number of tosses is even?
Hint: $x^{2k} = (x^2)^k$.

2. An experiment whose probability of success is .4 is repeated until a success is obtained.

(a)° What is the probability that at least three such experiments will be required?

(b)° Find a number $j$ such that the probability that more than $j$ trials will be required is less than .01.

(c) What is the probability that the number of trials will be an integral multiple of three?

(d) What is the probability that the number of trials will be of the form $3j + 1$ if $j$ is a non-negative integer?

(e) If the probability of success is $p$, what would you expect the answers for (c) and (d) to approach as $p \to 0$? Can you prove your conjecture?

3. In the gambler's ruin problem of Example 3.6, compute $p_6$ and $p_7$.

4.° In the gambler's ruin problem of Example 3.6, suppose the probability of the gambler's success on each play is $p$, and each sequence with $k$ wins and $n - k$ losses has probability $p^k(1 - p)^{n-k}$. Find $p_1, p_2, \ldots, p_7$.

## *3.9 PROBLEMS*

1. Suppose that the probability $p_k$ that $k$ phone calls are initiated in a time period of length $t$ hours is given by

$$p_k = e^{-10t}\frac{(10t)^k}{k!}.$$

Find a simple upper bound for:

(a)° the probability that more than $j$ phone calls are initiated in a six-minute period.

(b) the probability that more than $j$ phone calls are initiated in a one-hour period.

2. In Example 3.4, we constructed an interval $[p_u; 1] = [1 - \alpha^{1/(j_0-1)}; 1]$ such that $p$ values in this interval furnished poor explanations for the required number of tosses $j_0$. The value $p_u = 1 - \alpha^{1/(j_0-1)}$ is called an upper $1 - \alpha$ confidence limit for $p$. Can you construct an interval $[0; p_l]$ with similar properties, where all $p$ in this interval give the event "observed number of trials $\leq j_0$" a probability less than or equal to $\alpha$?

## 3.10 EXERCISES

The following exercises are intended for those with access to a high-speed digital computer; however, they may be "set up" profitably even by those without such facilities.

1. In launching a satellite aimed at Mars, it is given that the probability of success is .7. If it is decided that satellite launches will be attempted until three successes have been obtained, what is the probability that fewer than 10 attempts will be needed (see Example 3.5)?

2.° What should the probability of success on one trial of an experiment be so that the probability of needing fewer than 10 trials for three successes be approximately .95?

3. Suppose that the probability of no flat tires in a time period of $t = 5$ hours is .9 (see Example 3.7). What is the probability of fewer than 20 flat tires in a time period of 200 hours?

4.° It is not unreasonable to assume that the probability $p_k$ of $k$ arrivals at a service counter in a time period of length $t$ hours is given by

$$p_k = e^{-\lambda t} \frac{(\lambda t)^k}{k!}.$$

(a) We asserted earlier that $\lambda$ represents the expected number of (in this case) arrivals per hour. Compute the probability of fewer than $j$ arrivals in one hour for $j = 1, 2, 3, \ldots, 10$ and $\lambda = .1, .5, 2.5, 12.5$ to convince yourself of the reasonableness of our assertion.

(b) In order to determine whether or not to add another cash register, the number of arrivals at the counter is counted for a three-hour period. Another register will be added if the number of arrivals exceeds 15. Compute the probability of installing an additional register as a function of $\lambda$ for $\lambda = 1, 2, \ldots, 9$ per hour. The above procedure may be thought of as a statistical test of whether or not $\lambda$ is "small"; if $\lambda$ is small then an additional register is not required. (We will go thoroughly into the theory of such tests in Chapter 11.)

5. Compute the probability that more than $j$ phone calls are initiated in a six-minute time period for $j = 0, 1, 2, 3, 4, 5, 6$ corresponding to part (a) of Problem 3.9(1) and compare your result with the upper bound obtained there.
Hint: Use the complementary event in these computations so as to work with a finite series.

# 3

# General Probability Spaces

## SECTION 1
## General Sample Spaces

In all experiments considered so far, the sample space $\mathcal{S}$ consisted at most of the set of positive integers or of a set which may be put into one-to-one correspondence with the positive integers. Despite the fact that in reality we know that our measuring instruments limit the sample space to consist of a finite number of elements, we have already seen that infinite sample spaces are often needed to set up an appropriate model. At this point, we present some examples to show that even sample spaces as large as the set of positive integers may not always be adequate for theoretical treatment.

### *1.1 Example: Spinning Pointer*

A pointer is spun and we measure the angle $\theta$ at which it comes to rest, as indicated in Figure 3.1. For theoretical purposes it is simplest to let the sample space consist of all possible angles, that is, all real numbers $x$ satisfying $0 \leq x < 2\pi$. It can be proved that this sample space, the interval $[0, 2\pi)$, is "too big" to be put into one-to-one correspondence with the set of positive integers; the essential idea behind this demonstration will be given in the next example.

This particular sample space arises frequently in electronics for describing the time elapsed from the switching on of a sine wave generator to the start of the next cycle. △

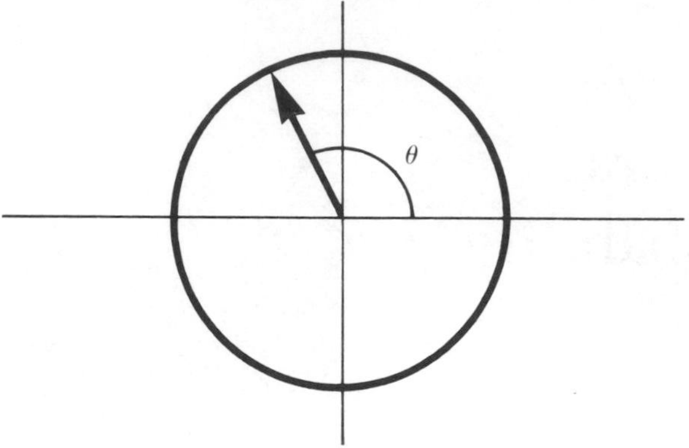


*FIGURE* **3–1.** Spinning pointer.

### *1.2 Example: Coin Tosses*

For the generality we are likely to require in coin toss problems, we may take the sample space, $\mathcal{S}$, to be the set of all infinite sequences of $H$'s and $T$'s. It is instructive to show that this set is too big to be put into one-to-one correspondence with the set of positive integers. We do this as follows:

> Suppose that a one-to-one correspondence between $\mathcal{S}$ and the set of positive integers were possible. Then we would be able to arrange the set of all infinite sequences of $H$'s and $T$'s in a single list that might look as follows:
>
> 1. $S_1 = (H, T, H, \ldots$
> 2. $S_2 = (T, H, \ldots$
> 3. $S_3 = (T, T, H, \ldots$
>
> . . .
> . . .
> . . .
>
> But we can show that no matter what is in this list, or how it is arranged, at least one sequence of $H$'s and $T$'s must be missing; that is, no such list can be exhaustive. To do this, define $M$ as that sequence whose $i$th coordinate value is opposite to the $i$th coordinate value of $S_i$; that is, for each $i$, if we find $H$ in the $i$th coordinate of $S_i$, then we put $T$ in the $i$th coordinate of $M$, and if $T$ appears in the $i$th coordinate of $S_i$ then we put $H$ in the $i$th coordinate of $M$. We see that $M$ is different from each $S_i$ in our proposed list, and thus $M$ is not in our list. Thus we see that we can never hope to put the set of all infinite sequences of $H$'s and $T$'s into a one-to-one correspondence with the set of positive integers. This technique of proof, called *diagonalization*, was developed by Georg Cantor (1845–1918), the founder of modern set theory.
>
> Here we have given an example of a sample space that is useful for handling a wide class of problems but that is larger than the set of positive integers. If we now identify $H$ with 1 and $T$ with 0, we see that the set of infinite decimals consisting solely of 0's and 1's is too large to be put into a one-to-one correspondence with any set of positive integers. From this it follows that the larger set, the interval $[0, 2\pi)$, cannot be put into a one-to-one correspondence with the set of positive integers, as asserted in the previous example. △

### 1.3 *Example: Height, Blood Pressure, Component Life*

In problems concerned with measurements of people's height, of blood pressure, or of the length of operating time of a component prior to failure, the most manageable sample space is the set of all non-negative real numbers. As in Example 1.2, this set is too large to be put into one-to-one correspondence with the positive integers. Thus, the notion of a probability system as developed in the previous chapter does not apply to these situations. △

### 1.4 *Example: Electrocardiogram*

An electrocardiogram is a graph of electrical activity of the heart (in volts) over some time interval. Each electrocardiogram may be thought of as a continuous real valued function whose domain is some time interval of positive duration. For the experiment of recording a cardiogram, the sample space might be the set of all continuous real valued functions with domain $[0, T]$, $T > 0$. It can be shown that this sample space is too large to be put into a one-to-one correspondence with the set of positive integers. △

### 1.5 *Example: Target*

In an experiment of shooting at a target whose radius is 1 yard, it is reasonable to let the sample space consist of the inside of a circle of unit radius. Here each point of the sample space corresponds to a possible point of impact. This sample space is again too large to be put into a one-to-one correspondence with the positive integers. In this example, it might be reasonable in certain cases to assign equal probabilities to areas of equal size. Thus the probability of a subset $A$ would be

$$P(A) = \frac{\text{area of } A \text{ (sq. yds.)}}{\pi \text{ (sq. yds.)}}$$

(since $\pi$ square yards is the area of a circle of radius 1 yard). △

Notice that in this case the elementary events are simply the points inside the unit circle. But since the area of a point is 0, we see that all elementary events would have probability 0. From this it is evident that simply assigning reasonable probabilities (in this case 0) to all elementary events may not generally be sufficient to determine the probabilities of other events of interest. This becomes clear if we consider that assigning each point area 0 is of no use in determining the area of sets in general. The basic reason for this difficulty lies in the fact that *points are not the fundamental objects* needed to develop the concept of area. We shall see that ***in the cases being treated here, the elementary events are no longer the fundamental sets to which probabilities are assigned.*** We shall return to these problems subsequently.

---

### *1.6 EXERCISES*

1.° The voltage across a resistor is measured with a meter whose scale goes from 0 to 10.

(a) List several possible sample spaces.

(b) In each of these sample spaces, conjecture some sets to which you believe probabilities should be assigned in order to compute probabilities of interesting sets.

2. The breaking strength of steel rods 1 inch in diameter is measured. List two sample spaces for this experiment.

3.° Diastolic (heart relaxed) and systolic (heart pumping) blood pressures are measured. Diastolic pressure may vary from 30 to 250 mm. of mercury and systolic pressure from 50 to 300 mm. of mercury, with systolic pressure always exceeding diastolic pressure.

(a) Give some reasonable sample spaces for this experiment.

(b) What sets in these sample spaces would you conjecture to require probabilities in order for other interesting sets to have computable probabilities?

Hint: The area of a set is determined by assigning area to rectangles.

4. Give a reasonable sample space for the experiment of measuring the lifetimes of the $n$ components of a television set.

5.° Seismographic activity is measured from time 0 to time $T$ on a device which can record values between 0 and 10 (Richter scale). What is a reasonable "large" sample space for this situation?

6. Give a reasonable "large" sample space for the experiment of recording electrical brain activity (electroencephalogram). You may think of the experiment as recording the voltage from some number $j$ of electrodes for a fixed period of time.

7. Give a reasonable sample space for measuring the age ($n$ years), height, and body temperature:

(a) for one individual.

(b) for 10 individuals.

## SECTION 2
## Elementary Set Manipulations and Their Use

As we mentioned before, the basic problem of probability is that of computing probabilities of sets which are related to other sets whose probabilities are known. We are therefore led to investigate how sets can be related to each other.

### 2.1 Membership and Subsets

Let $\mathcal{S}$ be an arbitrary set of objects. If $x$ is any one of the objects in $\mathcal{S}$, we write $x \in \mathcal{S}$, which is read "$x$ is a member of $\mathcal{S}$." If $x$ is not a member of $\mathcal{S}$, we write $x \notin \mathcal{S}$.

If $A$ is a set each of whose members is also in $\mathcal{S}$, that is, if whenever $x \in A$ then $x \in \mathcal{S}$, then we say $A$ is a *subset* of $\mathcal{S}$ and denote this by $A \subseteq \mathcal{S}$. More generally, if $A \subseteq \mathcal{S}$ and $B \subseteq \mathcal{S}$, and if whenever $x \in A$ then $x \in B$, we write $A \subseteq B$ ($A$ is a subset of $B$).

### 2.2 *Example: Coin Toss Sequences*

Let $\mathcal{S}$ consist of sequences of length 100 of $H$'s and $T$'s. The sequence $x$ whose first 10 coordinate values are $H$'s and whose remaining 90 coordinate values are $T$'s is an element of $\mathcal{S}$; that is, $x \in \mathcal{S}$. The collection $A_{10}$ of all sequences of length 100 with exactly 10 $H$'s and 90 $T$'s is a subset of $\mathcal{S}$, that is, $A_{10} \subseteq \mathcal{S}$. △

### 2.3 *Example: Inclusion Versus Membership*

It is necessary to distinguish inclusion ($\subseteq$) from membership ($\in$). If $\mathcal{S}$ stands for the United States Senate, then Senator Smith $\in \mathcal{S}$. On the other hand, the Foreign Relations Committee $\subseteq \mathcal{S}$. The whole Senate $\mathcal{S} \subseteq \mathcal{S}$. ($\mathcal{S}$ is a *subset* of $\mathcal{S}$, but *not* a member of $\mathcal{S}$, since the entire Senate is *not* a Senator.) △

Frequently a set may be described in two apparently different ways. For example, the set of sequence of length $n$ of alternate $H$'s and $T$'s, that is,

$$\{(H, T, H, T, \ldots), (T, H, T, H, \ldots)\},$$

is the same as the set of sequences of $H$'s and $T$'s of length $n$ which have no two $H$'s in succession and no two successive $T$'s.

Sometimes our initial description of a set may not be the most convenient one for computing its probability. We often want to show that two different descriptions yield the same set; that is, that two sets which may be described differently are in fact the same (equal). To show the equality of two sets, we must show that they have exactly the same membership. If $A$ and $B$ are two such sets, then to show that $A = B$, we must show that

$$\text{if } x \in A, \text{ then } x \in B \quad \textit{and} \quad \text{if } x \in B, \text{ then } x \in A,$$

that is,

$$A \subseteq B \quad \textit{and} \quad B \subseteq A.$$

### 2.4 *Example: Sum on Dice*

Let $\mathcal{S}$ be the set of ordered pairs $(i, j)$ of integers, with $1 \leq i \leq 6$ and $1 \leq j \leq 6$. Then the subset $A$ of $\mathcal{S}$ corresponding to a sum of 8 is the same as the subset $B$ consisting of the ordered pairs $(6, 2)$, $(5, 3)$, $(4, 4)$, $(3, 5)$ and $(2, 6)$. That is, $A = B$. △

To help simplify our set descriptions, we introduce the following notation.

## 2.5 Set Notation

When a set consists of a small number of elements, it may be possible to describe the set by listing its elements within braces. The set whose elements are 1, 2, and 4 would be written $\{1, 2, 4\}$.

Let $\mathcal{S}$ be a set and let $q$ be a property possessed by some (but not necessarily all) elements of $\mathcal{S}$. Then we often denote the subset of $\mathcal{S}$ consisting of those elements of $\mathcal{S}$ having property $q$ by

$$\{x \in \mathcal{S}\colon x \text{ has property } q\}.$$

For the set in Example 2.4, we have

$$A = \{(i, j) \in \mathcal{S}\colon i + j = 8\}.$$

## 2.6 *Examples: Set Descriptions*

(a) The set consisting of the integers strictly between 0 and 7 is $\{1, 2, 3, 4, 5, 6\}$. If $\mathscr{I}$ is the set of integers, we may also write this set as $\{x \in \mathscr{I}\colon 1 \le x \le 6\}$.

(b) Let $\mathcal{S}$ be the set of positive integers and $A$ be the set of even positive integers. Then $A = \{x \in \mathcal{S}\colon \text{for some } n \in \mathcal{S},\ x = 2n\}$.

(c) Let $\mathcal{S}$ be the set of sequences $x$ of length $n$ of $H$'s and $T$'s, and let $A$ consist of those sequences $x$ with $n/2$ or more $H$'s. Let $x_i$ be the number of $H$'s (1 or 0) in the $i$th coordinate of the sequence $x$. Then

$$A = \left\{x \in \mathcal{S}\colon 2 \sum_{i=1}^{n} x_i \ge n\right\}.$$

(d) Let $\mathscr{R}$ be the set of real numbers and let $\mathscr{I}r$ be the set of irrational numbers. Then

$$\mathscr{I}r = \{x \in \mathscr{R}\colon \text{for every pair } m, n \text{ of integers } x \ne m/n\}. \qquad \triangle$$

## 2.7 Complementation

Let $A \subseteq \mathcal{S}$. The set of those $x \in \mathcal{S}$ with $x \notin A$ is called the *complement of A* (relative to $\mathcal{S}$) and is denoted by $A^c$. Here it will usually be clear which set we are taking as the sample space $\mathcal{S}$. It is sometimes quite convenient to illustrate these concepts geometrically, in a Venn diagram. In Figure 3.2 we have $B \subseteq A$, $A \subseteq \mathcal{S}$, $x \in A$, $x \notin B$, $y \in A^c$, $x \in B^c$, $y \notin A$, and $A^c \subseteq B^c$.

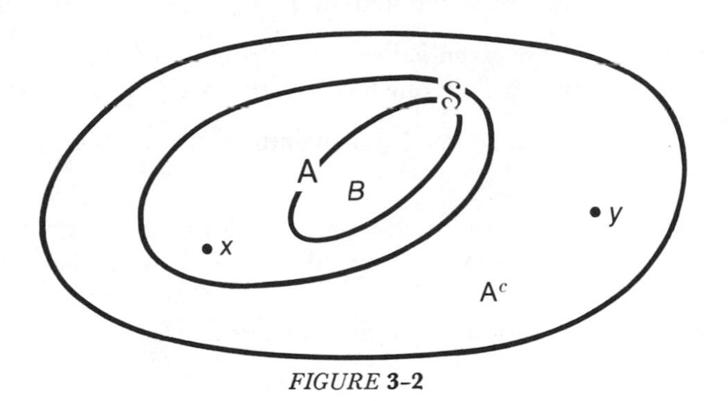


*FIGURE* **3-2**

## 2.8 *Examples: Complements*

(a) In the toss of two dice with sample space $\mathcal{S} = \{2, 3, 4, 5, 6, 7, 8, 9, 10, 11, 12\}$, the event "even" is

$$E = \{2, 4, 6, 8, 10, 12\}.$$

Its complement, "odd," is

$$E^c = \{3, 5, 7, 9, 11\}.$$

(b) In tossing a coin 100 times, we may let $\mathcal{S}$ consist of all sequences of length 100 of $H$'s and $T$'s. If $A_i$ is the subset of $\mathcal{S}$ in which the $i$th coordinate value is an $H$ (the event "heads on the $i$th toss"), then $A_i^c$ consists of those sequences in $\mathcal{S}$ whose $i$th coordinate value is a $T$. Let $B_{10}$ consist of all those elements of $\mathcal{S}$ which have at least 10 $H$'s. Then $B_{10}^c$ consists of all sequences of length 100 of $H$'s and $T$'s with fewer than 10 $H$'s. Also,

$$\{(H, T, H, T, \ldots), (T, H, T, H, \ldots)\}$$
$$= \{x \in \mathcal{S}: x \text{ has no run of length } 2\}$$
$$= \{x \in \mathcal{S}: x \text{ has at least one run of length } 2\}^c. \quad \wedge$$

## 2.9 Unions and Intersections

Let $A_1, A_2, \ldots$ be subsets of $\mathcal{S}$. Then we define $\bigcup_n A_n$ to be the set of all elements of $\mathcal{S}$ which are in at least one of the sets $A_n$. We call $\bigcup_n A_n$ the *union* of the sets $A_n$.

We define $\bigcap_n A_n$ to be the set of those elements of $\mathcal{S}$ which are simultaneously in each of the sets $A_n$. We call $\bigcap_n A_n$ the *intersection* of the sets $A_n$. If there are only finitely many sets being considered, we sometimes write

$$\bigcup_{n=1}^{k} A_n \quad \text{or} \quad A_1 \cup A_2 \cup \cdots \cup A_k \quad \text{for} \quad \bigcup_n A_n.$$

Occasionally we may only be interested in $\bigcup_n A_n$, where $A_n$ is defined only for $n_0 \leq n \leq k$; here $n_0$ and $k$ are given values. Referring to the experiment in Example 2.8, we see that the event "at least one heads in the last 20 tosses" can be described as $\bigcup_{n=81}^{100} A_n$, where $A_n$ is the event "heads on the $n$th toss."

Similar remarks pertain to intersections. Additionally, the notation $AB$ is sometimes used in place of $A \cap B$. Any other variations in the notation will be rather obvious from the context.

We illustrate the concepts of union and intersection in Figure 3.3.

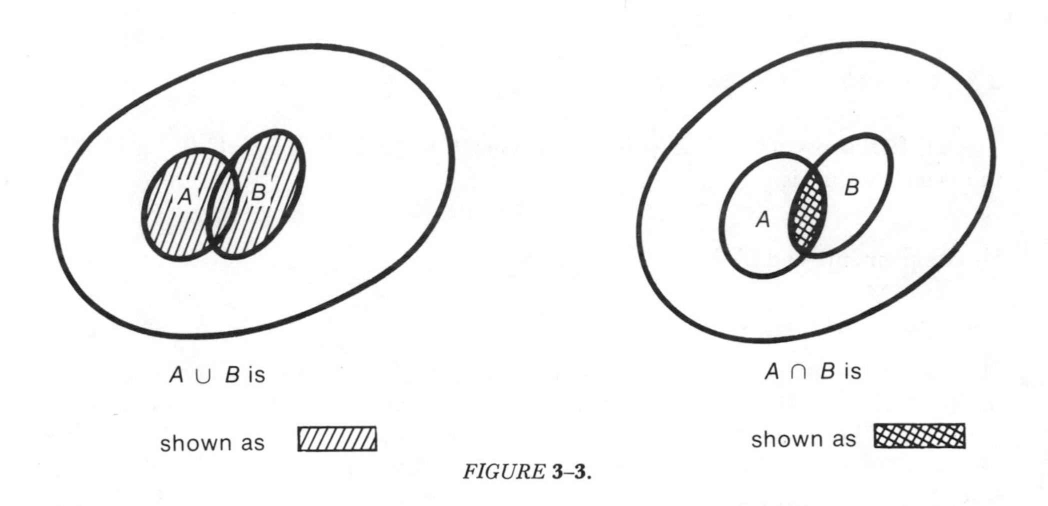


*FIGURE* **3–3.**

### *2.10 Example: Union and Intersection*

In the experiment of tossing a coin 100 times, let $\mathcal{S}$ be the set of all sequences of length 100 of $H$'s and $T$'s, and for $x \in \mathcal{S}$ let $x_i$ be the $i$th coordinate value of $x$ (that is, $x_i = H$ if there is an $H$ in the $i$th coordinate of the sequence $x$). For $i = 1, 2, \ldots, 100$, let $A_i = \{x \in \mathcal{S}\colon x_i = H\}$, so that $A_i$ is the event "heads on the $i$th toss." Then $\bigcup_n A_n$ is the event "at least one heads in 100 tosses," that is, $\bigcup_n A_n$ is the set of all sequences in $\mathcal{S}$ in which heads appears at least once. We see that $\bigcap_n A_n$ is the event "all heads in 100 tosses," that is, the event consisting of the sequence in $\mathcal{S}$ with *all* heads. △

### *2.11 Example: Reliability*

An electronic system consisting of $N$ components is said to be a *series system* if failure of at least one component causes system failure. It is called a *parallel system* if the system fails only when all components fail. (In general, system failure cannot be described in such simple terms. However, such descriptions are useful as first approximations, and do describe certain types of switching circuits rather well.) If $A_i$ denotes the event "failure of the $i$th component," then for the series case, system failure is the event $\bigcup_n A_n$, and for the parallel case, system failure is the event $\bigcap_n A_n$.

△

**2.12 Definition: Disjoint Sets.** If two sets $A$ and $B$ have no elements in common, we say that they are *disjoint*. More generally, if $A_1, A_2, \ldots$ is a sequence of sets, then we say that the sequence is *pairwise disjoint* if, whenever $i \neq j$, the pair $A_i$ and $A_j$ are disjoint.

It is very convenient to insist that the intersection of each pair of sets be again a set, even when the pair is disjoint. Thus we introduce the symbol $\varnothing$, and write $A \cap B = \varnothing$ whenever $A$ and $B$ are disjoint. We call $\varnothing$ the *empty set* or, in a probability context, the *impossible event*. The empty set is simply the set with no members at all (and recall that it is the members which determine a set). We see that $\varnothing$ is a subset of every set, since it is impossible to exhibit a member of $\varnothing$ which fails to be a member of any other set; that is, it is impossible to show $\varnothing \subseteq A$ to be false. Figure 3.4 illustrates disjoint sets.

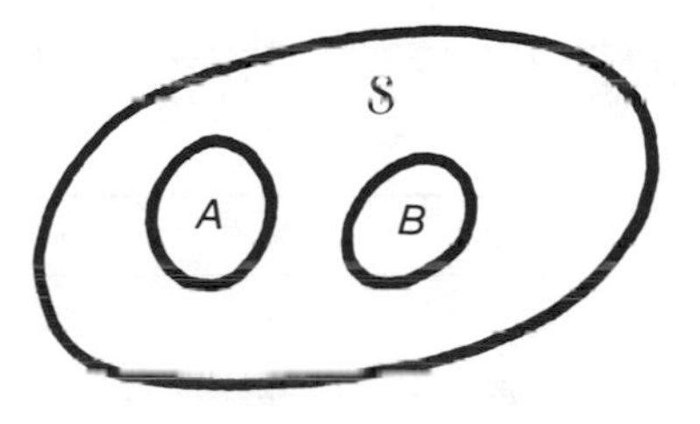


FIGURE 3–4. Two (pairwise) disjoint sets.

*2.13 Example: Disjoint Sets*

If a coin is tossed three times, the sample space may be taken as

$$\mathcal{S} = \{(H, H, H), (H, H, T), (H, T, H), (T, H, H), (T, T, H), (T, H, T), (H, T, T), (T, T, T)\}.$$

The event "heads on the first toss" is

$$H_1 = \{(H, H, H), (H, H, T), (H, T, H), (H, T, T)\}.$$

"Tails on the first toss" is

$$T_1 = \{(T, H, H), (T, H, T), (T, T, H), (T, T, T)\}.$$

Note that $T_1$ and $H_1$ are disjoint. "Tails on second toss" is

$$T_2 = \{(H, T, H), (T, T, H), (H, T, T), (T, T, T)\}.$$

Note that $H_1$ and $T_2$ are not disjoint. In fact,

$$H_1 \cap T_2 = \{(H, T, T), (H, T, H)\}.$$

In the experiment of tossing a pair of dice, we may take $\mathcal{S}$ to be the set of ordered pairs $(i, j)$ of integers $1 \leq i \leq 6$ and $1 \leq j \leq 6$, where $i$ represents the result appearing on the first die and $j$ the result appearing on the second one. Then the event

$$\text{“sum on faces} \leq 6\text{”} = \{(i,j) \in \mathcal{S}\colon\ i + j \leq 6\}$$

is disjoint from the event

$$\text{“sum on faces} > 8\text{”} = \{(i,j) \in \mathcal{S}\colon\ i + j > 8\}. \qquad \triangle$$

### 2.14 *Example: Set Operations* $A = (A \cap B) \cup (A \cap B^c)$

The preceding equation is intuitively true, since it merely asserts that $A$ consists of two parts: those points of $A$ inside $B$ and those points of $A$ outside $B$. This is illustrated in Figure 3.5.

To provide precise verification of the type needed when the result is not so clear, we must show that the sets $A$ and $(A \cap B) \cup (A \cap B^c)$ have exactly the same members. A method for demonstrating such set equalities is to make a table in which the truth and falsity of statements concerning membership in related sets is given in terms of the truth and falsity indicated for the statements $x \in A$ and $x \in B$.

**TABLE 3–1**

| $x \in A$ | $x \in B$ | $x \in A \cap B$ | $x \in A \cap B^c$ | $x \in (A \cap B) \cup (A \cap B^c)$ |
|---|---|---|---|---|
| $T$ | $T$ | $T$ | $F$ | $T$ |
| $T$ | $F$ | $F$ | $T$ | $T$ |
| $F$ | $T$ | $F$ | $F$ | $F$ |
| $F$ | $F$ | $F$ | $F$ | $F$ |

Note that the entries in the last column are determined from those of the previous two columns.

Thus, we see that in all cases $x \in A$ is true if and only if $x \in (A \cap B) \cup (A \cap B^c)$ is true. $\triangle$

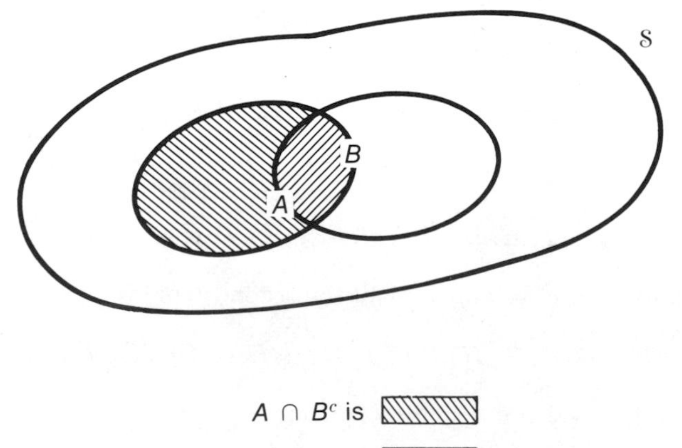


$A \cap B^c$ is [hatched pattern]

$A \cap B$ is [hatched pattern]

*FIGURE* **3–5.**

### 2.15 *Example: Set Operations* $\bigcup_n [AB_n]$

Let $B_1, B_2, \ldots$ be pairwise disjoint and $\bigcup_n B_n = \mathcal{S}$, and let $A$ be any subset of $\mathcal{S}$. We show that $A = \bigcup_n [AB_n]$.

For each $x$ in $\mathcal{S}$, we know that $x$ is in precisely one $B_i$—say $B_{i(x)}$. If $x \in A$, then $x \in A \cap B_{i(x)}$. Hence $x \in \bigcup_n [AB_n]$. Thus, whenever $x \in A$, we have $x \in \bigcup_n [AB_n]$. Similarly, if $x \in \bigcup_n [AB_n]$, then for some $n$ (say $n_0$) $x \in AB_{n_0}$, yielding $x \in A$. The sequence $AB_1, AB_2, \ldots$ is a partition of $A$, just as $B_1, B_2, \ldots$ is a partition of $\mathcal{S}$. △

The student should spend a little bit of time convincing himself of the legitimacy of equalities like the following, which are used continually, and some of which permit us to discard parentheses.

$$(A \cap B) \cap C = A \cap (B \cap C) \equiv A \cap B \cap C$$

$$A \cap B = B \cap A$$

$$A \cup B = B \cup A$$

$$(A \cup B) \cup C = A \cup (B \cup C) \equiv A \cup B \cup C.$$

---

### 2.16 *EXERCISES*

1. Referring to the sample space $\mathcal{S} = \{(H, H, H), \ldots, (T, T, T)\}$ of the first part of Example 2.13, for three tosses of a coin, let $H_i$ be defined as the event "heads on the $i$th toss" and $T_i$ as the event "tails on the $i$th toss." List the elements of:
   (a) $H_1 \cup H_2 \cup H_3$.
   (b)° $T_1 \cap T_2 \cap T_3$.
   (c) $T_1 \cap T_2 \cap T_3^c$.
   (d)° $H_1 \cup T_3$.
   (e) $(H_1 \cup T_2) \cap H_3$.
   (f)° $(H_1 \cap T_2) \cup H_3$.

2. In terms of the events defined in the previous question, write the events:
   (a)° tails at least once in the three tosses.
   (b) tails at least twice in the three tosses.
   (c)° heads on at least one of the first two trials.
   (d) tails on exactly one of the three trials.

3.° Throw two dice, and let $A_j$ be the event "die 1 comes up $j$" and $B_i$ be the event "die 2 comes up $i$." Write the event "sum comes up 9" in terms of the events $A_j$ and $B_i$.

4. Two bets are placed on successive horse races. If $A_{ij}$ is the event "winning \$$i$ in $j$th race," $i = -20, -19, \ldots, 18, 19, 20$, $j = 1, 2$, write the event $T_{10}$: total winnings \$10 in both races, in terms of the events $A_{ij}$.

5. Let $\mathcal{S}$ consist of all heads–tails sequences of length 100. Let $A_i$ be the event "heads on the $i$th toss."

(a)° Describe $A_i$ as a set of sequences from $\mathcal{S}$.

Hint: If $x \in \mathcal{S}$, let $x_i$ stand for the $i$th coordinate value of $x$. Then define $v(x_i) = 1$ if $x_i$ is heads, $v(x_i) = 0$ if $x_i$ is tails.

(b)° Write the event "exactly five heads" as a set of sequences.

(c) Write the event "at least five heads" as a set of sequences.

(d)° In terms of the $A_i$, write the event "heads from toss 70 on."

(e) Write the event "no heads from toss 70 on" in terms of the $A_i$.

(f)° Write the event "at least two heads" in terms of the $A_i$.

Hints: (i) Look at the complement of this event. (ii) The event "exactly one head" can be written as $\bigcup_{i=1}^{100} B_i$, where $B_i$ is the event "heads on $i$th trial, tails on all others."

6. Let $\mathcal{S}$ be the set of all sequences of length 100 of 0's and 1's, where a 1 in the $i$th coordinate indicates a success on the $i$th of 100 experiments. Let $x_i$ be the value in the $i$th coordinate of the sequence $x \in \mathcal{S}$.

(a)° Write the event "at least 70 successes" in terms of the $x_i$.

(b) Write the event "between 60 and 80 successes" in terms of the $x_i$.

7. Show that:

(a)° $A \cap (B \cup C) = (A \cap B) \cup (A \cap C)$.

(b) $A \cup (B \cap C) = (A \cup B) \cap (A \cup C)$.

Hint: See Example 2.14.

8. (a)° Write $A \cup B$ as a union of two disjoint sets.

Hint: Draw a picture.

(b) Write $\bigcup_n A_n$ as a union of disjoint sets.

9. Show that if $A \subseteq B$, then $A^c \supseteq B^c$ (that is, $B^c \subseteq A^c$).

10. A system consists of $N$ components and $A_n$ is the event "failure of $n$th component," $n = 1, 2, \ldots, N$.

(a)° Explain in terms of a series system (see Example 2.11) why $\left(\bigcup_n A_n\right)^c = \bigcap_n (A_n^c)$.

(b) Explain in terms of a parallel system (see Example 2.11) why $\left(\bigcap_n A_n\right)^c = \bigcup_n (A_n^c)$.

(c) Prove that the equations in (a) and (b) hold in general for infinitely many sets. These equations are called De Morgan's laws.

Hint: If $x \in \left(\bigcup_n A_n\right)^c$, then $x \notin \bigcup_n A_n$. Thus, for each $n$, $x \notin A_n$.

11. Given $A$, $B$, $A \cup B$, $A \cap B$, $A^c$, $B^c$, $A^c \cup B$, $A \cup B^c$, $A^c \cup B^c$, $A^c \cap B^c$, $A^c \cap B$, and $A \cap B^c$, which sets are subsets of which others?

12.° Two parallel systems of three elements each are run in series as shown:

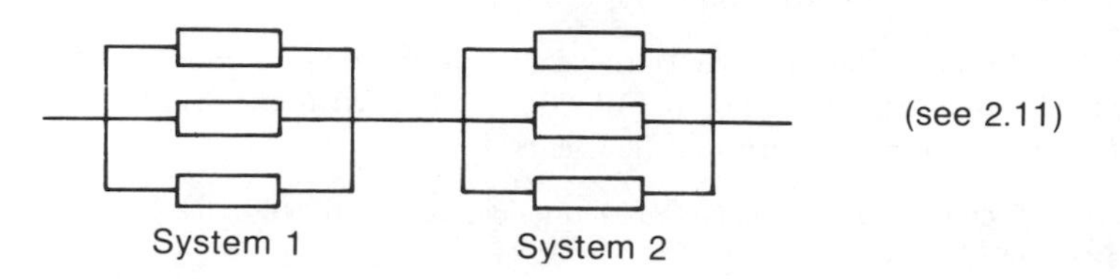

Let $A_{ij}$ be the event "failure of unit $i$ in system $j$," $i = 1, 2, 3$; $j = 1, 2$. Write the event unnumbered (fig.3.19) "failure of the combined system" in terms of the $A_{ij}$.

13.° Two series systems of four units each are run in parallel (see Example 2.11). Let $B_{ij}$ be the event "failure of unit $i$ in system $j$," $i = 1, 2, 3, 4$; $j = 1, 2$. Write the event "failure of the combined system" in terms of the $B_{ij}$.

14. Two series systems, the first with $n$ components and the second with $m$ components, are run in parallel (see Example 2.11). If $A_{ji}$ is the event "failure of component $j$ in system $i$" ($i = 1, 2$), how do we express the event "failure of the combined system"?

15. Let $A_{ji}$ represent the event "failure of component $j$ in system $i$," where system 1 is a series system, system 2 is a parallel system, and system 3 is a parallel system (see Example 2.11). These three systems are run in series. Describe the event "overall system failure."

## SECTION 3
## Axioms for a Probability System

When the sample space is "larger" than the set of positive integers, it can be shown* that it is often impossible to associate a probability with every subset of $\mathcal{S}$ and at the same time to satisfy some of the natural conditions that we impose, for example to make computations simpler or to reflect realistic conditions easily.

Given an initial reasonable assignment of probabilities to some collection of sets (such as the collection of all rectangles in the plane whose sides are parallel to the coordinate axes), which sets can be assigned probabilities consistent with the rules governing probabilities and the initial assignment? That is, if we say an *event* is any set to which a probability is attached, then we want to know what the class of events should consist of, given an initial assignment of probabilities to certain selected events. Our answer will be in line with the *frequency interpretation of* $P(A)$, in which $P(A)$ is thought of as the limiting proportion of repetitions of an experiment in which the outcome belongs to $A$, that is, in which $A$ occurs.

Surely $\mathcal{S}$ should be assigned a probability, and we should have $P(\mathcal{S}) = 1$, since every possible outcome is in $\mathcal{S}$. If $A$ is an event, then $A^c$ should also be an event, with $P(A^c) = 1 - P(A)$, since whenever the outcome is not in $A$ it is in $A^c$. Furthermore, suppose that $A_1, A_2, \ldots, A_j$ is any finite sequence of pairwise disjoint events. The proportion of outcomes in the union of these events is just the sum of the proportion of outcomes in each of these events. Hence, we shall want $\bigcup_n A_n$ to be an event and to have probability

**3.1**
$$P\left(\bigcup_n A_n\right) = \sum_n P(A_n).$$

It is also extremely convenient to include as events the union of any finite or infinite sequences of events, whether or not they are pairwise disjoint, and to insist that Equation 3.1 hold for all pairwise disjoint sequences of events. Using De Morgan's Law (Exercise 2.16(10)), it can be seen that the above assumptions imply that if $A_1, A_2, \ldots$ is any sequence of events, then $\bigcap_n A_n$ is an event.

---

* The methods needed for such a demonstration are, unfortunately, beyond the scope of this text.

We summarize the essence of the preceding discussion in our definition of an event class; this class is the type of collection of subsets of the sample space to which we shall want to attach probabilities.

**3.2 Definition: Event Class.** An *event class** $\mathscr{E}$ is a class of subsets of a sample space $\mathcal{S}$ such that:

(a) $\mathcal{S}$ is in $\mathscr{E}$ ($\mathcal{S}$ is an event).

(b) if $A$ is in $\mathscr{E}$, then $A^c$ is in $\mathscr{E}$ (the complement of an event is an event).

(c) if $A_1, A_2, \ldots$ is any finite or infinite sequence of sets in $\mathscr{E}$, then $\bigcup_n A_n$ is in $\mathscr{E}$ (the union of any sequence of events is an event).

(d) if $A_1, A_2, \ldots$ is any finite or infinite sequence of sets in $\mathscr{E}$, then $\bigcap_n A_n$ is in $\mathscr{E}$ (the intersection of any sequence of events is an event).

Note that (d) is a direct result of (b) and (c).

The definition of an event class is most useful for the general results and other definitions that follow from it. Despite the fact that it is not always possible to attach a probability to every subset of $\mathcal{S}$, in most problems of practical or theoretical importance this restriction is not a handicap. We shall not be greatly concerned with membership in the event class; it really belongs to a more advanced course.

**3.3 Definition: Probability System.** A *probability system*† is a triple, ($\mathcal{S}$, $\mathscr{E}$, $P$), where $\mathcal{S}$ is an arbitrary set (the sample space which includes all possible outcomes), $\mathscr{E}$ is an event class of subsets of $\mathcal{S}$, and $P$ is a real valued function defined for each $A \in \mathscr{E}$ such that:

(a) $0 \leq P(A) \leq 1$ for all $A$ in $\mathscr{E}$.

(b) $P(\mathcal{S}) = 1$.

(c) if $A_1, A_2, \ldots$ is any sequence of pairwise disjoint sets in $\mathscr{E}$, then

$$P\left(\bigcup_n A_n\right) = \sum_n P(A_n).$$

This property is called *countable additivity*. A function $P$ satisfying (a), (b), and (c) is called a *probability measure*. The elements of $\mathscr{E}$ are referred to as *events*.

It sometimes helps to think of $A$ as a physical object and of $P(A)$ as its mass, since the rules followed by a probability measure are essentially the same as those for assigning mass to parts of objects.

The reader should realize that it is possible for the outcome to be in $A$ even if $P(A) = 0$, as illustrated in the following example.

### *3.4 Example: Certainty of Occurrence of Event of Probability 0*

If $\mathcal{S}$ is the interval $[0, 1]$ and the probability assigned to each subinterval is simply the length of the subinterval, then each one-point set has probability 0, yet

* An event class is also called a *sigma algebra*, a *sigma field*, or a *Borel field*.

† Originally defined by A. N. Kolmogorov.

the outcome is bound to be in precisely one of these one-point sets. Nevertheless, the outcome of the first experiment is so unlikely to ever appear again that we may be confident that the limiting proportion of experiments in which it arises will be 0. Notice that this result seems unpleasant only if you are viewing the elementary events as the fundamental sets for computing probabilities (recall Example 1.5). △

We remark that $P(A)$ may be thought of as a measure of how confident we are that $A$ will occur, that is, that the outcome will be in $A$. There are those who prefer this subjective interpretation of probability to the frequency interpretation we gave earlier.

***The structure of a probability system is taken to be universal to all probability problems treated here.*** Every problem in the probability theory that we develop will involve a probability system. One of the main reasons for the difficulty of elementary probability stems, paradoxically, from the fact that this framework is so simple—it is difficult to accomplish much when very few tools are available.

We may think of probability as consisting of two parts, *mathematical probability* and *applied probability*. In applied probability we build specific models to reflect given real situations; we assign probabilities and make definitions which are useful for application. In mathematical probability we may attempt to relate given events $A_i$ to other events and derive various consequences which follow logically from our definitions and axioms. More precisely, we try to prove theorems about various kinds of probability systems (which often first arise in applied probability).

The following theorem contains some immediate consequences of the definition of a probability system.

### 3.5 Theorem

Let $A, B, A_1, A_2, \ldots$ all be in the event class $\mathscr{E}$ of some probability system. Then the following relations hold:

(a) $P(A^c) = 1 - P(A)$.

(b) if $B \subseteq A$, then $P(B) \leq P(A)$.

(c) $P(A \cup B) = P(A) + P(B) - P(A \cap B)$.

(d) $P\left(\bigcup_n A_n\right) \leq \sum_n P(A_n)$.

(e) $P(A \cap B) \geq 1 - P(A^c) - P(B^c)$.

(f) $P\left(\bigcap_n A_n\right) \geq 1 - \sum_n P(A_n^c)$.

*Proof.* Throughout the proof, although we do not do this explicitly, it can be shown that every subset considered is in $\mathscr{E}$.

(a) Since $A$ is an event, so is $A^c$, by Definition 3.2(b). Now $\mathcal{S} = A \cup A^c$ and $A$, $A^c$ is a disjoint pair of events. It follows from Definition 3.3(b) and 3.3(c) that

$$1 = P(\mathcal{S}) = P(A \cup A^c) = P(A) + P(A^c).$$

(b) If $B \subseteq A$, then (from Example 2.14)

$$A = (A \cap B) \cup (A \cap B^c) = B \cup (A \cap B^c).$$

Note that $B$ and $A \cap B^c$ are disjoint events. Hence, by Definition 3.3(c),

$$P(A) = P(B) + P(A \cap B^c)$$

and by Definition 3.3(a)

$$P(A) \geq P(B).$$

(c) As in Example 2.14, $B = (A \cap B) \cup (A^c \cap B)$, and therefore, due to the disjointness of $A \cap B$ and $A^c \cap B$,

$$P(B) = P(A \cap B) + P(A^c \cap B),$$

or

$$(*) \qquad P(A^c \cap B) = P(B) - P(A \cap B).$$

But, as in part (b), $A \cup B = A \cup (A^c \cap B)$ with $A$ and $A^c \cap B$ disjoint. Hence, by Definition 3.3(c),

$$P(A \cup B) = P(A) + P(A^c \cap B)$$

and by (*)

$$P(A \cup B) = P(A) + P(B) - P(A \cap B).$$

(d) We write $\bigcup_n A_n$ as a disjoint union

$$\bigcup_n A_n = A_1 \cup (A_2 \cap A_1^c) \cup (A_3 \cap A_2^c \cap A_1^c) \cup \cdots.$$

From Definition 3.3(c),

$$\begin{aligned} P\left(\bigcup_n A_n\right) &= P(A_1) + P(A_2 \cap A_1^c) + P(A_3 \cap A_2^c \cap A_1^c) + \cdots \\ &\leq P(A_1) + P(A_2) + P(A_3) \cdots \quad \text{from part (b) of this theorem} \\ &= \sum_n P(A_n). \end{aligned}$$

(e) Since $(A \cap B)^c = A^c \cup B^c$ (see answer to Exercise 2.16(10)(b)), we have by part (a) of this theorem,

$$\begin{aligned} P(A \cap B) &= 1 - P[(A \cap B)^c] \\ &= 1 - P(A^c \cup B^c) \\ &\geq 1 - P(A^c) - P(B^c) \end{aligned}$$

by part (d) of this theorem.

(f) The proof is essentially the same as in (e).

3.6 *Remarks.* Certain parts of the preceding theorem, part (b) in particular, seem so intuitively reasonable that it may seem superfluous to provide a proof. The purpose of a proof here is to demonstrate the adequacy of our model; if a proof could not be given, then surely the model would need substantial alteration to make it reflect reality. Parts (c) and (d) of this theorem are highly important for theory; parts (e) and (f) are useful when $A_i^c$ are all low probability events and the probabilities of $A$ and $B$ (or $P(A_i)$) are the the only known probabilities

We now present some illustrations of the uses of Theorem 3.5.

### *3.7 Example: Intersection of High Probability Events*

If the probability of fair weather is .97 and the probability of temperature above 65° is .92, what can be said about the probability of both fair weather and temperature above 65° occurring? Let $A$ be the event "fair weather" and $B$ be the event "temperature above 65°." Then $P(A^c) = 1 - P(A) = .03$ and $P(B^c) = 1 - P(B) = .08$. Hence, from Theorem 3.5(e), we have

$$P(A \cap B) \geq 1 - .03 - .08 = .89.$$

That is, the probability of both "fair weather" and "temperature above 65°" occurring is at least .89.

Note that if, instead of the given probability assignment, we had $P(A) = .2$, $P(B) = .3$, then this theorem, yielding the result $P(A \cap B) \geq 1 - .8 - .7 = -.5$, would be useless, although still correct. △

### *3.8 Example: Even on at Least One Toss*

Suppose a pair of fair dice are thrown. What is the probability that at least one of the dice comes up even? To solve this problem by a method simpler than mere enumeration, let $A_i$ be the event "$i$th die comes up even." Then we desire to compute $P(A_1 \cup A_2)$. Now $P(A_1) = 1/2$, $P(A_2) = 1/2$ (in our 36-point uniform sample space of pairs $(i, j)$), and

$$P(A_1 \cap A_2) = \frac{3 \times 3}{36} = \frac{1}{4}$$

(nine outcomes with even in both tosses). Hence, by Theorem 3.5(c),

$$P(A_1 \cup A_2) = 1/2 + 1/2 - 1/4 = 3/4.$$

The 1/4 must be subtracted because the probability of outcomes like (2, 4) and (4, 6) are added in both in $P(A_1)$ and in $P(A_2)$. This can also be seen from the following:

$$P(A_1 \cup A_2) = 1 - P[(A_1 \cup A_2)^c] = 1 - P(A_1^c \cap A_2^c) = 1 - 1/4 = 3/4. \quad \triangle$$

---

## *3.9 EXERCISES*

1. Suppose $A$, $B$, and $C$ are events in some sample space $\mathcal{S}$ such that $P(A) = .5$, $P(B) = .2$, and $P(C^c) = .3$. If possible, find each of the following probabilities. If it is not possible to do this, find the best possible upper and lower bounds consistent with the given information, and justify your conclusions.

(a)° $P(A^c)$.
(b)° $P(A \cup B)$.
(c) $P(A \cap C) + P(A^c \cap C)$.
(d) $P(A^c \cup B^c)$.
(e)° $P(A \cap B)$.
(f) $P(A^c \cap B)$.
(g)° $P(A \cup B \cup C)$.
(h) $P(A \cup B \cup C^c)$.

2. What is the probability of obtaining a 4 on at least one die, in tossing a pair of fair dice?

3.° Suppose that 10 components are connected in a series circuit, and that each component has probability .0001 of failure. What can be said about the probability of failure of the system (see Example 2.11)?

4.° Suppose that 15 components are connected in a parallel system and that each component has probability 1/20 of working properly. Is there any possibility that the probability of the system's working is .95? (See Example 2.11.)

5. Suppose that $P(A_i) \geq 1 - (.01/2^i)$, $i = 1, 2, \ldots$. Find a lower bound for $P\left(\bigcap_n A_n\right)$.

6. Suppose that $P(A_n) = 1/n^2$ for $n = 1, 2, \ldots$. Find an upper bound for:
(a)° the probability that for at least one $n \geq 10{,}000$ the outcome lies in $A_n$.
(b) the probability that for at least one $n \geq 1{,}000{,}000$ the outcome lies in $A_n$.

7.° Show that

$$P(A \cup B \cup C) = P(A) + P(B) + P(C) - P(AB) - P(AC) - P(BC) + P(ABC).$$

8. Find the probability that in the toss of three fair dice, at least one will come up even.
Hint: See Exercise 7.

## 3.10 PROBLEMS

1.° Show that

$$P(A_1 \cup A_2 \cup \cdots \cup A_n) = \sum_{1 \le i \le n} P(A_i) - \sum_{i,j:1 \le i < j \le n} P(A_i \cap A_j) + \sum_{i,j,k:1 \le i < j < k \le n} P(A_i \cap A_j \cap A_k) - \cdots + (-1)^{n+1} P\left(\bigcap_{1 \le j \le n} A_j\right)$$

where the notation $i, j\colon 1 \le i < j \le n$ indicates that the sum includes all subsets $A_i$, $A_j$ such that $1 \le i \le n$, $1 \le j \le n$, and $i < j$.

2.° Suppose that balls labeled $1, 2, \ldots, j$ are to be assigned to $j$ boxes (also numbered $1, 2, \ldots, j$), one ball per box, with all distinct assignments equally likely. What is the probability that at least one ball will be assigned to a box having the same number? Hint: See the result of the previous problem.

*Remarks.* This problem is called a *matching problem.* It also appears in the form, "If the hatcheck girl gives back hats at random to the $n$ customers, what is the chance that at least one customer gets his own hat?" We conclude this section with several examples of probability spaces, some of which have been introduced previously.

### 3.11 Example: Finite Set or Denumerable Set

Suppose that $\mathcal{S}$ is either a finite set or one that can be put into a one-to-one correspondence with the set of positive integers. If each element $e_j$ of $\mathcal{S}$ determines a member $\{e_j\}$ of the event class, $\mathcal{E}$, with $P\{e_j\} = p_j \ge 0$, $\sum_j p_j = 1$, then all subsets of $\mathcal{S}$ are in $\mathcal{E}$ with

$$P(A) = \sum_{j:e_j \text{ in } A} p_j.$$

### 3.12 Example: Roundoff Model

*Roundoff error* is error due to either limited readability of a measuring instrument or the limited accuracy of a digital computer (because of its finite "word length"). A model is: $\mathcal{S}$ is some interval $[a, b]$. A possible outcome $s \in \mathcal{S}$ is a number from $[a, b]$ and stands for the actual error = true value − stored value. For instance, if numbers were stored only to two decimals, then the error in storing $\pi$ would be $\pi - 3.14 = .00159\ldots$. Note that we may have available only an upper bound for this error. In cases like this, when we may not be able to observe the error, we may have to infer the usefulness of such a model from accurate prediction of related but observable events, like the total error in some computation in which the true answer is known and the computed answer is observable. (In such cases, we may need to make further assumptions concerning the relation between errors made in separate operations.)

The event family $\mathscr{E}$ includes all subintervals of $\mathcal{S}$, where for $[c, d]$ in $\mathscr{E}$ we assign

$$P[c, d] = \frac{d - c}{b - a}$$

if we believe that in some sense all outcomes are equally likely. In the measuring instrument case, when the measuring instrument has been properly calibrated (the so-called *unbiased* case), we take $a = -b < 0$. In this case, we think of the error as "all random," that is, none is systematic. In a computer, $[a, b]$ might be of the form $[-2^{-n}, 2^{-n}]$, or $[0, 2^{-n}]$, depending upon how the computer does its rounding off. △

### 3.13 *Example: Two-Dimensional Uniform Model*

A model useful in describing the joint behavior of a pair of roundoff errors is one that has as its sample space $\mathcal{S}$ the set of points $(x, y)$ in the plane for which $a \leq x \leq b$ and $a \leq y \leq b$. The probability assigned to any subsquare $A$ of $\mathcal{S}$ is (see Figure 3.6)

$$P(A) = \frac{\text{area of } A}{(b - a)^2}.$$

In fact, for this example the equation can be shown to hold for each subset $A$ of $\mathcal{S}$ that has an area. △

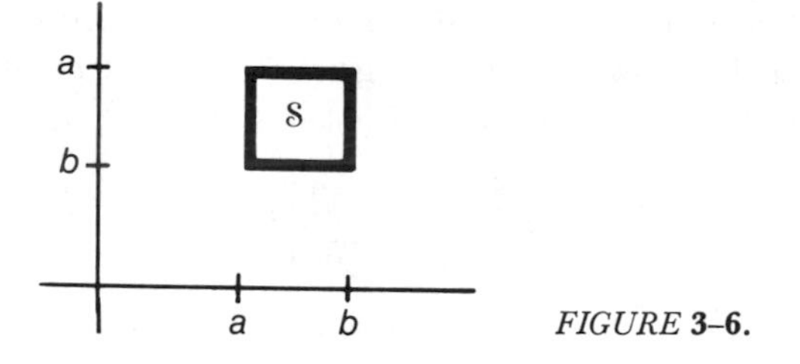


*FIGURE* **3–6.**

### 3.14 *Example: General One-Dimensional Density Model*

Let $f$ be any non-negative Riemann integrable† function with

$$\lim_{a \to \infty} \int_{-a}^{a} f(x)\, dx = 1$$

(this is usually written as $\int_{-\infty}^{\infty} f(x)\, dx = 1$). We let $P[a, b] = \int_a^b f(x)\, dx$ for each interval $[a, b]$. From this assignment we can compute the probabilities of all subsets

† That is, a function whose ordinary calculus integral $\int_\alpha^\beta f(x)\, dx$ is well defined for every pair $(\alpha, \beta)$ with $\alpha \leq \beta$.

of the real line of interest to us. A specific example is

$$f(x) = \begin{cases} \dfrac{3x^2}{2a^3} & \text{for} \quad -a \leq x \leq a, \\ 0 & \text{elsewhere,} \end{cases}$$

where $a$ is any fixed positive number. Another such example is

$$(x) = \begin{cases} \dfrac{\sin x}{2} & \text{for} \quad 0 \leq x \leq \pi, \\ 0 & \text{elsewhere.} \end{cases}$$

△

### *3.15 Example: Rainfall Model*

Suppose $\mathcal{S}$ to be the set of non-negative real numbers, and let the outcome $s \in \mathcal{S}$ stand for the amount of rainfall in some given time period. Let $f$ be a Riemann integrable function (see Example 3.14) such that:

(a) $f \geq 0$ and $f(x) = 0$ for $x < 0$.

(b) $\lim_{a \to \infty} \int_0^a f(x)\, dx \equiv \int_0^\infty f(x)\, dx = 1 - p < 1.$

We then assign each interval $[c, d] \subseteq \mathcal{S}$ with $c > 0$ the probability

$$P[c, d] = \int_c^d f(x)\, dx$$

and assign

$$P[0, d] = p + \int_0^d f(x)\, dx.$$

Here the event $[c, d]$ corresponds to total rainfall (in a fixed time period) between $c$ and $d$. The model seems to be a reasonable one to describe rainfall, since there is a non-zero probability $p$ of no rainfall in any given time period, but generally probability 0 of any other specific amount of rainfall. As in the previous example, the above information is sufficient to enable us to compute all probabilities of interest concerning total rainfall. △

---

### *3.16 EXERCISES*

1. Given the two-dimensional uniform roundoff model of Example 3.13, with $a = -.5$ and $b = .5$, what is the probability that both roundoff errors will be:
   (a)$^\circ$ positive?
   (b) in the same direction (that is, both positive or both negative)?

2. Given the one-dimensional uniform roundoff model of Example 3.12, with $a = -.5$, $b = .5$,

(a)° What is the probability that the first and second digits of the outcome are both even? (Zero is considered even.)

(b) What is the probability that the first and second digits of the outcome are both the same?

(c) Repeat parts (a) and (b) with $a = 0$ and $b = 1$.

3. In the two-dimensional uniform model of Example 3.13, suppose that a possible outcome $(x, y)$ represents the possible arrival times of a pair of individuals at some meeting place. If $a = 0$ and $b = 1$, what is:

(a)° the probability that the two individuals will arrive within 1/12 unit (5 minutes) of each other?

Hints: Draw a picture to help visualize this event, which can be written as

$$\{(x, y) \in \mathcal{S}: |x - y| \leq 1/12\} = \{(x, y) \in \mathcal{S}: -1/12 \leq x - y \leq 1/12\}$$
$$= \{(x, y) \in \mathcal{S}: -1/12 \leq x - y\} \cap \{(x, y) \in \mathcal{S}: x - y \leq 1/12\}$$
$$= \{(x, y) \in \mathcal{S}: y \leq x + 1/12\} \cap \{(x, y) \in \mathcal{S}: y \geq x - 1/12\}.$$

(b) the probability that the two individuals will arrive more than 1/4 unit (15 minutes) apart?

*Remark.* The extension of this problem to more than two variables occurs in problems in which near simultaneous occurrence of many events leads to something unusual, e.g., turning on many appliances in a short period of time leads to breakdown of the power transformer due to the surges of current. Chain reactions are a similar phenomenon, as well as floods due to local showers in many places simultaneously.

4. With the same model as in the preceding question, what is the probability of:

(a)° $\{(x, y) \in \mathcal{S}: y \leq x^2\}$.

(b) $\{(x, y) \in \mathcal{S}: y \leq 1/12\}$

5.° A target has three concentric circles, with radii of 1 yard, 3 yards, and 5 yards. Set up and solve the problem of finding the probability of a bullseye (a hit in the smallest circle). Please state your assumptions explicitly. It may be that your assumptions depend on the nature of the marksman.

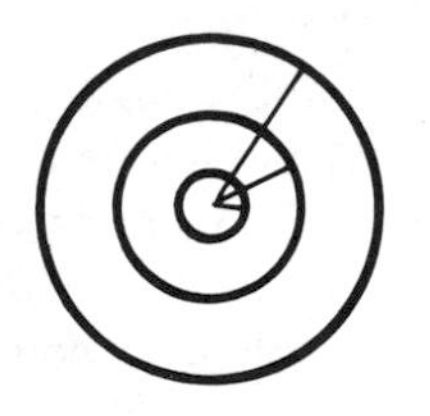

6. In the two-dimensional uniform model of Example 3.13, with $a = 0$ and $b = 1$, let

$$A_t = \{(x, y) \in \mathcal{S}: x \geq t\} \quad \text{and} \quad B_t = \{(x, y) \in \mathcal{S}: y \geq t\}.$$

(a) Compute $P(A_{.01} \cap B_{.01})$ and compare it with the lower bound for this event obtained via Theorem 3.5(e).

(b) Do the same as in part (a) for $P(A_{.2} \cap B_{.2})$.

(c)° If $(x, y)$ represent arrival times, find the probability that the first individual arrives later than the second one.

(d) As in (c), find the probability that the first individual arrives at least .2 units later than the second one.

## SECTION 4
## Conditional Probability

Consider a general probability model, $(\mathcal{S}, \mathcal{E}, P)$. For any given event $A$, we may think of $P(A)$ as a measure of our belief *before performing the experiment* that the outcome of the experiment will be in $A$. It happens in many situations that partial information concerning the outcome becomes available while the experiment is in progress. For example, complete election returns may come in from some states while the polls are still open in others, or data may be transmitted back to earth continuously during the orbiting of a space vehicle, or the face of one die of a pair that has been tossed may be known. It is thus reasonable to ask the question of *how we should modify our probability model to take account of partial information* obtained during the performance of the experiment or possibly after its conclusion. In fact, we must also inquire as to what type of information can reasonably be taken into account.

To be more specific, let $B$ be some given event and suppose that in every performance of the experiment we are told whether the actual outcome is in $B$ or in $B^c$ as soon as the information becomes available. (Note that in many experiments such information can become available prior to the termination of the experiment.) How should this information influence our feelings about the chances of occurrence of some other event $A$? How can we reasonably use this information to reassign a probability to $A$? In order to handle this partial information, we should modify the sample space to take into account the fact that it is now known whether the outcome is in $B$ or in $B^c$. We lose no generality by considering only the case in which the outcome is in $B$, since the other case would be handled in the same way. Under these circumstances, it seems reasonable to consider the event $B$ as our new sample space. Now, if we think of

$$P(A) = \frac{P(A \cap \mathcal{S})}{P(\mathcal{S})}$$

as the "proportion of space which $A$ occupies in $\mathcal{S}$," then in the new sample space we should take "the proportion of space which $A$ occupies in $B$" as the new probability of $A$, as illustrated in Figure 3.7. That is, the modified probability of $A$,

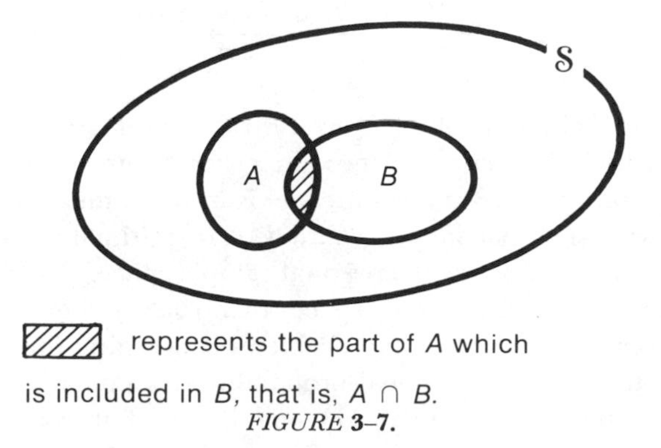


*FIGURE* **3–7.**

when the outcome is known to be in $B$, should equal $P(A \cap B)/P(B)$. Note that we divide by $P(B)$ so that the modified probability of $B$ (the probability of the new sample space) is 1. We are thus led to the following definition.

**4.1 Definition: Conditional Probability.** In the probability system ($\mathcal{S}$, $\mathcal{E}$, $P$), let $A$ and $B$ be events with $P(B) > 0$. We define the *conditional probability of $A$ given $B$*, denoted by $P(A \mid B)$, as

$$P(A \mid B) = \frac{P(A \cap B)}{P(B)} .$$

We read "$P(A \mid B)$" as the conditional probability of $A$ subject to the information that $B$ has occurred; it is *not* read as the probability of some event $A \mid B$, since $A \mid B$ is not any event in $\mathcal{S}$. $P(A \mid B)$ is our newly assigned or updated probability of $A$. Actually, we should think of all probabilities as conditional, with $P(A)$ being $P(A \mid \mathcal{S})$.

*4.2 Remarks.* We see that for any given $B$ with $P(B) > 0$, the function associating the value $P(A \mid B)$ with each event $A$ is a probability measure on $\mathcal{E}$ (see Problem 4.17(5)). We also notice that $P(A \mid B)$ will be large if "$A$ takes up a large proportion of $B$." Finally, it should be pointed out that conditional probability applies to a wide variety of situations. Examples would be the behavior of vacuum tubes or transistors given that they have functioned for a certain period of time, the height of a child given the height of its parents, and the incidence of a disease in two months given its present incidence.

The motivation given for our definition of $P(A \mid B)$ has the merit of simplicity. For those who prefer an operational approach, we will now show that the frequency interpretation of probability furnishes another argument that makes this a reasonable definition. Suppose we repeat the experiment some large number of times ($n$ times) and restrict consideration to precisely those experiments for which the outcome is in $B$. The conditional probability of $A$ given $B$ should be the limiting proportion of *these* experiments, such that the outcome is in $A$ (it is in $B$ already). But the number of the $n$ experiments in which the outcome is in $B$ should be "approximately" $nP(B)$, while the number of these $n$ experiments in which the outcome is in $A$ as well as in $B$ should be "approximately" $nP(A \cap B)$. Hence, the limiting proportion we are looking for should be approximately

$$\frac{nP(A \cap B)}{nP(B)} = \frac{P(A \cap B)}{P(B)} .$$

Note that the definition of $P(A \mid B)$ restricts consideration to precisely those experiments in which $B$ occurs. Hence, in order to use $P(A \mid B)$ legitimately in practical applications, whenever the outcome is in $B$ we must be so informed.

In dealing with such partial information, it is legitimate to convert to the new sample space, $B$, and to discard the original sample space, $\mathcal{S}$. If the only problems to be solved were in the new sample space, then such a course of action would be reasonable. However, there are many problems concerning the entire sample space $\mathcal{S}$ which require the use of conditional probabilities; to mention just two, there are the Stratified Sampling Theorem (4.11) and Bayes' Theorem (4.12). Therefore, we usually will not discard the original sample space $\mathcal{S}$ when dealing with partial information, but will adhere to Definition 4.1.

### 4.3 *Example: Conditional Probabilities with Dice*

We toss two fair dice and assume a uniform model on the 36 ordered pairs in the sample space. What is the probability that the sum is 9, given that the first die comes up 5?

Let $F_5$ be the event that the first die comes up 5, and $S_9$ be the event that the sum is 9. From the definition,

$$P(S_9 \mid F_5) = \frac{P(S_9 \cap F_5)}{P(F_5)}.$$

But $S_9 \cap F_5 = \{(5, 4)\}$ and

$$F_5 = \{(5, 1), (5, 2), (5, 3), (5, 4), (5, 5), (5, 6)\}.$$

Hence $P(S_9 \cap F_5) = 1/36$ and $P(F_5) = 6/36$. Thus

$$P(S_9 \mid F_5) = \frac{1/36}{6/36} = 1/6 .$$

Similarly, we may compute $P(F_5 \mid S_9)$, that is, the probability that the first die comes up 5 given that the sum is 9. By definition,

$$P(F_5 \mid S_9) = \frac{P(F_5 \cap S_9)}{P(S_9)},$$

and since $S_9 = \{(6, 3), (5, 4), (4, 5), (3, 6)\}$, we see that $P(S_9) = 4/36$. Thus

$$P(F_5 \mid S_9) = \frac{1/36}{4/36} = 1/4. \qquad \triangle$$

### 4.4 *Example: Conditional Probabilities with Coins*

If three coins are tossed, and it is known that precisely two heads come up (but we do not know on which coins), what is the "new" probability of the event $A$, that the first coin came up $H$ and the second one $T$?

We assume a uniform model over the eight possible sequences that make up the sample space. Let $B$ be the event that precisely two heads come up. Then

$$B = \{(H, H, T), (H, T, H), (T, H, H)\},$$

and

$$A = \{(H, T, H), (H, T, T)\}.$$

Hence

$$A \cap B = \{(H, T, H)\}$$

and

$$P(A \mid B) = \frac{P(A \cap B)}{(PB)} = \frac{1/8}{3/8} = 1/3. \qquad \triangle$$

### 4.5 *Example: Conditional Probability of Lifetime*

Suppose that the probability that a given component in some system lasts for at least $x$ hours is $e^{-7x}$. Such a model is related to the Poisson distribution in Problem 5.7, Chapter 5. (The theoretical justification of the Poisson model is given in Chapter 7.) We want to find the probability that the component will last at least $y$ additional hours, given that it has already survived $z$ hours. Here we let $B$ be the event that the component has survived at least $z$ hours, and $A$ the event that the unit survives at least $z + y$ hours (that is, that it survives at least $y$ additional hours). Note that

$$B = \{x \in \mathscr{R}: x \geq z\}, \qquad A = \{x \in \mathscr{R}: x \geq z + y\}$$

where $\mathscr{R}$ denotes the set of real numbers.

Because in this case $A = A \cap B$, the desired probability

$$P(A \mid B) = \frac{P(A \cap B)}{P(B)} = \frac{P(A)}{P(B)} = \frac{e^{-7(z+y)}}{e^{-7z}} = e^{-7y}.$$

Notice the rather striking result that the chance of surviving $y$ additional hours is the same as the chance of surviving the *initial* $y$ hours of life. That is, in this example there appears to be no "wearout" of the component during the time period being considered. Such a model is actually quite realistic when the cause of failure is unrelated to wearout—for example, an accident may demolish a car that had not begun to wear out, or a generator may fail due to a temporary overload. △

### 4.6 *Example: Arrivals under Conditions*

Suppose that arrivals of a pair $\mathsf{x}, \mathsf{y}$ of individuals are governed by the model of Example 3.13 (see also Exercise 3.16(3)), the uniform model on the square with vertices (0, 0) hour and (1, 1) hour. Knowing that individual $\mathsf{y}$ arrived after 12:30 (.5 on the $\mathsf{y}$ scale), what is the probability that individual $\mathsf{x}$ arrives later than individual $\mathsf{y}$?

Let $B$ be the event that individual $\mathsf{y}$ arrives after 12:30 and let $A$ be the event that individual $\mathsf{x}$ arrives later than individual $\mathsf{y}$. Examine Figure 3.8. Since

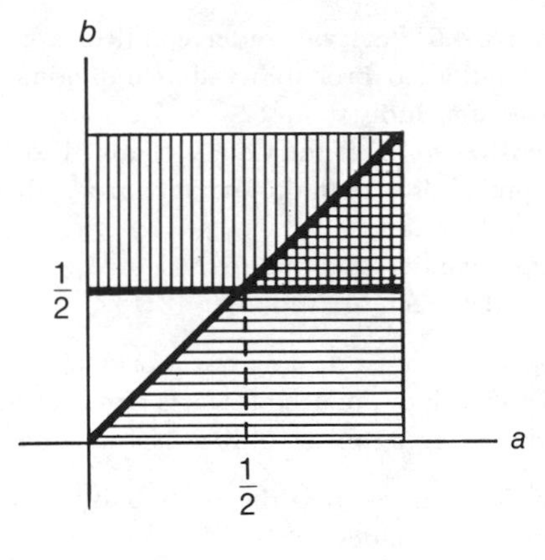


a is

b is

*FIGURE 3.8*

that individual $x$ arrives later than individual $y$. Examine Figure 3.8. Since probabilities in this case are simply areas, we see that

$$P(A \mid B) = \frac{P(A \cap B)}{P(B)} = \frac{1/8}{1/2} = 1/4. \qquad \triangle$$

---

### 4.7 EXERCISES

1. Suppose $A$, $B$, $C$, and $D$ are events with

$$P(A) = .1 \qquad P(B) = .3 \qquad P(C) = .8 \qquad P(D) = .95$$

and $P(A \cap C) = .05$. Find the best possible upper and lower bounds for the conditional probabilities below consistent with the given information and justify:

(a)° $P(A \mid C)$.
(b) $P(C \mid A)$.
(c)° $P(A \mid B)$.
(d)° $P(A \mid B^c)$.
(e) $P(B \mid A)$.
(f)° $P(D \mid A)$.
(g) $P(A \mid A \cap C)$.
(h) $P(A \cap C \mid A)$.
(i)° $P(A^c \mid C)$.
(j) $P(A \mid C^c)$.

2. If two dice are tossed and the sum is given to be 7, what is the probability:
(a)° that the smaller of the two values is less than 4?
(b) that the amount on the first die is less than 4?

3. Given that in three tosses of a fair coin there are a total of two heads, what is the probability that the first toss was heads?

4. Referring to Example 4.6, "arrival under conditions":
   (a)° What is the conditional probability that individual $\mathbf{y}$ arrives after 12:30, given that he arrives after individual $\mathbf{x}$?
   (b) What is the answer to (a) if the time is changed to 12:45?
   (c) What is the probability that the arrivals are within $\mathbf{z}$ hours, $0 \leq \mathbf{z} \leq 1$, of each other?
   (d) What is the probability that the arrivals are within $\mathbf{z}$ hours, $0 \leq \mathbf{z} \leq 1$, of each other, given that each arrived after 12:15?

5. What is the probability that in dealing out a bridge hand:
   (a)° the person who receives the ace of hearts also receives the king of hearts?
   (b) the person who receives the ace of hearts receives no other hearts?

6. In a five-card poker hand, what is the probability of a flush in spades, given that the first four cards dealt are spades?

7. Suppose that the probability of any given sequence of $k$ successes and $n - k$ failures is $p^k(1 - p)^{n-k}$, where $p$ is given (see 2.19(2) in Chapter 2). What is the probability that in 18 repetitions all of the successes will be among the last nine trials:
   (a)° given that there is one success?
   (b) given that there are two successes?
   (c) given that there are eight successes?

8. The probability that a given component in a system lasts for at least $x$ hours is $e^{-(x^2)}$. What is the conditional probability that it will last at least one more hour, given that it has already survived one hour? Does this unit show evidence of wearout?

9. Given two fair dice, when is the set of probabilities for die one unaltered by knowledge of the sum? In other words, when is $P(D_{1i} \mid S_j) = P(D_{1i})$, where $D_{1i}$ is the event "die 1 comes up $i$," and $S_j$ is the event "sum is $j$"?

10. Recalling Example 4.5, suppose that the probability that a unit lasts for at least $t$ hours is $e^{-\lambda t}$, where $\lambda > 0$ is given. Let $A_t$ be the event "unit lasts at least $t$ hours." Show that $P(A_{z+y} \mid A_z) = P(A_y)$.

11. Show that a discrete analog of the probability assignment in the previous problem is given by letting $P(B_k) = (1 - p)^k p$, where $p \in (0, 1)$ is some given number and $B_k$ is the event "survival for exactly $k$ units." (That is, the *geometric distribution* is a discrete analog of the exponential distribution, in regard to lack of wearout.)

Hint: $A_k = \bigcup_{j=k}^{\infty} B_j$. See the geometric series in the Appendix.

In the remainder of this section we go somewhat deeper into the study of conditional probability. We first illustrate the fact that in many circumstances it is natural to assign both unconditional and conditional probabilities in setting up an appropriate model.

### *4.8 Example: Urn Model*

The following type of model is used in studying the spread of epidemics. An urn contains 20 white balls with red spots and 200 pure white balls. A ball is selected at random from the urn. If it is pure white, it is returned to the urn with two additional pure white balls; if it has red spots, it is returned with five additional red spotted balls. The process is repeated indefinitely. The pure white balls may correspond to healthy people and the red spotted balls to people having some contagious disease, such as measles.

Let $R_j$ be the event "red spotted ball selected on the $j$th drawing." We ask how $P(R_1)$ and $P(R_2)$ compare, where $P(R_2) > P(R_1)$ would mean that initially the incidence of the disease tends to rise. By the hypothesis of selection at random (see 2.3(6) in Chapter 2), we know that $P(R_1) = 1/11$. But notice that here, because we are drawing at random, we also know $P(R_2 \mid R_1)$, $P(R_2 \mid R_1^c)$, $P(R_3 \mid R_1 \cap R_2)$, and so forth. For example, since $R_1^c$ means that the first ball chosen is pure white, and we therefore put in three pure white balls to replace the first one, we see that $P(R_2 \mid R_1^c) = 20/202$ (20 is the number of red spotted balls in the urn just prior to the second draw and 202 is the total number of balls in the urn just prior to the second draw if a pure white ball was selected on the first drawing). To find $P(R_2)$, we make use of the fact that

$$R_2 = R_2 \cap \mathcal{S} = R_2 \cap [R_1 \cup R_1^c] = [R_2 \cap R_1] \cup [R_2 \cap R_1^c].$$

That is, we break up the event $R_2$ into the two possibilities that can occur on the first drawing. Since the union just given is disjoint, we have

$$P(R_2) = P(R_2 \cap R_1) + P(R_2 \cap R_1^c).$$

Now

$$P(R_2 \mid R_1) = \frac{P(R_2 \cap R_1)}{P(R_1)},$$

and therefore $P(R_2 \cap R_1) = P(R_1)P(R_2 \mid R_1)$. But if $R_1$ occurs (that is, the first ball drawn has red spots), the urn's contents are altered to contain 25 red spotted balls and 200 pure white balls. So $P(R_2 \mid R_1) = 25/225$, and therefore $P(R_2 \cap R_1) = (1/11) \cdot (25/225)$. A similar computation yields $P(R_2 \cap R_1^c) = (10/11) \cdot (20/222)$, and consequently

$$P(R_2) = \tfrac{1}{11} \cdot \tfrac{25}{225} + \tfrac{10}{11} \cdot \tfrac{20}{222} = \tfrac{337}{3663} > \tfrac{1}{11} = P(R_1).$$

We conclude that in this model the disease initially has a tendency to become more widespread. △

We remark that in this last example we may view the urn mechanism simply as a convenient description of how to alter the probabilities governing disease spread. However, suppose we had started off with an urn containing exactly as many balls as the population being considered, and suppose that whenever a spotted ball was drawn, we put spots on five white balls in the urn (returning the spotted ball to the urn). When a white ball was drawn it was simply returned to the urn. Finally, any ball which had been spotted for $k$ drawings had its spots removed with probability $p_k$. Then we might view this as a simulation of the behavior of the disease in the population. Much more realism can be built into an epidemic model. For example, we could build models in which there was a greater contagion in certain seasons.

### *4.9 Example: Voting Preference*

Let the proportion of women who vote Republican be $p_w$ and the proportion of men who vote Republican be $p_m$. The proportion of men in the population is $\mathcal{M}$. What is the probability that a person chosen at random from the voter population is a Republican?

Let $R$ be the event that a person selected at random from the voter population votes Republican and let $M$ be the event that this person is male. Hence

$$\begin{aligned} P(R) &= P(R \cap M) + P(R \cap M^c) \\ &= P(R \mid M)P(M) + P(R \mid M^c)P(M^c) \\ &= p_m \cdot \mathcal{M} + p_w(1 - \mathcal{M}). \end{aligned}$$

△

### *4.10 Example: Gambler's Ruin*

Recall Example 3.2, Chapter 2, and assume that after each play the probability that the gambler will be ruined depends solely on the remaining amount of his fortune. Let $p_a$ denote the probability that the gambler will be ruined if his initial fortune is $a$ dollars, where the total amount of money owned by the gambler and his adversary is $b$ dollars. On each play, the gambler wins \$1 with probability $p$ or loses \$1 with probability $1 - p$. Let $R$ be the event that the gambler is ruined and let $W$ be the event that the gambler wins the first play. Then

$$\begin{aligned} p_a = P(R) &= P(R \cap [W \cup W^c]) \\ &= P[(R \cap W) \cup (R \cap W^c)] \\ &= P(R \cap W) + P(R \cap W^c) \\ &= P(R \mid W)P(W) + P(R \mid W^c)P(W^c) \\ &= p_{a+1} \cdot p + p_{a-1}(1 - p), \end{aligned}$$

since if the gambler wins the first play, his fortune increases by \$1 to $\$(a + 1)$, and the problem is approached similarly if he loses. Thus

$$(*) \qquad p_a = pp_{a+1} + (1 - p)p_{a-1} \qquad \text{for} \qquad a = 1, 2, \ldots b - 1.$$

Since no credit is allowed, we have

$$(**) \qquad p_0 = 1 \qquad \text{and} \qquad p_b = 0.$$

We may think of (*) and (**) as a system of $b + 1$ linear equations in $b + 1$ unknowns $p_0, p_1, \ldots, p_b$ and solve them by the usual Gaussian elimination method of adding and subtracting equations. There are also other specialized techniques available for their solution (see Feller [8]).

As a final remark, we note that we never specified a sample space for this problem. There is, however, a natural one, namely the set of all infinite sequences of wins and losses. △

It sometimes happens that a population divides into non-overlapping subsets in a rather natural way. For instance, we may classify a population according to economic groups, religious or ethnic groups, age groups, and so on. Quite frequently it is not difficult to closely approximate the size of the various groups $B_j$ forming the population (that is, to estimate $P(B_j)$) when the experiment is that of choosing an individual at random from the population. Such estimation will be carefully treated in Chapter 9. Similarly, it may be rather easy to closely approximate the frequency of occurrence of some attribute $A$ in each of the $B_j$ (that is, to estimate $P(A \mid B_j)$ closely for each $j$) when the experiment is that of choosing a person at random from group $B_j$. For example, we may know fairly precisely the proportion of people in each age group owning a color television set. From the given information, it is possible to determine accurately the frequency of the attribute $A$ in the overall population, that is, the probability of $A$ in the experiment of drawing a random sample from the population. A technique that uses this idea is of major importance in the rather rapid and (sometimes) accurate election forecasts made as partial returns come in, and is known as stratified sampling. The basic result of stratified sampling is given in the following theoretical counterpart of the foregoing discussion.

### *4.11 Theorem: Stratified Sampling*

Let $A, B_1, B_2, \ldots$ be events and suppose that:
(a) $B_i$ and $B_j$ are disjoint whenever $i \neq j$.
(b) $\bigcup_k B_k = \mathcal{S}$, the sample space.
(c) $P(B_k) > 0$ for each $k$.
Then

$$P(A) = \sum_k P(A \mid B_k) P(B_k).$$

*Proof:*

$$\begin{aligned} P(A) &= P(A \cap \mathcal{S}) \\ &= P\Big(A \cap \Big[\bigcup_j B_j\Big]\Big) \\ &= P\Big(\bigcup_j \big[A \cap B_j\big]\Big) \\ &= \sum_j P(A \cap B_j) && \text{(by disjointness)} \\ &= \sum_j P(A \mid B_j) P(B_j) && \text{(by the definition of conditional probability),} \end{aligned}$$

concluding the proof.

Note that although this theorem is motivated by sampling a population, it holds even if the probability space $\mathcal{S}$ in question is not related to picking a person at random from some population. We actually made use of a special version of the stratified sampling theorem in Example 4.8, where we did essentially the following:

$$\begin{aligned} P(R_2) &= P(R_2 \cap [R_1 \cup R_1^c]) \\ &= P(R_2 \cap R_1) + P(R_2 \cap R_1^c) \\ &= P(R_2 \mid R_1)P(R_1) + P(R_2 \mid R_1^c)P(R_1^c). \end{aligned}$$

The important idea here is that $\{R_1, R_1^c\}$ consistutes a disjoint partition of $\mathcal{S}$ and the terms $P(R_2 \mid R_1)$, $P(R_1)$, $P(R_2 \mid R_1^c)$, $P(R_1^c)$ in the last expression are all computable. In general, the stratified sampling theorem is useful if $P(A \mid B_i)$ and $P(B_i)$ are known.

A corollary of the stratified sampling theorem known as Bayes' Theorem provides a way to compute $P(B_i \mid A)$, knowing the $P(B_j)$ and $P(A \mid B_j)$ for all $j$. In practical situations, we often may have to estimate the $P(B_j)$ and $P(A \mid B_j)$ from data. The question of how much data is needed for accurate estimation will be treated in Chapters 10 and 12.

### 4.12 *Corollary to 4.11: Bayes' Theorem*

Under the hypotheses of 4.11, and with the additional assumption that $P(A) > 0$, we have for each $i$

$$P(B_i \mid A) = \frac{P(A \mid B_i)P(B_i)}{\sum_j P(A \mid B_j)P(B_j)}.$$

*Proof:*

$$\begin{aligned} P(B_i \mid A) &= \frac{P(A \cap B_i)}{P(A)} \\ &= \frac{P(A \mid B_i)P(B_i)}{\sum_j P(A \mid B_j)P(B_j)}, \end{aligned}$$

the denominator coming from use of 4.11 and the numerator from the definition of conditional probability, concluding the proof.

### 4.13 *Example: Medical Diagnosis*

Let $A$ be the event that an individual has a certain specified set of symptoms (that is, measurements associated with medical instruments) and let $B_j$ be the event that the person has disease $j$. From extensive data on the incidence of the disease in the population, we may "know" (that is, have accurate estimates of) the probabilities $P(B_j)$. This incidence may be determined from bacterial culture studies, microscopic tissue examinations, and autopsy reports. Furthermore, from such studies, coupled with careful records concerning symptoms, we may also "know" the $P(A \mid B_j)$. When a new patient comes in with symptoms corresponding to $A$, we would like to determine $P(B_j \mid A)$ for each $j$, in order to know which treatments or further diagnostic tools are called for (namely, those for which $P(B_j \mid A)$ is large). Bayes' Theorem yields the needed $P(B_j \mid A)$. So-called *computer diagnosis* is based on the ideas just presented, since the computation of $P(B_j \mid A)$ as given in Theorem 4.12 can be done rapidly by a properly programmed computer.

In order to take into account the possibility that a patient has several diseases, and to still keep the $B_j$ pairwise disjoint, we might be forced to redefine the $B_j$ to be "states," where a state is a fixed combination of diseases. For example, if there are three diseases, $D_1$, $D_2$, and $D_3$, then we might let

$$B_0 = D_1^c \cap D_2^c \cap D_3^c = \text{good health}$$

$$B_1 = D_1 \cap D_2^c \cap D_3^c = \text{disease 1}$$

$$B_2 = \cdots$$

$$B_3 = \cdots$$

$$B_4 = D_1 \cap D_2 \cap D_3^c = \text{both disease 1 and disease 2, but not disease 3,}$$

and so on.

Of course, such generality may not be needed if we know that the patient has precisely one of several possible diseases. The art of good diagnosis lies in knowing which measurements to make. The "shotgun approach"—in which as many measurements as possible are made—turns out to be self-defeating, because estimating the many probabilities $P(A \mid B_j)$ for each sympton set $A$ does not permit much accuracy in the estimates without an excessive amount of data. Hence, a good knowledge of the physical mechanism is vital in order to know which measurements will pay off by furnishing enough information to warrant their use. △

We now consider a classic problem that never fails to arouse a good deal of classroom controversy.

### 4.14 *Example: The Three Drawers Problem*

A chest has three drawers. The first contains two gold coins, the second contains a gold and a silver coin, and the third has two silver coins. A drawer is chosen at random (see 2.3(6), Chapter 2) and from it a coin is chosen at random. What is the probability that the coin still remaining in the chosen drawer is gold, given that the coin chosen is silver?

The solution to the problem hinges on a precise understanding of exactly what we have just stated. In order to include in each possible outcome the information concerning the choice of drawer and coin, we take as our sample space

$$\mathcal{S} = \{(1, g_1), (1, g_2), (2, g), (2, s), (3, s_1), (3, s_2)\},$$

where the first coordinate of the outcome is reserved for the drawer chosen and the second is reserved for the coin which is chosen (the coins in drawer 1 being labeled

$g_1$ and $g_2$, and so on). We let $A_i$ be the event that drawer $i$ was chosen. Then $A_1 = \{(1, g_1), (1, g_2)\}$. We let $G_1$ denote the event that the chosen coin is $g_1$, $G_2$ the event that the chosen coin is $g_2$, $G$ the event that the chosen coin is $g$, and so forth. Our assumptions then become

$$P(A_i) = \tfrac{1}{3} \text{ for } i = 1, 2, 3 \qquad \text{(drawer chosen at random)}$$

$$P(G_1 \mid A_1) = \tfrac{1}{2} \qquad \text{(coin chosen at random from the chosen drawer).}$$

$$P(G_2 \mid A_1) = \tfrac{1}{2}$$

We see then that

$$P\{(1, g_1)\} = P(A_1 \cap G_1) = P(A_1)P(G_1 \mid A_1) = \tfrac{1}{3} \cdot \tfrac{1}{2} = \tfrac{1}{6},$$

and in fact (by similar computations) all the elementary events have probability 1/6. Now let $B$ be the event that a silver coin was selected and let $C$ denote the event that the coin remaining in the chosen drawer is gold. Then

$$B = \{(2, s), (3, s_1), (3, s_2)\}$$

and

$$C = \{(1, g_1), (1, g_2), (2, s)\}.$$

We can now compute the answer:

$$\begin{aligned} P(C \mid B) &= \frac{P(C \cap B)}{P(B)} \\ &= \frac{P\{(2, s)\}}{P\{(2, s), (3, s_1), (3, s_2)\}} \\ &= \frac{\tfrac{1}{6}}{\tfrac{3}{6}} = \tfrac{1}{3}. \end{aligned}$$ △

Another classic puzzler is the following example:

### 4.15 *Example: The Three Prisoners Problem*

Of three prisoners, Jones, Smith, and Wilson, two have been selected at random for execution. Jones, being somewhat curious, would like to know whether or not he will be executed, but he realizes that the warden (who knows who will live) will not answer the question directly. So he says, "Warden, of the other two prisoners, name one who will be executed." The warden, who is always truthful in such puzzles, replies, "Smith." Now Jones is in a quandary. Has he really learned anything more about his chances of survival?

As usual, we must first choose a sample space that reflects the possible outcomes; to do this we must have the outcomes indicate both the possible execution situations and the possible answers of the warden. Hence we take as the sample space

$$\mathcal{S} = \{(J, S; S), (J, W; W), (S, W; S), (S, W; W)\}.$$

Here the first two coordinate values of a possible outcome indicate who are to be executed, and the last coordinate value indicates the warden's answer to Jones' question. Note that $P\{(J, S; S)\}$ is the probability that Jones and Smith were chosen, because the warden must give Smith's name in this case. Hence $P\{(J, S; S)\} = P\{\text{Jones and Smith chosen}\}$, which is 1/3 because of the assumption of random selection. Reasoning in this way, we see that our assumptions concerning the random selection of prisoners to execute tell us precisely that

$$P\{(J, S; S)\} = \tfrac{1}{3} \qquad P\{(J, W; W)\} = \tfrac{1}{3}$$

and

$$P\{(S, W; S), (S, W; W)\} = \tfrac{1}{3},$$

since this last is just the probability that Smith and Wilson will be executed. In particular, we have no way of assigning separately the probabilities $P\{(S, W; S)\}$ and $P\{(S, W; W)\}$ with the information provided; that is, we have no way of knowing the warden's strategy (if indeed he has any) when he must make a choice. Thus, for practical purposes, Jones may just as well not ask the question, since unless we make further assumptions he cannot use the information supplied by the warden.

If, however, Jones knows the warden to be methodical, so that he always answers with the name that comes first in alphabetical order, then we know that

$$P\{(S, W; S)\} = \tfrac{1}{3}$$

$$P\{(S, W; W)\} = 0.$$

If now we let $E_J$ be the event that Jones is executed

$$E_J = \{(J, S; S), (J, W; W)\}$$

and $N_S$ be the event that the warden names Smith

$$N_S = \{(J, S; S), (S, W; S)\},$$

then

$$P(E_J \mid N_S) = \frac{P(E_J \cap N_S)}{P(E_S)}$$

$$= \frac{P\{(J, S; S)\}}{P\{(J, S; S), (S, W; S)\}}$$

$$= \frac{P\{(J, S; S)\}}{P(\{J, S; S\} \cup \{S, W; S\})}$$

$$= \frac{\tfrac{1}{2}}{\tfrac{2}{3}} = \tfrac{3}{4}.$$

If, when he needs to, the warden tosses a fair coin to decide on his answer, then we would have

$$P\{(S, W; S)\} = 1/6$$

$$P\{(S, W; W)\} = 1/6$$

and

$$P(E_J \mid N_S) = \frac{1/3}{1/2} = 2/3.$$

Further variations on this puzzle will be explored in the problems. △

## 4.16 *EXERCISES*

1. Referring to the urn model of Example 4.8:

(a)° What is the probability of removing a red spotted ball on the first draw, given that a red spotted ball was chosen on the second draw?

(b) What is the probability of a red spotted ball on the first draw, given that a pure white ball was selected on the second draw?

2. In an urn model for diffusion (that is, for the passage of pollutant particles through the atmosphere), we might start out with two urns as shown, where the red balls represent the pollutants. A complete drawing consists of the following two actions: (1) Five balls are drawn at random from urn 1 and put in urn 2. (2) Ten balls are then drawn at random from urn 2 and placed in urn 1.

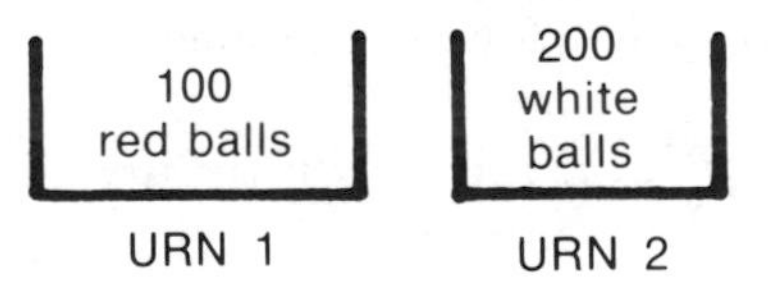


(a)° What is the probability that $j$ white balls are in urn 2 following the first complete drawing?

(b) What would be the answer to (a) if initially urn 2 had 150 white balls and 50 red balls, and urn 1 had 50 red balls and 50 white balls?

Hint: Use the Stratified Sampling Theorem (4.11) with $B_k$ the event "$k$ red balls chosen on the first drawing."

3. If men constitute 47 per cent of the population and tell the truth 78 per cent of the time, and women tell the truth only 63 per cent of the time, what is the probability that a person selected at random from the population will answer a question truthfully?

4. Solve the gambler's ruin problem, that is, find the probability that the gambler will be ruined (Example 4.10, equations (*) and (**)) for

(a)° $b = \$4$, $p = 1/2$, $a = \$1$. Does the answer seem reasonable? (Explain.) Recall that play will terminate with probability 1 (see 3.2 in Chapter 2).

(b) $b = \$4$, $p = 2/3$, $a = \$1$.

5.° A fair die is tossed, and if it comes up $j$ (with $1 \leq j \leq 6$) then we toss $j$ fair coins. What is the probability of getting exactly three heads in the second part of the experiment?

6. A pair of fair dice is tossed, and if the sum is $j$ ($2 \leq j \leq 12$) then we toss $2j$ fair coins. What is the probability of exactly $k$ heads in the second part of this experiment?

We remark that such two-stage procedures really arise in practice, when the results of the pilot run (represented by the dice toss) are used to determine the extent of the definitive final run (represented by the coin tosses).

7. If you classify people by three economic classes—poor, middle, and rich,—suppose you find that 10 per cent of the poor, 54 per cent of the middle, and 17 per cent of the rich have color television sets. Suppose further that the poor constitute 20 per cent of the population, the middle 75 per cent, and the rich 5 per cent.

(a)° A person selected at random from the population is found to be a color television addict. How likely is he to be rich?

(b) A person selected at random from the population does not have color television. How likely is he to be rich?

(c)° What is the probability that a person selected at random from the population has a color television set?

8.° A patient is known to have precisely one of two diseases, $B_1$ or $B_2$. There are two possible sets of symptoms,

$$\{\text{fever, rash}\} = C, \qquad \text{and} \qquad \{\text{sore throat, headache}\} = D.$$

It has been found that

$$P(C \mid B_1) = .6 \qquad P(D \mid B_1) = .4$$
$$P(C \mid B_2) = .7 \qquad P(D \mid B_2) = .3$$

Furthermore, $B_1$ occurs twice as often as $B_2$ in the population, and they never occur together in the same person. Assuming the patient to have been drawn at random from the population and to have a fever and rash, what are the probabilities of $B_1$ and of $B_2$ for this patient?

9. A patient is known to have precisely one of three diseases, $B_1$, $B_2$, or $B_3$. There are four possible sets of symptoms:

$$C = \{\text{fever, rash}\}, \qquad D = \{\text{headache, fever}\},$$
$$E = \{\text{upset stomach, rash}\}, \qquad F = \{\text{upset stomach, headache, rash, fever}\}.$$

It has been found that

$$P(C \mid B_1) = .2 \qquad P(C \mid B_2) = .1 \qquad P(C \mid B_3) = .4$$
$$P(D \mid B_1) = .3 \qquad P(D \mid B_2) = .2 \qquad P(D \mid B_3) = .3$$
$$P(E \mid B_1) = .3 \qquad P(E \mid B_2) = .3 \qquad P(E \mid B_3) = .2$$
$$P(F \mid B_1) = .2 \qquad P(F \mid B_2) = .4 \qquad P(F \mid B_3) = .1$$

and $B_1$ occurs twice as often as $B_2$ and three times as often as $B_3$ in the population (never together in the same person). What is the probablity that a person with symptom set $F$ has disease $B_1$? That he has disease $B_2$?

10.° Referring to the three drawers problem (4.14), add three more drawers, one with a gold and a copper coin, one with a silver and a copper coin and one with two copper coins. What is the probability that the remaining coin is copper if the coin drawn is silver?

11. Referring to the three drawers problem, let us add a fourth drawer with two platinum coins. What is the probability that the coin still remaining in the chosen drawer is gold, given that the chosen coin is silver? (This question was added because one student was sure that the answer to 4.14 was 1/3 because each drawer had probability 1/3 of being chosen.)

---

## 4.17 PROBLEMS

1. Referring back to Example 4.8 (the urn), what is an acceptable sample space for this experiment?

2. Referring to Example 4.14 (the three drawers), let $\mathscr{G}$ be the event that a gold coin is drawn and let $D_j$ be the event that drawer $j$ is opened.

(a) Find $P(D_j \mid \mathscr{G})$.

Hint: Since we know $P(\mathscr{G} \mid D_j)$, Bayes' Theorem seems applicable.

(b) Does this actually solve the three drawers problem of Example 4.14?

3.° (a) A given lot of 25 units has five defectives. Two units are selected at random, and from these two units, one is selected at random. What is the probability that the second selection is defective?

(b) Can you generalize this result?

4.° Suppose that the probability that a component survives at least $x$ hours is $\frac{1}{2}e^{-x} + \frac{1}{2}e^{-3x}$.

(a) What is the probability that it will survive at least another hour, given that it has survived at least one hour?

(b) How does this compare with the probability of survival at least one hour?

Hint: Use the fact that $e^{-t}$ is concave up.

(c) Can you explain this apparent increase in life expectancy?

5. Show that for $B$ with $P(B) \neq 0$, the function whose value at $A \in \mathscr{E}$ is $P(A \mid B)$ is a probability measure.

6. Show that

$$(a)^\circ \quad P(A \cap B \cap C) = P(A \mid B \cap C)P(B \mid C)P(C),$$

and that in general

$$(b) \quad P(A_1 \cap A_2 \cap \cdots \cap A_n) = P(A_1 \mid A_2 \cap \cdots \cap A_n)P(A_2 \mid A_3 \cap \cdots \cap A_n) \cdots P(A_{n-1} \mid A_n)P(A_n)$$

and

$$(c) \quad P(A_1 \cap A_2 \cap \cdots \cap A_n) = P(A_1)P(A_2 \mid A_1)P(A_3 \mid A_1 \cap A_2) \cdots P(A_n \mid A_{n-1} \cap \cdots \cap A_2 \cap A_1).$$

7. (a) It is intuitively reasonable that if the average income in each religious group of a population exceeds \$8000 a year, then the overall population average income exceeds \$8000 a year. This idea should be reflected in your answer to the following: If $B_1, B_2, \ldots$ are pairwise disjoint events with $\bigcup_n B_n = \mathcal{S}$ and $P(A \mid B_j) \geq .99$ for all $j$, what conclusion can be drawn about $P(A)$?

Hint: Theorem 4.11.

(b) Given a collection of pairs of people, if we are told that each person in the population is included in at least one pair and that the average income of each pair exceeds $10,000 per year, we couldn't conclude anything, since John D. Rockefeller could have been the first member of every pair. This should provide intuition for the following question: Suppose $\bigcup_n B_n = S$ and $P(A \mid B_j) \geq .99$ for each $j$. Show that it is still possible for $P(A)$ to be small.
Hint: Suppose $P(A) = .001$. Partition $A^c$ into a union $C_j$ of pairwise disjoint sets $C_j$, with $P(C_j) = .00001$. Now let $B_j = C_j \cup A$ and compute $P(A \mid B_j)$.

8.° Referring to the three prisoners problem (Example 4.15), what is the conditional probability of Jones' being executed, given the warden's two possible answers to the question, "Will Smith be executed?"

(b) Suppose that Jones knows that in answer to the question, "Name one of the two other prisoners who will be executed," the warden will either (i) choose the first name in alphabetical order (when there is a choice) or (ii) with probability .9 choose the first name in alphabetical order (when there is a choice). Can Jones do anything with the "information" he gets? In particular, can he legitimately assume his probability of execution to be 2/3?

9.° What relation (if any) is there between:

(a) $P(A \mid B)$ and $P(A \mid B^c)$?

(b) $P(A \mid B)$ and $P(A^c \mid B)$?

10. On election night, one radio station reported, "Senator Doaks is trailing by 5000 votes," whereas another station reported, "Senator Doaks is leading by 2000 votes." Is there any way to use this information?

(a) What would you do?

(b) What reason do you give for the discrepancy and what might be done to overcome it?
Hint: You have only one way of using partial information. If you could somehow determine a sample space in which each piece of information corresponded to the occurrence of some specific event, you would be in good shape.

'11. In a recent court trial, a man was convicted of a crime based on the following evidence. The probability of a man in his locality (Los Angeles, population 4,000,000) satisfying the description given of the criminal and his accomplice was 1/1,000,000. He (and his girl friend) satisfied this description. So unlikely an event must, beyond doubt, pin the rap on him. Is this reasoning correct? Support your assertion.

## SECTION 5
## Statistical Independence

An important special case arises when occurrence of $B$ (that is, the outcome of the experiment being in $B$) does not affect the probability assigned to $A$. If knowledge that the outcome is in $B$ does not cause us to change the probability assigned to $A$, then we say that $A$ and $B$ are independent events.

We are considering independence when we ask such questions as, "Is cigarette smoking related to lung cancer?" "Is exercise good for longevity?" "Are high school grades related to college performance?" "Does heat significantly increase wearout?" "Do primary elections significantly influence nominations?"

We are tempted to say that $A$ is independent of $B$ if $P(A \mid B) = P(A)$. Using the definition of conditional probability, we therefore want to say that $A$ is independent of $B$ if $P(A \cap B)/P(B) = P(A)$, that is, if $P(A \cap B) = P(A)P(B)$. Since this form looks so symmetrical, we say that "$A$ and $B$ are independent" rather than "$A$ is independent of $B$." We summarize our discussion in the following definition.

**5.1 Definition: Pairwise Independence.** *The pair $A$, $B$ of events is said to be independent* (or *statistically independent*) if $P(A \cap B) = P(A)P(B)$. If $A$ and $B$ are not independent, then we say they are *dependent*.

In order to extend the idea of independence to more than two events, we reason that if we are given events $A_1, A_2, \ldots, A_n$, we would like to call them independent if each event we form via set operations from any sub-collection $A_{i_1}, \ldots, A_{i_k}$ is independent of every event which may be formed via set operations from the remaining $A_j$'s. It can be shown that the following definition accomplishes this aim.

**5.2 Definition: Mutual Independence.** Let $A_1, A_2, \ldots, A_n$ be events in a probability space $(\mathcal{S}, \mathscr{E}, P)$. We shall say that these events are (mutually) *independent* (or *statistically independent*) relative to $P$ if the equation

$$(*) \qquad P(A_1^* \cap A_2^* \cap \cdots \cap A_n^*) = P(A_1^*)P(A_2^*)\cdots P(A_n^*)$$

holds for every possible choice of $A_1^*, \ldots, A_n^*$ where $A_i^*$ is either $A_i$ or $A_i^c$.

Note that this definition requires the satisfaction of $2^n$ equations. If $A_1, \ldots, A_n$ fail to be independent, they are said to be *dependent*.

*5.3 Remarks.* For a pair of events, 5.2 certainly implies 5.1. (We shall see in Exercise 5.8(2) that 5.2 and 5.1 are equivalent for $n = 2$.) Furthermore, if $P(A \cap B) = P(A)P(B)$ and $P(B) \neq 0$, then $P(A \mid B) = P(A)$, which agrees with our intuitive feeling that $A$ is independent of $B$. For three events, $A$, $B$, and $C$, to be independent, Definition 5.2 requires

$$P(A \cap B \cap C) = P(A)P(B)P(C)$$
$$P(A^c \cap B \cap C) = P(A^c)P(B)P(C)$$
$$P(A \cap B^c \cap C) = P(A)P(B^c)P(C)$$
$$P(A \cap B \cap C^c) = P(A)P(B)P(C^c)$$
$$P(A^c \cap B^c \cap C) = P(A^c)P(B^c)P(C)$$
$$P(A^c \cap B \cap C^c) = P(A^c)P(B)P(C^c)$$
$$P(A \cap B^c \cap C^c) = P(A)P(B^c)P(C^c)$$
$$P(A^c \cap B^c \cap C^c) = P(A^c)P(B^c)P(C^c).$$

Note that if we add the first two equations we obtain

$$P(B \cap C) = P(B)P(C),$$

so that mutual independence of $A$, $B$, and $C$ implies pairwise independence.

### 5.4 *Example: Toss of Two Coins*

In a uniform model for the toss of two coins, where $S = \{(H, H), (H, T), (T, H), (T, T)\}$, let $H_1$ be the event "heads on the first toss," that is,

$$H_1 = \{(H, H), (H, T)\},$$

and let $H_2$ be "heads on the second toss." Then

$$P(H_1) = P\{(H, H)\} + P\{(H, T)\} = 1/4 + 1/4 = 1/2.$$

Similarly, $P(H_2) = 1/2$. Furthermore,

$$P(H_1 \cap H_2) = P\{(H, H)\} = 1/4 = P(H_1)P(H_2).$$

Hence $H_1$ and $H_2$ are independent.

If we had instead assigned

$$P\{(H, H)\} = p^2, P\{(H, T)\} = p(1 - p) = P\{(T, H)\} \text{ and } P\{(T, T)\} = (1 - p)^2,$$

then we would still have found $H_1$ and $H_2$ to be independent, each having probability $p$. In fact, $H_1$ and $T_2 = \{(H, T), (T, T)\}$ are independent, since

$$\begin{aligned} P(H_1) &= P\{(H, H), (H, T)\} = P\{(H, H)\} + P\{(H, T)\} \\ &= p^2 + p(1 - p) = p^2 + p - p^2 = p \\ P(T_2) &= P\{(H, T), (T, T)\} = P\{(H, T)\} + P\{(T, T)\} \\ &= p(1 - p) + (1 - p)^2 = (p + 1 - p)(1 - p) \\ &= 1 - p \end{aligned}$$

and

$$P(H_1 \cap T_2) = P\{(H, T)\} = p(1 - p) = P(H_1)P(T_2). \qquad \triangle$$

You may wonder whether pairwise independence wouldn't be enough to guarantee mutual independence, or whether the equation $P(A_1 \cap A_2 \cap \cdots \cap A_n) = P(A_1)P(A_2) \cdots P(A_n)$ wouldn't be enough to ensure mutual independence. Life just is not that simple, as the following example shows.

### 5.5 *Example: Pairwise and Triple Independence*

If we consider the usual model for the toss of two fair dice (a uniform one on the 36 ordered pairs) and let

$E_1$ be the event "even on first die,"

$E_2$ be the event "even on second die," and

$E_S$ be the event "even sum,"

then it is not difficult to verify that these events are pairwise independent, that is,

$$P(E_1 \cap E_2) = 1/4 = 1/2 \cdot 1/2 = P(E_1)P(E_2)$$
$$P(E_1 \cap E_S) = 1/4 = 1/2 \cdot 1/2 = P(E_1)P(E_S)$$
$$P(E_2 \cap E_S) = 1/4 = 1/2 \cdot 1/2 = P(E_2)P(E_S).$$

But certainly $E_1$, $E_2$, and $E_S$ are not mutally independent, since if $x \in E_1 \cap E_2$, then $x \in E_S$. Hence

$$P(E_1 \cap E_2 \cap E_S) = P(E_1 \cap E_2) = 1/4 \neq P(E_1 \cap E_2)P(E_S),$$

since $P(E_S) = 1/2$.

*Thus, pairwise independence does not imply mutual independence.*

On the other hand, for $A = \varnothing$, we surely have

$$P(A \cap B \cap C) = P(\varnothing) = 0$$

$$P(A)P(B)P(C) = P(\varnothing)P(B)P(C) = 0 \cdot P(B)P(C) = 0.$$

Hence,

$$P(A \cap B \cap C) = P(A)P(B)P(C).$$

Clearly, this equation does not tell us *anything* about probabilities concerning $B$ and $C$. Hence, *"triple independence" does not imply pairwise independence.* △

We further illustrate the concept of independence with some familiar models.

### 5.6 *Example: Dice and Coins*

In the toss of two dice we can generate the familiar uniform model as follows: Let $A_i$ be the event "first die comes up $i$" and $B_j$ be the event "second die comes up $j$." For each $i$ and $j$, $P(A_i) = 1/6 = P(B_j)$. Make $A_i$ and $B_j$ independent. Then $P\{(i, j)\} = P(A_i \cap B_j) = P(A_i)P(B_j) = 1/6 \cdot 1/6 = 1/36$, which is the familiar uniform model.

Similarly, the uniform model for tossing a fair coin $n$ times (in which an outcome is a sequence of length $n$ of $H$'s and $T$'s) can be generated by the assignment

$$P(H_i) = 1/2$$

and

$$H_1, \ldots, H_n \text{ are independent,}$$

where $H_i$ is the event "heads on $i$th toss." This follows from the fact that every elementary event is of the form

$$H_1^* \cap H_2^* \cap \cdots \cap H_n^*$$

(where $H_i^*$ is either $H_i$ or $H_i^c$), and hence, by the foregoing, has probability $1/2^n$. △

It can be shown that if $A_1, \ldots, A_n$ are independent events, then from the values $P(A_1), P(A_2), \ldots, P(A_n)$, we can compute the probability of any event that can be formed from the $A_i$ by set operations.

### 5.7 *Example: $P(A \cup B)$ When A and B are Independent*

Recall from Theorem 3.5(c) that

$$P(A \cup B) = P(A) + P(B) - P(A \cap B).$$

From this we see that when $A$ and $B$ are independent,

$$P(A \cup B) = P(A) + P(B) - P(A)P(B). \qquad \triangle$$

***We see that independence makes our probability computations much simpler, and hence we go to great effort to introduce independence in many practical situations*** whenever it is feasible (just as we did by choosing a sample space leading to a uniform model). Because independence is of such great practical importance, it is useful to be able to make a test on data to help decide whether it is reasonable to assume that some given events are independent. A reasonable statistical test of independence of events $A$ and $B$ might be as follows: Run the experiment $n$ times, letting $P_n(E)$ stand for the observed proportion of the $n$ experiments in which the outcome is in any given event $E$; now assert the independence of $A$ and $B$ if

$$|P_n(A \cap B) - P_n(A)P_n(B)|$$

is small. Just what we mean by "small" is taken up in studies of such tests. We shall treat such problems in Chapter 12.

---

### *5.8 EXERCISES*

1. (a)° Suppose that $A$, $B$, and $C$ are independent, each with probability .1. Find $P(A \cup B \cup C)$. Try each of the given hints to see which you prefer. Hint (i) Use the result given in 3.9.7. Hint (ii) Use De Morgan's Law (Exercise 2.16.10(a)).

(b) Find a general expression for $P\left(\bigcup_n A_n\right)$ in terms of the $P(A_n)$ when the $A_n$ are independent (see hint (ii)).

(c) If $A_1, \ldots, A_n$ are independent with $0 < P(A_i) < 1$, could $P\left(\bigcup_k A_k\right) = 1$? Hint: See (b).

2. Prove that if $A$ and $B$ are an independent pair of events (see Definition 5.1), then they are mutually independent (5.2). That is, prove that $(A^c, B)$, $(A, B^c)$, and $(A^c, B^c)$ all satisfy Definition 5.1 if $(A, B)$ satisfies 5.1.

3. Suppose our experiment consists of drawing a number at random from $[0, 1]$, that is, if $I$ is any subinterval of $[0, 1]$, then $P(I)$ is just the length of $I$.

(a)° Let $A = [0, 1/2]$ and let $B = [0, 1/4] \cup [3/4, 1]$. Are $A$ and $B$ independent?

(b) Let $A = [0, 1/2]$ and $B = [1/4, 3/4]$. Are $A$ and $B$ independent?

(c)° Let $X_i$ be the $i$th digit in the decimal representation of the outcome. Are the events $A_i$ that $X_1 = i$ and $B_j$ that $X_2 = j$ independent, for a given $i$ and $j$?

4. (a)° Show that if $A$, $B$, and $C$ are independent, then $A \cup B$ and $C$ form an independent pair (see Definition 5.1) of events.

(b) Assuming (a), compute $P[(A \cup B) \cap C]$ in terms of $P(A)$, $P(B)$, and $P(C)$.

5.° Suppose that in the experiment of tossing a coin three times we let $H_i$ be the event "heads on $i$th toss" and know that $P(H_i) = p$ for $i = 1, 2, 3$, and that $H_1$, $H_2$, and $H_3$ are mutually independent. Compute the probability of exactly two heads in the three tosses.

6. An experiment is repeated four times. Let $S_i$ be the event "success on the $i$th trial" and suppose $S_1$, $S_2$, $S_3$, and $S_4$ to be mutually independent events.

(a)° If $P(S_i) = p$ for $i = 1, 2, 3, 4$, find the probability of exactly three successes in the four trials.

(b) If $P(S_i) = p_i$ for $i = 1, 2, 3, 4$, find the probability of exactly one success in the four trials.

7. In determining the relation between smoking and lung cancer, we might perform the following experiment. Choose a person at random from the population, observe whether or not this person is a heavy smoker, and see whether or not he develops lung cancer. (This experiment takes quite a while to perform.) Let $S$ be the event that the person is a heavy smoker, and $L$ be the event that this person develops lung cancer. (Actually, we must be quite careful about the definition of $S$—does it concern the lifetime total of cigarettes smoked, is inhaling a part of the definition, and so forth.)

(a) How would you phrase the assertion that smoking and lung cancer are related?

(b) How would you phrase the Surgeon General's assertion in this matter (that is, that a high incidence of lung cancer is associated with heavy smoking)?

8. Suppose two tosses of an unfair die are made. Let $\mathcal{S}$ be the usual 36-point sample space. Let $A_i$ be the event "$i$ on the first toss" and $B_j$ the event "$j$ on the second toss." Suppose that for each $i$ and $j$, $A_i$ and $B_j$ are independent, with

$$P(A_i) = P(B_i) = \begin{cases} .1 \text{ for } i = 1, 2, 3, 4 \\ .3 \text{ for } i = 5, 6. \end{cases}$$

Compute the probability that the sum is:

(a)° 7.

(b) 11.

(c)° even.

**5.9 Definition: Conditional Independence.** We say that $A$ and $B$ are *conditionally independent*, given $C \subseteq \mathcal{S}$ (that is, when $C$ is the sample space), if

$$P(A \cap B \mid C) = P(A \mid C)P(B \mid C).$$

In many practical situations it might be the case that two events are conditionally independent but fail to be independent. For example, let us choose a person at random from the population and let:

A be the event that the person is a smoker,

B be the event that the person contracts lung cancer, and

C be the event that the person lives in a smog-free environment (or an environment with some such attribute).

Then it might very well be that

$$P(A \cap B) \neq P(A)P(B) \text{—that is, that } P(A \cap B \mid \mathcal{S}) \neq P(A \mid \mathcal{S})P(B \mid \mathcal{S})$$

but that $P(A \cap B \mid C) = P(A \mid C)P(B \mid C)$—that is, that smoking has no effect in the "proper" type of environment. Alternatively, we might have $A$ and $B$ independent given $\mathcal{S}$, but not independent given $C$.

---

### *5.10 PROBLEMS*

1. Show by constructing a mathematical example that it is possible for $A$ and $B$ to be independent given $\mathcal{S}$ but not to be conditionally independent given $C$. Hint: Let $A, B$ be independent (in $\mathcal{S}$) with $0 < P(A) < 1$ and $0 < P(B) < 1$. Let $C = A \cup B$; now make use of the results of Exercise 5.8.1(c).

2. If $A$ and $B$ are not independent, show that it is possible for them to be conditionally independent, given $C$.
Hint: Start off with sample space $C$ containing two independent events $A$ and $B$. Now enlarge to a bigger sample space $\mathcal{S}$, destroying independence in this larger space by enlarging only one of the independent sets.

## 5.11 Combined Experiments, Independent Experiments

So far we have only discussed independent events. In this section we will define what is meant by independent experiments. This is the mathematical model for experiments that do not interact. Our definition will have to be formulated in terms of independent events in the same sample space. For this reason we must see how to find a common sample space for a sequence of experiments.

Suppose we have $n$ experiments, the $i$th one being defined by the probability system $(\mathcal{S}_i, \mathscr{E}_i, P_i)$. We may consider these $n$ experiments to constitute one "combined" experiment, whose sample space $\mathcal{S}$ consists of typical outcomes of the form $(x_1, \ldots, x_n)$ with $x_i \in \mathcal{S}_i$. Here the outcome $(x_1, \ldots, x_n)$ of the combined experiment indicates that the first of the original experiments resulted in $x_1$, the second in $x_2$, and so on. For instance, we might consider three experiments:

(i) measuring the systolic blood pressure of a person.

(ii) measuring the respiration rate of this person.

(iii) measuring the person's pulse.

These three experiments might be thought of as the combined experiment "measure vital signs," where $x_i$ is the outcome of the $i$th experiment and $(x_1, x_2, x_3)$ is the outcome of the combined experiment. Suppose $\tilde{A}_1$ is an event in the first (blood pressure) experiment, such as the event "blood pressure above 200."

We insist that if $\tilde{A}_i$ is an event in $\mathscr{E}_i$ (that is, an event associated with the $i$th of the original experiments), then the corresponding set in the sample space of the combined experiment must also be an event for the combined experiment. That is, $A_i = \{(x_1, \ldots, x_n) \in \mathcal{S}: x_i \in \tilde{A}_i\}$ must be an event in the combined experiment. Note that there is no restriction on $x_j$, $j \neq i$, here. The model that reflects non-interaction of the original experiments is simply that if we are given any $n$ events $A_1, A_2, \ldots, A_n$, where $A_i$ is associated solely with the $i$th experiment (as just described), then these events must be mutually independent. In practice, showing that a given model is that of independent experiments may be almost impossible. However, suppose we try to set up the experiments so that there is no interaction between them. Then we may reasonably postulate the mathematical description just given, and make use of this structure to compute probabilities.

An important illustration is given in the following example.

### 5.12 *Example: Binomial Distribution*

An experiment whose probability of success is $p$ is repeated independently $n$ times. We wish to determine the probability of exactly $k$ successes. If we consider these $n$ experiments to constitute one combined experiment, then the sample space consists of sequences of length $n$ of $S$'s (successes) and $F$'s (failures). Let $S_i$ be the event that the $i$th *trial* (that is, the $i$th of the constituent experiments) was successful. That is, $S_i = \{(x_1, \ldots, x_n) \in \mathcal{S}: x_i = S\}$. The assumption of repeated trials yields the relation

$$P(S_i) = p \text{ for all } i,$$

and the assumed independence of the trials yields the condition that $S_1, S_2, \ldots, S_n$ are independent events. We now note that the intersection

$$S_1 \cap S_2 \cap \cdots \cap S_k \cap S_{k+1}^c \cap \cdots \cap S_n^c$$

is simply the elementary event (in $\mathcal{S}$) in which the first $k$ trials are successes and the remaining $n - k$ trials are failures. By independence, this event has probability $p^k(1 - p)^{n-k}$, as does every other elementary event in the event (in $\mathcal{S}$), "$k$ successes in $n$ trials." Hence, we see that the probability of $k$ successes in $n$ independent trials, each having probability $p$ of success, is

$$\begin{pmatrix}\text{number of sequences of } n \\ S\text{'s and } F\text{'s having } k\ S\text{'s}\end{pmatrix} \times p^k(1 - p)^{n-k}.$$

But we know from 2.3(5) in Chapter 2 that the number of sequences with $k$ $S$'s and $n - k$ $F$'s is just $\binom{n}{k}$. Thus, *the probability of exactly $k$ successes in $n$ independent trials each having probability $p$ of success is*

$$\binom{n}{k} p^k(1 - p)^{n-k}, \qquad 0 \leq k \leq n.$$

The probability measure assigning probability

$$p_k = \begin{cases} \binom{n}{k} p^k (1-p)^{n-k} & \text{for} \quad 0 \leq k \leq n, \\ 0 & \text{otherwise} \end{cases}$$

is called the *Binomial Distribution, with parameters n and p.* △

Sometimes students confuse the concepts of independence and disjointness. We stress that *disjointness is a set theoretic concept*, whereas *independence is a property of the specific probability measure*. Except in trivial cases, disjoint sets are almost never independent, since if $A$ and $B$ are disjoint then $P(A \mid B) = 0$, and in most cases of interest $P(A) \neq 0$. Thus, in a sense, disjointness and independence are almost opposite in spirit. The following example illustrates why these concepts are sometimes confused.

### 5.13 *Example: Disjointness Versus Independence*

Suppose that a coin is tossed independently two times, and that $H_i$ is the event, "heads on the $i$th toss" and $T_i$ the event, "tails on the $i$th toss." Thus $T_i = H_i^c$, $H_1 = \{(H, H), (H, T)\}$, and $T_2 = \{(H, T), (T, T)\}$. Then $H_1$ and $T_2$ are not disjoint, but they are independent, whereas $H_1$ and $T_1$ are disjoint but not independent. △

*5.14 Remarks.* Recall that we say that an event $A$ has occurred if the outcome lies in $A$. When two events are disjoint, they have no outcomes in common, so that if one of them occurs, the other cannot occur. Hence, if two events, $A$ and $B$, can both occur on one trial of an experiment, then we know that they cannot be disjoint. This observation is especially helpful in preventing the error of assuming disjointness when the sample space has not been made explicit but when certain events have been assumed to exist and have been assigned probabilities.

In the following example we combine the ideas of independent trials and conditional probability to present a simple solution to a problem that looks difficult.

### 5.15 *Example: "Three" Before "Even" on Dice*

Let us imagine that we are engaged in an indefinitely long sequence of independent tosses of a fair die. Intuitively, both the event "3" and the event "even" are certain to take place eventually, since the die is a fair one and the trials are independent (see Problem 5.17(1)). One of these events will occur first in any such series of trials. Thus, it is reasonable to consider the event "3 before even," which we will denote by $T$. Let $A_i$ be the event "$i$ on the first toss." Then, using the Stratified Sampling Theorem (4.11), we have

$$(*) \qquad P(T) = \sum_{i=1}^{6} P(T \mid A_i) P(A_i).$$

Now $P(A_i) = 1/6$ for all $i$. We claim that it is intuitively clear that

$$P(T \mid A_i) = \begin{cases} P(T) & \text{for} \quad i = 1, 5 \quad \text{(by independent trial assumption)} \\ 0 & \text{for} \quad i = 2, 4, 6 \\ 1 & \text{for} \quad i = 3. \end{cases}$$

The cases in which $i = 1, 5$, though intuitively reasonable, can well use a rigorous justification. In the sample space $\mathcal{S}$ of infinite sequences of elements from $\{1, 2, 3, 4, 5, 6\}$, the event $T$ consists of all sequences of the form

$$\text{(i)} \qquad (a_1, a_2, \ldots, a_n, \ldots)$$

where $a_n = 3$, and $a_j \in \{1, 5\}$ for $j < n$, with no restrictions on the $a_j$ for $j > n$, for some $n = 1, 2, \ldots$. For $i = 1, 5$, the event $T \cap A_i$ consists of all sequences of the form

$$\text{(ii)} \qquad (b, a_1, a_2, \ldots, a_n, \ldots)$$

where $b = i$ and $a_1, a_2, \ldots$ are as described in (i). Now we notice that for any given $n$, the set of all sequences of the form (ii) has probability $(1/6)q_n$, where $q_n$ is the probability of all sequences of the form (i) for the same $n$. But, using 3.3(c) on the probability of a disjoint union, we see that

$$P(T) = \sum_{n=1}^{\infty} q_n$$

and

$$P(T \cap A_i) = \sum_{n=1}^{\infty} (1/6) q_n,$$

from which it follows that

$$P(T \mid A_i) = \frac{P(T \cap A_i)}{P(A_i)} = \frac{\sum_{n=1}^{\infty} (1/6) q_n}{1/6} = \sum_{n=1}^{\infty} q_n = P(T)$$

as asserted for $i = 1, 5$.

Substituting for $P(T \mid A_i)$ in (*) yields $P(T) = \frac{1}{6}P(T) + \frac{1}{6}P(T) + \frac{1}{6}$. Thus, $\frac{4}{6}P(T) = 1/6$ or $P(T) = 1/4$. △

---

## 5.16 *EXERCISES*

1.° In 20 independent tosses of a fair die, what is the probability that the first three tosses result in 1's, followed by five odd outcomes, followed by nine even outcomes?

2. In four independent tosses of a fair die, what is the probability that the sum of the first two tosses is seven and the last toss is even?

3. In tossing a pair of dice repeatedly and independently, what is the probability of a 7 before a 4 (see Example 5.15)?

4.° (a) An experiment with probability $p$ of success is repeated independently. What is the probability that exactly $k - 1$ failures will precede the first success? (This result justifies the model of 3.4 in Chapter 2, the *geometric distribution*.)

(b) From the answer to (a), write down the probability that at least $k$ failures will precede the first success.

(c) What is the probability that exactly $k$ trials are needed to obtain a success?

(d) Could the answer to (b) have been derived more simply by not using (a)?

5. (a) An experiment whose probability of success is $p$ is repeated independently until exactly $j$ successes are obtained. What is the probability that exactly $k$ trials are needed?
Hint: We must achieve exactly $j - 1$ successes on the first $k - 1$ trials and a success on the $k$th one. (This justifies the model of 3.5 in Chapter 2, the Pascal distribution.)

(b) What is the probability that at least $k$ trials are needed?
Hint: You could use part (a), or you could note that if fewer than $j$ successes are achieved on the first $k - 1$ trials, then more than $k - 1$ trials are needed for the $j$ successes.

6. In $n$ independent trials of an experiment whose probability of success is $p$, what is the probability that:

(a)° there are at least $k$ successes?

(b) the number of successes lies between $j$ and $k$?

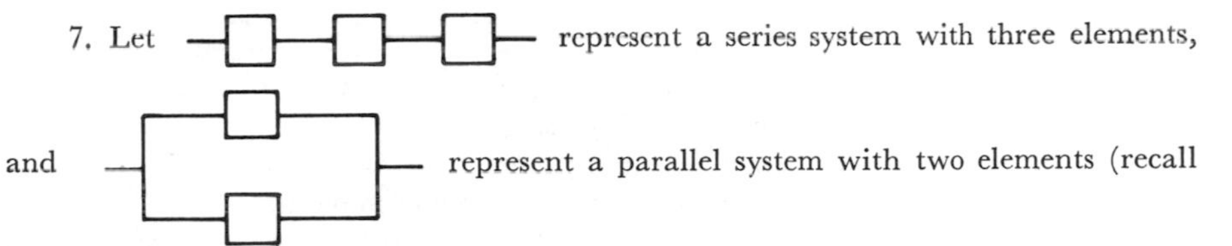

7. Let [series diagram] represent a series system with three elements, and [parallel diagram] represent a parallel system with two elements (recall Example 2.11). Suppose component $i$ of the system

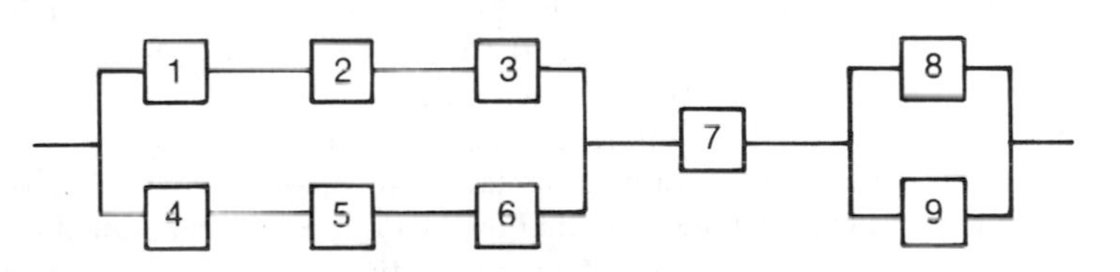


has probability $p_i$ of failure, and suppose failures to be independent (an unusual circumstance). What is the probability of failure of the system?

---

## 5.17 PROBLEMS

1. Show that in repeatedly tossing a die, the event "3" is practically certain to occur (that is, has probability 1). (This was assumed to be true in Example 5.15.)
Hint: Let $T_i$ be the event that the $i$th toss is a 3. The event "3 never occurs" is $\bigcap_{i=1}^{\infty} (T_i^c)$.

But

$$P\left[\bigcap_{i=1}^{\infty} (T_i^c)\right] \le P\left[\bigcap_{i=1}^{n} (T_i^c)\right]$$

2. Solve the problem in Example 5.15 (three before even) directly by computing $\sum_{n=1}^{\infty} q_n = \sum_{n=1}^{\infty} P(B_n)$, where $B_n = \{(a_1, \ldots) \in \mathcal{S}\colon\ a_n = 3,\ a_j \in \{1, 5\} \text{ for } j < n\}$.

3. What is the probability of two 7's before a 4 in repeated identical rolls of a pair of fair dice?

4. What is the probability of two successive 7's before a 4 in repeated identical rolls of a pair of fair dice?

5. Consider $n$ independent tosses of a fair die. For all $j \leq i \leq 6j$, let $u_{j,i}$ be the probability that on $j$ tosses a sum of $i$ is achieved. Write an equation relating $u_{j,i}$ and the various $u_{j-1,i}$ for $j \leq n$. If you have access to a digital computer, find the probability of a sum of 18 on six tosses of a die.
Hint: The $j$th toss must result in an element of $\{1, 2, \ldots, 6\}$. Now use the stratified sampling theorem.

6.° A sequence of $n$ independent experiments is performed, where $p_i$ is the probability of success on the $i$th experiment. Let $p_{j,k}$ be the probability of $k$ successes on the first $j$ of these experiments. Derive an expression for $p_{j,k}$ in terms of $p_{j-1,k}$, $p_{j-1,k-1}$ and $p_{j-1}$.
Hint: The $j$th experiment is either a success or a failure.

7.° Here is a challenge for compulsive gamblers. The game of craps is played as follows: You toss a pair of dice. *On the first toss:*

if the sum is 7 or 11, you win (and the game terminates).

if the sum is 2, 3, or 12, you lose (and the game terminates).

if the sum is 4, 5, 6, 8, 9, or 10 you continue to toss until the first occurrence of the following possibilities:

(a) a sum of 7 comes up, in which case you lose and the game terminates.

(b) the same sum as that of your first toss comes up, in which case you win and the game terminates. (This event is called *making your point.*)

What is your probability of winning, assuming fair dice?
Hint: The stratified sampling theorem is useful, as in Example 5.15. Let $W$ be the event that you win and let $A_i$ be the event that the first roll results in a sum of $i$, $2 \leq i \leq 12$. Now note, for example, that $P(W \mid A_4)$ is simply the probability of a 4 before a 7 in repeated rolls of a pair of fair dice.

8.° We have met many people who put forth the following argument with great conviction: "I've been extremely lucky in my life. I fly a lot and so far haven't been involved in any airplane accidents. I'm sure that my luck is running out and that there is a big chance that some accident will happen on one of my flights very soon." Assuming that the events $A_i$, $i = 1, 2, \ldots$, "accident on his $i$th flight," are independent, we know that $P(A_i \mid A_1^c \cap \cdots \cap A_{i-1}^c) = P(A_i)$, which directly contradicts his feelings.

(a) Can you supply a convincing intuitive argument in reply to his feelings?
Hint: What about the remaining passengers?

(b) The really tough argument to counter is the following:
"What you say may be true, but it seems to run counter to the frequency interpretation of probability, since we know that the proportion of accidents happening to me must settle down to $p = P$ (accident) and this means that my run of non-accidents must be countered by an accident, the likelihood of which increases the longer I stay healthy." Can you counter this argument? (It's not easy.)

## SECTION 6
## Systematic Procedure for Solving Probability Problems

In this section we summarize what we consider to be a systematic approach to most probability problems. This summary permits us to focus on the significant aspects of such problems.

(a) Choose an appropriate sample space $\mathcal{S}$ which includes as subsets all of the events to be considered.

(b) Identify all of those events to which reasonable probabilities or conditional probabilities can be assigned and make these assignments.

(c) Describe the event $A$ for which you wish to compute the probability.

(d) Relate $A$ to the events for which probability is assigned.

(e) Using the various laws of probability and the definitions of conditional probability and independence, compute $P(A)$, and see if your answer makes sense.

(f) If, as is often the case, you cannot compute $P(A)$ precisely, find a good approximation, or useful bounds for $P(A)$.

Notice that the first three steps are a part of applied probability, whereas the remaining ones are part of mathematical probability. Also notice that these steps are interrelated. For instance, we try to choose $\mathcal{S}$ to be as simple as possible while still satisfying the requirements that are imposed by the other steps. As an example, we note that in rolling a pair of dice (even if—indeed, *especially* if—we are interested only in the sum), we choose the 36-point sample space, since it permits easy and reasonable computation of probabilities. The following example illustrates the six steps.

### 6.1 *Example: Birthday Problem*

Given $n$ people selected at random, what is the probability that no two of these people have the same birthday (month and day)? Now, when this problem is posed in class, usually each person announces his birthday. This suggests that we take the sample space $\mathcal{S}$ to be the set of all sequences of length $n$ whose elements are chosen from $\{1, 2, \ldots, 365\}$. The value $j$ in the first coordinate indicates that the first person has his birthday on the $j$th day of the year, and so on. Step (a) is done. We let $A_i$ be the event that the $i$th person's birthday differs from all previously announced ones. That is, $A_i$ is the set of sequences in $\mathcal{S}$ whose $i$th entry differs from all entries in coordinates preceding the $i$th. We make the assignment

$$(*)\qquad P(A_1) = 1,\quad P(A_2 \mid A_1) = 364/365, \ldots,$$

$$P(A_i \mid A_{i-1} \cap A_{i-2} \cap \cdots \cap A_1) = \frac{365 - i + 1}{365}$$

based on our feelings that the probabilities assigned for the birthday in the $i$th coordinate ($i$th person) should not be changed with knowledge of birthdays for others in the class, and that the choice of the $i$th birthday follows a uniform model.† We see that (*) is a natural consequence of these feelings: In a sequence of uniform models that are independent from coordinate to coordinate, the probability that the $i$th birthday will differ from the $i - 1$ previous different ones is just the probability that the $i$th birthday falls in one of the $365 - (i - 1)$ unannounced days. But this is just

$$\frac{365 - i + 1}{365}$$

in a uniform model where each outcome has probability $1/365$. This finishes step (b).

For step (c), we note that the event $A$ is "no two people have the same birthday." Next note that $A = A_1 \cap A_2 \cap \cdots \cap A_n$, which accomplishes step (d). For step (e), we use the result of Problem 4.17(6)(c), namely, that

$$\begin{aligned} P(A) &= P(A_1 \cap A_2 \cap \cdots \cap A_n) \\ &= P(A_1)P(A_2 \mid A_1)P(A_3 \mid A_2 \cap A_1) \cdots P(A_n \mid A_{n-1} \cap A_{n-2} \cap \cdots \cap A_1) \\ &= 1 \cdot \frac{364}{365} \cdot \frac{363}{365} \cdots \frac{365 - n + 1}{365}. \end{aligned}$$

At this stage we see that, even for a moderate-size $n$, this expression may not be too useful in its given form, so the time is ripe for step (f). Let us take $n = 26$ and try to find a useful approximation for

$$\begin{aligned} P(A) &= \frac{354}{365}\frac{363}{365} \cdots \frac{340}{365} \\ &= (1 - 1/365)(1 - 2/365) \cdots (1 - 25/365). \end{aligned}$$

Computation of a product of this sort suggests the use of logarithms:

$$\ln P(A) = \sum_{i=1}^{25} \ln\left(1 - \frac{i}{365}\right).$$

(Here we use the logarithm to the base $e$ in order to facilitate our theoretical work.) We note that in all of the foregoing terms, $i/365$ is small compared to 1, which suggests the tangent line approximation

$$\begin{aligned} \ln(1 - x) &\cong \ln 1 + (\ln' 1)(-x) \\ &= 0 + 1/1(-x) = -x. \end{aligned}$$

† This latter assumption appears to be a somewhat crude first approximation. The non-uniformity that is likely to be present would seem to increase the probability of multiple birthdays over the answer we will obtain.

Thus we have the approximation

$$\begin{aligned}
\ln P(A) &\simeq \sum_{i=1}^{25} \left(\frac{-i}{365}\right) \\
&= -\frac{1}{365} \sum_{i=1}^{25} i \\
&= -\frac{1}{365} \cdot \frac{25 \cdot 26}{2} \quad \left(\text{by the well-known formula for the sum of the first } n \text{ integers, namely, } \sum_{i=1}^{n} i = \frac{n(n+1)}{2}\right) \\
&= -65/73 \\
&\simeq -.89.
\end{aligned}$$

Hence, the probability that of 26 people, no two will have the same birthday, is approximately

$$P(A) \simeq e^{-.89} \simeq .412,$$

a result which is somewhat surprising to many. △

We suggest that the student look over previous exercises and problems to see how the procedure just outlined might be applied to them.

# 4

# Random Variables and Their Distributions

## SECTION 1
## Random Variables As Functions or As Outcomes

Up to now our discussions of probability spaces have been concerned with situations in which the sample space $\mathcal{S}$ is arbitrary. For many purposes, it is convenient to convert the outcome of a probabilistic experiment into a numerical outcome, in particular to be able to use the familiar structure of the real numbers, and to simplify by filtering out needless information. Such filtering permits a focus on what is essential, much as a blueprint allows great precision because of its limited scope.

In order to convert to a numerical outcome, we associate a number $X(s)$ with each elementary event $s$ in $\mathcal{S}$. This association is simply the familiar process of defining a real valued function $X$ on $\mathcal{S}$. Now suppose that instead of observing the original outcome $s$, we are permitted to observe only the numerical value $X(s)$. Then, in effect, we have a new sample space, which we can take to be the real line $\mathscr{R}$. The fundamental objects in this sample space are intervals. Therefore, it is reasonable to insist that each interval in this new sample space be assigned a probability in a manner consistent with the probabilities assigned to events of the original probability system, $(\mathcal{S}, \mathscr{E}, P)$. But when we refer to "the probability of the interval $I$ in the new sample space"—that is, "the probability that the observed numerical result lies in the interval $I$"—we mean simply the "probability that the original outcome $s$ satisfies the condition $X(s) \in I$." In order to talk meaningfully about this probability, we must insist that the set $\{s \in \mathcal{S}\colon X(s) \in I\}$ be an event in the original event class $\mathscr{E}$ associated with $\mathcal{S}$. (Note that the set $\{s \in \mathcal{S}\colon X(s) \in I\}$ is the subset of $\mathcal{S}$ consisting of those original possible outcomes that get converted by $X$ to numerical possible outcomes lying in $I$.)

### *1.1 Abbreviations* $\{X \in I\}$, $P_X(I)$

We abbreviate the event $\{s \in \mathcal{S}: X(s) \in I\}$ by $\{X \in I\}$. We sometimes denote $P\{s \in \mathcal{S}: X(s) \in I\} = P\{X \in I\}$ by $P_X(I)$. We read $P\{X \in I\}$ as "the probability that $X$ assumes a value in $I$."

**1.2 Definition: Random Variable.** Let $(\mathcal{S}, \mathcal{E}, P)$ be a probability system and let $X$ be a real valued function on $\mathcal{S}$, satisfying the condition that for each interval $I$

$$(*) \qquad \{X \in I\} \equiv \{s \in \mathcal{S}: X(s) \in I\} \text{ is in } \mathcal{E}.$$

We call $X$ a *random variable* on $(\mathcal{S}, \mathcal{E}, P)$.

In all cases that we shall consider, condition (*) will be met for each real valued function we define on $\mathcal{S}$; therefore we will not attempt to verify it in any specific situations. ***We may think of a random variable as being a measuring instrument that is used to measure a certain attribute of the outcome.***

Keep this interpretation in mind in the examples that follow.

### *1.3 Example: Random Variables*

Let $\mathcal{S}$ consist of sequences of length $n$ of heads and tails. Let $H_i$ be the event "heads on the $i$th toss" (that is, the set of sequences in $\mathcal{S}$ with $H$ in the $i$th coordinate). Let $s$ denote any sequence in $\mathcal{S}$, and define the random variables $X_i$ by

$$X_i(s) = \begin{cases} 1 \text{ if } H \text{ is in the } i\text{th coordinate of } s \\ 0 \text{ otherwise.} \end{cases}$$

Thinking of $s$ as a possible outcome when tossing a coin $n$ times, we notice that $X_i(s)$ is the number of heads on the $i$th toss when the outcome is $s$; hence $\sum_{i=1}^{n} X_i(s)$ must be the total number of heads in the $n$ tosses. We refer to $\sum_{i=1}^{n} X_i$ as the total number of heads in $n$ tosses.

If we take $\mathcal{S}$ to be the set of all sequences of length $n$ of integers $1, 2, \ldots, 6$, then we have the familiar sample space for the toss of $n$ dice. If we let $X_i(s)$ be the value in the $i$th coordinate of the sequence $s$, then the value of $\sum_{i=1}^{n} X_i$ is the sum on the $n$ dice. △

### *1.4 Example: Coordinate Variables*

Let $\mathcal{S}$ consist of all ordered pairs of real numbers corresponding to pressure and temperature measurements on a given object. Then two important and natural random variables are given by

$$X(x, y) = x \qquad \text{and} \qquad Y(x, y) = y.$$

$X$ is the "pressure" random variable and $Y$ the "temperature" random variable. △

### 1.5 *Example: Disease Diagnosis*

If $n$ medical measurements are made on an individual, our sample space might consist of sequences $(x_1, x_2, \ldots, x_n)$. We would probably include as events all sets of the form $A_j = \{(x_1, \ldots, x_n) \in \mathcal{S}: a_j \leq x_j \leq b_j\}$. If we have sufficient information on the symptoms caused by the disease, then we might be able to develop a useful model for diagnosis in which the value $X(x_1, \ldots, x_n)$ of some random variable $X$ is likely to be close to 1 when the disease is present and close to 0 when the disease is absent. Such a random variable $X$ would be a good "diagnostic" function. △

### 1.6 *Example: Speed Function*

Suppose that the sample space is the same as that in Example 1.4, but with $(x, y)$ representing velocity. Here the velocity (3, 7) mph means 3 mph north, 7 mph east, whereas (−3, 7) mph means 3 mph south, 7 mph east. Then a natural random variable is the one that associates with velocity $(x, y)$ its magnitude

$$s(x, y) = \sqrt{x^2 + y^2},$$

which represents the speed. △

### 1.7 *Example: Maximum Temperature*

The routine of a hospital calls for measurement of a patient's temperature every hour. A typical full day's outcome is a sequence $(x_1, \ldots, x_{24})$. A natural random variable is the one that associates with these measurements their maximum, that is,

$$M(x_1, \ldots, x_{24}) = \max_{1 \leq i \leq 24} x_i. \qquad △$$

### 1.8 *Example: Electric Current*

If electrical current—amount of electrical charge per unit time passing a given point—is measured continuously for a period of two hours, then a typical outcome is a real valued function $f$ with domain [0, 2]. Here $f(t)$ is the current at time $t$. A natural random variable is the average current $X$ given by

$$X(f) = \tfrac{1}{2}\int_0^2 f(t)\, dt. \qquad △$$

*1.9 Remarks.* In many cases it is natural to define more than one random variable on a given probability space. As an example, when information is telemetered back from a satellite, we may have a random variable $X_t$ (radiation at time $t$) associated with each time point. We shall investigate such cases in the next chapter.

In quite a few of the situations that we have examined, the outcome of the original experiment is already a real number or a sequence of real numbers. In these cases we can think of the actual outcome as the value of a random variable or of a sequence of random variables. Since the conversion by means of a random variable is often done prior to presenting the results for analysis, we often see a random variable defined as the outcome of a numerical experiment—that is, as a symbol that assumes different numerical values with various probabilities. This usage is not inconsistent with our definition of a random variable.

---

## *1.10 EXERCISES*

In these exercises, several experiments will be described. Define some random variables that you feel might be useful or interesting (supply reasons if you think they might be needed).

1.° The experiment consists of recording the point count of each card in the sequence of 13 cards dealt to you in the game of bridge. (Here the point counts are as follows: Ace, 4; King, 3; Queen, 2; Jack, 1; all others, zero.)

2. The experiment consists of drawing a random sample of 2000 people from the population of the United States.

3.° The experiment consists of recording the grades made by each member of a class of 22 on five tests.

4. The experiment consists of measuring the three side lengths of a rectangular parallelepiped.
Hint: The measurements may be for the purpose of determining volume or for the purpose of determining area.

5.° The experiment consists of recording:
(a) the number of telephone calls in progress at time zero.
(b) all times in [0, 12] at which telephone calls are initiated in some trunk line.
(c) all times in [0, 12] at which telephone calls are terminated.

6. The experiment consists of recording the values of 100 selected stocks at hourly intervals during the stock market day.

7.° The experiment consists of recording the location and time of discovery of cases of a given disease.

8. The experiment consists of recording the voltages across various resistors in a working electronic device.

9.° The experiment consists of measuring the amount of vitamins $A$, $B$, $C$, $D$, and $E$ and the caloric content of each of the foods in a hospital menu.

## SECTION 2
## Univariate Distributions, Distribution Functions, Discrete Probability Functions, and Densities

Since random variables are real valued functions, it is possible to use the structure of the real numbers and methods of calculus to study them. In fact, we can describe the probabilities associated with random variables by means of ordinary real valued functions of a real variable. In order to do this, we must give several definitions.

**2.1 Definition: Univariate Distribution.** A *univariate distribution* is a probability measure whose domain consists of subsets of the real line. There are various ways to describe univariate distributions by means of ordinary functions. The most important one for theoretical purposes is the distribution function.

**2.2 Definition: Univariate Distribution Function.** Let $(-\infty, b]$ stand for the set of real numbers $x$ satisfying $x \leq b$. If $P$ is a univariate distribution, then the function $F$ defined by

$$F(b) = P(-\infty, b] \text{ for each real } b$$

is called the *univariate distribution function* of the distribution $P$.

**2.3 Definition: Distribution of a Random Variable.** The univariate distribution $P_X$ determined by

$$P_X(I) = P\{X \in I) = P\{s \in \mathcal{S}: X(s) \in I\} \text{ for each interval } I$$

is called the *distribution of the random variable X*. Its distribution function is denoted by $F_X$, so that

$$F_X(b) = P\{X \leq b\}.$$

Note that the probability $P_X(I_1 \cup I_2)$ is determined by this assignment, as are the probabilities of more complicated sets, in a manner similar to that to be given in Example 2.5.

*Remark.* It can be shown that a knowledge of the distribution function $F$ is equivalent to knowledge of the corresponding distribution $P$. We shall indicate why this is so in Example 2.5.

### 2.4 Properties of Univariate Distribution Functions

It is easy to show that every univariate distribution function is monotone non-decreasing and satisfies

$$\lim_{b \to -\infty} F(b) = 0, \qquad \lim_{b \to \infty} F(b) = 1,$$

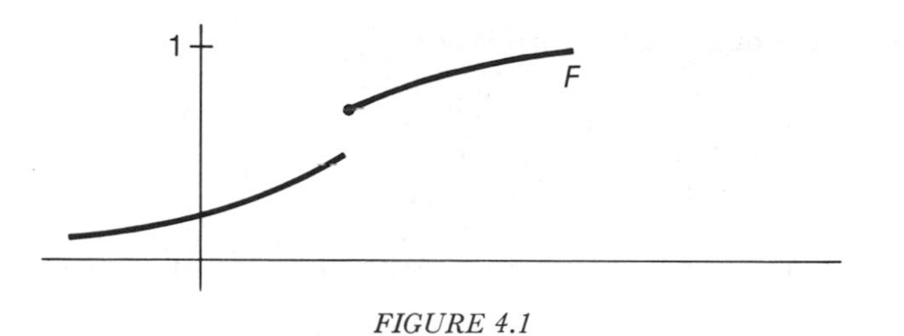


FIGURE 4.1

for each real $a$

$$\lim_{\substack{b \to a \\ b > a}} F(b) = F(a) \qquad \text{(right continuity).}$$

These properties are illustrated in Figure 4.1. The proof of monotonicity follows in Example 2.5. The remainder of the proof is left to Problem 2.9. *Any function with these properties is a distribution function.*

### 2.5 *Example: Finding P(a, b] from F*

If $F$ is the distribution function of the distribution $P$, then for $a < b$, the probability $P(a, b]$ that the outcome $x$ satisfies $a < x \leq b$ is given by

$$(*) \qquad P(a, b] = F(b) - F(a).$$

This follows from the fact that we may write $(-\infty, b]$ as the disjoint union $(-\infty; a] \cup (a, b]$. Hence

$$P(-\infty, a] + P(a, b] = P(-\infty, b],$$

that is,

$$P(a, b] = P(-\infty, b] - P(-\infty, a],$$

which is the same as (*). Figure 4.2 illustrates Example 2.5. △

It is intuitively reasonable that we can build most subsets of the real line from sets of the form $(a, b]$. For example, we can obtain the probability $P(-\infty, a)$ that the outcome $x < a$ as follows:

$$(-\infty, a) = (-\infty, a - 1] \cup \left(\bigcup_{n \geq 1} \left(a - \frac{1}{n}, a - \frac{1}{n+1}\right]\right).$$

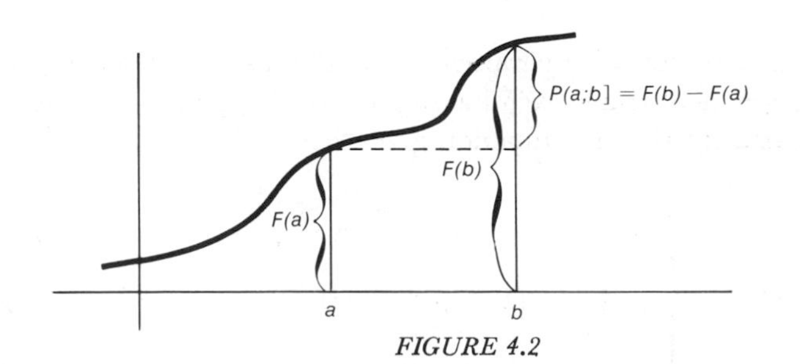


FIGURE 4.2

The right side is a disjoint union, and hence

$$
\begin{aligned}
P(-\infty, a) \\
&= F(a-1) + \sum_{n\geq 1}\left[F\left(a - \frac{1}{n+1}\right) - F\left(a - \frac{1}{n}\right)\right] \\
&= F(a-1) + \left[F\left(a-\frac{1}{2}\right) - F(a-1)\right] + \left[F\left(a-\frac{1}{3}\right) - F\left(a-\frac{1}{2}\right)\right] \\
&\qquad + \left[F\left(a-\frac{1}{4}\right) - F\left(a-\frac{1}{3}\right)\right] + \cdots \\
&= \lim_{n\to\infty} F\left(a - \frac{1}{n}\right).
\end{aligned}
$$

## 2.6 *Example: Exponential Distribution Function*

A distribution function that is used to describe the intervals between the arrivals of buses is given by

$$
F(x) = \begin{cases} 1 - e^{-x/\theta} & \text{for } x \geq 0 \\ 0 & \text{for } x < 0, \end{cases}
$$

where $\theta > 0$ is a given constant. We refer to the associated univariate distribution as the *exponential distribution with mean value* $\theta$. (The significance of the term "mean value" will be explained in the next chapter.)

This distribution is related to the Poisson model in Chapter 5, Problems 4.9(3) and 5.7. The Poisson model is justified in Chapter 7.

If $T$ is a random variable with this distribution and represents the time elapsed between successive bus arrivals, then recall from Example 4.5 and Exercise 4.7(10) in Chapter 3 that

$$
P\{T \geq z + y \mid T \geq z\} = P\{T \geq y\}.
$$

Because of this property, we say that *an exponential random variable is memoryless.* If $T$ represents the lifetime of a unit, then we say that *this unit exhibits no wearout.* △

## 2.7 *Example: Binomial Distribution*

The binomial distribution with parameters $n$ and $p$, used to describe the number of successes in $n$ independent trials, each having probability $p$ of success (see Example 5.12, Chapter 3), has a distribution function given by

$$
F(x) = \begin{cases} 0 & \text{if } x < 0 \\ \displaystyle\sum_{j\leq x} \binom{n}{j} p^j (1-p)^{n-j}, & \text{if } 0 \leq x \leq n \\ 1 & \text{if } x > n. \end{cases}
$$

△

---

## 2.8 *EXERCISES*

1. If the time between bus arrivals is governed by the exponential distribution with mean value 10 minutes, what is:

(a)° the probability that the interval between the arrival of the first and second bus exceeds 25 minutes (furnish a numerical answer accurate to three decimals).

(b) the probability that the second bus arrives more than 10 minutes after the first?

(c) the probability that the second bus arrives between 5 and 15 minutes after the first one (furnish a numerical answer accurate to three decimals)?

2. An experiment whose probability of success is .1 is repeated five times. What is:

(a)° the probability of not more than two successes?

(b) the probability of not more than one success?

(c) the probability of no successes?

3. Given the exponential distribution function

$$F(x) = \begin{cases} 1 - e^{-x/\theta} & \text{for} \quad x \geq 0 \\ 0 & \text{for} \quad x < 0 \end{cases}$$

which values of $\theta$ correspond to long-lived populations? Justify.

---

## 2.9 *PROBLEM*

Prove the assertions of 2.4 concerning the properties of univariate distribution functions.

Although a distribution function can be used to represent any distribution on the real line, it is mainly a theoretical tool. The two special cases that arise most commonly in practical applications are given in the following definition.

**2.10 Definition: Discrete Probability Function, Density.** Let $P$ be a univariate distribution. If there is a set $\{x_0, x_1, \ldots\}$ such that for each event $A$

$$P(A) = \sum_{x_i \in A} P\{x_i\},$$

then the function $p$ defined by

(a) $$p_i = P\{x_i\}$$

is called the *discrete probability function for P*. Note that $P$ is defined for every event $A$, whereas $p$ is defined only for specific one-point events. Nonetheless, for each event $A$ we can find $P(A)$ from $p$. Clearly, $P\{x_i\}$ is the probability of the one-point set $\{x_i\}$.

If $P$ is a univariate distribution, and if we can represent the distribution function $F$ of $P$ in the form

$$F(x) = \int_{-\infty}^{x} f(t)\, dt \quad \text{for all real } x, \tag{b}$$

then we call $f$ the *density of the distribution P*. (Recall that $\int_{-\infty}^{x} f(t)\, dt = \lim_{a\to\infty} \int_{-a}^{x} f(t)\, dt$.) The density of the distribution $P_X$ of a random variable $X$ is denoted by $f_X$ and is called simply the *density of X*. The value $F(t)$ represents the *probability per unit length at t*. That is, $f(t)\,\Delta t \simeq P[t, t + \Delta t]$ for $\Delta t$ small.

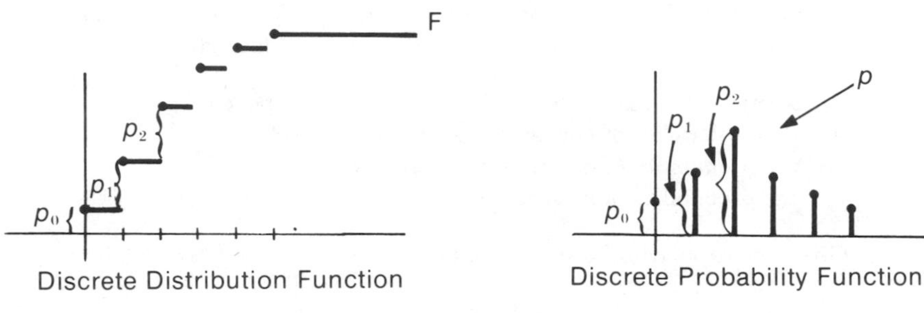

*FIGURE 4.3*

We illustrate the concepts of distribution function and corresponding discrete probability function in Figure 4.3, and Figure 4.4 shows a distribution function with its corresponding density.

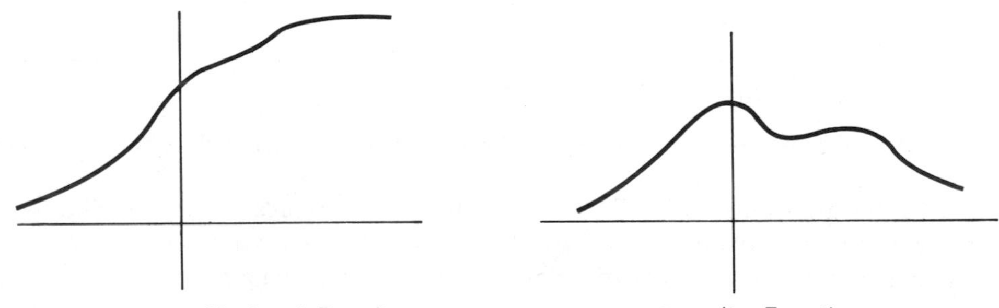

*FIGURE 4.4*

When there is a density, we can see from the Fundamental Theorem of Calculus that we usually have $F'(x) = f(x)$. In fact, the following result can be established.

### 2.11 *Theorem: Recovery of Density from Distribution Function*

Suppose $F'(x)$ is defined for all but a finite number of values $x = a_1, \ldots, x = a_k$. Let us arbitrarily put $F'(a_i) = 0$ for all $a_i$. If $F'$ so defined is integrable with $\int_{-\infty}^{\infty} F'(x)\, dx = 1$, then $F'$ is the density for the associated distribution $P$.

We note that a density cannot possibly be negative at all points of any non-zero length interval, as follows: If $f(t) < 0$ for $a < t < b$, then

$$P(a, b] = F(b) - F(a) = \int_{-\infty}^{b} f(t)\, dt - \int_{-\infty}^{a} f(t)\, dt = \int_{a}^{b} f(t)\, dt < 0.$$

That is, if $f(t) < 0$ on $(a, b)$, then we would have $P(a, b] < 0$, which is impossible. The following result is important in helping us construct densities.

### 2.12 *Theorem: Criterion for a Density*

If $f$ is a real valued function of a real variable such that

(a) $f(t) \geq 0$ for all real $t$,

(b) $\int_{-\infty}^{\infty} f(t)\, dt = \lim_{a \to \infty} \int_{-a}^{a} f(t)\, dt = 1$,

then $f$ is a probability density.

The proof follows from the fact that the function $F$ given by

$$F(x) = \int_{-\infty}^{x} f(t)\, dt$$

is easily seen to be a distribution function.

The following analogous result follows in the discrete case.

### 2.13 *Theorem: Criterion for Discrete Probability Function*

Any sequence $p_1, p_2, \ldots$ is a discrete probability function if

(a) $p_k \geq 0$ for each $k$, and

(b) $\sum_{k=1}^{\infty} p_k = 1$.

### 2.14 *Example: Exponential Density*

Recalling the exponential distribution function $F(t) = 1 - e^{-t/\theta}$, $t \geq 0$ from Example 2.6, we see that

**2.15**
$$F'(t) = \begin{cases} \dfrac{1}{\theta} e^{-t/\theta} & \text{for} \quad t \geq 0 \\ 0 & \text{elsewhere.} \end{cases}$$

Since $\int_{-\infty}^{\infty} F'(t)\, dt = 1$, we see that the exponential density with mean value $\theta$ is given by Equation 2.15. △

### 2.16 *Example: Binomial Probability Function*

From Example 2.7 and the binomial theorem $\left(\sum_{k=0}^{m} \binom{m}{k} a^k b^{m-k} = (a+b)^m\right)$, we see that the binomial distribution has a discrete probability function given by

$$p_k = \begin{cases} \binom{n}{j} p^j (1-p)^{n-j} & j = 0, 1, \ldots, n \\ 0 & \text{otherwise.} \end{cases}$$

△

### 2.17 *Example: Poisson Distribution*

The Poisson distribution, used in models for the number of phone calls in a time interval of duration $t$, is given by the discrete probability function

$$p_k = P\{k\} = e^{-\mu t} \frac{(\mu t)^k}{k!}, \qquad k = 0, 1, 2, \ldots.$$

***The parameter $\mu > 0$ represents the expected number of phone calls initiated per unit time, as we will show later.*** △

### 2.18 *Example: Normal Distribution*

The density given by

$$f(x) = \frac{1}{\sqrt{2\pi\sigma^2}} e^{-(x-\mu)^2/(2\sigma^2)} \quad \text{for } x \text{ real,}$$

where $\mu$ is a real constant and $\sigma$ is a positive constant, is called the normal density with mean $\mu$ and variance $\sigma^2$. It is the familiar bell-shaped curve (Figure 4.5) so often associated with statistics. Its great theoretical and practical importance will become apparent in Chapter 7. ***The density is largest at $x = \mu$, and if $\sigma$ is small, most of the area under the density is concentrated near $\mu$.*** Proof that $f$ is a density follows from Example 2.25 and from A.12 in the Appendix. △

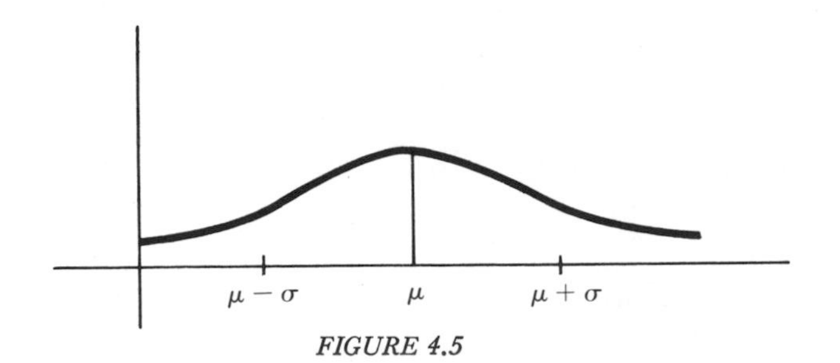


*FIGURE 4.5*

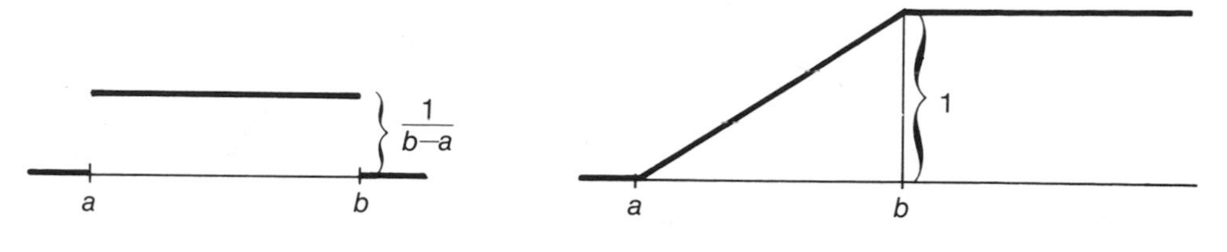


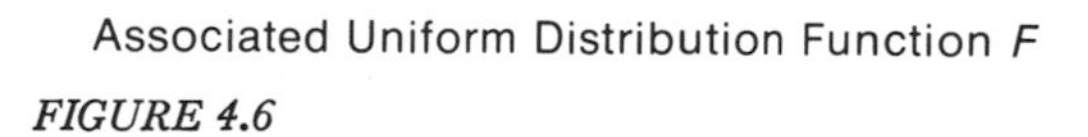

Uniform Density on $[a;b]$ Associated Uniform Distribution Function $F$

*FIGURE 4.6*

### 2.19 *Example: Uniform Distribution*

The density given by

$$f(x) = \begin{cases} \dfrac{1}{b-a} & \text{for} \quad a \le x \le b \\ 0 & \text{otherwise} \end{cases}$$

is called the uniform density on $[a, b]$. As we know, it is used in models of simple roundoff error (see Example 3.12, Chapter 3). We call $f$ the *uniform* or *rectangular* density on $[a, b]$ (see Figure 4.6). △

### 2.20 Computing Probabilities from a Density

Let $F_X$ have density $f_X$, and let $A$ be any event on the real line. Then

$$P_X(A) = P\{X \in A\} = \int_A f_X(t)\, dt.$$

Here, $\int_A f_X(t)\, dt$ has the geometric interpretation of the area under $f$ and above $A$, as illustrated in Figure 4.7. This follows because Definition 2.10 of a density and Definition 2.3 of the distribution function of a random variable imply that

$$P\{X \le x\} = F_X(x) = \int_{-\infty}^{x} f_X(t)\, dt.$$

Hence, using the reasoning of Example 2.5, where we found $P(a,b]$, we have

$$P\{a < X \le b\} = \int_a^b f_X(t)\, dt.$$

Since $\lim_{n\to\infty} \int_{a-(1/n)}^{a} f_X(t)\, dt = 0$, we also see that

$$P\{a \le X \le b\} = P\{a < X \le b\} = P\{a \le X < b\}$$

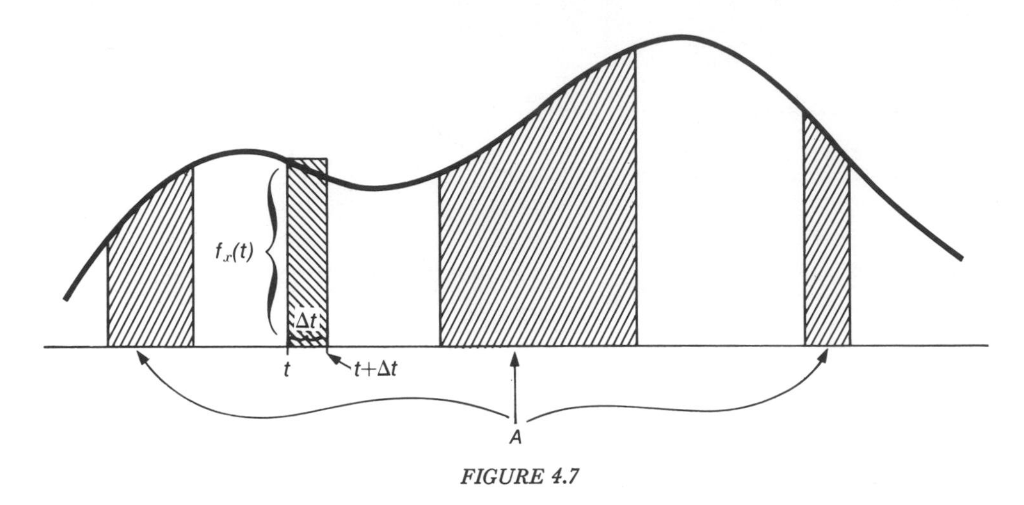


*FIGURE 4.7*

in the density case. ***We note that if $\Delta t$ is "small," then we may write the approximation*** $P\{t \leq X \leq t + \Delta t\} \simeq f_X(t)\,\Delta t$, as illustrated in Figure 4.7. Of course this is only justifiable if $f_X$ is continuous at $t$, by the mean value theorem for integrals.

### 2.21 *Example:* $P\{X^3 < a\}$

Let $X$ have density $f_X(t) = \begin{cases} e^{-t} & \text{for } t > 0 \\ 0 & \text{otherwise} \end{cases}$ and let $a$ be real. Find $P\{X^3 < a\}$.

The event $\{X^3 < a\} \equiv \{X \in A_a\}$, where $A_a = \{x \in \mathscr{R}\colon x < \sqrt[3]{a}\}$. Clearly, $P(A_a) = 0$ for $a \leq 0$, since then $f_X(t) = 0$ for all $t \in A_a$. For $a > 0$,

$$P\{X^3 < a\} = P\{X \in A_a\} = \int_{A_a} f_X(t)\,dt$$

$$= \int_{-\infty}^{\sqrt[3]{a}} f_X(t)\,dt = \int_0^{\sqrt[3]{a}} e^{-t}\,dt = 1 - e^{-\sqrt[3]{a}}. \qquad \triangle$$

### 2.22 *Example: Probability that the Digit to the Right of the Decimal is 1*

Let $X$ be a random variable with $f_X(t) = \begin{cases} e^{-t} & \text{for } t > 0 \\ 0 & \text{otherwise.} \end{cases}$ What is the probability that the first digit of $X$ to the right of the decimal point is 1; that is, what is

$P\{X \in A\}$ for $A = \{x \in \mathscr{R}\colon$ first digit to the right of the decimal point in $x$ is $1\}$?

We see that

$$A = \bigcup_{n=0}^{\infty} [n + .1,\, n + .2) \qquad (\text{since } f_X(t) = 0 \quad \text{for} \quad t \leq 0).$$

Hence

$$P\{X \in A\} = \int_A e^{-t}\,dt = \int_{\bigcup_{n=0}^{\infty}[n+.1,\,n+.2)} e^{-t}\,dt$$

$$= \sum_{n=0}^{\infty} \int_{n+.1}^{n+.2} e^{-t}\,dt$$

$$= \sum_{n=0}^{\infty} [e^{-(n+.1)} - e^{-(n+.2)}]$$

$$= e^{-.1} \sum_{n=0}^{\infty} \left(\frac{1}{e}\right)^n - e^{-.2} \sum_{n=0}^{\infty} \left(\frac{1}{e}\right)^n$$

$$= \frac{1}{e^{.1}} \frac{1}{1 - \frac{1}{e}} - \frac{1}{e^{.2}} \frac{1}{1 - \frac{1}{e}} = \frac{e^{.9} - e^{.8}}{e - 1}$$

(here we used the result that for

$$|x| \leq 1, \qquad \sum_{n=0}^{\infty} x^n = \frac{1}{1 - x};$$

see Appendix). △

*2.23 Remarks on Discrete versus Continuous Models.* In building a model for any particular situation, it is usually convenience that determines whether to use a density or a discrete probability function, since most practical situations can be well described either way. In fact, suppose $X$ is a random variable with a density. Let $X_n$ be the random variable obtained by discarding all the digits following the digit which is $n$ places to the right of the decimal point in the expansion of $X$. (For simplicity, we assume that the value of $X$ is never written with an infinite string of 9's.) Then $X_n$ has a discrete probability function, since it can only assume values that are integral multiples of $10^{-n}$. Furthermore, for large enough $n$, we see that $X_n$ approximates $X$ quite well because $|X - X_n| \leq 10^{-n}$. Hence we see that *a random variable with a density can be approximated quite well by a random variable with a discrete probability function.* Such an approximation will not usually greatly change the answers to most practical questions.

We can use a density to approximate a discrete distribution in which the points are spaced an equal distance apart as follows: Simply replace each discrete probability $p_k$ of $\{k\}$ by a uniform density with total area

$$p_k \quad \text{on} \quad \left[k - \frac{1}{n},\, k + \frac{1}{n}\right],$$

as shown in Figure 4.8.

For $n$ quite large, the approximation should be adequate for most practical applications. In general, when the integration is easy to perform, a density might be preferred. When addition is easy, if a digital computer is readily available, a discrete probability function might furnish a preferable description.

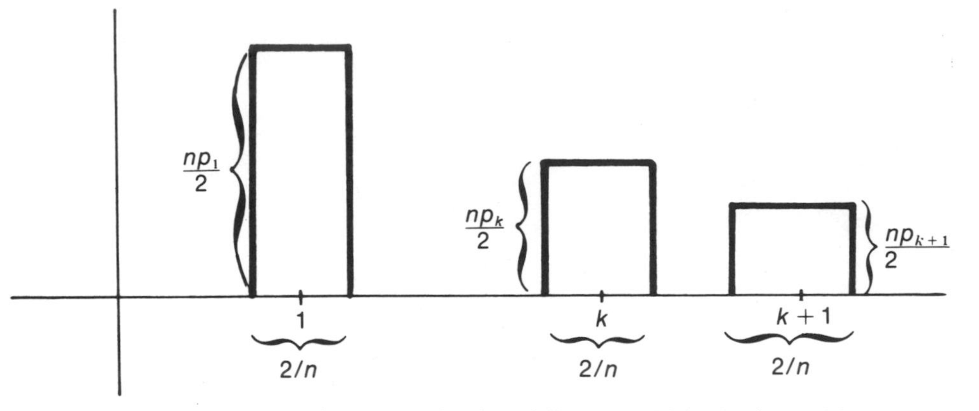


*FIGURE 4.8* Approximation of discrete model by density model.

### *2.24 Example: Roundoff*

In describing roundoff to the nearest dollar, we might use a discrete uniform model,

$$p_j = P\left\{\frac{j}{100}\right\} = \frac{1}{101} \quad \text{for} \quad j = -50, -49, \ldots, 0, 1, \ldots, 50.$$

However, a model that is somewhat simpler to handle and that gives almost the same answers to most of the problems likely to arise concerning roundoff uses the uniform density defined by

$$f(x) = \begin{cases} 1 & \text{if} \quad -.5 \le x \le .5 \\ 0 & \text{otherwise.} \end{cases}$$

△

As we shall see in Chapter 7, a normal distribution is a frequently used approximation to other distributions. Tables of the normal density and distribution function (see Example 2.18) are usually made only for the case $\mu = 0$ and $\sigma^2 = 1$. In order to make use of such tables to compute the desired probabilities, we prove the following result.

### *2.25 Example: Density of* $\dfrac{X - \mu}{\sigma}$

Let $X$ be a normal random variable with mean $\mu$ and variance $\sigma^2 > 0$; that is,

$$f_X(x) = \frac{1}{\sqrt{2\pi}\,\sigma} e^{-(x-\mu)^2/(2\sigma^2)}$$

for each real $x$. Then the related random variable

$$Y = \frac{X - \mu}{\sigma}$$

has a *standard normal distribution*. That is,

$$f_Y(y) = \frac{1}{\sqrt{2\pi}} e^{-y^2/2}$$

for each real $y$. To see this, we first find $F_Y(y)$. But

$$F_Y(y) = P\{Y \le y\} = P\left\{\frac{X-\mu}{\sigma} \le y\right\} = P\{X \le \sigma y + \mu\}$$

$$= \int_{-\infty}^{\sigma y+\mu} \frac{1}{\sqrt{2\pi}\,\sigma} e^{-(x-\mu)^2/(2\sigma^2)}\, dx.$$

Now

$$f_Y(y) = F_Y'(y) = \frac{1}{\sqrt{2\pi}\,\sigma} e^{-(\sigma y+\mu-\mu)^2/(2\sigma^2)} \cdot \sigma$$ (using the Fundamental Theorem of Calculus and the chain rule: see Appendix)

$$= \frac{1}{\sqrt{2\pi}} e^{-y^2/2},$$

as asserted.

To use this result, since

$$Y = \frac{X-\mu}{\sigma},$$

we have $X = \sigma Y + \mu$. Thus, for example,

$$P\{X \le a\} = P\{\sigma Y + \mu \le a\} = P\left\{Y \le \frac{a-\mu}{\sigma}\right\} = \int_{-\infty}^{(a-\mu)/\sigma} \frac{1}{\sqrt{2\pi}} e^{-y^2/2}\, dy.$$

This value is readily available in tables. △

---

## 2.26 EXERCISES

1. Which of the following (a) are probability densities, (b) are not probability densities, but could be converted into densities by multiplying by a suitable constant? (Justify, and supply the constant if possible.)

(i)° $f(x) = \begin{cases} \sin x \text{ for } 0 \le x \le \pi/2 \\ 0 \quad \text{otherwise.} \end{cases}$

(ii)° $f(x) = \begin{cases} \sin x \text{ for } 0 \le x \le 2\pi \\ 0 \quad \text{otherwise.} \end{cases}$

(iii)° $f(x) = \begin{cases} x^2 e^{-x^3} \text{ for } x > 0 \\ 0 \quad \text{otherwise.} \end{cases}$

(iv) $f(x) = x^3 e^{-x^4}$ for $x$ real.

(v) $f(x) = |x^3|\, e^{-x^4}$ for $x$ real.

(vi) $f(x) = x^2 e^{-x^3}$ for $x$ real.

(vii) $f(x) = e^{-|x|}$ for $x$ real.

(viii) $f(x) = \begin{cases} e^{x} \text{ for } x < 0 \\ 0 \text{ otherwise.} \end{cases}$

(ix) $f(x) = \begin{cases} e^{-x} \text{ for } x \ge 0 \\ \dfrac{1}{1+x^2} \text{ for } x < 0. \end{cases}$

2. Which of the following (*a*) are discrete probability functions, (*b*) are not discrete probability functions but could be converted into discrete probability functions by multiplying by a suitable constant? Justify.

(i)° $p_k = \begin{cases} 1/k \text{ for } k = 1, 2, \ldots \\ 0 \text{ otherwise.} \end{cases}$

(ii)° $p_k = \begin{cases} 1/k \text{ for } k = 1, 2, \ldots, 50 \\ 0 \text{ otherwise.} \end{cases}$

(iii)° $p_k = \begin{cases} 1/k^2 \text{ for } k = \pm 1, \pm 2, \ldots \\ 0 \text{ otherwise.} \end{cases}$

(iv)° $p_k = \begin{cases} 1/k^3 \text{ for } k = \pm 1, \pm 2, \ldots \\ 0 \text{ otherwise.} \end{cases}$

(v) $p_k = 1/2^k$ for $k = 0, 1, 2, \ldots$ .

(vi) $p_k = 1/2^k$ for $k = 0, \pm 1, \pm 2, \ldots$ .

(vii) $p_k = 1/2^{|k|}$ for $k = 0, \pm 1, \pm 2, \ldots$

(viii) $p_k = \begin{cases} 1/2^k \text{ for } k = 0, 2, 4, 6, \ldots . \\ 1/3^k \text{ for } k = 1, 3, 5, 7, \ldots \\ 0 \text{ otherwise.} \end{cases}$

3. Suppose $f_X(x) = \begin{cases} 0 & \text{for } x < 0 \\ 3x^2 & \text{for } 0 \leq x \leq 1 \\ 0 & \text{for } x < 1. \end{cases}$

Find:

(a)° $P\{X < 1/3\}$.
(b) $\{1/2 \leq X \leq 2/3\}$.
(c) $P\{|X - 1/2| \leq 1[2\}$.
(d)° $P\{|X - 1/2| \leq 1/10\}$.
(e) $P\{X \in [.1, .2] \cup [.7, .8]\}$.
(f)° $P\{1/3 \leq X \leq 2\}$.
(g) $P\{-.3 \leq X \leq -.2\}$.
(h) $P\{X \text{ is rational}\}$.

4. Suppose that $P\{X = k\} = p_k = \begin{cases} \dfrac{1}{k(k+1)} & \text{for } k = 1, 2, \ldots \\ 0 & \text{otherwise.} \end{cases}$

(a) Show that $p_k$ is a discrete probability function.

Hint: $$\sum_{k=m}^{n} \frac{1}{k(k+1)} = \frac{1}{m(m+1)} + \frac{1}{m(m+1)(m+2)} + \cdots + \frac{1}{n(n+1)}$$
$$= \left[\frac{1}{m} - \frac{1}{m+1}\right] + \left[\frac{1}{m+1} - \frac{1}{m+2}\right] + \cdots + \left[\frac{1}{n} - \frac{1}{n+1}\right]$$
$$= \frac{1}{m} - \frac{1}{n+1}.$$

(b)° Find $P\{X \leq j\}$.
(c) Find $P\{j \leq X \leq m\}$.
(d) Find an expression for $P\{X \text{ is even}\}$.
(e) Find an expression for $P\{X^2 \text{ is even}\}$.

5. Dropout is a type of defect in magnetic tape recording in which the signal disappears for a brief period of time. It can arise from a variety of factors, such as a defective oxide coating, a buildup of static electricity, and badly cut or stretched tape. It is especially prevalent and disturbing on today's narrow track, slow speed stereo tapes. When it occurs, you get the feeling of having gone deaf in one ear. Let $N$ be a random variable whose value is the number of dropout defects in an 1800-foot reel of tape. It is common to assume that $N$ is Poisson with some mean $\mu$, that is,

$$P\{N = k\} = \frac{e^{-\mu}\mu^k}{k!}, \; k = 0, 1, 2, \ldots .$$

(We shall see a theoretical justification for this assumption in Chapter 7.) Assuming this

(a)° Find the probability of no dropout defects in a reel of tape if $\mu = 1$.
(b) Find the probability of no dropout defects in a reel of tape if $\mu = 2$.
(c)° If the probability of no dropout defects in a reel of tape is .4, find the probability of more than three defects in a reel.

6. Let the random variable $X$ have a uniform density on $[-1, 1]$ (see 2.19).
(a)° Find $P\{X^2 > .5\}$.

Hint: This is $P\{X \in A\} = \int_A f_X(x)\, dx$, where

$$A = \{x \in [-1, 1]\colon\; x^3 > .5\} = [-1, -\sqrt{.5}) \cup (\sqrt{.5}, 1].$$

(b) Find $P\{X^2 > .5 \mid X > 0\}$.

7. Let $X$ be uniformly distributed on $[1, 2]$ (see 2.19).

(a)° Find $P\left\{\dfrac{1}{X} < \tfrac{2}{3}\right\}$.

(b) Find $P\left\{\dfrac{1}{X^2} < \tfrac{2}{3}\right\}$.

8. Let $X$ be uniformly distributed on $[-.5, .5]$. (See 2.19.) Find $P\left\{\dfrac{1}{X} < 10\right\}$.

9. Suppose that $X$ has density $f_X(x) = \begin{cases} e^{-x} & \text{for } x > 0 \\ 0 & \text{otherwise.} \end{cases}$

(a) What is the probability that the first digit of $X$ to the *left* of the decimal point is a 1?
(b) What is the probability that the first digit to the right of the decimal point in $X$ is a 2?

10. Let $X$ have an exponential distribution with mean value 2 (see Example 2.6).
(a) Find $P\{X \leq 1\}$.
(b)° Find $P\{X \leq 1 \mid X \leq 2\}$.

11. Let $X$ represent the error in some experimental measurement, and suppose that $X$ has a normal distribution (see Example 2.18) with $\mu = 0$ and $\sigma^2 = 1$.
(a)° Find a number $c$ such that the probability of an error exceeding $c$ is less than .01, that is, that $P\{X > c\} < .01$.

Hint: $e^{-x^2}/2 \leq \begin{cases} e^{-x/2} \text{ for } x > 1 \\ 1 \text{ for } 0 \leq x \leq 1 \end{cases}$ and $P\{X \leq 0\} = \frac{1}{2}$ by the symmetry of the normal density. (Problems like this one arise in certain statistical tests.)

(b) Find $c$ such that $P\{|X| > c\} < .01$.

---

*2.27 PROBLEMS*

1. Suppose that $X$ is the number of phone calls initiated in the time period [0, 2], and that it has a Poisson distribution (see Example 2.17 ) with $\mu = 1$, $t = 2$. Find a number $c$ such that the chances that more phone calls than $c$ are initiated in [0, 2] is less than .1, that is, find $c$ satisfying $P\{X > c\} \leq .1$. (Such computations are frequently needed in statistical tests.)

Hints: (a) $e^{-2}\frac{2^k}{k!} \leq \frac{2^k}{k!} \leq \frac{1}{2^k}$ for $k > 10$. (b) $\sum_{k=n}^{\infty} x^k = \frac{x^n}{1-x}$ for $|x| < 1$.

2. When an electronic sinusoidal signal generator (generating the voltage $\sin t$ at time $t$) is switched on, the time $\theta$ from the switch-on, $t_0$, until the start of the next cycle (the next 0 of the sine function having positive slope) is called a *phase angle.* It is not an angle in the geometrical sense. Examine Figure 4.9. If the total length of

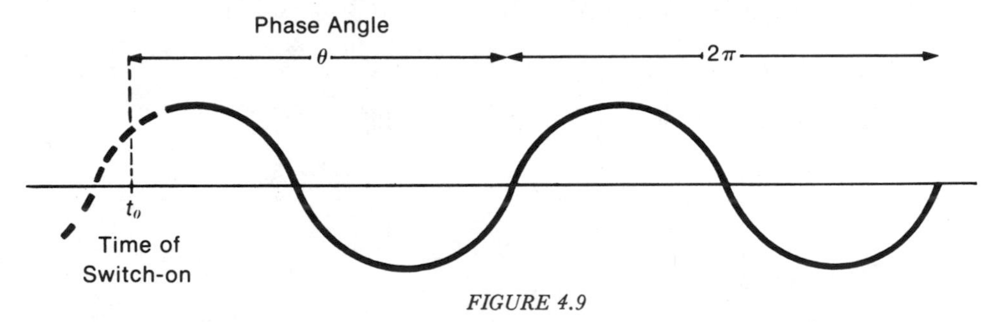


*FIGURE 4.9*

time of one cycle is $2\pi$ units, we often assume that $\theta$ has a uniform distribution on $[0, 2\pi]$ (see Theorem 2.12). Under this assumption, find $P\{\sin \theta \leq \gamma\}$.
Hint: Draw the sine function and examine $A = \{\theta \in [0, 2\pi]: \sin \theta \leq \gamma\}$. Once you have determined $A$, use Theorem 2.13.

The notion of *failure rate* is sufficiently important in reliability to warrant a brief discussion here. Let $T$ be the lifetime of some component, and suppose that

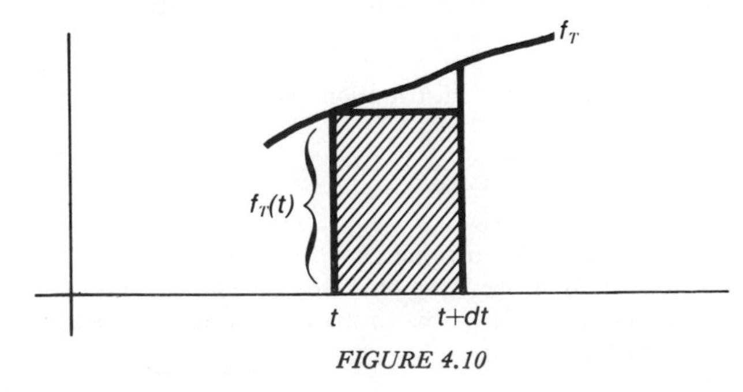


*FIGURE 4.10*

$T$ has distribution function $F_T$ and density $f_T$. Then we know that the probability that the component's lifetime falls in the interval $(t, t + dt]$ is approximately $f_T(t)\,dt$: see Figure 4.10. Hence the probability of failure in the interval $(t, t + dt]$, given survival past time $t$, is

$$\frac{P\{\text{failure in } [t, t + dt] \text{ and survival past time } t\}}{P\{\text{survival past time } t\}}$$

$$= \frac{P\{\text{failure in } [t, t + dt]\}}{P\{\text{survival past time } t\}}$$

(since if the unit does not survive past time $t$ it it must have failed in $[0, t]$, and hence cannot fail in $(t, t + dt]$)

$$\simeq \frac{f_T(t)\,dt}{1 - F_T(t)}.$$

Thus, the probability of failure per unit time "at $t$," given survival up to $t$, is

$$\frac{f_T(t)}{1 - F_T(t).}$$

We summarize our discussion in the following definition.

**2.28 Definition: Failure Rate.** Let $T$ represent the lifetime of some component. The quantity

$$q(t) = \frac{f_T(t)}{1 - F_T(t)}$$

is called the *failure rate of the component at time t*;

$$\frac{f_T(t)}{1 - F_T(t)}\,dt$$

is approximately the conditional probability of failure in $[t, t + dt]$, given survival up to time $t$. If $q(t)$ is increasing with $t$, the component exhibits wearout, and $F_T$ is called an i.f.r. (*increasing failure rate*) distribution. If $q(t)$ is decreasing, $F_T$ is called a d.f.r. (*decreasing failure rate*) distribution.

The exponential distribution $F_T(t) = 1 - e^{-t/\theta}, t > 0$ (see Example 2.6) is seen to have a constant failure rate, since

$$\frac{f_T(t)}{1 - F_T(t)} = \frac{\frac{1}{\theta}e^{-t/\theta}}{e^{-t/\theta}} = \frac{1}{\theta}.$$

Because of this, it plays a crucial role in reliability theory. Furthermore, if $q(t)$ is constant—that is,

$$\frac{F'(t)}{1 - F(t)} = c, \qquad \text{or} \qquad (1 - F)' = -c(1 - F)$$

then, from the theory of differential equations and the fact that $F$ is a distribution function, it follows that $1 - F(t) = e^{-ct}$, that is, $F(t) = 1 - e^{-ct}$ for $t \geq 0$.

## 2.29 *EXERCISES*

1. For which numbers $\alpha > 0$ does the distribution given by

$$F_T(t) = \begin{cases} 1 - e^{-(t^\alpha)} & \text{if } t > 0 \\ 0 & \text{otherwise} \end{cases}$$

have:

(a) an increasing failure rate?
(b) a decreasing failure rate?
(c) a constant failure rate?

2. Show that a distribution of the form

$$F_T(t) = p(1 - e^{-t/\theta_1}) + (1 - p)(1 - e^{-t/\theta_2})$$

where $0 < p < 1$, and $\theta_1 > 0$, $\theta_2 > 0$, has a decreasing failure rate, and give an intuitive reason.
Hint: Show that the derivative of the failure rate with respect to $t$ is negative, using $a^2 - 2ab + b^2 = (a - b)^2 \geq 0$.

# SECTION 3
# Functions of Random Variables

In the previous section, we encountered random variables that were determined by other random variables. For instance, in Example 2.25 the random variable

$$Y = \frac{X - \mu}{\sigma}$$

was determined by $X$, and in Example 2.21, $X^3 = Z$ is certainly determined by $X$. Now, any outcome $s$ that determines the value $X(s)$ also determines the values

$$Y(s) = \frac{X(s) - \mu}{\sigma}$$

and $Z(s) = [X(s)]^3$, respectively. But recall that we are assuming that each function $W$ encountered here satisfies the condition that $\{s \in \mathcal{S}: W(s) \in I\}$ is an event for each interval $I$. Hence, we conclude that all functions of random variables (such as $Y$ and $Z$) that we encounter are themselves random variables.

Up to now, our approach to finding the distribution of a function of a random variable has been somewhat on an ad hoc basis. In this section, we treat this problem more systematically. We examine the case in which we are given a real valued function $g$ of a real variable, and a random variable $X$ with density $f_X$. Let $W = g(X)$. We ask how we can find the density of $W$ (provided, of course, that $W$ has a density). The method we propose for finding $f_W$ is actually quite simple. We first find $F_W$, the distribution function of $W$, by integration of $f_X$ over $\{x \in \mathscr{R}: g(x) \leq w\}$. Then define

$$q_W(w) = \begin{cases} 0 & \text{if } F'_W(w) \text{ fails to exist} \\ F'_W(w) & \text{otherwise.} \end{cases}$$

(See Problem 2.9.) If

$$\int_{-\infty}^{\infty} q_W(w)\, dw \equiv \lim_{a\to\infty} \int_{-a}^{a} q_W(w)\, dw = 1,$$

then $W$ has a density, $f_W = q_W$. That is, we first find

**3.1** $$F_W(w) = P\{W \leq w\} = P\{g(X) \leq w\} = \int_{\{x \in \mathscr{R}: g(x) \leq w\}} f_X(x)\, dx$$

and then proceed to find $f_W$. A few illustrations follow.

### 3.2 *Example:* $W = \sqrt[3]{X}$

Given $X$ with density $f_X(x) = \frac{1}{2}e^{-|x|}$, find the density of $W = \sqrt[3]{X}$.

$$F_W(w) = P\{W \leq w\} = P\{\sqrt[3]{X} \leq w\} = P\{X \leq w^3\} = \int_{-\infty}^{w^3} \frac{e^{-|x|}}{2}\, dx\ .$$

Now, we could evaluate this integral and then differentiate with respect to $w$. However, it is simpler to use the Fundamental Theorem of Calculus and the chain rule (see the Appendix) to obtain

$$F'_W(w) = \frac{e^{-|w^3|}}{2}\, 3w^2,$$

that is,

$$f_W(w) = \frac{3}{2}\, w^2 e^{-|w^3|}$$

Actually, it would have been simpler to use the symbol $f_X(x)$ rather than the specific $e^{-|x|}/2$ to derive the general result that if $X$ has density $f_X$, then $W = \sqrt[3]{X}$ has density $f_W$, given by $f_W(w) = f_X(w^3) \cdot 3w^2$, as follows:

$$f_W(w) = F'_W(w) = \frac{d}{dw} \int_{-\infty}^{w^3} f_X(x)\, dx = f_X(w^3)\, \frac{d}{dw}\, (w^3) = f_X(w^3) \cdot 3w^2. \qquad \triangle$$

### 3.3 *Example:* $W = X^2$

Given $X$ with density $f_X$, find the density of $W = X^2$. Here

$$
\begin{aligned}
F_W(w) = P\{W \le w\} &= P\{X^2 \le w\} \\
&= \begin{cases} P\{|X| \le \sqrt{w}\} & \text{for } w \ge 0 \\ 0 & \text{for } w < 0 \end{cases} \\
&= \begin{cases} P\{-\sqrt{w} \le X \le \sqrt{w}\} & \text{for } w \ge 0 \\ 0 & \text{for } w < 0 \end{cases} \\
&= \begin{cases} \displaystyle\int_{-\sqrt{w}}^{\sqrt{w}} f_X(x)\, dx & \text{for } w \ge 0 \\ 0 & \text{for } w < 0 \end{cases} \\
&= \begin{cases} \displaystyle\int_{-\infty}^{\sqrt{w}} f_X(x)\, dx - \int_{-\infty}^{-\sqrt{w}} f_X(x)\, dx & \text{for } w > 0 \\ 0 & \text{for } w \le 0. \end{cases}
\end{aligned}
$$

Therefore, from the Fundamental Theorem of Calculus and the chain rule (see the Appendix), we find that for $W = X^2$ we have

$$
f_W(w) = \begin{cases} \dfrac{1}{2\sqrt{w}} f_X(\sqrt{w}) + \dfrac{1}{2\sqrt{w}} f_X(-\sqrt{w}) & \text{for } w \ge 0 \\ 0 & \text{otherwise.} \end{cases}
$$

Note here that $f_W$ may be unbounded near 0, so that by $\int_{-a}^{a} f_W(w)\, dw$ we mean

$$
\lim_{\substack{\varepsilon \to 0 \\ \varepsilon > 0}} \left[ \int_{-a}^{-\varepsilon} f_W(w)\, dw + \int_{\varepsilon}^{a} f_W(w)\, dw \right].
$$

That is, $\int_{-a}^{a} f_W(w)\, dw$ is an improper Riemann integral (see the Appendix). △

---

### 3.4 *EXERCISES*

1.° Suppose that $X$ is a positive random variable with density $f_X$, that is, $f_X(x) = 0$ for $x \le 0$. Let $W = 1/X$. Find $f_W$ in terms of $f_X$.

2. Let $W = 1/X$. Find $f_W$ in terms of $f_X$.

3. Find $f_W$ in terms of $f_X$ when $W = X^3$. (We may write this as "find $f_{X^3}$ in terms of $f_X$.")

4.° Let $F_X(x) = x^2$ for $0 \le x \le 1$. Find the density of $W = X^2$.

5. Let $F_X(x) = \sin x$ for $0 \le x \le \pi/2$. Find the density of $W = \sin X$.

## *3.5 PROBLEMS*

1.° Give an example in which $W = g(X)$ does not have a density even though $X$ does have a density.

2. Establish a general result about $f_W$ when $W = g(X)$ can be solved as $X = g^{-1}(W)$, with $g$ monotone increasing.

3. A pointer is spun on a circle, but the outcome is the location on the horizontal line, as shown in Figure 4.11. Assuming a uniform distribution on the circle, what is $f_X$? We call $f_X$ the standard Cauchy density.

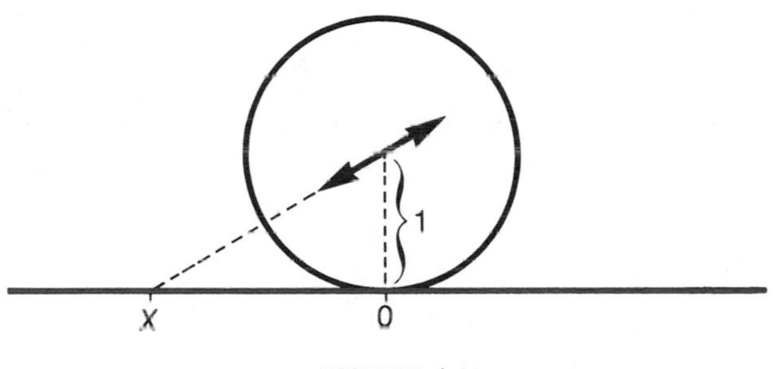


*FIGURE 4.11*

4. Let $W = \sin X$. Find an expression for $f_W$ in terms of $f_X$.

Exercises 3.4(4) and 3.4(5) illustrate a very important concept in statistical theory and in the theory of simulation of experiments. For this reason, we present a general result in this direction. It is wise to keep Exercises 3.4(4) and 3.4(5) in mind throughout this proof.

To establish this result, let $X$ be a random variable with distribution function $F_X$. For each real number $t$, we know that $F_X(t)$ is a real number in $[0, 1]$. However, if the random variable $X$ appears in place of the real number $t$, then some reflection should convince you that $F_X(X)$ is again a random variable whose value is $F_X(t)$ when $X$ takes on the value $t$. Resist the urge to write $F_X(X) = P\{X \leq X\}$, since it is almost sure to confuse you.* $F_X(X)$ is a random variable. For example, in our previous exercises, $F_X(x) = \sin x$ yields $F_X(X) = \sin X$, a nice random variable (function whose domain is $\mathcal{S}$).

Concerning the random variable $F_X(X)$, we have the following theorem.

### 3.6 *Theorem: Probability Transformation*

Let $X$ have distribution function $F_X$ which is continuous, that is, for all $t$,

$$\lim_{h \to 0} F_X(t + h) = F_X(t).$$

* Recall that $P\{X \in A\}$ is simply an abbreviation for $P\{s \in \mathcal{S}: X(s) \in A\}$. Hence $F_X(X) = P\{s \in \mathcal{S}: X(s) \leq X\}$. It is a random variable whose value at the outcome $s_0$ is $P\{s \in \mathcal{S}: X(s) \leq X(s_0)\}$.

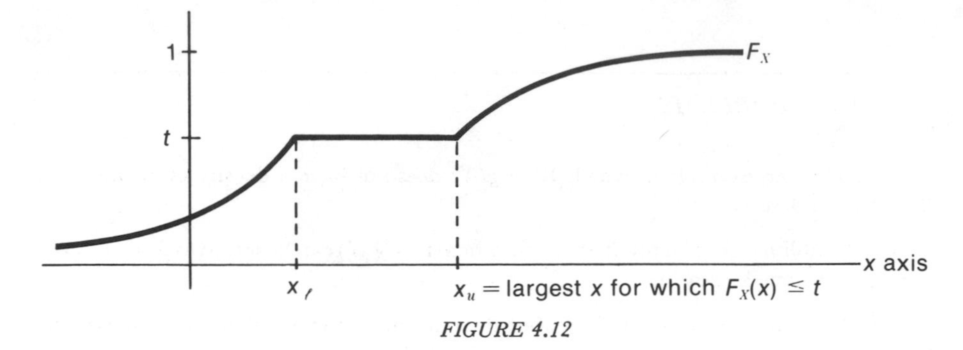


*FIGURE 4.12*

Let $Y = F_X(X)$ (that is $Y = g(X)$ where $g$ happens to be chosen to be $F_X$; note that the subscript is just for labeling). Then $Y$ is uniformly distributed on [0, 1], that is, for all $x$ in [0, 1],

$$P\{Y \leq x\} = x.$$

*Proof:* Figure 4.12 is a picture representative of a distribution function $F_X$. From it we see that the events $\{F_X(X) \leq t\}$ and $\{X \leq x_u\}$ are the same. Hence

$$P\{Y \leq t\} = P\{F_X(X) \leq t\} = P\{X \leq x_u\} = F_X(x_u) = t.$$

Since this argument applies to every $t$ in (0, 1), the theorem is proved.

Note that if $F_X$ is not continuous, then (see Figure 4.13) there is no largest $x$ such that $F_X(x) \leq t$. Hence, $F_X(X) \leq t$ is not equivalent to $X \leq x_u$.

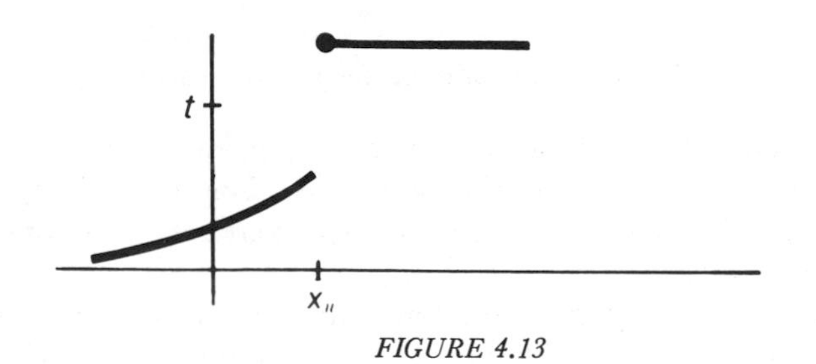


*FIGURE 4.13*

This theorem permits us to generate values of most random variables from values of uniform random variables as follows: If $F$ is a given continuous and increasing distribution function, we know that if $X$ has distribution function $F$, then $F(X)$ is uniformly distributed on [0, 1] by Theorem 3.6. But $X = F^{-1}(F(X))$ ($F^{-1}$ being well defined because $F$ is strictly increasing). Thus we have the following corollary.

### *3.7 Corollary to Theorem 3.6: Generation of Random Variables*

If $Y$ is a random variable with the uniform distribution on [0, 1] and $F$ is a continuous strictly increasing distribution function, then

$$Z = F^{-1}(Y)$$

has distribution function $F$.

Thus, if we have available random variables which are uniform on [0, 1], in theory we can generate random variables with any desired distribution function. The ability to do this is of great importance when we try to simulate the behavior of various systems. For example, in order to simulate telephone call initiation, we usually require the values of independent random variables* $T_1, T_2, \ldots$, each with an exponential distribution. Here $T_j$ represents the time between the initiation of the $j$th and the $j + 1$st call. It can be shown (see Problems 4.9(3) and 5.7, both in Chapter 5) that from these we can obtain the values of the Poisson variables representing the number of phone calls in a given time interval.

We must point out here that the method of generating random variables starting with uniform ones, given by Corollary 3.7, is of more theoretical than practical interest. The main obstacle is the difficulty of evaluating $F^{-1}$ on a computer. Even when $F^{-1}$ is relatively simple, for example when $F$ is the exponential distribution function so that $F^{-1}$ is a logarithm, there are usually much less time-consuming methods of generating the desired random variables from uniform ones.

## *3.8 EXERCISES*

1.° Suppose that $X$ is uniform on [0, 1] (see Example 2.19). How should $g$ be chosen so that $Y = g(X)$ has an exponential distribution with mean $\theta$?

2.° Suppose that $X$ is uniform on [0, 1] (see Example 2.19). How should $g$ be chosen so that $Y = g(X)$ has density $f_Y$, given by

$$f_Y(y) = \begin{cases} 5y^4 & \text{for} \quad 0 \le y \le 1 \\ 0 & \text{otherwise?} \end{cases}$$

3. Suppose that $X$ is uniform on [0, 1]. How should $g$ be chosen so that $Y = g(X)$ has density $f_Y$, given by

$$f_Y(y) = \begin{cases} \dfrac{7}{2^7} y^6 & \text{for} \quad 0 \le y \le 2 \\ 0 & \text{otherwise?} \end{cases}$$

4. Suppose that $Z$ is normal, with mean 0 and variance 1. How should $H$ be chosen so that $Y = H(Z)$ has the density given in the previous question?

---

* The concept of independent random variables will be formally defined in Chapter 5.

# 5

# Random Vectors and Multivariate Distributions

## SECTION 1
## Introduction

Frequently it is necessary to define more than one random variable on a given probability space. For example, in the experiment of recording the vital signs of a hospital patient, we might let $T_i$ be a random variable representing the patient's average temperature over the time period $[i, i + 1]$, $i = 1, 2, \ldots, n$.

We refer to a sequence $\boldsymbol{T} = (T_1, \ldots, T_n)$ of random variables all defined on the same probability space as a *random vector*. In this chapter, we will investigate the joint probabilistic behavior of such sequences of random variables. First, we introduce some definitions.

## SECTION 2
## Distributions, Densities and Discrete Probabilities

**2.1 Definition: $n$-Dimensional Space, $\mathscr{R}_n$; $n$-Dimensional Distribution.** The set $\mathscr{R}_n$ of all sequences of length $n$ of real numbers is called *n-dimensional space*. That is, $\mathscr{R}_n$ consists of elements $(r_1, \ldots, r_n)$ where each $r_i$ is real. A probability measure whose domain consists of subsets of $n$-dimensional space is called an *n-dimensional distribution*. An $n$-dimensional distribution is called *multivariate* if $n > 1$.

As a single random variable induces a distribution on the real line, so $n$ random variables $X_1, \ldots, X_n$ all defined on the one probability space $(\mathcal{S}, \mathscr{E}, P)$ induce a distribution $P_{X_1,\ldots,X_n}$ on $n$-dimensional space, defined by

$$P_{X_1,\ldots,X_n}(A) = P\{s \in \mathcal{S}: (X_1(s), \ldots, X_n(s)) \in A\}$$

where $A$ is an event in $n$-dimensional space. Just as we looked for a real valued function of a real variable to represent a one-dimensional distribution, here we try to find a real valued function of $n$ real variables to represent an $n$-dimensional distribution.

In generalizing distributions to $n$ dimensions, we find it useful to extend the notion of interval. In one dimension, if we write $(-\infty, b] = \{s \in \mathcal{R}: s \leq b\} \equiv A(b)$, then $F(b) = P\{A(b)\}$, where $F$ is the distribution function of the one-dimensional distribution $P$. For each sequence $(b_1, \ldots, b_n)$ of real numbers, it is natural to let $A(b_1, \ldots, b_n)$ be the set of real sequences $(x_1, \ldots, x_n)$ such that for all $i = 1, \ldots, n$, $x_i \leq b_i$. In two dimensions, $A(b_1, b_2)$ is the set indicated in Figure 5.1. We call $A(b_1, \ldots, b_n)$ an *$n$-dimensional semi-infinite interval.*

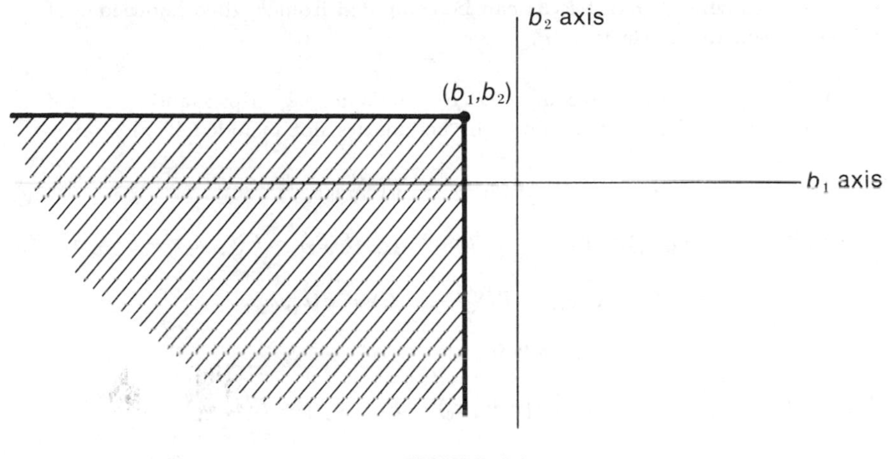


*FIGURE 5.1*

**2.2 Definition: Distribution Function.** For each sequence $(b_1, \ldots, b_n)$ of real numbers, let $A(b_1, \ldots, b_n)$ be the semi-infinite interval just defined. If $P$ is an $n$-dimensional distribution, then the function $F$ defined by

$$F(b_1, \ldots, b_n) = P[A(b_1, \ldots, b_n)] \quad \text{for all real} \quad b_1, \ldots, b_n$$

is called the *distribution function of the distribution* $P$. The distribution function $F_{X_1,\ldots,X_n}$ corresponding to the distribution $P_{X_1,\ldots,X_n}$ induced by $X_1, \ldots, X_n$ is called the *joint distribution function of the random variables* $X_1, \ldots, X_n$, or the *distribution function of the random vector* $\mathbf{X} = (X_1, \ldots, X_n)$.

It can be shown that knowledge of $F$ is equivalent to knowledge of the probability measure $P$ that determines $F$.

We briefly outline the way in which we can use the distribution function to compute probabilities of fairly general sets. For simplicity, we shall work in two dimensions. In Figure 5.2, let $B$ be the indicated rectangle, including only the parts of the upper and right-hand boundaries not included in the other boundaries. It is reasonable to believe that most sets of interest can be built up from countable

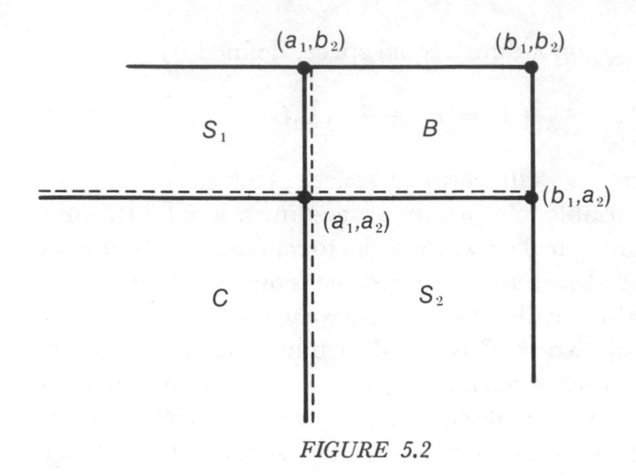


*FIGURE 5.2*

unions and complements of countable unions of such sets. Thus, we can be reasonably convinced that if $P(B)$ can be computed from $F$, then knowledge of $F$ is equivalent to knowledge of $P$.

We will now show how to compute $P(B)$. In Figure 5.2, strips $S_1$ and $S_2$ include only the parts of their upper and right boundaries not included in any other boundary. Now

$$P(S_1) = F(a_1, b_2) - F(a_1, a_2)$$

(since $P(S_1) + F(a_1, a_2) = F(a_1, b_2)$). Similarly, $P(S_2) = F(b_1, a_2) - F(a_1, a_2)$. But

$$P(B) + P(S_1) + P(S_2) + P(C) = F(b_1, b_2),$$

that is, substituting for $P(S_1)$ and $P(S_2)$,

$$P(B) + [F(a_1, b_2) - F(a_1, a_2)] + [F(b_1, a_2) - F(a_1, a_2)] + F(a_1, a_2) = F(b_1, b_2).$$

Hence,

$$P(B) = F(b_1, b_2) - F(b_1, a_2) - F(a_1, b_2) + F(a_1, a_2).$$

Once we can compute probabilities of sets like $B$, we can construct much more complicated sets, such as countable disjoint unions and complements of such sets, and use the general rules of a probability measure to compute their probabilities. Similar arguments apply to the general $n$-dimensional case.

Especially in $n$ dimensions, the distribution function is used mainly for theoretical purposes.

Just as in the univariate case, the two most frequently encountered situations involve a discrete probability function or a density.

**2.3 Definition: Multivariate Discrete Probability Function, Joint Density, $\{X_1 = b_i, X_2 = b_j\}$, $\{X_1 \leq b_1, \ldots, X_n \leq b_n\}$ Notations.** For simplicity, we define discrete probability functions only for two dimensions, the $n$-dimensional case being a natural extension that is easily recognized in practice. For each pair $i, j$ of non-negative integers, let $(b_i, b_j)$ be a point in the plane such that $P\{(b_i, b_j)\} = p_{ij}$, where $p_{ij} \geq 0$ and $\sum_{\text{all } i,j} p_{ij} = 1$. We call the function given by the $p_{ij}$ a *two-dimensional discrete probability function* and note that

$$P(A) = \sum_{\substack{\text{all } i,j \text{ such} \\ \text{that } (b_i, b_j) \in A}} p_{ij}.$$

If the discrete probability function is induced by random variables $X_1, X_2$, we may write $p_{ij} = P\{X_1 = b_i, X_2 = b_j\}$. That is, we write $\{X_1 = b_i, X_2 = b_j\}$ rather than $\{X_1 = b_i\} \cap \{X_2 = b_j\}$. Similarly, we write $\{X_1 \leq b_1, X_2 \leq b_2, \ldots, X_n \leq b_n\}$ rather than $\{X_1 \leq b_1\} \cap \{X_2 \leq b_2\} \cap \cdots \cap \{X_n \leq b_n\}$. Then

$$F_{X_1,\ldots,X_n}(b_1, \ldots, b_n) = P\{X_1 \leq b_1, \ldots, X_n \leq b_n\}.$$

If we can write the distribution function $F$ of the distribution $P$ in the form

$$F(b_1, \ldots, b_n) = \int_{-\infty}^{b_n} \int_{-\infty}^{b_{n-1}} \cdots \int_{-\infty}^{b_1} f(x_1, \ldots, x_n)\, dx_1 \cdots dx_n$$

$$= \lim_{a \to \infty} \int_{-a}^{b_n} \int_{-a}^{b_{n-1}} \cdots \int_{-a}^{b_1} f(x_1, \ldots, x_n)\, dx_1 \cdots dx_n$$

for all $b_n, b_{n-1}, \ldots, b_1$, then we call $f$ the *joint density of the distribution P*. (A review of multiple integrals is given in the Appendix.) If the distribution $P$ is induced by the random variables $X_1, \ldots, X_n$, then we denote the density by $f_{X_1,\ldots,X_n}$. Note that these subscripts are used merely for the purpose of identification.

### 2.4 Properties of Densities

The density, if it exists, can be determined from the distribution function by the formula

$$f(x_1, \ldots, x_n) = \frac{\partial^n F}{\partial x_1\, \partial x_2 \cdots \partial x_n}(x_1, \ldots, x_n).$$

Conversely, if this mixed partial derivative integrates out to 1, it can be shown that $P$ does have a density. Any non-negative function $f$ of $n$ variables whose $n$-fold integral over $\mathscr{R}_n$ equals 1 is a probability density on $\mathscr{R}_n$.

### 2.5 Computing Probabilities from a Joint Density

As in one dimension,

$$P(A) = \int \cdots \int_A f(x_1, \ldots, x_n)\, dx_1 \cdots dx_n$$

for each event $A$ in $n$-dimensional space (see the Appendix for a discussion of evaluation of multiple integrals). In the case $n = 2$, this integral represents the volume lying between the set $A$ and the surface determined by the density $f$. Except for a few simple cases, the integration needed to evaluate $P(A)$ must be done by some approximation technique. Often this will involve a digital computer, and great care must be taken to keep the error of approximation small and to avoid consumption of too much computer time.

As in the one-dimensional case, we note that if $\Delta x$ and $\Delta y$ are small and $f$ is continuous, then $P\{x \leq X \leq x + \Delta x,\ y \leq Y \leq y + \Delta y\} \cong f_{X,Y}(x, y)\ \Delta x\ \Delta y$. We have great flexibility in our choice of $f$. For instance, in the case of a target, if we wanted sets near the bullseye to be more probable than those near the border, we might let $f$ be some bell-shaped surface centered over the bullseye.

We now present a few examples of multivariate distributions.

### 2.6 *Example: Multinomial Discrete Probability Function*

An experiment has $k$ possible outcomes, say $1, 2, \ldots, k$, where the probabilities $p_j, j = 1, \ldots, k$, are arbitrary. The experiment is repeated independently $m$ times. Each possible outcome is a sequence of length $m$ with each coordinate value from the set $\{1, 2, \ldots, k\}$. We want to compute the probability $p_{j_1, j_2, \ldots, j_k}$ of observing

the value 1 in $j_1$ coordinates
the value 2 in $j_2$ coordinates
.
(*) .
.
the value $k$ in $j_k$ coordinates

where $j_1 + j_2 + \cdots + j_k = m$. It follows from the independence assumption that each specific possible outcome satisfying conditions (*) has probability

$$p_1^{j_1} p_2^{j_2} \cdots p_k^{j_k}.$$

Thus, in order to compute the desired probability, we need only multiply this expression by the number of distinct sequences satisfying (*). From Corollary 2.11, Chapter 2, we know that this number is the multinomial coefficient

$$\frac{m!}{j_1!\, j_2! \cdots j_k!}.$$

Thus, the desired probability is

$$p_{j_1, j_2, \ldots, j_k} = \frac{m!}{j_1!\, j_2! \cdots j_k!} p_1^{j_1} p_2^{j_2} \cdots p_k^{j_k}$$

and the members of the associated family of distributions are called *multinomial distributions*. Note that in order to specify a member of this family we must know $m$, $k$, and $p_1, \ldots, p_k$. △

### 2.7 *Example: Two-Dimensional Uniform Density*

In Example 3.13, Chapter 3, we introduced a uniform model in which the probability of a subset of a rectangle was proportional to the area of the subset. This model admits the density

$$f(x, y) = \begin{cases} \dfrac{1}{(b-a)^2} & \text{for } a \le x \le b \text{ and } a \le y \le b \\ 0 & \text{otherwise.} \end{cases}$$

△

### 2.8 *Example: Total Lifetime*

Suppose that a typical outcome $(t_1, t_2)$ represents the lifetimes of two components, an original and its replacement. Then $t_1 + t_2$ is the total lifetime of the so-called *standby system* consisting of this pair. It is therefore of interest to determine $P(A_a)$, where

$$A_a = \{(t_1, t_2) \in \mathcal{S}: t_1 + t_2 \le a\}.$$

In such a case, we often assume a joint density of the form

$$f(t_1, t_2) = \begin{cases} \dfrac{1}{\theta^2} e^{-t_1/\theta - t_2/\theta} & \text{for both } t_1 \ge 0 \text{ and } t_2 \ge 0 \\ 0 & \text{otherwise.} \end{cases}$$

This model is related to the Poisson model in Problems 4.9(3) and 5.7. The Poisson model is given a theoretical basis relating it to the binomial distribution in Chapter 7.

Figure 5.3 helps us to evaluate

$$P(A_a) = \iint_{A_a} f(t_1, t_2)\, dt_1\, dt_2.$$

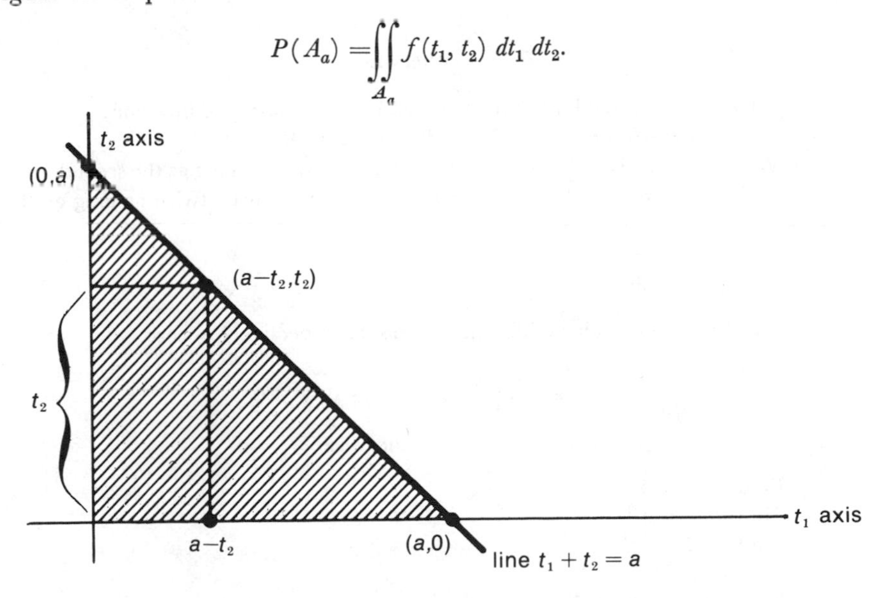


*FIGURE 5.3*

We see that for fixed $t_2$ in $[0, a]$, in order that $(t_1, t_2)$ be in $A_a$, we must have $0 \leq t_1 \leq a - t_2$. Hence, using evaluation of multiple integrals (see the Appendix),

$$\iint_{A_a} f(t_1, t_2)\,dt_1\,dt_2 = \int_0^a \int_0^{a-t_2} f(t_1, t_2)\,dt_1\,dt_2$$

$$= \int_0^a \int_0^{a-t_2} \frac{1}{\theta^2} e^{-t_1/\theta} e^{-t_2/\theta}\,dt_1\,dt_2$$

$$= \int_0^a \frac{1}{\theta^2} e^{-t_2/\theta} \left[\int_0^{a-t_2} e^{-t_1/\theta}\,dt_1\right] dt_2$$

$$= \int_0^a \frac{1}{\theta^2} e^{-t_2/\theta} [\theta - \theta e^{-(a-t_2)/\theta}]\,dt_2$$

$$= \int_0^a \left[\frac{1}{\theta} e^{-t_2/\theta} - \frac{1}{\theta} e^{-a/\theta}\right] dt_2$$

$$= 1 - e^{-a/\theta} - \frac{a}{\theta} e^{-a/\theta}. \qquad \triangle$$

## 2.9 EXERCISES

In all of these exercises, it is extremely useful to draw a picture of the set over which you must integrate.

1.° A pair of components have lifetimes $X$ and $Y$, respectively, where

$$f_{X,Y}(x, y) = \begin{cases} \frac{1}{2} e^{-x-y/2} & \text{for } x > 0 \text{ and } y > 0 \\ 0 & \text{otherwise.} \end{cases}$$

(a) What is the probability that the second unit outlasts the first one?
Hint: We want to determine $P\{Y > X\} = P\{Y - X > 0\}$.
(b) What is the probability that the first unit lasts twice as long as the second one?
(c) What is the probability that the first unit lasts at least twice as long as the second one?
(d) Find $P\{X + Y \leq a\}$.
(e) Find the density of $X + Y$.

2. A pair of components have lifetimes $X$ and $Y$, respectively, where

$$f_{X,Y}(x, y) = \begin{cases} x e^{-x} \frac{1}{2} e^{-y/2} & \text{for } x > 0, y > 0 \\ 0 & \text{otherwise.} \end{cases}$$

(a) Find $P\{X > Y\}$.
(b) Find $P\{Y > X\}$.
(c) Find $P\{2X > Y\}$.
(d) Find $P\{X + Y \leq a\}$.
(e) Find the density of $X + Y$.

3.° Referring to Example 2.8:

(a) find the density of the distribution of the total lifetime.

(b) find the probability that the first component outlasts the second one.

4.° Referring to Example 2.8:

(a) find the probability that the maximum of the two lifetimes is less than or equal to $b$.
Hint: $\max(X, Y) \leq b$ if and only if $X \leq b$ and $Y \leq b$.

(b) Find the probability that the minimum of the two lifetimes exceeds $c$.
Hint: $\min(X, Y) > c$ if and only if $X > c$ and $Y > c$.

(c) Find the density of the distribution of the maximum of the two lifetimes.

(d) Find the density of the minimum of the two lifetimes.

5. A pair of components have lifetimes $X$ and $Y$, respectively, with

$$f_{X,Y}(x, y) = \begin{cases} \dfrac{1}{\theta^2} e^{-x/\theta - y/\theta} & \text{for } x > 0 \text{ and } y > 0, \\ 0 & \text{otherwise,} \end{cases}$$

as in Example 2.8.

(a) Find the density of the lifetime of the system formed by connecting these units in series (see Example 2.11 in Chapter 3).

(b) Find the density of the lifetime of the system formed by connecting these units in parallel (see Example 2.11 in Chapter 3).

---

## 2.10 PROBLEMS

1. Suppose that $f_{X,Y}(x, y) = \begin{cases} 1 & \text{if } 0 \leq x \leq 1 \text{ and } 0 \leq y \leq 1 \\ 0 & \text{otherwise.} \end{cases}$

What is the probability that

$$a = \min(X, Y), \qquad b = \max(X, Y) - \min(X, Y), \qquad c = 1 - \max(X, Y)$$

are the side lengths of a triangle? (Here $\max(X, Y)$ is the larger of the two variables $X$ and $Y$, and $\min(X, Y)$ is the smaller of the two.)
Hint: The conditions required in order that $a$, $b$, and $c$ be the side lengths of a triangle are: (i) $a + b > c$; (ii) $a + c > b$; (iii) $b + c > a$. Now suppose that $X = \min(X, Y)$. Then, examining Figure 5.4, we see that condition (i) becomes $Y > 1/2$; condition (ii) becomes $X + (1 - Y) > Y - X$, that is, $Y - X < 1/2$; and condition (iii) becomes $X < 1/2$. Calling the event of interest $D$, we want to compute

$$P(D) = P(D \cap \{\min(X, Y) = X\}) + P(D \cap \{\min(X, Y) = Y\}).$$

But

$$D \cap \{\min(X, Y) = X\} = \{X < \tfrac{1}{2}, Y > \tfrac{1}{2}, Y - X < \tfrac{1}{2}\}$$

with a similar representation for $D \cap \{\min(X, Y) = Y\}$.

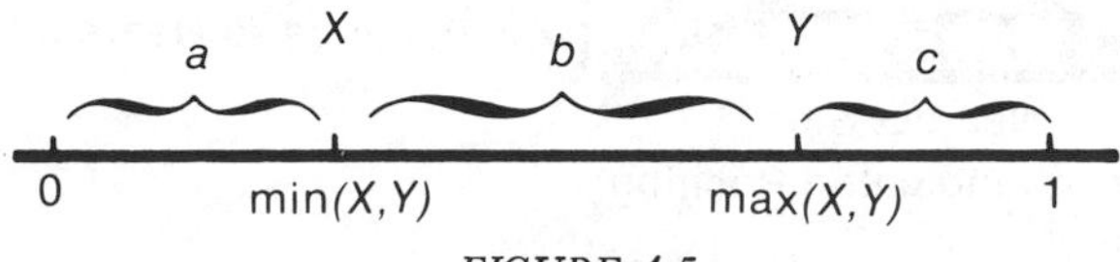

*FIGURE 4.5*

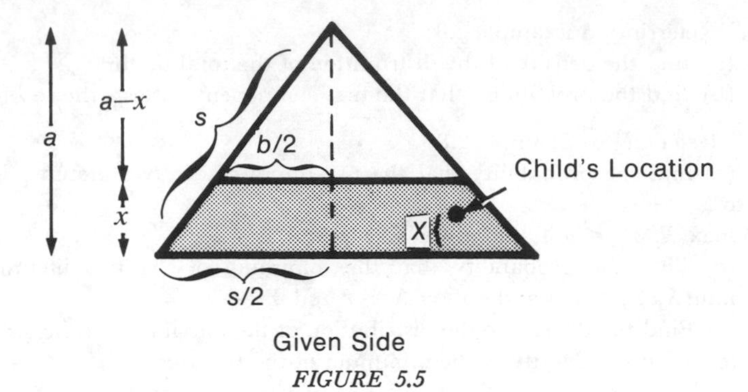


*FIGURE 5.5*

2.° A child is lost in a forest whose sides form an equilateral triangle. If the probability of the child's being in any given subset of the triangle is proportional to the area of this subset:

(a) what is the distribution of the child's distance to a given side?

Hint: See Figure 5.5. Let $X$ be the distance to the given side. Then $X \leq x$ if and only if the child is in the shaded area.

(b) what is the distribution of the child's distance to the nearest side?

Hint: From Figure 5.6, it should be evident that the child is equally likely to be in one of the three indicated subtriangles, and the conditional probability of a set in the subtriangle is proportional to its area, that is,

$$P(X \leq x) = \sum_{i=1}^{3} P(X \leq x \mid T = i)P(T = i).$$

$T$ is the subtriangle in which the child is located. The computation of $P(X \leq x \mid T = i)$ is similar to that in part (a).

## SECTION 3
## Marginal Distributions and Densities

We will now investigate how to obtain the joint distribution function and density of a *subsequence* of a sequence $X_1, \ldots, X_n$ of random variables from the corresponding distribution function and density of the full set $X_1, \ldots, X_n$.

Let $F_{X_1,\ldots,X_k,X_{k+1},\ldots,X_n}$ be the joint distribution function of $X_1, \ldots, X_n$ and $F_{X_1,\ldots,X_k}$ be the joint distribution of its subsequence $X_1, \ldots, X_k$. Since the condition $X_{k+1} \leq b_{k+1}, \ldots, X_n \leq b_n$ is really no restriction when $b_{k+1}, \ldots, b_n$ are unboundedly large, we see the following relationship.

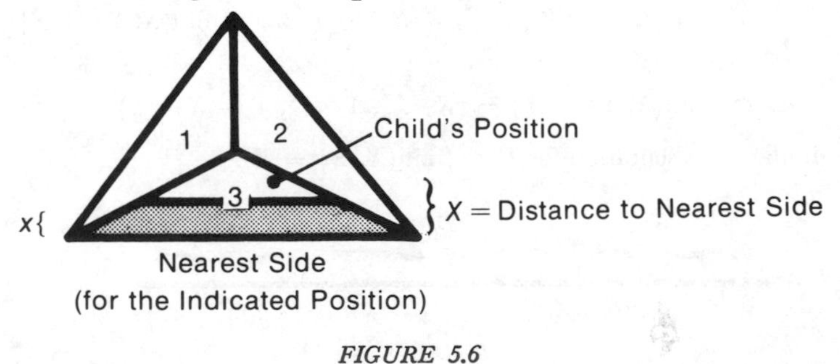


*FIGURE 5.6*

### 3.1 Distribution Function of a Subsequence

$$F_{X_1,\ldots,X_k}(b_1,\ldots,b_k) = \lim_{\substack{b_{k+i}\to\infty \\ \text{for all } i=1,\ldots,n-k}} F_{X_1,\ldots,X_n,X_{k+1},\ldots,X_n}(b_1,\ldots,b_k,b_{k+1},\ldots,b_n).$$

Note that it doesn't matter whether we choose to determine the distribution function of $X_1, \ldots, X_k$ or of some other subset $X_{i_1}, \ldots, X_{i_k}$, except that the notation corresponding to our choice is simpler. That is to say, if $i_1, i_2, \ldots, i_k$ are any $k$ integers with $1 \le i_i < i_2 < \cdots < i_k \le n$, then

$$F_{X_{i_1},\ldots,X_{i_k}}(b_{i_1},\ldots,b_{i_k}) = \lim_{\substack{b_j\to\infty \\ j\neq i_1,\ldots,i_k}} F_{X_1,\ldots,X_n}(b_1,\ldots,b_n).$$

To see how to obtain the density of $X_1, \ldots, X_k$ from that of $X_1, \ldots, X_k, X_{k+1}, \ldots, X_n$, we first see how to find $f_{X_1}$ from $f_{X_1,X_2}$. Now

$$F_{X_1}(b_1) = \int_{-\infty}^{b_1}\left[\int_{-\infty}^{\infty} f_{X_1,X_2}(x_1, x_2)\,dx_2\right]dx_1$$

from 2.3. Hence, by the definition of $f_{X_1}$ as the function satisfying

$$F_{X_1}(b_1) = \int_{-\infty}^{b_1} f_{X_1}(x_1)\,dx_1 \qquad \text{for all } b_1,$$

we see that

$$f_{X_1}(x_1) = \int_{-\infty}^{\infty} f_{X_1,X_2}(x_1, x_2)\,dx_2.$$

In the same fashion, we can derive the following result.

### 3.2 Marginal Density of $X_1, \ldots, X_k$

$$f_{X_1,\ldots,X_k}(x_1,\ldots,x_k)$$
$$= \int_{-\infty}^{\infty}\cdots\int_{-\infty}^{\infty} f_{X_1,\ldots,X_k,X_{k+1},\ldots,X_n}(x_1,\ldots,x_k,x_{k+1},\ldots,x_n)\,dx_{k+1}\cdots dx_n.$$

The distribution, distribution function, and density function of random variables are referred to as *marginal* when they are computed from the corresponding functions of a larger set of random variables. That is, the marginal density is computed by integrating out all unwanted variables in the joint density.

### *3.3 Example: Bivariate Normal Marginals*

Let $X$ and $Y$ have joint density

**3.4** $$f_{X,Y}(x, y) = \frac{1}{2\pi\sqrt{1-\rho^2}}\exp\left[\frac{-1}{2(1-\rho^2)}(x^2 - 2\rho xy + y^2)\right],$$

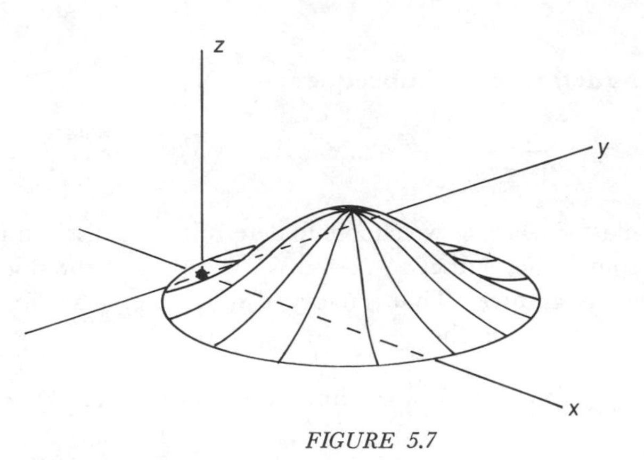


*FIGURE 5.7*

for $x$ and $y$ real, where $-1 < \rho < 1$. The function $f_{X,Y}$ is a "bell-shaped surface" (Figure 5.7). In the following optional part, we will show that

$$f_Y(y) = \frac{1}{\sqrt{2\pi}} e^{-y^2/2} \quad \text{and} \quad f_X(x) = \frac{1}{\sqrt{2\pi}} e^{-x^2/2} \quad \text{for all } \rho,$$

thus showing that *knowledge of the marginal densities $f_X$ and $f_Y$ does not in general determine the joint density $f_{X,Y}$*.

We use the technique of completion of the square to write

$$x^2 - 2\rho xy + y^2 = x^2 - 2\rho xy + \rho^2 y^2 - \rho^2 y^2 + y^2 = (x - \rho y)^2 + (1 - \rho^2)y^2.$$

Hence

$$f_Y(y) = \frac{1}{2\pi\sqrt{1-\rho^2}} e^{-y^2/2} \int_{-\infty}^{\infty} e^{-(x-\rho y)^2/[2(1-\rho^2)]} \, dx.$$

Referring to Example 2.18 in Chapter 4, we see that $g$ given by

$$g(x) = \frac{1}{\sqrt{2\pi\sigma^2}} e^{-(x-\mu)^2/(2\sigma^2)} \qquad \text{all real } x$$

is a density for each real $\mu$ and each $\sigma^2 > 0$; its integral must therefore equal 1. Now let us identify $\mu$ with $\rho y$ and $\sigma^2$ with $1 - \rho^2$. Then we have

$$\int_{-\infty}^{\infty} \frac{1}{\sqrt{2\pi(1-\rho^2)}} e^{-(x-\rho y)^2/[2(1-\rho^2)]} \, dx = 1$$

for each fixed $y$, and hence

$$f_Y(y) = \frac{1}{\sqrt{2\pi}} e^{-y^2/2},$$

as asserted. By symmetry, we obtain the desired result for $f_X$.

The *general bivariate normal density* is given by

**3.5** $$f_{X,Y}(x, y) = \frac{1}{2\pi\sigma_x\sigma_y\sqrt{1-\rho^2}} \exp -\frac{1}{2(1-\rho^2)} \times \left[\left(\frac{x-\mu_x}{\sigma_x}\right)^2 - 2\rho\left(\frac{x-\mu_x}{\sigma_x}\right)\left(\frac{y-\mu_y}{\sigma_y}\right) + \left(\frac{y-\mu_y}{\sigma_y}\right)^2\right].$$

By the same technique, we can find

**3.6** $$f_X(x) = \frac{1}{\sqrt{2\pi}\,\sigma_x} \exp -\frac{1}{2}\left(\frac{x-\mu_x}{\sigma_x}\right)^2, \qquad f_Y(y) = \frac{1}{\sqrt{2\pi}\,\sigma_y} \exp -\frac{1}{2}\left(\frac{y-\mu_y}{\sigma_y}\right)^2.$$

The bivariate normal density is frequently used as a model for describing pairs of measurements, such as parent height and child height, or high school performance and college performance. Its theoretical justification will be explored in Chapter 7.
△

For completeness, we state the discrete analog to 3.2. We let

$$p_{X_1,\ldots,X_n}(x_1, \ldots, x_n) = P\{X_1 = x_1, \ldots, X_n = x_n\}.$$

### 3.7 Marginal Probability Function of $X_1, \ldots, X_k$

If $p_{X_1,\ldots,X_n}$ is the joint discrete probability function of $X_1, \ldots, X_n$, then for $k < n$ the joint discrete probability function $p_{X_1,\ldots,X_k}$ is obtained from $p_{X_1,\ldots,X_n}$ by the formula

$$p_{X_1,\ldots,X_k}(x_1, \ldots, x_k) = \sum_{\substack{\text{all possible values} \\ (x_{k+1},\ldots,x_n) \text{ of} \\ (X_{k+1},\ldots,X_n)}} p_{X_1,\ldots,X_n}(x_1, \ldots, x_k, x_{k+1}, \ldots, x_n).$$

That is, we sum over all possible values of the unwanted variables.

---

*3.8 EXERCISES*

1. Let $f_{X,Y}(x, y) = \begin{cases} 1/\pi & \text{if } x^2 + y^2 \leq 1 \\ 0 & \text{otherwise.} \end{cases}$

Geometrically, $f_{X,Y}$ is a circular cylinder section of height $1/\pi$ above the unit circle. The probability of a subset of the unit circle is just the proportion of area taken up by this set inside the unit circle. Find:

(a)° $f_X(x)$.
(b) $f_Y(y)$.
(c)° $P\{\sqrt{X^2 + Y^2} \leq a\}$.
(d) $f_{\sqrt{X^2+Y^2}}$.

2.° Let $f_{X,Y}(x, y) = \begin{cases} x + y & \text{for } 0 \le x \le 1 \text{ and } 0 \le y \le 1 \\ 0 & \text{otherwise.} \end{cases}$
Find $f_X(x)$.

3. Let $f_{X,Y}(x, y) = \begin{cases} x^2 + y^2 & \text{if } -\sqrt[4]{3/8} \le x \le \sqrt[4]{3/8} \text{ and } -\sqrt[4]{3/8} \le y \le \sqrt[4]{3/8} \\ 0 & \text{otherwise.} \end{cases}$
Find $f_Y(y)$.

4. Let $f_{X,Y}(x, y) = \begin{cases} \frac{1}{2}e^{-x/2} \dfrac{1}{\sqrt{2\pi}} e^{-y^2/2} & \text{for } x > 0, y \text{ real} \\ 0 & \text{otherwise.} \end{cases}$

(a)° Find $f_X(x)$.
(b) Find $f_Y(y)$.

## SECTION 4
## Independent Random Variables

We would like to extend the concept of independence to random variables $X_1, \ldots, X_n$. Surely, if these random variables are to be considered independent, we would like the following to hold: If $A$ is any event determined by any $k$ of these random variables, and $B$ is any event determined by the remaining ones, then $A$ and $B$ should be independent.

In particular, let $A_i = \{X_i < b_i\}$ for $i = 1, 2, \ldots, n$, where $b_1, \ldots, b_n$ are arbitrary numbers. Then we want $A_1, \ldots, A_n$ to be mutually independent. In fact, we would have

**4.1**
$$F_{X_1,\ldots,X_n}(b_1, \ldots, b_n) = F_{X_1}(b_1) \cdots F_{X_n}(b_n)$$

for all possible choices of $b_1, \ldots, b_n$. Satisfying 4.1 is sufficient to guarantee all of the conditions mentioned previously. Hence, we give the following definition.

**4.2 Definition: Independent Random Variables, Random Sample.** The random variables $X_1, \ldots, X_n$ with multivariate distribution function $F_{X_1,\ldots,X_n}$ are said to be (*mutually*) *independent* if

$$F_{X_1,\ldots,X_n}(b_1, \ldots, b_n) = F_{X_1}(b_1) \cdots F_{X_n}(b_n)$$

for all choices of $b_1, \ldots, b_n$. If $X_1, \ldots, X_n$ are independent random variables all having the same univariate distribution function $F$, then we refer to the observed values $x_1, \ldots, x_n$ of $X_1, \ldots, X_n$ as a *random sample of size n from the distribution function F*.

An immediate consequence of the definition of independent random variables is the following lemma.

### *4.3 Lemma: Independence*

If $X_1, \ldots, X_n$ are random variables with a discrete probability function, then they are independent if and only if for all $x_1, \ldots, x_n$ we have

$$P\{X_1 = x_1, X_2 = x_2, \ldots, X_n = x_n\} = P\{X_1 = x_1\}P\{X_2 = x_2\} \cdots P\{X_n = x_n\}.$$

If $X_1, \ldots, X_n$ have a multivariate density, then it can be shown that they are independent if and only if

$$(*) \qquad f_{X_1,\ldots,X_n}(x_1, \ldots, x_n) = f_{X_1}(x_1) \cdots f_{X_n}(x_n)$$

for all choices of $x_1, \ldots, x_n$.†

It follows that if $X_1, \ldots, X_n$ are independent random variables, then the probability of any event determined by $X_1, \ldots, X_n$ can be computed solely from the marginal densities $f_{X_1}, f_{X_2}, \ldots, f_{X_n}$ or the marginal discrete probability functions $p_{X_1}, \ldots, p_{X_n}$.

### *4.4 Example: Independent Interarrival Times*‡

Suppose that $X$ is the time between the arrival of the first customer and the arrival of the second one, and that $Y$ is the time between the arrival of the second customer and the arrival of the third. Suppose also that $X$ and $Y$ are independent, with

$$f_X(x) = \begin{cases} e^{-x} & \text{for } x > 0 \\ 0 & \text{otherwise} \end{cases} \quad \text{and} \quad f_Y(y) = \begin{cases} e^{-y} & \text{for } y > 0 \\ 0 & \text{otherwise.} \end{cases}$$

Let us call $t = 0$ the arrival time of the first customer. We want to compute the probability of the event $B$, that fewer than two customers arrive in the time interval $(0, a]$. A little reflection shows that $B = \{X + Y > a\}$, since $X + Y > a$ means that the third customer arrives after time $a$, and hence only one customer (the second one) could possibly have arrived in $(0, a]$. Now, from the independence of $X$ and $Y$, we have

$$f_{X,Y}(x, y) = \begin{cases} e^{-x-y} & \text{for } x > 0 \quad \text{and} \quad y > 0 \\ 0 & \text{otherwise.} \end{cases}$$

$$\begin{aligned} P(B) = P\{X + Y > a\} &= 1 - P\{X + Y \leq a\} \\ &= 1 - (1 - e^{-a} - ae^{-a}) \text{ from the results of Example 2.8} \\ &= e^{-a} + ae^{-a}. \end{aligned}$$

△

† We might have to redefine $f_{X_1,\ldots,X_n}$ and $f_{X_1}, \ldots, f_{X_n}$ slightly in order for (*) to hold. The difficulty arises because we can change the value of any of these functions at a finite number of points without affecting any probabilities.

‡ This model is related to the Poisson distribution in Problems 4.9(3) and 5.7. We will give the Poisson model a theoretical justification in Chapter 7.

### 4.5 *Example: Sum of Independent Poisson Variables*

Suppose that $X$ and $Y$ are independent random variables, $X$ representing the number of phone calls initiated in the time interval $[0, t_1]$, and $Y$ representing the number of calls initiated in $(t_1, t_2]$. If

$$P\{X = k\} = e^{-\lambda t_1}(\lambda t_1)^k/k!$$

and

$$P\{Y = j\} = e^{-\lambda(t_2 - t_1)}[\lambda(t_2 - t_1)]^j/j!,$$

let us find $P\{X + Y = m\}$, the probability that a total of $m$ phone calls are initiated in $[0, t_2]$. To simplify notation, let $\lambda t_1 = \mu$ and $\lambda(t_2 - t_1) = \nu$. Then

$$\begin{aligned}
P\{X + Y = m\} &= \sum_{k=0}^{m} P\{X = k,\ Y = m - k\} \\
&= \sum_{k=0}^{m} P\{X = k\}P\{Y = m - k\} \quad \text{by the independence of } X \text{ and } Y \\
&= \sum_{k=0}^{m} e^{-\mu}\frac{\mu^k}{k!}\, e^{-\nu}\frac{\nu^{m-k}}{(m-k)!} \\
&= \frac{e^{-(\mu+\nu)}}{m!}\sum_{k=0}^{m}\frac{m!}{k!\,(m-k)!}\,\mu^k\nu^{m-k} \qquad \text{multiplying by } m!/m! \\
&= \frac{e^{-(\mu+\nu)}}{m!}\sum_{k=0}^{m}\binom{m}{k}\mu^k\nu^{m-k} \\
&= \frac{e^{-(\mu+\nu)}}{m!}(\mu + \nu)^m \qquad \text{by the Binomial Theorem (see Problem 2.19(4) in Chapter 2)} \\
&= \frac{e^{-\lambda t_2}}{m!}(\lambda t_2)^m.
\end{aligned}$$

Thus, the number of calls initiated in $[0, t_2]$ is Poisson with mean value $\lambda t_2$. We have shown that *if $X$ and $Y$ are independent with means $\mu$ and $\nu$, respectively, then $X + Y$ is Poisson with mean $\mu + \nu$.* △

### 4.6 *Example: Sum of Independent Normal Random Variables*

Let $X$ and $Y$ be independent random variables with

$$f_X(x) = \frac{1}{\sqrt{2\pi\sigma_x^2}}\, e^{-(x-\mu_x)^2/(2\sigma_x^2)} \quad \text{and} \quad f_Y(y) = \frac{1}{\sqrt{2\pi\sigma_y^2}}\, e^{-(y-\mu_y)^2/(2\sigma_y^2)}.$$

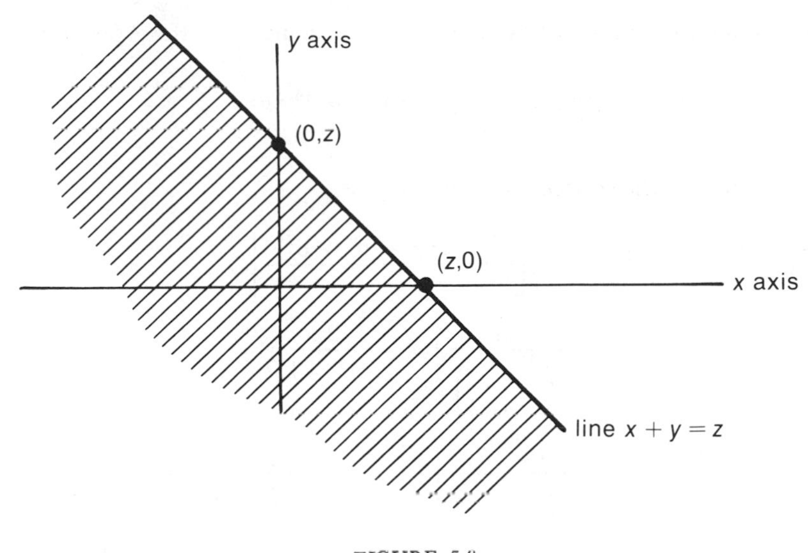


*FIGURE 5.8*

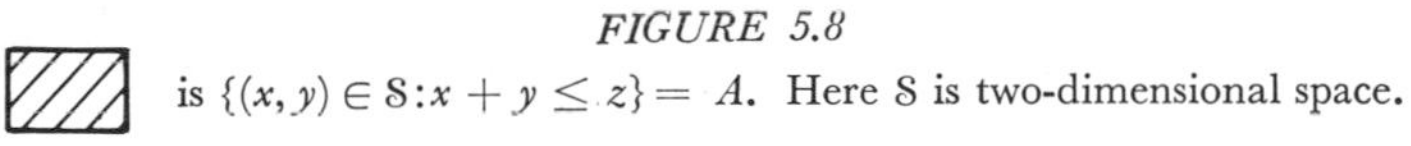
is $\{(x, y) \in \mathcal{S} : x + y \leq z\} = A$. Here $\mathcal{S}$ is two-dimensional space.

It can be shown that

$$f_{X+Y}(w) = \frac{1}{\sqrt{2\pi\sigma_w^2}}\, e^{-(w-\mu_w)^2/(2\sigma_w^2)}$$

where $\mu_w = \mu_x + \mu_y$ and $\sigma_w^2 = \sigma_x^2 + \sigma_y^2$. Since this result will be proved by simpler methods in the next chapter (see 7.16 of Chapter 6), we shall prove this result only for the special case $\mu_x = 0 = \mu_y$, $\sigma_x^2 = \sigma^2$, $\sigma_y^2 = 1$.

Letting $Z = X + Y$, we determine first that

$$F_Z(z) = P\{Z \leq z\} = P\{X + Y \leq z\}.$$

To find this probability, we must integrate $f_{X,Y}(x, y) = f_X(x) f_Y(y)$ over the region indicated in Figure 5.8. From the evaluation of double integrals (see the Appendix), we see that

$$F_Z(z) = \iint_A f_{X,Y}(x, y)\, dy\, dx = \frac{1}{2\pi\sigma} \int_{-\infty}^{\infty} \int_{-\infty}^{z-y} e^{-(1/2)[x^2/\sigma^2 + y^2]}\, dx\, dy.$$

Then

$$f_Z(z) = \frac{d}{dz} F_Z(z) = \frac{1}{2\pi\sigma} \int_{-\infty}^{\infty} \left[\frac{\partial}{\partial z} \int_{-\infty}^{z-y} e^{-(1/2)[x^2/\sigma^2 + y^2]}\right] dx\, dy.$$

(See the Appendix for the conditions that permit this interchange of the order of limiting operations. Note that the ability to differentiate "under" an integral sign may be considered an extension of the fact that the derivative of a sum is the sum of the derivatives, since an integral is essentially a sum.)

Now we use the Fundamental Theorem of Calculus (see the Appendix) to find

$$f_Z(z) = \frac{1}{2\pi\sigma}\int_{-\infty}^{\infty} e^{-(1/2)[(z-y)^2/\sigma^2+y^2]}\,dy.$$

Using completion of the square, we may write

$$-\frac{1}{2}\left[\frac{(z-y)^2}{\sigma^2}+y^2\right] = -\frac{1}{2}\left[\frac{z^2-2zy+y^2+\sigma^2y^2}{\sigma^2}\right]$$

$$= \frac{-1}{2\left(\dfrac{\sigma^2}{\sigma^2+1}\right)}\left[y^2-\frac{2z}{\sigma^2+1}y+\frac{z^2}{\sigma^2+1}\right]$$

$$= \frac{-1}{2\left(\dfrac{\sigma^2}{\sigma^2+1}\right)}\left[\left(y-\frac{z}{\sigma^2+1}\right)^2-\frac{z^2}{(\sigma^2+1)^2}+\frac{z^2}{\sigma^2+1}\right]$$

$$= -\frac{1}{2\left(\dfrac{\sigma^2}{\sigma^2+1}\right)}\left(y-\frac{z}{\sigma^2+1}\right)^2-\frac{z^2}{2(\sigma^2+1)}.$$

Hence

$$f_Z(z) = \frac{1}{2\pi\sigma}\int_{-\infty}^{\infty}\exp-\left[\frac{1}{2\left(\dfrac{\sigma^2}{\sigma^2+1}\right)}\left(y-\frac{z}{\sigma^2+1}\right)^2-\frac{z^2}{2(\sigma^2+1)}\right]dy$$

$$= \frac{\exp\dfrac{-z^2}{2(\sigma^2+1)}}{\sqrt{2\pi(\sigma^2+1)}}\int_{-\infty}^{\infty}\frac{1}{\sqrt{2\pi\dfrac{\sigma^2}{\sigma^2+1}}}\exp-\frac{1}{2\left(\dfrac{\sigma^2}{\sigma^2+1}\right)}\left(y-\frac{z}{\sigma^2+1}\right)^2 dy.$$

But the integrand is simply the normal density with mean

$$\frac{z}{\sigma^2+1}$$

and variance

$$\frac{\sigma^2}{\sigma^2+1},$$

and hence the integral is 1 (see Example 2.18, Chapter 4). Thus

$$f_Z(z) = \frac{1}{\sqrt{2\pi(\sigma^2+1)}}\exp-\frac{z^2}{2(\sigma^2+1)}.$$

This proves the desired result for $\sigma_x^2 = \sigma^2$, $\sigma_y^2 = 1$, $\mu_x = 0 = \mu_y$. △

The following is a classic example relating probability to the number $\pi$.

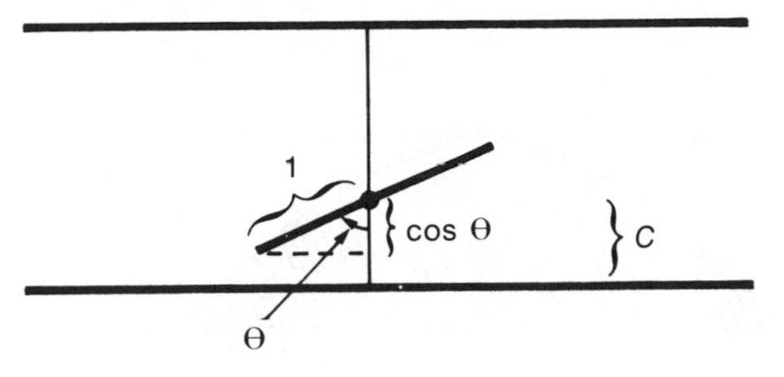


*FIGURE 5.9* Buffon needle.

### 4.7 *Example: The Buffon Needle Problem and π*

We first state the problem in a rather vague fashion. A 2-inch long needle is "dropped at random" on a grid consisting of parallel lines spaced 2 inches apart. What is the probability that the needle intersects one of the lines? In Figure 5.9, let $C$ denote the distance from the needle's center to the nearest of the parallel lines (viewing this line as horizontal and below the center of the needle) and let $\Theta$ denote the angle that the needle makes with the vertical direction, as indicated.

As a model for the experiment, we take $C$ to be uniformly distributed on [0, 1], $\Theta$ to be uniformly distributed on

$$\left[-\frac{\pi}{2}, \frac{\pi}{2}\right],$$

and $C$ and $\Theta$ to be independent, that is,

$$f_{C,\Theta}(c, \theta) = f_C(c) f_\Theta(\theta) = \begin{cases} 1 \cdot \dfrac{1}{\pi} & \text{for } 0 \le c \le 1 \quad \text{and} \quad -\pi/2 \le \theta \le \pi/2 \\ 0 & \text{otherwise.} \end{cases}$$

The event of interest is (see Figure 5.9)

$$\{\cos \Theta \ge C\}.$$

But $P\{\cos \Theta \ge C\}$ is the integral of $f_{C,\Theta}$ over the set indicated in Figure 5.10. That is (see the Appendix on multiple integrals),

$$\begin{aligned} P\{\cos \Theta \ge C\} &= \int_{-\pi/2}^{\pi/2} \int_0^{\cos\theta} f_{C,\Theta}(c, \theta)\, dc\, d\theta \\ &= \int_{-\pi/2}^{\pi/2} \int_0^{\cos\theta} \frac{1}{\pi}\, dc\, d\theta = \int_{-\pi/2}^{\pi/2} \frac{\cos\theta}{\pi}\, d\theta \\ &= \frac{\sin\theta}{\pi}\Bigg|_{-\pi/2}^{\theta=\pi/2} = \frac{2}{\pi}. \end{aligned}$$

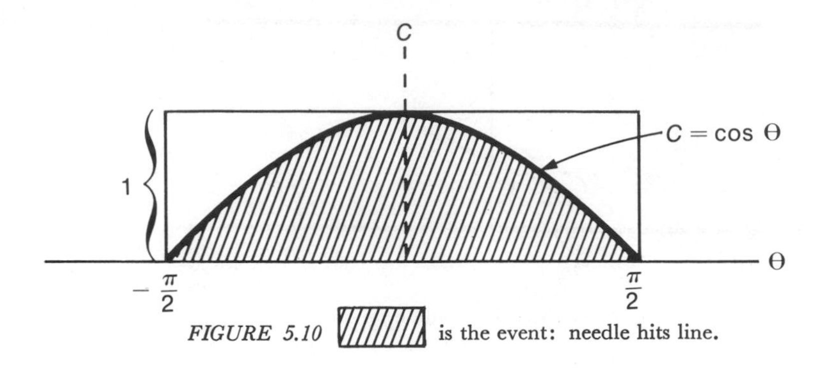


*FIGURE 5.10* [hatched box] is the event: needle hits line.

That is, the probability that the needle touches the line is $2/\pi$.

This result furnishes one of the poorest methods available for determining $\pi$, since (as we shall see in Chapter 7) the rate of approach of the proportion of crossings in $n$ drops of the needle is of the order of $1/\sqrt{n}$, so that the needle would have to be dropped 1,000,000 times to estimate $2/\pi$ to about .001 accuracy. Actually, we might use this result more as a check on the correctness of the model: If the proportion of crossings in dropping the needle 1,000,000 times is near $2/\pi$, then we have faith in the model's validity. △

---

## *4.8 EXERCISES*

1.° Let $X$ and $Y$ be independent random variables, each uniformly distributed on [0, 1]; that is

$$f_X(x) = \begin{cases} 1 & \text{if } 0 \leq x \leq 1 \\ 0 & \text{otherwise} \end{cases}.$$

(a) Find the density of $Z = X + Y$.
(b) Find the density of $W = X \cdot Y$.
In this problem it is helpful to draw pictures.

2. Let $X$ and $Y$ be independent random variables, each uniformly distributed on $[-1/2, 1/2]$; that is

$$f_X(x) = \begin{cases} 1 & \text{if } -\frac{1}{2} \leq x \leq \frac{1}{2} \\ 0 & \text{otherwise} \end{cases}.$$

(a) Find the density of $Z = X + Y$.
(b) Find the density of $W = X \cdot Y$.
Again, it is helpful to draw pictures.

3. Let $X$ and $Y$ be independent, with discrete probability functions given by the binomial distributions

$$P\{X = k\} = \binom{n}{k} p^k (1 - p)^{n-k} \qquad 0 \leq k \leq n$$

$$P\{Y = j\} = \binom{m}{j} p^j (1 - p)^{m-j} \qquad 0 \leq j \leq m.$$

Recall that $X$ is the number of successes in $n$ independent repetitions of an experiment whose probability of success is $p$, and view $Y$ similarly.

(a) From the meaning of $X$ and $Y$, guess the distribution of $X + Y$.

(b) Verify your guess analytically.

4. In the Buffon Needle problem (4.7), generalize the result to any needle of length $2l \leq 2$.

---

## *4.9 PROBLEMS*

1.° Let $X_1, \ldots, X_n$ be independent random variables, each with a uniform distribution on [0, 1].

(a) Find the density of $Y = \max\limits_{1 \leq i \leq n} X_i$ (that is, $Y$ is the largest of these random variables).

Hint: Max $X_i \leq a$ if and only if $X_i \leq a$ for all $i$.

(b) Find the joint density of $Y = \max\limits_{1 \leq i \leq n} X_i$ and $Z = \min\limits_{1 \leq i \leq n} X_i$.

Hints: First find $P\{a \leq \min X_i, \max X_i \leq b\}$. Then use the equality $P\{a \leq Z, Y \leq b\} + P\{Z < a, Y \leq b\} = P\{Y \leq b\}$. But here, $P\{Z < a, Y \leq b\} - P\{Z \leq a, Y \leq b\} = F_{Z,Y}(a, b)$. From $F_{Z,Y}$ it is easy to find $f_{Z,Y}$.

2. Suppose $n$ people are to meet on a street corner, and that their arrival times are independent, each uniformly distributed on [0, 1] (12 o'clock to 1 o'clock). Each person will wait 5 minutes: If all $n$ persons arrive in this time, they go to lunch; otherwise, no lunch. What is the probability that they will go to lunch?

Hint: The desired condition is that max $X_i$ − min $X_i \leq 1/12$. In the previous problem we determined $f_{\min X_i, \max X_i}$.

3. Suppose that $T_1, \ldots, T_n$ represent interarrival times, that is, $T_j$ is the time from the arrival of the $j$th customer to the arrival of the $j + 1$st customer. Further suppose that $T_1, \ldots, T_n$ are independent, with

$$f_{T_i}(x) = \begin{cases} e^{-x} & \text{for } x \geq 0 \\ 0 & \text{otherwise} \end{cases}.$$

Then $Y_n = T_1 + T_2 + \cdots + T_n$ represents the time from the arrival of the first customer to the arrival of the $n + 1$st customer.

(a) Show that

$$P\{Y_n \leq t\} = 1 - \left(e^{-t} + te^{-t} + \frac{t^2}{2!}e^{-t} + \cdots + \frac{t^{n-1}}{(n-1)!}e^{-t}\right).$$

Hint: You must use induction.

(b) Let $N_t$ be the number of customers who have arrived in the time interval $(0, t)$ (customer number 1, the first arrival, arrives at time 0 and is not included). Show that

$$P\{N_t \leq n - 1\} = e^{-t}\left(1 + t + \frac{t^2}{2!} + \cdots + \frac{t^{n-1}}{(n-1)!}\right).$$

Hint: $\{N_t \le n-1\} = \{Y_n > t\}$. Thus, $N_t$ has a Poisson distribution with mean $t$.

(c) Solve parts (a) and (b) if

$$f_{T_i}(x) = \begin{cases} \dfrac{1}{\theta}\, e^{-x/\theta} & \text{if } x \ge 0 \\ 0 & \text{otherwise.} \end{cases}$$

In particular, show that

$$P\{Y_n \le t\} = 1 - e^{-t/\theta}\left(1 + \frac{t}{\theta} + \frac{1}{2!}\left(\frac{t}{\theta}\right)^2 + \cdots + \frac{1}{(n-1)!}\left(\frac{t}{\theta}\right)^{n-1}\right).$$

## SECTION 5 Analytic Approach to Computing the Distribution of a Function of Random Variables

In the previous section, we considered such problems as finding the distribution of $X + Y$ or $X \cdot Y$ when $f_{X,Y}$ was given. Now we tackle the general problem of finding the distribution of $W = g(X_1, \ldots, X_n)$ when we are given the density $f_{X_1,\ldots,X_n}$. We use the same basic idea as in Section 3 of the previous chapter. Namely, we find

$$F_W(w) = P\{W \le w\} = P\{g(X_1, \ldots, X_n) \le w\}$$

$$= \int_B \cdots \int f_{X_1,\ldots,X_n}(x_1, \ldots, x_n)\, dx_1 \cdots dx_n$$

where

$$B = \{(x_1, \ldots, x_n) \in \mathscr{R}_n\colon g(x_1, \ldots, x_n) \le w\}$$

and $\mathscr{R}_n$ is $n$-dimensional space.

However, the evaluation of this $n$-dimensional integral, in the few cases in which closed form evaluation is possible, usually requires writing it as an iterated integral. We have done this with the help of appropriate pictures when $n = 2$. Now we must develop a method that works for $n > 2$. To do this, we state the following theorem.

### *5.1 Theorem: Setting Limits of Integration*

Suppose we can rewrite the set

$$B = \{(x_1, \ldots, x_n) \in \mathscr{R}_n\colon g(x_1, \ldots, x_n) \le w\}$$

in the *sequential form*

$$B = \{(x_1, \ldots, x_n) \in \mathscr{R}_n\colon a \le x_1 \le b,\ L_1(x_1) \le x_2 \le U_1(x_1), \ldots,$$
$$L_{n-1}(x_1, \ldots, x_{n-1}) \le x_n \le U_{n-1}(x_1, \ldots, x_{n-1})\}.$$

(The inequality $L_1(x_1) \leq x_2 \leq U_1(x_1)$ gives the allowable values of the second coordinate $x_2$ for any given allowable value $x_1$ of the first coordinate, and so forth.) Then

$$\int \cdots \int_B f_{X_1,\ldots,X_n}(x_1, \ldots, x_n)\, dx_1 \cdots dx_n$$
$$= \int_a^b \int_{L_1(x_1)}^{U_1(x_1)} \cdots \int_{L_{n-1}(x_1,\ldots,x_{n-1})}^{U_{n-1}(x_1,\ldots,x_{n-1})} f_{X_1,\ldots,X_n}(x_1, \ldots, x_n) \times dx_n\, dx_{n-1} \cdots dx_1$$

(where this latter is an iterated integral).

The following examples show how to write a given set in sequential form.

### 5.2 *Example:*

$$\{(x, y) \in \mathscr{R}_2 \colon x^2 + y^2 \leq 1\} = \{(x, y) \in \mathscr{R}_2 \colon -1 \leq x \leq 1, \; -\sqrt{1 - x^2} \leq y \leq \sqrt{1 - x^2}\}$$

To verify this equality, we notice that the set of all possible first coordinates is precisely $-1 \leq x \leq 1$, since for each $x$ in $[-1, 1]$ there is a real value $y$ such that $x^2 + y^2 \leq 1$; if $|x| > 1$ there is no value $y$ with $x^2 + y^2 \leq 1$. Now, for each possible $x$ in $[-1, 1]$, a condition equivalent to $x^2 + y^2 \leq 1$ is $y^2 \leq 1 - x^2$, that is, $-\sqrt{1 - x^2} \leq y \leq \sqrt{1 - x^2}$, which proves the asserted equality. This example is illustrated in Figure 5.11. △

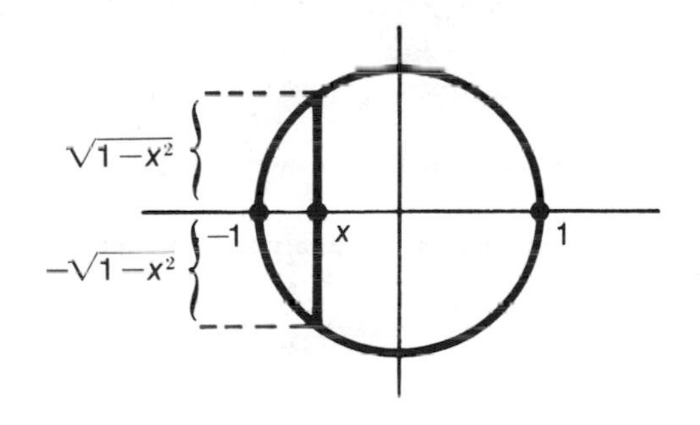


*FIGURE 5.11.* Illustrating restriction on second coordinate for given first coordinate.

### 5.3 *Example:*

$$A = \{(w, x, y, z) \in \mathscr{R}_4 \colon w + x + y + z \leq a\}$$
$$= \{(w, x, y, z) \in \mathscr{R}_4 \colon -\infty < w < \infty, \quad -\infty < x < \infty,$$
$$-\infty < y < \infty, \quad -\infty < z \leq a - w - x - y\}$$

To understand this set equality, we note that for any given choice of $w$, $x$, and $y$, if $z \leq a - w - x - y$, then $(w, x, y, z) \in A$. Conversely, if $(w, x, y, z) \in A$, then $z \leq a - w - x - y$.

Notice that *here* only the last limits of integration depend on the previous ones. △

### 5.4 *Example: Total Lifetime*

Let $X$, $Y$, and $Z$ be the lifetimes of a transistor and two spares. Suppose that these lifetimes are independent, with joint density

$$f_{X,Y,Z}(x, y, z) = \begin{cases} \frac{1}{6}e^{-x-y/2-z/3} & \text{for } x \geq 0 \text{ and } y \geq 0 \text{ and } z \geq 0 \\ 0 & \text{otherwise.} \end{cases}$$

(When we justify the Poisson model in Chapter 7, the meaning of this model becomes clearer from the results of Problems 4.9(3) and 5.7.)

We want to determine $P\{X + Y + Z \leq a\}$, the probability that the total lifetime does not exceed $a$. Reasoning as in the previous example, we see that

$$\{(x, y, z) \in \mathscr{R}_3\colon x + y + z \leq a\} = \{(x, y, z) \in \mathscr{R}_3\colon -\infty < x < \infty,$$
$$-\infty < y < \infty, \qquad -\infty < z \leq a - x - y\}.$$

Thus, for $a > 0$,

$$P\{X + Y + Z \leq a\} = \int_{-\infty}^{\infty}\int_{-\infty}^{\infty}\int_{-\infty}^{a-x-y} f_{X,Y,Z}(x, y, z)\, dz\, dy\, dx$$
$$= \int_0^a \int_0^{a-x} \int_0^{a-x-y} \tfrac{1}{6}e^{-x-y/2-z/3}\, dz\, dy\, dx.$$

Note that in setting the limits of integration for $y$ we had to eliminate those values of $y > a - x$, in order to replace $f_{X,Y,Z}(x, y, z)$ by $\frac{1}{6}e^{-x-y/2-z/3}$. This is because for any $(x, y, z)$ such that $y > a - x$, we would have to have $z < 0$ in order for $x + y + z \leq a$. But when $z < 0$, $f_{X,Y,Z}(x, y, z) = 0$. Routine but rather tedious evaluation of this iterated integral yields

$$P\{X + Y + Z \leq a\} = 1 - \tfrac{1}{2}e^{-a} + 4e^{-a/2} - \tfrac{9}{2}e^{-a/3}$$

for $a \geq 0$. Note that the answer is reasonable at $a = 0$ and as $a \to \infty$. △

The essence of our analytic approach to setting limits of integration is as follows.

### 5.5 Systematic Procedure for Setting Limits of Integration to Evaluate $\int \cdots \int_A f(x_1, \ldots, x_n)\, dx_n \ldots dx_1$

(1) Determine for which values of $x_1$ there are points $(x_1, \ldots, x_n)$ in $A$. The interval so obtained furnishes the limits of integration for $x_1$.

(2) For any given value of $x_1$ in the limits determined by (1), determine those values of $x_2$ for which there are $(x_1, x_2, \ldots, x_n)$ in $A$. These are the limits of integration for $x_2$ and in general will depend on $x_1$.

⋮

(n) For any given values of $x_1, \ldots, x_{n-1}$ in their limits of integration as determined by the previous steps, determine the values of $x_n$ such that $(x_1, \ldots, x_n)$ is in $A$. These are the limits of integration for $x_n$.

Note that the limits of integration for $x_j$ will generally depend on $x_1, \ldots, x_{j-1}$. Also note that since we can set the limits of integration in any order we desire, it may pay to investigate which order yields the simplest results (see the exercises that follow).

We note that if $f$ is the joint density of $X_1, \ldots, X_n$, then the integral in 5.5 is simply $P\{(X_1, \ldots, X_n) \in A\}$. Thus, the technique just presented may be helpful for calculating the probabilities of events defined in terms of several random variables.

---

## *5.6 EXERCISES*

1.° Let $X$ and $Y$ be independent and represent the lifetimes of two components, where

$$f_X(x) = \begin{cases} e^{-x} & \text{for } x \geq 0 \\ 0 & \text{otherwise} \end{cases}$$

and

$$f_Y(y) = \begin{cases} \frac{1}{2}e^{-y/2} & \text{for } y \geq 0 \\ 0 & \text{otherwise.} \end{cases}$$

What is the probability that the first component outlasts the second $(X > Y)$?

2. Let $X$ and $Y$ be two random variables with density $f_{X,Y}$.

(a) Write an expression for $P\{X + Y \leq a\}$, using the analytic method for setting limits of integration.

(b) Find $P\{X + Y \leq a\}$ where

$$f_{X,Y}(x, y) = \begin{cases} 1 & \text{if } x \text{ and } y \text{ are both in } [0, 1] \\ 0 & \text{otherwise.} \end{cases}$$

(c)° Write an expression for $P\{\max(X, Y) \leq a\}$, using the analytic method for setting limits of integration.

3.° Let $X$, $Y$, and $Z$ represent lifetimes, with

$$f_{X,Y,Z}(x, y, z) = \begin{cases} \frac{1}{6}e^{-x-y/2-z/3} & \text{if } x \geq 0,\ y \geq 0,\ z \geq 0 \\ 0 & \text{otherwise.} \end{cases}$$

What is the probability that the total lifetime of the first two components $(X + Y)$ exceeds that of the third $(Z)$?

4. Let $X$, $Y$, and $Z$ be as in Exercise 3. Find $P\{\max(X, Y, Z) > 5\}$.

5.° Suppose that

$$f_{X,Y}(x, y) = \begin{cases} 1 & \text{if } 0 \leq x \leq 1 \text{ and } 0 \leq y \leq 1 \\ 0 & \text{otherwise.} \end{cases}$$

Find $P\{XY \leq 1/2\}$.

6. Let $f_{X,Y}$ be some given density. Write an expression for $P\{XY \leq 5\}$.
Hint: Write $\{(x, y) \in \mathscr{R}_2\colon xy \leq 5\}$ as

$$\{(x, y) \in \mathscr{R}_2\colon x > 0, xy \leq 5\} \cup \{(x, y) \in \mathscr{R}_2\colon x < 0, xy \leq 5\}$$
$$\cup \{(x, y) \in \mathscr{R}_2\colon x = 0\}.$$

---

### *5.7 PROBLEM*

Suppose that the initiation of telephone calls into a given exchange is governed by the following model:

(a) In any time interval of length $t$, the probability that $k$ phone calls are initiated is $e^{-\lambda t}(\lambda t)^k/k!$ , where $\lambda > 0$ is given.

(b) If $[a_1, b_1], [a_2, b_2], \ldots, [a_n, b_n]$ are any $n$ non-overlapping time intervals, and if $N_i$ is the number of phone calls initiated in $[a_i, b_i]$, then $N_1, \ldots, N_n$ are mutually independent random variables.

Let $T_j$ be the length of time between the initiations of the $j$th and the $j + 1$st calls. Find the joint density of $T_1, \ldots, T_k$ where $k$ is a positive integer. (We are assuming that $T_1, \ldots, T_k$ possess a joint density—the proof of which is not trivial.)
Hints: Compute $P\{t_i \le T_i \le t_i + h_i \text{ for all } i = 1, \ldots, k\}$ in two ways. (i) from the given assumptions, where the event $\{t_i \le T_i \le t_i + h_i \text{ for all } i = 1, \ldots, k\}$ can be viewed as $\{1 \text{ phone call in } [t_i, t_i + h_i] \text{ for } i = 1, \ldots, k, \text{ no phone calls in } [t_i + h_i, t_{i+1}] \text{ for } i = 1, \ldots, k - 1\}$; and (ii) from the joint density of $f_{T_1,\ldots,T_k}$ as

$$\int_{t_1}^{t_1+h_1} \cdots \int_{t_k}^{t_k+h_k} f_{T_1,\ldots,T_k}(x_1, \ldots, x_k)\, dx_1 \cdots dx_k,$$

which you approximate by $f_{T_1,\ldots,T_k}(t_1, \ldots, t_k) \cdot h_1 \cdots h_k$.

## SECTION 6
## Discrete Conditional Probability Distributions

In this section we introduce the main results concerning conditional probability distributions in the simpler setting of discrete distributions. An understanding of this material will make the next section much easier to follow.

Suppose that $X$ and $Y$ are two related random variables having a discrete probability function

$$P_{X,Y}(i, j) = p_{ij} = P\{X = i,\ Y = j\}.$$

In many cases, it will happen that we can observe the value of $Y$ prior to that of $X$. In this case we would want to update our description of $X$; in fact, if we observed $Y = j$, we would want to use as our probability that $X$ assumes the value $i$ the expression

$$P\{X = i \mid Y = j\} = \frac{P\{X = i,\ Y = j\}}{P\{Y = j\}}$$

$$= \frac{P\{X = i,\ Y = j\}}{\sum\limits_{\substack{\uparrow \\ \text{all values} \\ x \text{ of } X}} P\{X = x,\ Y = j\}} \qquad \text{(by 3.7)}$$

provided that $P\{Y = j\} \neq 0$.

**6.1 Definition: Conditional Probability Function.** The function whose value at $i$ is $P\{X = i \mid Y = j\}$ is called the *conditional probability function of $X$ given $Y = j$.*

## 6.2 Claim: Conditional Probability Function Represents a Distribution

It is rather easy to prove that *the conditional probability function of $X$ given $Y = j$ is a bona fide discrete probability function.*

This follows from the facts that

(a) $P\{X = i \mid Y = j\} \geq 0$

(b)
$$\sum_{\substack{\text{all values} \\ i \text{ of } X}} P\{X = i \mid Y = j\} = \sum_{\substack{\text{all values} \\ i \text{ of } X}} \frac{P\{X = i,\ Y = j\}}{\sum_{\substack{\text{all values} \\ x \text{ of } X}} P\{X = x,\ Y = j\}}$$

$$= \frac{\sum_{\substack{\text{all values} \\ i \text{ of } X}} P\{X = i,\ Y = j\}}{\sum_{\substack{\text{all values} \\ x \text{ of } X}} P\{X = x,\ Y = j\}}$$

$$= 1.$$

## 6.3 *Example: Drawing Without Replacement*

Suppose we draw an ordered sample of 2 items at random from the population $M = \{1, 2, \ldots, m\}$. Because the sample is chosen at random, we know that each elementary event $\{(i, j)\}$ has probability

$$P\{(i, j)\} = \frac{1}{m(m-1)}, \qquad i \neq j,\ 1 \leq i \leq m,\ 1 \leq j \leq m.$$

Let $X$ be a random variable whose value is the first object and $Y$ one whose value is the second object. Then

$$P\{X = i,\ Y = j\} = \frac{1}{m(m-1)} \qquad \text{for} \qquad i \neq j,\ 1 \leq i, j \leq m.$$

Then for the non-trivial $i$ and $j$,

$$P\{Y = j\} = \sum_{\text{all } x} P\{X = x,\ Y = j\} = \sum_{\substack{1 \leq x \leq m \\ x \neq j}} \frac{1}{m(m-1)} = \frac{1}{m}.$$

Hence

$$P\{X=i \mid Y=j\} = \frac{P\{X=i, Y=j\}}{P\{Y=j\}} = \frac{\dfrac{1}{m(m-1)}}{\dfrac{1}{m}}.$$

Thus

$$P\{X=i \mid Y=j\} = \frac{1}{m-1} \qquad \text{for} \qquad 1 \leq i \leq m, i \neq j, \qquad \text{any } j \text{ in } M.$$

We see that $X$ has a discrete uniform distribution over $\{1, 2, \ldots, j-1, j+1, \ldots, m\}$ given $Y=j$. △

To conclude this section, we state the following reasonable result.

### *6.4 Theorem: Conditional Distribution and Independence*

Suppose that $X$ and $Y$ are random variables with a discrete probability function. Then $X$ and $Y$ are statistically independent if and only if for each $j$ at which $P\{Y = j\} \neq 0$, we have

$$P\{X=i \mid Y=j\} = P\{X=i\}.$$

If $P\{X=i \mid Y=j\}$ is a constant function of $j$ for each $i$, then $X$ and $Y$ are independent.

The student may do the proof as an exercise.

---

### *6.5 EXERCISES*

1.° Prove Theorem 6.4.

2. Suppose that an ordered sample of 3 items is drawn from $\{1, 2, \ldots, m\}$ with replacement. Let $X_i$ be the random variable representing the result of the $i$th drawing. Show that $X_1$, $X_2$, and $X_3$ are independent random variables.

3. In Example 6.3, show that $X$ and $Y$ are not independent random variables. Doesn't this seem to conflict with the final result of 6.4?

## SECTION 7 Conditional Densities

In this section, we will show how we must modify the density of some random variables when we are given the values of other related random variables. To be specific, we will define the conditional density of $X_1, \ldots, X_k$ given $X_{k+1} = x_{k+1}, \ldots, X_n = x_n$ when $X_1, \ldots, X_n$ have a known joint density $f_{X_1,\ldots,X_n}$, since the partial information given us is often of the form $X_{k+1} = x_{k+1}, \ldots, X_n = x_n$. To simplify our discussion, we will confine our motivation to the conditional density of $X$ given $Y = y$. It is natural to define this conditional density as the derivative of the conditional distribution function of $X$ given $Y = y$, where we would be tempted to define this conditional distribution function by the relation

$$F_{X|Y}(x \mid y) = \frac{P\{X \le x,\ Y = y\}}{P\{Y = y\}}.$$

However, when $Y$ has a density, this expression is meaningless, since both numerator and denominator are 0. We overcome this difficulty by use of a limiting process, namely, by letting

$$F_{X|Y}(x \mid y) = \lim_{h\to 0} \frac{P\{X \le x, y \le Y \le y + h\}}{P\{y \le Y \le y + h\}},$$

provided the right side is meaningful. But for small positive $h$, the expression

$$\frac{P\{X \le x, y \le Y \le y + h\}}{P\{y \le Y \le y + h\}} = \frac{\displaystyle\int_{-\infty}^{x}\int_{y}^{y+h} f_{X,Y}(t, u)\, du\, dt}{\displaystyle\int_{-\infty}^{\infty}\int_{y}^{y+h} f_{X,Y}(t, u)\, du\, dt}$$

is approximately

$$\frac{\displaystyle\int_{-\infty}^{x} hf_{X,Y}(t, y)\, dt}{\displaystyle\int_{-\infty}^{\infty} hf_{X,Y}(t, y)\, dt} = \frac{\displaystyle\int_{-\infty}^{x} f_{X,Y}(t, y)\, dt}{\displaystyle\int_{-\infty}^{\infty} f_{X,Y}(t, y)\, dt} = \frac{\displaystyle\int_{-\infty}^{x} f_{X,Y}(t, y)\, dt}{f_Y(y)}$$

by 3.2. Thus, we see that we should define $F_{X|Y}(x \mid y)$ by the formula

$$F_{X|Y}(x \mid y) = \frac{\displaystyle\int_{-\infty}^{x} f_{X,Y}(t, y)\, dt}{f_Y(y)}.$$

We now observe from the Fundamental Theorem of Calculus that we should define $f_{X|Y}(x \mid y)$ by the formula

$$f_{X|Y}(x \mid y) = \frac{d}{dx} F_{X|Y}(x \mid y) = \frac{f_{X,Y}(x, y)}{f_Y(y)}.$$

Thus we have the motivation for the following definition.

**7.1 Definition: Conditional Distribution Function, Conditional Density.** For the case in which $X$ and $Y$ have a joint density and in which $f_Y(y) \ne 0$, the *conditional distribution function* $F_{X|Y}(\ |y)$ of $X$ given $Y = y$ is defined by

$$F_{X|Y}(x \mid y) = \frac{\int_{-\infty}^{x} f_{X,Y}(t, y)\, dt}{f_Y(y)}$$

and the *conditional density* $f_{X|Y}(\ |y)$ of $X$ given $Y = y$ is defined by

$$f_{X|Y}(x \mid y) = \frac{f_{X,Y}(x, y)}{f_Y(y)}.$$

In general, the conditional density of $X_1, \ldots, X_k$ given $X_{k+1} = x_{k+1}, \ldots, X_n = x_n$ is defined by the formula

$$f_{X_1,\ldots,X_k|X_{k+1},\ldots,X_n}(x_1, \ldots, x_k \mid x_{k+1}, \ldots, x_n) = \frac{f_{X_1,\ldots,X_n}(x_1, \ldots, x_n)}{f_{X_{k+1},\ldots,X_n}(x_{k+1}, \ldots, x_n)}$$

provided the denominator is not 0. (Again, we remark that there is no significance in the choice of $X_1, \ldots, X_k$ and $X_{k+1}, \ldots, X_n$ except for notational convenience.)

It is not difficult to prove the following theorem.

### 7.2 *Theorem: Conditional Density Is a Density*

The conditional density of $X_1, \ldots, X_k$ given $X_{k+1} = x_{k+1}, \ldots, X_n = x_n$ is a bona fide probability density for each fixed $x_{k+1}, \ldots, x_n$.

*Proof:* We will supply the proof only for the case of two variables. That is, we will show that $f_{X|Y}(\ \mid y)$ is a probability density function for each $y$ at which it is defined.

(a) $f_{X|Y}(x \mid y) = \dfrac{f_{X,Y}(x, y)}{f_Y(y)} \geq 0$

(b)
$$\begin{aligned}
\int_{-\infty}^{\infty} f_{X|Y}(x \mid y)\, dx &= \int_{-\infty}^{\infty} \frac{f_{X,Y}(x, y)}{f_Y(y)}\, dx \\
&= \int_{-\infty}^{\infty} \frac{f_{X,Y}(x, y)}{\int_{-\infty}^{\infty} f_{X,Y}(x^*, y)\, dx^*}\, dx \qquad \text{by Theorem 3.2} \\
&= \frac{\int_{-\infty}^{\infty} f_{X,Y}(x, y)\, dx}{\int_{-\infty}^{\infty} f_{X,Y}(x^*, y)\, dx^*} \\
&= 1.
\end{aligned}$$

Thus $\int_{-\infty}^{\infty} f_{X|Y}(x \mid y)\, dx = 1$, proving that $f_{X|Y}(\ \mid y)$ is a probability density.

The proof for the general case is left to the student.

We can also prove the following theorem.

### 7.3 *Theorem: Conditional Densities and Independence*

If $X$ and $Y$ have a joint density, then $X$ and $Y$ are statistically independent if and only if for each $y$ where $f_Y(y) \neq 0$,

**7.4** $$f_{X|Y}(x \mid y) = f_X(x) \qquad \text{for all } x.$$

If $f_{X|Y}(x \mid y)$ is a constant function of $y$ for each $x$, then $X$ and $Y$ are statistically independent.

*Proof:* First we suppose $X$ and $Y$ to be statistically independent, and try to prove that 7.4 holds. But under the independence hypothesis, it follows from Lemma 4.3 that

$$f_{X|Y}(x \mid y) = \frac{f_{X,Y}(x, y)}{f_Y(y)} = \frac{f_X(x) f_Y(y)}{f_Y(y)} = f_X(x).$$

Hence, 7.4 holds when $X$ and $Y$ are independent random variables.

Now, to prove the converse, we will show that if for each $x$, $f_{X|Y}(x \mid y)$ does not vary with $y$, then $X$ and $Y$ are independent.

Since we are assuming that $f_{X|Y}(x \mid y)$ does not vary with $y$, we may let $g(x)$ denote $f_{X|Y}(X \mid Y)$. Now

$$g(x) = f_{X|Y}(x \mid y) = \frac{f_{X,Y}(x, y)}{f_Y(y)}.$$

Hence,

$$f_{X,Y}(x, y) = g(x) f_Y(y).$$

Now, integrating both sides with respect to $y$, using 3.2 and the fact that $\int_{-\infty}^{\infty} f_Y(y)\, dy = 1$, we see that

$$f_X(x) = g(x).$$

Hence, if for each fixed $x$ the function $f_{X|Y}(x \mid y)$ does not vary with $y$, we see that

$$f_{X|Y}(x, y) = f_X(x) f_Y(y).$$

Thus, by Lemma 4.3, $X$ and $Y$ are independent random variables, as asserted.

There are many situations in which it is easier to assign conditional rather than unconditional densities. For example, we may run an experiment in which we observe the value $x$ of a random variable $X$, and using this result as a basis, we perform an experiment in which the density of the outcome $Y$ is determined to be some given function $g_x$. From its meaning, we interpret $g_x$ to be the conditional density of $Y$ given $X = x$, that is, $g_x(y) = f_{Y|X}(y \mid x)$. Therefore, as a model to describe the joint behavior of $X$ and $Y$, we have

**7.5** $$f_{X,Y}(x, y) = f_X(x) g_x(y).$$

This result is similar to the results we obtained in two-stage sampling from an urn, when the mixture in the urn on the second drawing depended on the results of the first drawing. The general version of this result is given below.

### 7.6 Model for a Pair of Linked Experiments

Suppose that $X_1, \ldots, X_k$ have joint density $f_{X_1,\ldots,X_k}$. We proceed as follows: First we observe the values $x_1, \ldots, x_k$ of $X_1, \ldots, X_k$. Now, for each possible choice of $x_1, \ldots, x_k$, let $g_{x_1,\ldots,x_k}$ be a density in $n - k$ dimensions. That is, the value of

$$g_{x_1,\ldots,x_k} \text{ at } x_{k+1}, \ldots, x_n \text{ is } g_{x_1,\ldots,x_k}(x_{k+1}, \ldots, x_n) \geq 0,$$

and

$$\int_{-\infty}^{\infty} \cdots \int_{-\infty}^{\infty} g_{x_1,\ldots,x_k}(x_{k+1}, \ldots, x_n)\, dx_{k+1} \cdots dx_n = 1.$$

Next, we perform experiments with outcomes $X_{k+1}, \ldots, X_n$ having joint density $f_{X_{k+1},\ldots,X_n} = g_{x_1,\ldots,x_k}$. Then we interpret $g_{x_1,\ldots,x_k}$ as the conditional density of $X_{k+1}, \ldots, X_n$, given $X_1 = x_1, \ldots, X_k = x_k$. That is,

$$g_{x_1,\ldots,x_k}(x_{k+1}, \ldots, x_n) = f_{X_{k+1},\ldots,X_n|X_1,\ldots,X_k}(x_{k+1}, \ldots, x_n \mid x_1, \ldots, x_k).$$

Hence

$$f_{X_1,\ldots,X_n}(x_1, \ldots, x_n) = f_{x_1,\ldots,x_k}(x_1, \ldots, x_k) g_{x_1,\ldots,x_k}(x_{k+1}, \ldots, x_n).$$

We now compute a conditional density in the important case of a bivariate normal distribution.

### 7.7 *Example: Conditional Normal Density*

Suppose that $X$ and $Y$ are a pair of random variables—say, the value of $X$ is the score on the college board mathematics test and the value of $Y$ is some measure of the same student's performance in his college mathematics courses. A model for the behavior of $X$ and $Y$ that has proved reasonably useful is given by

**7.8**
$$f_{X,Y}(x, y) = \frac{1}{2\pi\sigma_x\sigma_y\sqrt{1-\rho^2}} \exp \frac{-1}{2(1-\rho^2)} \times \left[\left(\frac{x-\mu_x}{\sigma_x}\right)^2 - 2\rho\left(\frac{x-\mu_x}{\sigma_x}\right)\left(\frac{y-\mu_y}{\sigma_y}\right) + \left(\frac{y-\mu_y}{\sigma_y}\right)^2\right].$$

(The theoretical basis for such a model is treated in Chapter 7.)

In order to try to predict college performance, $Y$, from the observed value $x$ of the college board score $X$, it is first useful to determine

$$_{Y|X}(y \mid x) = \frac{f_{X,Y}(x, y)}{f_X(x)}.$$

We shall carry out the computations only for the specific case $\mu_x = 0 = \mu_y$, $\sigma_x = 1 = \sigma_y$, and state the general result (which can be derived by more tedious computations of the same type). In this case,

$$f_{Y|X}(y \mid x) = \frac{f_{X,Y}(x, y)}{f_X(x)} = \frac{\dfrac{1}{2\pi\sqrt{1-\rho^2}} \exp \dfrac{-1}{2(1-\rho^2)} \cdot (x^2 - 2\rho xy + y^2)}{\dfrac{1}{\sqrt{2\pi}} \exp(-x^2/2)}$$

(see Example 3.3)

$$= \frac{\exp\left(\dfrac{-1}{2(1-\rho^2)}[x^2 - 2\rho xy + y^2] + \dfrac{x^2}{2}\right)}{\sqrt{2\pi(1-\rho^2)}}$$

$$= \frac{1}{\sqrt{2\pi(1-\rho^2)}} \exp \frac{-1}{2(1-\rho^2)} [x^2 - 2\rho xy + y^2 - (1-\rho^2)x^2]$$

$$= \frac{1}{\sqrt{2\pi(1-\rho^2)}} \exp \frac{-1}{2(1-\rho^2)} [\rho^2 x^2 - 2\rho xy + y^2]$$

$$= \frac{1}{\sqrt{2\pi(1-\rho^2)}} \exp - \frac{(y - \rho x)^2}{2(1-\rho^2)} .$$

That is, given $X = x$, the random variable $Y$ is normally distributed with mean value $\rho x$ and variance $1 - \rho^2$.

We note the following for reference purposes. In the general case given in 7.8, we can show by means of a computation similar to this optional one that *the conditional density of $Y$ given $X = x$ is normal with mean*

$$\mu = \rho \frac{\sigma_y}{\sigma_x}(x - \mu_x) + \mu_y$$

*and variance* $\sigma^2 = (1 - \rho^2)\sigma_y^2$. Note that if $\rho = 0$, the conditional distribution of $Y$ given $X = x$ does not depend on $x$. This is reasonable in view of the fact that when $\rho = 0$, we have from 7.8 that

$$\begin{aligned} f_{X,Y}(x, y) &= \frac{1}{2\pi\sigma_x\sigma_y} \exp - \frac{1}{2}\left[\left(\frac{x - \mu_x}{\sigma_x}\right)^2 + \left(\frac{y - \mu_y}{\sigma_y}\right)^2\right] \\ &= \left[\frac{1}{\sqrt{2\pi\sigma_x^2}} \exp - \frac{1}{2}\left(\frac{x - \mu_x}{\sigma_x}\right)^2\right]\left[\frac{1}{\sqrt{2\pi\sigma_y^2}} \exp - \frac{1}{2}\left(\frac{y - \mu_y}{\sigma_y}\right)^2\right] \\ &= f_X(x) f_Y(y). \end{aligned}$$

That is, when $\rho = 0$, $X$ and $Y$ are independent random variables. The quantity $\rho$ is called the *coefficient of linear correlation between $X$ and $Y$* and is a measure of the dependence between $X$ and $Y$. It will be further investigated in Chapter 6 (Section 5), but its fundamental significance will become apparent in the next example. △

### 7.9 *Example: Prediction with Bivariate Normal Random Variables*

Continuing the discussion of Example 7.7, suppose that the college board score $X$ and the college performance score $Y$ have a bivariate normal density (see Equation 7.8) with $\mu_x = 600$, $\sigma_x = 70$, $\mu_y = 75$, $\sigma_y = 8$, and correlation coefficient $\rho = .7$. We know from our determination of bivariate normal marginals in 3.5 and 3.6 that the college performance score $Y$ has marginal density given by

$$f_Y(y) = \frac{1}{\sqrt{2\pi}\,8} e^{-(y-75)^2/(2\cdot 64)}$$

From the results of Example 2.25 in Chapter 4, we know that $(Y - 75)/8$ has a standard normal distribution. From the table of the standard normal distribution function in the Appendix, we find that

$$P\left\{-1.96 \leq \frac{Y - 75}{8} \leq 1.96\right\} \cong .95.$$

Hence, $P\{59.32 \leq Y \leq 90.68\} \cong .95$. That is, we can predict with 95 per cent assurance that the value of $Y$ will lie in $[59.32, 90.68]$, assuming we have no knowledge of $X$.

Now, what should our 95 per cent assurance prediction for $Y$ be if it is known that $X = 670$ $(= \mu_x + 70)$? From the previous example, we know that the conditional density of $Y$ given $X = 670$ is normal, with mean

$$\mu = \rho \frac{\sigma_y}{\sigma_x}(x - \mu_x) + \mu_y = .7 \times \tfrac{8}{70}(670 - 600) + 75 = 80.6$$

and variance

$$\sigma^2 = (1 - \rho^2)\sigma_y^2 = (1 - .49)\cdot 64 = 32.64 \cong (5.713)^2.$$

Thus, given $X = 670$, we see that

$$\frac{Y - 80.6}{5.713}$$

has a standard normal density. Hence

$$P\left\{-1.96 \leq \frac{Y - 80.6}{5.713} \leq 1.96 \mid X = 670\right\} = .95,$$

that is, $P\{69.4025 \leq Y \leq 91.7975 \mid X = 670\} = .95$. That is, with 95 per cent assurance, we can predict the value of $Y$ to be in $[69.4025, 91.7975]$ when we are given that $X = 670$.

Notice that the length of the interval obtained without benefit of knowledge of the value of $X$ is 31.36 units, whereas that of the interval obtained with knowledge of the value of $X$ is 22.395 units, an improvement in precision of

$$100 \cdot \frac{31.36 - 22.395}{31.36} \% \cong 28\%.$$

Note that the percentage improvement was not influenced by the actual value of $X$, but was determined solely by the value of $\rho$ (in this case, $\rho = .7$), since the length of a 95 per cent assurance prediction interval for a random variable $Y$ with given $\sigma_y$ is $2 \cdot 1.96 \cdot \sigma_y$, whereas the *conditional prediction interval* for $Y$ has length $2 \cdot 1.96 \cdot \sqrt{1 - \rho^2}\, \sigma_y$. Thus, larger values of $|\rho|$ yield shorter conditional prediction intervals, that is, better prediction from $Y$. △

---

## 7.10 *EXERCISES*

1.° Let

$$f_{X,Y}(x, y) = \begin{cases} x + y & \text{for } 0 \le x \le 1 \text{ and } 0 \le y \le 1 \\ 0 & \text{otherwise.} \end{cases}$$

Find $f_{Y|X}(y \mid x)$.

2. Let

$$f_{X,Y}(x, y) = \begin{cases} 1/\pi & \text{if } x^2 + y^2 \le 1 \\ 0 & \text{otherwise.} \end{cases}$$

Find $f_{Y|X}(y \mid x)$.

3. Let

$$f_{X,Y}(x, y) = \begin{cases} x^2 + y^2 & \text{if } -\sqrt[4]{3/8} \le x \le \sqrt[4]{3/8} \text{ and } -\sqrt[4]{3/8} \le y \le \sqrt[4]{3/8} \\ 0 & \text{otherwise.} \end{cases}$$

Find $f_{X|Y}(x \mid y)$.

4.° Suppose that you have a random number generator that you can adjust to generate successively the values of normal random variables with any mean values and variances you desire. (It is assumed that for any single generation, only your setting has any influence on the value generated.) With such a device, how would you generate values assumed by a pair of random variables, $X$ and $Y$, with

$$f_{X,Y}(x, y) = \frac{1}{2\pi\sqrt{1 - \rho^2}} \exp \frac{-1}{2(1 - \rho^2)} [x^2 - 2\rho xy + y^2]$$

where $\rho$ is some given number between $-1$ and 1?
Hint: Use Equation 7.5 and the results established in Example 7.7.

5. With the device described in the previous question, how could you generate values assumed by a random variable with an arbitrary bivariate normal distribution (see Equation 7.8)?

6. Suppose that a pair of measurements $T$ and $Y$ on temperature and yield have a bivariate normal density (see Equation 3.5) with $\mu_t = 65$, $\sigma_t = 12$, $\mu_y = 800$, $\sigma_y = 15$, and $\rho = .95$.

(a) Predict the value of $Y$ with 90 per cent assurance (as in Example 7.9), without knowing the value of $T$.

(b) Predict the value of $Y$ with 90 per cent assurance (as in Example 7.9), knowing that the temperature is 60 degrees.

---

## *7.11 PROBLEMS*

1.° Referring to Exercise 7.10(6), find a number that you are 95 per cent sure will be a lower bound for the value of $Y$:

(a) knowing $T = 60$ degrees.
(b) not knowing $T$.

2. Repeat part (1) for an upper bound for the value of $Y$.

3. Suppose that

$$f_{X,Y}(x, y) = \begin{cases} 1 & \text{if } 0 \le x \le 1 \text{ and } 0 \le y \le 1 \\ 0 & \text{otherwise.} \end{cases}$$

Let $Z = \max(X, Y)$ and $W = \min(X, Y)$ (that is, if $X$ takes on the value $x$ and $Y$ the value $y$, then $Z$ is the larger of $x$ and $y$, and $W$ is the smaller). Find $f_{Z|W}(z \mid w)$. Hint: To find $f_{Z,W}$ see the hint in Problem 4.9(1)(b).

4.° Let $X$ and $Y$ be as in part (3) and let $Z = X + Y$. Find $f_{Z|Y}(z \mid y)$. Hint: Find $f_{Z,Y}$ by computing $P\{X + Y \le z, Y \le y\}$, using geometrical reasoning.

5. Justify the model for high school and college board performance of Example 7.7 in light of the fact that it assigns non-zero probabilities to the impossible set of negative outcomes.

# 6

# Expectations, Moment Generating Functions, and Quantiles

## SECTION 1
## Introduction

Most of us automatically feel that the average of 50 stock prices is a better measure of market behavior than is just one stock price. In the physics laboratory we are told that the average of $n$ measurements on the speed of sound is likely to be a better estimate of the speed of sound than any single such measurement. The habit of averaging is likely to be so deeply ingrained that many of us feel that no justification is required. However, throughout this text we have insisted that any operations on data must be done in a meaningful framework. Our purpose in this chapter is to construct the theory that makes averaging and other such operations meaningful. In fact, we shall see that there are useful alternatives to averaging to improve accuracy, and that sometimes averaging does not increase our accuracy.* Indeed, it may even decrease accuracy.

## SECTION 2
## Expectations as Limiting Averages

Suppose that $X$ is a random variable whose value is the outcome of some experiment with $P\{X = b_j\} = p_j$, for $j = 1, \ldots, k$. Imagine repeating this experiment, letting $X_1, X_2, \ldots$ be the sequence of random variables whose values are the

* See the last problem in the chapter for such an example.

respective outcomes. Then

**2.1** $$P\{X_i = b_j\} = p_j \quad \text{for} \quad j = 1, 2, \ldots, k, \quad \text{with} \quad \sum_{j=1}^{k} p_j = 1.$$

Let us suppose that these random variables arise in a gambling situation in which the value of $X_i$ is the payoff of the $i$th game. (Note that $X_i$ may assume negative as well as positive values.) Then the value of

$$\bar{X}_{(n)} = \frac{1}{n} \sum_{i=1}^{n} X_i$$

is the average gain per trial in the first $n$ trials.

In practice, it has been observed that as $n$ increases, this average gain per trial appears to settle down to some fixed number. What should we expect this number to be? In the simplest non-trivial case, when $P\{X_i = 1\} = p$, $P\{X_i = 0\} = 1 - p$ for all $i$, then the value of

$$\frac{1}{n} \sum_{i=1}^{n} X_i$$

is just the proportion of 1's observed in the first $n$ trials, and thus we should certainly anticipate from statistical regularity arguments that the value of the random variable

$$\frac{1}{n} \sum_{i=1}^{n} X_i$$

would be close to $p$ when $n$ is large.

### 2.2 *Example: Expected Sum on Dice*

Let $X_i$ be a random variable whose value is the sum on the $i$th independent toss of a pair of dice. Let us determine what the average

$$\frac{1}{n} \sum_{i=1}^{n} X_i$$

of the sums on $n$ tosses should tend to when $n$ is large. We know the probabilities for the sum of two dice (Table 6.1).

**TABLE 6-1**

| Value of sum | 2 | 3 | 4 | 5 | 6 | 7 | 8 | 9 | 10 | 11 | 12 |
|---|---|---|---|---|---|---|---|---|---|---|---|
| Probability | $1/36$<br>$p_2$ | $2/36$<br>$p_3$ | $3/36$<br>$p_4$ | $4/36$<br>$p_5$ | $5/36$<br>$p_6$ | $6/36$<br>$p_7$ | $5/36$<br>$p_8$ | $4/36$<br>$p_9$ | $3/36$<br>$p_{10}$ | $2/36$<br>$p_{11}$ | $1/36$<br>$p_{12}$ |

Hence, we see from the frequency interpretation of probabilities that in $n$ repeated tosses, the value

2 should result approximately $n \cdot 1/36 = np_2$ times
3 should result approximately $n \cdot 2/36 = np_3$ times
.
.
.
12 should result approximately $n \cdot 1/36 = np_{12}$ times.

That is, approximately

$n \cdot 1/36 = np_2$ of the $X_i$ will take on the value 2
$n \cdot 2/36 = np_3$ of the $X_i$ will take on the value 3
.
.
.
$n \cdot 1/36 = np_{12}$ of the $X_i$ will take on the value 12.

Thus

the $X_i$ taking on the value 2 will contribute about $n \cdot 1/36 \cdot 2 = np_2 \cdot 2$ to $\sum_{i=1}^{n} X_i$

the $X_i$ taking on the value 3 will contribute about $n \cdot 2/36 \cdot 3 = np_3 \cdot 3$ to $\sum_{i=1}^{n} X_i$
.
.
.
the $X_i$ taking on the value 12 will contribute about $n \cdot 1/36 \cdot 12 = np_{12} \cdot 12$ to $\sum_{i=1}^{n} X_i$.

Thus, $\sum_{i=1}^{n} X_i$ will have a value close to

$$n \cdot \frac{1}{36} \cdot 2 + n \cdot \frac{2}{36} \cdot 3 + \cdots + n \cdot \frac{1}{36} \cdot 12 = \sum_{j=2}^{12} np_j \cdot j$$

and

$$\frac{1}{n} \sum_{i=1}^{n} X_i$$

should have a value close to

$$\frac{1}{36} \cdot 2 + \frac{2}{36} \cdot 3 + \cdots + \frac{1}{36} \cdot 12 = \sum_{j=2}^{12} p_j j = \sum_{j=2}^{12} P\{X_i = j\} \cdot j. \qquad \triangle$$

In the more general case specified by 2.1, arguments identical to the ones just given lead us to conclude that for large $n$, the value of

$$\frac{1}{n} \sum_{i=1}^{n} X_i$$

should be approximately

$$\frac{1}{n}\sum_{j=1}^{k} np_j b_j = \sum_{j=1}^{k} b_j p_j = \sum_{j=1}^{k} \underset{\substack{\uparrow \\ \text{possible value assumed by the } X_i}}{b_j} \underbrace{P\{X_i = b_j\}}_{\substack{\uparrow \\ \text{probability of this value}}}.$$

Based on this intuition, we give the following definition.

**2.3 Definition: Expectation in the Finite Case.** Let $X$ be a random variable with $P\{X = b_j\} = p_j$ for $j = 1, \ldots, k$ where $\sum_{j=1}^{k} p_j = 1$. We define the *expectation* of $X$, denoted by $E(X)$, by the formula

$$E(X) = \sum_{j=1}^{k} b_j\, p_j. \leftarrow \text{probability of this value}$$

sum over *all* possible values of the random variable $X$ ↗ ↖ possible value of the random variable $X$

$E(X)$ is also commonly referred to as:

*the first moment of the distribution of X,*
*the mean value of X,*
*the mathematical expectation of X,*
*the mean of X,*
*the expected value of X,*
*the average value of X,*
*the average of X,* or
*the ensemble mean of X.*

Note that ***a random variable need never assume its expected value.*** This can be seen by looking at 2.1 in the case in which $X_i$ can only take on the values 0 or 1, but $E(X) = p$, which generally does not equal either 0 or 1. ***Rather,*** $E(X)$ ***is the value we expect***

$$\frac{1}{n}\sum_{i=1}^{n} X_i$$

***to settle down to as*** $n$ ***gets large, where*** $X_1, X_2, \ldots$ ***represent the results of repeated experiments each governed by the distribution function*** $F_X$.

Now, our mathematical model for repeated experiments, each of which is governed by $F_X$, consists of independent random variables $X_1, X_2, \ldots$ each having distribution function $F_X$. Within this model, there should be a mathematical result reflecting the approach of

$$\frac{1}{n}\sum_{i=1}^{n} X_i$$

to $E(X)$ as $n$ grows large. This is in fact the case, as we shall see when we prove the weak law of large numbers (Theorem 6.5).

### 2.4 *Example: Discrete Uniform Expectation*

In problems of rounding off to the nearest dollar, we may have

$$P\left\{X = \frac{j}{2m}\right\} = \frac{1}{2m+1} \qquad \text{for} \qquad j = -m, -m+1, \ldots, 0, 1, \ldots, m.$$

That is, the roundoff $X$ has a discrete uniform distribution on the set of points of the form $j/(2m)$, where $j$ goes from $-m$ to $m$. The expected roundoff

$$E(X) = \sum_{j=-m}^{m} \frac{j}{2m} \cdot \frac{1}{2m+1} = \frac{1}{2m \cdot (2m+1)} \sum_{j=-m}^{m} j = 0$$

because of the cancellation. This result could hardly be called unexpected. More generally, if

$$P\left\{X = \alpha + \frac{j}{2m}\right\} = \frac{1}{2m+1}$$

for $m = -m, -m+1, \ldots, 0, 1, \ldots, m$, then

$$E(X) = \sum_{j=-m}^{m} \left(\alpha + \frac{j}{2m}\right) \frac{1}{2m+1} = \alpha \sum_{j=-m}^{m} \frac{1}{2m+1} + \sum_{j=-m}^{m} \frac{j}{2m} \cdot \frac{1}{2m+1}$$

$$= \alpha + 0 = \alpha. \qquad \triangle$$

### 2.5 *Example: Expected Number of Successes in One Trial*

Let $Y$ be the number of successes (0 or 1) in one trial of an experiment whose probability of success is $p$. Then, since $\{Y = 1\}$ means "1 success," or success, and $\{Y = 0\}$ means "0 successes," or failure, we have $P\{Y = 1\} = p$, $P\{Y = 0\} = 1 - p$. Hence

$$E(Y) = \sum_{j=0}^{1} \underset{\substack{\uparrow \\ \text{possible} \\ \text{value}}}{j} \; \underset{\substack{\uparrow \\ \text{probability of this} \\ \text{possible value}}}{\underbrace{P\{Y = j\}}}$$

$$= 0 \cdot (1 - p) + 1 \cdot p = p.$$

Hence, we see that if the probability of success on a trial is $p$ and the probability of failure is $1 - p$, then the expected number of successes on one trial is simply $p$. $\triangle$

### 2.6 *Example: Expectation of Binomial Random Variable*

When $X$ is the number of successes in $n$ independent trials, each of which has probability $p$ of success, then we know from Example 5.12 of Chapter 3 that

$$P\{X = j\} = \binom{n}{j} p^j (1 - p)^{n-j}.$$

In this case,

$$E(X) = \sum_{j=0}^{n} \underset{\substack{\uparrow \\ \text{possible} \\ \text{value} \\ \text{of } X}}{j} \ \underbrace{\binom{n}{j} p^j (1-p)^{n-j}}_{\substack{\uparrow \\ \text{probability of} \\ \text{this possible} \\ \text{value}}}$$

$$= \sum_{j=0}^{n} j \frac{n!}{j!\,(n-j)!} p^j (1-p)^{n-j}$$

$$= \sum_{j=0}^{n} j \frac{n!}{j!(n-j)!} p^j (1-p)^{n-j} \qquad \text{(since the term with } j = 0 \text{ is 0, recall that } 0! = 1\text{)}$$

$$= \sum_{j=1}^{n} \frac{n!}{(j-1)!\,(n-j)!} p^j (1-p)^{n-j}$$

$$= np \sum_{j=1}^{n} \frac{(n-1)!}{(j-1)!\,(n-j)!} p^{j-1} (1-p)^{n-j}$$

$$= np \sum_{j=1}^{n} \binom{n-1}{j-1} p^{j-1} (1-p)^{n-j}$$

$$= np \sum_{k=0}^{n-1} \binom{n-1}{k} p^k (1-p)^{(n-1)-k}$$

$$= np(p + [1-p])^{n-1}$$

$$= np \qquad \text{(by the Binomial Theorem, Problem 2.19(4), Chapter 2).}$$

Thus $E(X) = np$, which should not be too surprising: We already know from Example 2.5 that the expected number of successes on one trial is $p$, and thus it is reasonable that the expected number of successes on $n$ trials is $np$. △

In the case in which the set of possible values of $X$ is in one-to-one correspondence with the integers, it is natural to extend the definition of expectation.

**2.7 Definition: Expectation in the Infinite Discrete Case.** Let $X$ be a random variable with $P\{X = b_j\} = p_j$, where $\sum_{j=0}^{\infty} p_j = 1$. If $\sum_{j=0}^{\infty} |b_j|\, p_j$ is finite, we define

$$E(X) = \sum_{j=0}^{\infty} b_j p_j.$$

The condition on $\sum_{j=0}^{\infty} |b_j|\, p_j$ is imposed in order to guarantee that the main theorems on the uses of expectation hold true, that is, that in some sense the average,

$$\frac{1}{n} \sum_{i=1}^{n} X_i,$$

tends to $E(X)$ as $n$ increases.

### 2.8 *Example: Expectation of Poisson Variables*

We have previously asserted that if

$$P\{X = k\} = e^{-\lambda t}\frac{(\lambda t)^k}{k!}$$

for $k = 0, 1, \ldots,$ where $X$ represents the number of phone calls initiated in a time interval of duration $t$, then $\lambda t$ is the expected number of phone calls in this interval, and hence $\lambda$ is the expected number of phone calls per unit time. We are now in a position to show that $E(X) = \lambda t$.

$$E(X) = \sum_{k=0}^{\infty} k e^{-\lambda t}\frac{(\lambda t)^k}{k!} = \sum_{k=1}^{\infty} e^{-\lambda t}\frac{(\lambda t)^k}{(k-1)!} = \lambda t e^{-\lambda t}\sum_{k=1}^{\infty}\frac{(\lambda t)^{k-1}}{(k-1)!}$$

$$= \lambda t e^{-\lambda t}\sum_{j=0}^{\infty}\frac{(\lambda t)^j}{j!}$$

$$= \lambda t e^{-\lambda t} e^{\lambda t} \qquad \text{using the well-known expansion } e^x = \sum x^j/j!$$

$$= \lambda t.$$

△

The following example shows that there are times when the expectation is not defined, even though the random variable under consideration is quite meaningful.

### 2.9 *Example: Digging for Gold*

Suppose that we are digging for gold. With probability $1/2^j$, we find $2^j$ dollars worth of gold, $j = 1, 2, \ldots$. Let $X$ represent the amount of gold we actually dig up. Then $X$ does not have an expectation, since

$$\sum_{\text{all values } x \text{ of } X} xP\{X = x\} = \sum_{j=1}^{\infty} 2^j \frac{1}{2^j} = 1 + 1 + 1 + \cdots.$$

It is reasonable to say that the expected amount of gold is infinite, despite the fact that the amount of gold that will actually be dug out must have a finite value. The explanation is as follows. Suppose we repeat the experiment independently and let $X_i$ represent the result of the $i$th trial. It can be shown that the average,

$$\frac{1}{n}\sum_{i=1}^{n} X_i,$$

will tend to grow unboundedly large as $n$ becomes large. △

In the next example, the expectation of $X$ is not defined. However, as we shall see, it is not reasonable here to say that the expectation is infinite.

### 2.10 Counterexample

$$p_j = \frac{3}{\pi^2 j^2}$$

It can be shown that

$$\sum_{j=1}^{\infty} \frac{1}{j^2} = \frac{\pi^2}{6}.$$

Let

$$P\{X = j\} = \frac{3}{\pi^2 j^2} \qquad \text{for} \qquad j = \pm 1, \pm 2, \ldots,$$

and note that this assignment is a probability function (provided you accept the assertion

$$\sum_{j=1}^{\infty} \frac{1}{j^2} = \frac{\pi^2}{6},$$

whose proof can be found in texts on Fourier Series). By the symmetry of the distribution of $X$, it might be thought that $X$ has expectation 0. This is *not* the case, since it can be shown that

$$\sum_{\text{all } j \neq 0} |j|\, p_j = \sum_{\text{all } j \neq 0} \frac{3}{\pi^2 |j|}$$

is not finite. (Essentially,

$$\sum_{j=1}^{k} \frac{1}{j}$$

behaves like

$$\int_1^k \frac{dx}{x} = \log k,$$

which becomes unbounded for large $k$.) It can be shown that in this case, the average,

$$\frac{1}{n} \sum_{i=1}^{n} X_i,$$

does not settle down to any fixed number as $n$ becomes large, but tends to fluctuate, becoming unbounded both positively and negatively. This behavior is consistent with the fact that $E(X)$ is undefined in this case. △

---

### 2.11 EXERCISES

1.° In a lottery, a number is drawn at random from the set $\{1, 2, \ldots, 100\}$. If $X$ is a random variable whose value is the number drawn, find $E(X)$:

(a) by guessing intelligently.

(b) using the definition.

2. Suppose that in a given year the probability that no dividend is declared on a given stock is 1/8, while the probability of a dividend of $\$2^j$ is $1/2^j$, $j = 1, 2, 3$. What is the expected dividend?

3. Find $E(X)$ for the following cases:

(a)° $p_j = \dfrac{j}{55}, j = 1, 2, \ldots, 10.$

Hint: $\displaystyle\sum_{j=1}^{n} j^2 = \frac{j(j+1)(2j+1)}{6}.$

(b) $p_j = \dfrac{j^2}{385}, j = 1, 2, \ldots, 10.$

Hint: $\displaystyle\sum_{j=1}^{n} j^3 = \left[\frac{n(n+1)}{2}\right]^2.$

(c) $p_j = \dfrac{c}{j(j+1)(j+2)}, j = 1, 2, \ldots$ . Here $c$ is chosen so that $\displaystyle\sum_{j=1}^{\infty} p_j = 1.$

Hint: $\dfrac{1}{(j+1)(j+2)} = \left(\dfrac{1}{j+1} - \dfrac{1}{j+2}\right).$

4.° Suppose that in the game of roulette (see Example 4.3, Chapter 1), we bet a dollar on "reds." If "reds" occurs, we receive \$2.00 (the dollar we bet plus another dollar); we lose the dollar we bet if "reds" fails to occur. Assuming a uniform model over the 38 possible outcomes, what is the expected gain? (Here, if $X$ is the gain, we gain either \$1.00 or $-$\$1.00.) This expected loss (negative gain) might be thought of as the cost of admission to play.

5. In roulette (see Example 4.3, Chapter 1) if we bet \$1.00 on a particular number and it comes up, we receive that dollar plus 35 dollars more. If another number comes up, the dollar that we bet is lost. What is the expected gain, assuming a uniform distribution over $S$? (Here, if $X$ is the gain, $X = 35$ or $X = -1$.)

6.° What is the expected number of aces in a bridge hand? Could you guess this answer without doing any computation?

7. (a) A lot of 100 units contains seven defectives. We choose $j$ units at random from the 100 (without replacement). Find an expression for the expected number of defectives among the $j$ chosen. Can you guess the value of this expression without computation?

(b) Evaluate $E(X)$.

8.° Let $P\{X = j\} = (1 - p)^{j-1}p$ for $j = 1, 2, \ldots$ ($X$ is the number of trials needed until exactly one success is obtained when the trials are independent, each with probability $p$ of success). What is $E(X)$?

Hint: $\dfrac{d}{dp}(1 - p)^j = -j(1 - p)^{j-1}.$

9. (a) Suppose an experiment whose probability of success is $p$ is repeated until precisely $k$ successes are obtained. Let $X$ be the required number of trials. Find an expression for $E(X)$ (see Example 3.5, Chapter 2). Can you guess the value of this expression from the answer to the previous question?

(b) Evaluate $E(X)$.

10.° Suppose that $P\{X = j\} = p_j = c/|j^{3/2}|$ for $j = \pm 1, \pm 2, \ldots$, where $c$ is chosen so that $\sum_{j=1}^{\infty} p_j = 1/2$, which guarantees that the $p_j$'s do form a discrete probability function. Determine whether or not $X$ has an expectation.

11. Perform the previous exercise for $p_j = c/|j^{5/2}|$.

Now suppose that $X$ is a random variable that can assume more than a discrete set of values. Recall (Remarks 2.23, Chapter 4) that such a random variable can be approximated to an accuracy of $10^{-n}$ by a discrete random variable $X_n$. Such a random variable arises naturally, since even if $X$ yields the "true value" of some measured quantity, most measuring instruments permit only a limited accuracy. We might therefore define $E(X)$ to be $\lim_{n\to\infty} E(X_n)$. This approach is very much like the one used in advanced probability. Rather than develop the advanced machinery needed for this approach, we give the following definition, which is adequate for our purposes.

**2.12 Definition: Expectation in the Density Case.** Let $X$ be a random variable with density $f_X$. If

$$\int_{-\infty}^{\infty} |x| f_X(x)\, dx = \lim_{a\to\infty} \int_{-a}^{a} |x| f_X(x)\, dx$$

is finite, we define $E(X)$ by

$$E(X) = \int_{-\infty}^{\infty} x f_X(x)\, dx.$$

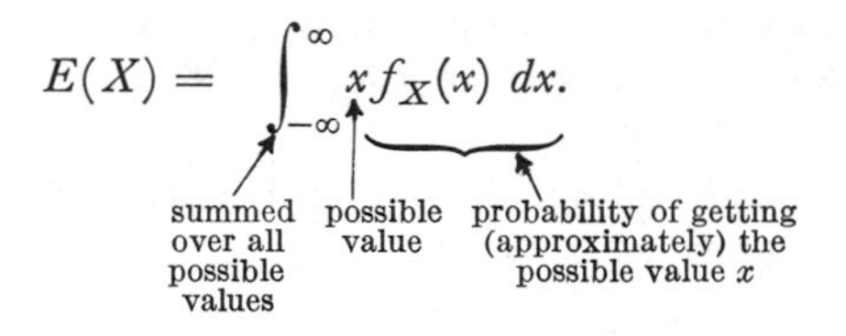


### *2.13 Example: Expected Waiting Time*

The time $T$ between initiation of successive telephone calls is often assumed to have the exponential density

$$f_T(t) = \begin{cases} \frac{1}{\theta} e^{-t/\theta} & \text{for} \quad t \geq 0 \\ 0 & \text{otherwise.} \end{cases}$$

Then the expected waiting time between phone call initiations

$$E(T) = \int_0^\infty t \cdot \frac{1}{\theta} e^{-t/\theta}\, dt$$

$$= \int_0^\infty ue^{-u}\, \theta\, du \qquad \left(\text{letting } u = \frac{t}{\theta}\right)$$

$$= \theta.$$

Thus $E(T) = \theta$. △

---

## 2.14 EXERCISES

1. Find $E(X)$ for the following densities:

(a)° $f_X(x) = \begin{cases} 3x^2 & \text{for } 0 \le x \le 1 \\ 0 & \text{otherwise.} \end{cases}$

(b) $f_X(x) = \begin{cases} x^3 & \text{for } 0 \le x \le \sqrt{2} \\ 0 & \text{otherwise.} \end{cases}$

(c)° $f_X(x) = \begin{cases} x & \text{for } 0 \le x \le 1 \\ 1/(3x^2) & \text{for } 1 \le x \le \sqrt[3]{3/2} \\ 0 & \text{otherwise.} \end{cases}$

(d) $f_X(x) = \begin{cases} 1 & \text{for } 0 \le x \le 1/2 \\ x & \text{for } 1/2 < x \le \sqrt{5/4} \\ 0 & \text{otherwise.} \end{cases}$

(e) $f_X(x) = \begin{cases} \sin x & \text{for } 0 \le x \le \pi/2 \\ 0 & \text{otherwise.} \end{cases}$

(f) $f_X(x) = \begin{cases} \dfrac{\sin x}{2} & \text{for } 0 \le x \le \pi \\ 0 & \text{otherwise.} \end{cases}$

2.° Let $f_X(x) = \dfrac{1}{\sqrt{2\pi}} e^{-(x-\mu)^2/2}$. Find $E(X)$.

3. Let $f_X(x) = \dfrac{1}{\sqrt{2\pi\sigma^2}} e^{-(x-\mu)^2/(2\sigma^2)}$ where $\sigma^2 > 0$ and $\mu$ is real. Show that $E(X) = \mu$.

4. Let $f_X(x) = \begin{cases} \dfrac{n-1}{(x+1)^n} & \text{for } x \ge 0 \\ 0 & \text{otherwise.} \end{cases}$

Find $E(X)$ when $n > 2$.

5.° The waiting time $T$ required from the initiation of the first telephone call until the initiation of the $k + 1$st call is often assumed to have a *Gamma density* with parameters $k$ and

$$\alpha = \frac{1}{\theta}$$

where

$$f_T(t) = \begin{cases} \dfrac{1}{(k-1)!}\dfrac{t^{k-1}}{\theta^k}e^{-t/\theta} & \text{for } t \geq 0 \\ 0 & \text{otherwise.} \end{cases}$$

From the answer to 4.9(3), Chapter 5, it is not difficult to see that $T = \sum_{i=1}^{k} T_i$ where $T_1, \ldots,$ $T_k$ are independent with density

$$f_{T_i}(t) = \begin{cases} \dfrac{1}{\theta}e^{-t/\theta} & \text{for } t \geq 0 \\ 0 & \text{otherwise,} \end{cases}$$

where $T_i$ represents the time from the initiation of the $i$th call to the initiation of the $i + 1$st call. From Definition 2.12, we know that $E(T_i) = \theta$. Find $E(T)$:

(a) By guessing intelligently.

(b) Using integration by parts "enough" times (legitimately by induction).

(c) By assuming that it is legitimate to differentiate and antidifferentiate under the integral sign (as if the integral were a sum). In this case it helps to replace $\theta$ by $-1/\alpha$ and differentiate and antidifferentiate $G(\alpha) = \int_0^\infty t^k e^{t\alpha}\,dt$ with respect to $\alpha$. Conditions under which such differentiation is legitimate are given in the Appendix.

---

## *2.15 PROBLEMS*

1. (a) Does $X$ have an expectation when $f_X(x) = 1/[\pi(1 + x^2)]$?

(b) It can be shown that if $X$ and $Y$ are independent with this density, then

$$f_{\frac{X+Y}{2}}(t) = \frac{1}{\pi(1+t^2)}.$$

How does this result help justify the result of part (a)?

2. Let $X$ be a random variable that can assume values $b_1, \ldots, b_k$, with $P\{X = b_i\} = p_i$, where $\sum_{i=1}^{k} p_i = p < 1$. Suppose further that for any interval $[a, b]$ not including any of the $b_i$, $P\{a \leq X \leq b\} = \int_a^b f(x)\,dx$, where $f \geq 0$ is a given function, $\int_{-\infty}^{\infty} f(x)\,dx = 1 - p$. Extend the definition of $E(X)$ to such random variables in a meaningful way, that is, so that you would expect $\frac{1}{n}\sum_{i=1}^{n} X_i$ to approach $E(X)$ as $n$ gets large.

## SECTION 3
## Expectation of Functions of a Random Variable, Chebyshev's Inequality

If $X$ is a random variable and $g$ a given real valued function of a real variable, then $Z = g(X)$ is a random variable for any "well-behaved" $g$. The problem of determining $E(Z)$ when we know the density $f_X$ or the discrete probability function of $X$ arises frequently. For instance, if $X$ is the electric current across a one ohm resistor, then $Z = g(X) = X^2$ is the heat generated per unit time by this current. The average heat generated per unit time, $E(Z)$, is sure to be of interest in many problems. The straightforward approach to determining $E(Z)$ is simply to find the density or the discrete probability function of $Z$ and then to make use of the definition of expectation. Fortunately, there is a method that is usually much simpler for computing $E(Z)$. We shall derive the result in the finite discrete case and then state the general theorem.

**3.1**

$$E(Z) = \sum_{\substack{\text{all possible}\\ \text{values } z \text{ of } Z}} zP\{Z = z\}$$

$$= \sum_{\substack{\text{all possible}\\ \text{values } z \text{ of } g(X)}} zP\{g(X) = z\}$$

$$= \sum_{\substack{\text{all possible}\\ \text{values } z \text{ of } g(X)}} z\Big(\sum_{\substack{\text{all values } x \text{ of } X\\ \text{such that } g(x) = z}} P\{X = x\}\Big)$$

$$= \sum_{\substack{\text{all possible}\\ \text{values } z \text{ of } Z}} \left(\sum_{\substack{\text{all values } x \text{ of } X\\ \text{such that } g(x) = z}} g(x)P\{X = x\}\right)$$

$$= \sum_{\substack{\text{all values}\\ x \text{ of } X}} g(x)P\{X = x\}.$$

The previous sum is just a particular arrangement of the last sum based on the values $z$ of $g(x)$.

Hence $E[Z] = E[g(X)] = \sum_x g(x)P\{X = x\}$. The general result in this direction is the following theorem.

### *3.2 Theorem: Evaluation of E[g(X)]*

Let $X$ be a random variable and let $g$ be a real valued function of a real variable. Let $Z = g(X)$.

If $X$ is discrete and $\sum_{\substack{\text{all values} \\ x \text{ of } X}} |g(x)| P\{X = x\}$ is finite, then $Z$ has an expectation

$$E(Z) = E(g(X)) = \sum_{\substack{\text{all values} \\ x \text{ of } X}} g(x) P\{X = x\}.$$

If $X$ has a density $f_X$ and $g$ is any function of the type encountered in practice such that $\int_{-\infty}^{\infty} |g(x)| f_X(x)\, dx$ is finite, then $Z$ has an expectation

$$E(Z) = E(g(x)) = \int_{-\infty}^{\infty} g(x) f_X(x)\, dx.$$

***Note that even here we may view the expectation as the sum of all possible values multiplied by their corresponding probabilities.***

In addition to being useful in providing a convenient description of distributions, the following definitions are required in the theoretical development.

**3.3 Definition: Higher Moments, Variance.** Let $X$ be a random variable. If $E(X^n)$ exists, it is called the *nth moment of X about the origin*, or the *nth moment of the distribution of X about the origin*. The quantity $E[(X - a)^n]$ is called the *nth moment of X about a;* if we denote $E(X)$ by $\mu$, then $E[(X - \mu)^2]$ is called the *variance of X*, and is written Var $X$ or $\sigma_X^2$. The non-negative square root $\sqrt{\sigma_X^2}$ denoted by $\sigma_X$ is called the *standard deviation of X*.

From Theorem 3.2, we see that in the discrete case,

**3.4**
$$\text{Var } X = \sum_{\substack{\uparrow \\ \text{all values} \\ x \text{ of } X}} (x - \mu)^2 P\{X = x\},$$

whereas in the density case,

**3.5**
$$\text{Var } X = \int_{-\infty}^{\infty} (x - \mu)^2 f_X(x)\, dx.$$

*3.6 Remarks on Moments.* It is not difficult to see that if $E(X^n)$ exists for some given positive integer $n$, then $E(X^k)$ exists for all positive integers $k < n$. In the discrete case, this is proved as follows: Since $E(X^n)$ exists, $\sum_x |x^n|\, P\{X = x\}$ is finite. But

$$\begin{aligned}\sum_x |x^n|\, P\{X = x\} &\geq 1 + \sum_{|x| \geq 1} |x^n|\, P\{X = x\} \\ &\geq \sum_{|x| < 1} |x^k|\, P\{X = x\} + \sum_{|x| \geq 1} |x^k|\, P\{X = x\} \\ &= \sum_x |x^k|\, P\{X = x\}.\end{aligned}$$

So $\sum_x |x^k|\, P\{ X = x \}$ is finite.

Now, by the comparison test (Appendix A1.7(2)), $\sum_x x^k P\{X = x\}$ converges, so that $E(X^k)$ exists.

In many cases, knowledge of all of the moments of $X$ is equivalent to knowledge of the distribution of $X$. To make this plausible, suppose that $X$ is discrete, with $P\{X = j\} = p_j$ for $j = 1, \ldots, k$, where $\sum_{j=1}^{k} p_j = 1$. Suppose we are given numbers $m_1, \ldots, m_{k-1}$, being told that $E(X) = m_1, E(X^2) = m_2, \ldots, E(X^{k-1}) = m_{k-1}$. That is,

$$
\begin{aligned}
p_1 + p_2 + \cdots + p_k &= 1 \\
p_1 + 2p_2 + \cdots + kp_k &= m_1 \\
p_1 + 4p_2 + \cdots + k^2 p_k &= m_2 \\
&\vdots \\
p_1 + 2^{k-1}p_2 + \cdots + k^{k-1}p_k &= m_{k-1}.
\end{aligned}
$$

We can actually solve this system of linear equations by the standard Gaussian elimination technique (in which you are allowed to multiply any of the equations by a constant and add it to any of the other equations). You are urged to do this for $k = 3$, and to examine this system to see that you can use the first equation to eliminate $p_1$ from the remaining ones, then you can use the second equation to eliminate $p_2$ from those that follow, and so on. Hence, at least in this case, it seems that the moments determine the distribution. It is thus reasonable to believe that this may hold fairly generally, and therefore it is interesting to ask, "When do all the moments of a distribution determine that distribution?" There are examples to show that even if all moments exist, there may be two distinct distributions with a given set of moments, so that this question has a non-trivial answer. One may ask the more general question as to when a given sequence of numbers can be the moments of some distribution. There is a good deal of mathematical literature on this subject, known as the *moment problem*.

In light of the previous discussion, it is plausible that knowledge of the first several moments of a distribution might furnish valuable information about this distribution. We shall see that this is in fact the case; for example, the variance is, in a sense (we will be more precise later), an easily computed measure of the "spread" or variability of our measurements. For the moment, we simply notice that *if $X$ has all of its mass concentrated at $\mu$, then its variance is 0; in any other case the variance is positive.* The fact that if $\operatorname{Var} X = 0$ then $P\{X = \mu\} = 1$ is not hard to see in the discrete case, since $\operatorname{Var} X = \sum_x (x - \mu)^2 P\{X = x\}$, which is positive if for some $x \neq \mu$, we have $P\{X = x\} > 0$. The continuous case requires a more delicate proof, which we will omit.

*The reader is cautioned that generally $E(X^n) \neq [E(X)]^n$.* For instance, just examine the case $n = 2$. If the density of $X$ is symmetric about the origin and $E(X)$ exists, then $E(X) = 0$, and hence $[E(X)]^2 = 0$. However, $E(X^2)$ is the expectation of the almost certainly positive random variable $X^2$, and hence $E(X^2) > 0$.

### 3.7 *Example: Second Moment Computations in the Poisson Case*

We see that when $X$ is Poisson with $E(X) = \lambda t$, then using Theorem 3.2 for evaluating $E[g(X)]$, with $g(X) = X^2$,

$$
\begin{aligned}
E(X)^2 &= \sum_{j=0}^{\infty} j^2 e^{-\lambda t} \frac{(\lambda t)^j}{j!} = \sum_{j=1}^{\infty} j^2 e^{-\lambda t} \frac{(\lambda t)^j}{j!} \\
&= \sum_{j=1}^{\infty} [j(j-1) + j] e^{-\lambda t} \frac{(\lambda t)^j}{j!} \\
&= \sum_{j=2}^{\infty} e^{-\lambda t} \frac{(\lambda t)^j}{(j-2)!} + \sum_{j=1}^{\infty} e^{-\lambda t} \frac{(\lambda t)^{j-1}}{(j-1)!} \\
&= (\lambda t)^2 e^{-\lambda t} \sum_{j=2}^{\infty} \frac{(\lambda t)^{j-2}}{(j-2)!} + (\lambda t) e^{-\lambda t} \sum_{j=1}^{\infty} \frac{(\lambda t)^{j-1}}{(j-1)!} \\
&= (\lambda t)^2 + (\lambda t),
\end{aligned}
$$

using the fact that both infinite series are of the form

$$1 + x + \frac{x^2}{2!} + \frac{x^3}{3!} + \cdots = e^x.$$

Thus, if $X$ is Poisson with $E(X) = \lambda t$, then $E(X^2) = (\lambda t)^2 + (\lambda t)$. Furthermore, since

$$
\begin{aligned}
\text{Var } X &= E[(X - E(X))^2] \\
&= E[X^2 - 2XE(X) + (E(X))^2] \\
&= E(X^2) - 2E(X)E(X) + (E(X))^2 \quad \text{(note that } E(X) \text{ is a constant) by Theorem 3.2, evaluation of } E[g(X)] \\
&= E(X^2) - (E(X))^2
\end{aligned}
$$

we see that Var $X = \lambda t$. △

The following lemma, which we already established in the Poisson example above, sometimes aids in computing the variance.

### 3.8 *Lemma: Relation Between the Variance and the First Two Moments*

If we let Var $X$ denote the variance of $X$, then

$$\text{Var } X = E(X^2) - [E(X)]^2.$$

### 3.9 *Example: Variance of Bernoulli Random Variable*

Let $P\{X = 1\} = p = 1 - P\{X = 0\}$. We know that $X$ represents the number of successes in one trial of an experiment whose probability of success is $p$. Since $X^2 = X$ in this case, and $E(X) = p$ is known from Example 2.5, we have from Lemma 3.8 that

$$\text{Var } X = E(X^2) - (E(X))^2 = p - p^2 = p(1 - p).$$

Since

$$\frac{d}{dp}(p - p^2) = 1 - 2p,$$

we see from calculus that Var $X$ is largest for $p = 1/2$. This is not surprising, since the outcome is most variable (least predictable) when $p = 1/2$. △

### 3.10 *Example: Variance of Normal Random Variable*

If

$$f_X(x) = \frac{1}{\sqrt{2\pi\sigma^2}} e^{-(x-\mu)^2/(2\sigma^2)},$$

then since $E(X) = \mu$ (see Exercise 2.14(3)), it follows from Equation 3.5 that

$$\text{Var } X = \int_{-\infty}^{\infty} \frac{(x-\mu)^2}{\sqrt{2\pi\sigma^2}} e^{(x-\mu)^2/(2\sigma^2)}\, dx.$$

Using the substitution

$$u = \frac{x - \mu}{\sigma},$$

we find that

$$\text{Var } X = \sigma^2 \int_{-\infty}^{\infty} \frac{u^2}{\sqrt{2\pi}} e^{-u^2/2}\, du.$$

Using integration by parts, with

$$f(u) = u \quad \text{and} \quad g(u) = \frac{u}{\sqrt{2\pi}} e^{-u^2/2},$$

the fact that

$$\lim_{u\to\infty} ue^{-u^2/2} = \lim_{u\to\infty} \frac{u}{e^{u^2/2}} = 0,$$

which follows from L'Hospital's rule (see the Appendix), and the fact that

$$\int_{-\infty}^{\infty} \frac{1}{\sqrt{2\pi}} e^{-u^2/2}\, du = 1,$$

(see A.12, in the Appendix), we find that

$$\operatorname{Var} X = \sigma^2. \qquad \triangle$$

### *3.11 Lemma: Normalization*

Suppose that $X$ is a random variable with expectation $\mu$ and variance $\sigma^2$. Then

$$E\left(\frac{X-\mu}{\sigma}\right) = 0, \qquad \operatorname{Var}\left(\frac{X-\mu}{\sigma}\right) = 1$$

*Remark.* The transformation from $X$ to

$$\frac{X-\mu}{\sigma} = 1$$

is called *normalization* or *standardization.*

*Proof:* We will supply proofs for the density case only. From Theorem 3.2 (evaluation of $E[g(X)]$), with

$$g(X) = \frac{X-\mu}{\sigma},$$

we see that

$$\begin{aligned}
E\left[\frac{X-\mu}{\sigma}\right] &= \int_{-\infty}^{\infty} \frac{x-\mu}{\sigma} f_X(x)\, dx \\
&= \frac{1}{\sigma}\left[\int_{-\infty}^{\infty} x f_X(x)\, dx - \mu \int_{-\infty}^{\infty} f_X(x)\, dx\right] \\
&= \frac{1}{\sigma}[\mu - \mu] \\
&= 0,
\end{aligned}$$

proving the first part. Again by Theorem 3.2,

$$\begin{aligned}
\operatorname{Var}\left(\frac{X-\mu}{\sigma}\right) &= E\left[\left\{\left(\frac{X-\mu}{\sigma}\right) - E\left(\frac{X-\mu}{\sigma}\right)\right\}^2\right] \\
&= E\left[\left(\frac{X-\mu}{\sigma}\right)^2\right] \\
&= \int_{-\infty}^{\infty} \left(\frac{x-\mu}{\sigma}\right)^2 f_X(x)\, dx \\
&= \frac{1}{\sigma^2}\int_{-\infty}^{\infty} (x-\mu)^2 f_X(x)\, dx \\
&= 1,
\end{aligned}$$

using Equation 3.5. This establishes the second result.

## 3.12 EXERCISES

1. Find $E(X^2)$ and Var $X$ when $X$ has the density:

(a)° $f_X(x) = \begin{cases} 3x^2 \text{ for } 0 \leq x \leq 1 \\ 0 \quad \text{otherwise.} \end{cases}$

(b) $f_X(x) = \begin{cases} 4x^3 \text{ for } 0 \leq x \leq 1 \\ 0 \quad \text{otherwise.} \end{cases}$

(c)° $f_X(x) = \begin{cases} \sin x \text{ for } 0 \leq x \leq \pi/2 \\ 0 \quad \text{otherwise.} \end{cases}$

(d) $f_X(x) = \begin{cases} (\sin x)/2 \text{ for } 0 \leq x \leq \pi \\ 0 \quad \text{otherwise.} \end{cases}$

2.° Find the variance of a random variable that is uniformly distributed on $[-.5, .5]$ (see Example 2.19, Chapter 4).

3. Find the variance of a random variable with the uniform density on $[a, b]$ (see Example 2.19, Chapter 4).

4. Find the variance of $X$ when

$$f_X(t) = \begin{cases} \frac{1}{\theta} e^{-t/\theta} \text{ for } t \geq 0 \\ 0 \quad \text{otherwise.} \end{cases}$$

5. Find $E(X^n)$ when $X$ is uniformly distributed on $[a, b]$ (see Example 2.19, Chapter 4).

6. If $X$ is a normal random variable with $E(X) = 0$, Var $X = \sigma^2$—that is, $f_X(x) = \frac{1}{\sqrt{2\pi\sigma^2}} e^{-x^2/(2\sigma^2)}$—find $E(X^n)$ for $n = 3, 4, 5$.

7.° Suppose independent experiments, each with probability $p$ of success, are repeated until a success is obtained. Let $M$ be the number of trials required. Find Var $M$.

8. If Var $X = \sigma^2$, find Var $(aX + b)$ where $a$ and $b$ are real numbers.

## 3.13 PROBLEMS

1.° To show that the first three moments do not determine the distribution, find two distinct distributions for $X$ such that $E(X) = 0$, $E(X^2) = 1$, $E(X^3) = 0$ in both cases.

2. Show that in general

$$E\left(\frac{1}{X}\right) \neq \frac{1}{E(X)}.$$

Hint: Try the simplest case, where $X$ can only take on two values 1 and $-1$.

3. Find a distribution for which:

(a) $E(X)$ exists but $E(X^2)$ does not.

(b) $E(X), E(X^2), \ldots, E(X^n)$ exist but $E(X^{n+1}), \ldots$ fail to exist.

4.° Recall that if an experiment is performed independently until a total of $k$ successes result, and if $N$ is the needed number of experiments, then

$$P\{N = j\} = \binom{j-1}{k-1} p^{k-1}(1-p)^{j-k}p,$$

where $p$ is the probability of success on any one experiment. Find Var $N$.

5.° Find the variance of a binomial random variable with parameters $n$ and $p$. Hint: Use Remark 3.8 and Example 2.6.

Earlier we intoduced the variance as an intuitively reasonable measure of variability. At this point we now show that it has a useful operational meaning.

### *3.14 Theorem: Chebyshev Inequality*

Let $X$ be a random variable with $E(X) = \mu$ and Var $X = \sigma^2$, and let $\sigma = \sqrt{\sigma^2}$ (the non-negative square root of $\sigma^2$). Then for each $t \geq 1$, we have

$$P\{|X - \mu| \geq t\sigma\} \leq \frac{1}{t^2}.$$

A remark is in order before giving the proof. This theorem says that if we view $X$ in terms of its distance from its expectation $\mu$, then it is useful to measure this distance in "$\sigma$-units." For example, in Figure 6.1, the probability that $X$ assumes a value outside the shaded region (3 $\sigma$-units to each side of $\mu$) is less than or equal to 1/9.

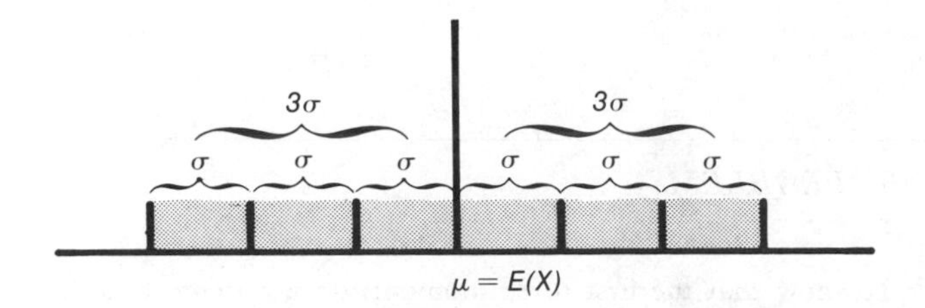


*FIGURE* 6–1.

*Proof of 3.14:* We give the proof for the density case only:

$$\sigma^2 = \text{Var } X$$

$$= E[(X-\mu)^2]$$

$$= \int_{-\infty}^{\infty} (x-\mu)^2 f_X(x)\, dx \qquad \text{by Equation 3.5}$$

$$\geq \int_{\{x \in \mathscr{R}: |x-\mu| \geq t\sigma\}} (x-\mu)^2 f_X(x)\, dx \qquad \text{(we discarded only non-negative contributions to the integral)}$$

$$\geq \int_{\{x \in \mathscr{R}: |x-\mu| \geq t\sigma\}} t^2\sigma^2 f_X(x)\, dx \qquad \text{(since in the set over which we integrate, we know that } (x-\mu)^2 \geq t^2\sigma^2\text{)}$$

$$= t^2\sigma^2 \int_{\{x \in \mathscr{R}: |x-\mu| \geqq t\sigma\}} f_X(x)\, dx$$

$$= t^2\sigma^2 P\{|X-\mu| \geq t\sigma\}.$$

Looking at the first and last terms of this sequence of relations, we have

$$\sigma^2 \geq t^2\sigma^2 P\{|X-\mu| \geq t\sigma\},$$

from which the desired result follows.

---

## 3.15 EXERCISES

1. High performance transistors have an expected lifetime $E(T) = 5$ years with $\text{Var } T = .16 \text{ yr}^2$.

(a) What can be said about the probability that these transistors will live between 3 and 7 years?

(b) What can be said about the probability that the lifetime of these transistors is between 2 and 8 years?

2. The expected percentage yield $Y$ in a certain organic chemistry reaction is 63 per cent and the variance of the percentage yield is 4. Find an upper bound for:

(a) the probability that the percentage yield exceeds 75 per cent.
Hint: $P\{Y > 75\} \leq P\{|Y - 63| \geq 12\}$.

(b) the probability that the percentage yield falls below 61.

3. Let $X$ be a Poisson random variable with mean $E(X) = 225$. Give a simple upper bound for $P\{X > 390\}$.
Hint: $P\{X > 390\} \leq P\{|X - 225| > 165\}$.

4. Let $N$ be the number of trials needed to obtain 16 successes in an experiment whose probability of success is $2/(1 + \sqrt{17}) \cong 1/2.56$. Find a simple upper bound for $P\{N \geq 72.96\}$, that is, for $P\{N \geq 73\}$.

5.° Assume that the Nielson survey includes 1600 ($40^2$) homes. On the basis of the Chebyshev inequality, does this appear to be a large enough sample to obtain a reasonably close estimate of the proportion of television sets tuned to a particular program:

(a) when $p = .25$?
(b) when $p = .5$?

Hint: Use the result of 3.13(5) that if $X$ is a binomial random variable with parameters $n$ and $p$, then $\text{Var } X = np(1 - p)$, and use Example 2.6, that $E(X) = np$.

---

### *3.16 PROBLEMS*

1.° Suppose you know that $P\{X \in [a, b]\} = 1$ where $a$ and $b$ are finite. Find a simple upper bound for $\text{Var } X$.

2.° Suppose you were given a random variable $X$ such that $E(X) = 10$ and $\text{Var } X = 1$. Would you be greatly surprised at the occurrence of the event $|X - 10| \leq .0001$? Justify your reasoning.
Hint: We know that a very small standard deviation means that $X$ is very likely to be close to the mean. Does a large standard deviation necessarily mean that $X$ is likely to be far from the mean?

3. If $X$ is a random variable, such as a lifetime for which $P\{X < 0\} = 0$ and $E(X) = 3$, find an upper bound for $P\{X > 30\} \equiv p$.
Hint: $3 = E(X) \geq p \cdot 30 + (1 - p) \cdot 0$.

4. Let $Y$ be a random variable for which $M_n \equiv E(|Y|^n)$ is finite. Show that

$$P\{|Y| \geq t\sqrt[n]{M_n}\} \leq 1/t^n.$$

Notice that when this result is applicable, it is more powerful than the Chebyshev inequality for large $t$ when $n > 2$. Note that if $Y = X - E(X)$ and $n = 2$, then this result is the Chebyshev inequality.

## SECTION 4
## Expectation of Functions of a Random Vector, Covariance

Just as we want to determine $E[g(X)]$ when $X$ is a random variable and $g$ is a real valued function of a real variable, we need to be able to compute $E[g(X)]$

when $X$ is a random $n$-dimensional vector and $g$ is a real valued function of $n$ real variables. The following is the $n$-dimensional version of Theorem 3.2.

### 4.1 *Theorem:* $E[g(\boldsymbol{X})]$

Let $\boldsymbol{X} = (X_1, \ldots, X_n)$ be a random vector and let $g$ be a real valued function of $n$ real variables. Let $Z = g(\boldsymbol{X})$.

If $\boldsymbol{X}$ is discrete, let $\{\boldsymbol{X} = \boldsymbol{x}\}$ represent the event $\{X_1 = x_1, \ldots, X_n = x_n\}$. If

$$\sum_{\substack{\uparrow \\ \text{all values} \\ \boldsymbol{x} \text{ of } \boldsymbol{X}}} |g(\boldsymbol{x})| \, P\{\boldsymbol{X} = \boldsymbol{x}\}$$

is finite, then $Z$ has an expectation

$$E(Z) = E(g(\boldsymbol{X})) = \sum_{\substack{\uparrow \\ \text{all values} \\ \boldsymbol{x} \text{ of } \boldsymbol{X}}} g(\boldsymbol{x}) P\{\boldsymbol{X} = \boldsymbol{x}\}.$$

If $\boldsymbol{X}$ has a joint density $f_{\boldsymbol{X}} = f_{X_1, \ldots, X_n}$ and $g$ is any function of the type encountered in practice such that the multiple integral

$$\int_{-\infty}^{\infty} \cdots \int_{-\infty}^{\infty} |g(x_1, \ldots, x_n)| f_{X_1, \ldots, X_n}(x_1, \ldots, x_n) \, dx_1 \cdots dx_n$$

$$= \lim_{c \to \infty} \int_{-c}^{c} \cdots \int_{-c}^{c} |g(x_1, \ldots, x_n)| f_{X_1, \ldots, X_n}(x_1, \ldots, x_n) \, dx_1 \cdots dx_n$$

is finite, then $Z$ has an expectation and

$$E(Z) = E[g(\boldsymbol{X})] = \int_{-\infty}^{\infty} \cdots \int_{-\infty}^{\infty} g(x_1, \ldots, x_n) f_{X_1, \ldots, X_n}(x_1, \ldots, x_n) \, dx_1 \cdots dx_n.$$

You can convince yourself that this assertion is reasonable for the finite discrete case by going through the same manipulations as in 3.1 but with $\boldsymbol{X}, \boldsymbol{x}, g, P\{\boldsymbol{X} = \boldsymbol{x}\}$ and $g$ interpreted in the $n$-dimensional setting.

It is tempting to give some examples and exercises illustrating this theorem, but they are always artificial. The main use of this theorem is in the theoretical development, and there we shall see its power. One such theoretical result is that if $X$ and $Y$ are two random variables defined on the same probability space and $a$ and $b$ are real numbers, then $E(aX + bY) = aE(X) + bE(Y)$. The following intuitive arguments make this plausible: Suppose that $Z_1, Z_2, \ldots$ are independent random variables having the same distribution and with $E(Z_i) = E(Z)$. From our original motivation for expectation, we expect that when $n$ is large, it is highly probable that

$$\frac{1}{n} \sum_{i=1}^{n} Z_i \quad \text{is close to} \quad E(Z).$$

Now it is evident that

$$(*) \qquad \frac{1}{n}\sum_{i=1}^{n}(aX_i + bY_i) = a\frac{1}{n}\sum_{i=1}^{n}X_i + b\frac{1}{n}\sum_{i=1}^{n}Y_i.$$

By this argument, it is highly probable *the left side is close to* $E(aX + bY)$ for large $n$. Similarly,

$$a\frac{1}{n}\sum_{i=1}^{n}X_i$$

is very likely to be close to $aE(X)$ and

$$b\frac{1}{n}\sum_{i=1}^{n}Y_i$$

is very likely to be close to $bE(Y)$ when $n$ is large. Hence, using the fact that the intersection of two high probability events is again an event of high probability (see Theorem 3.5(e), Chapter 3) we see that for large $n$ *the right side should be close to* $aE(X) + bE(Y)$. That is, for large $n$ it is highly probable that the left side of (*) is close to $E(aX + bY)$ and the right side of (*) is close to $aE(X) + bE(Y)$. Hence, we should not be surprised that $E(aX + bY) = aE(X) + bE(Y)$. We now state this result formally and supply a "proof" based on Theorem 4.1.

### 4.2 *Theorem:* $E(aX + bY) = aE(X) + bE(Y)$

Let $X$ and $Y$ be random variables defined on the same probability space with $E(X)$ and $E(Y)$ defined. Then whenever $a$ and $b$ are real numbers, $E(aX + bY)$ is defined, and

$$E(aX + bY) = aE(X) + bE(Y).$$

*Proof:* We give the proof only for the density case, but that for the discrete case is similar. Let $g(X, Y) = aX + bY$. Then

$$\int_{-c}^{c}\int_{-c}^{c}|ax + by| f_{X,Y}(x, y)\, dx\, dy$$

$$\le |a|\int_{-c}^{c}\int_{-c}^{c}|x| f_{X,Y}(x, y)\, dy\, dx + |b|\int_{-c}^{c}\int_{-c}^{c}|y| f_{X,Y}(x, y)\, dx\, dy$$

$$\le |a|\int_{-\infty}^{\infty}\int_{-\infty}^{\infty}|x| f_{X,Y}(x, y)\, dy\, dx + |b|\int_{-\infty}^{\infty}\int_{-\infty}^{\infty}|y| f_{X,Y}(x, y)\, dx\, dy$$

$$= |a|\int_{-\infty}^{\infty}|x| f_X(x)\, dx + |b|\int_{-\infty}^{\infty}|y| f_Y(y)\, dy \quad \text{by 3.2, Chapter 5.}$$

It follows that

$$\int_{-\infty}^{\infty}\int_{-\infty}^{\infty}|ax + by| f_{X,Y}(x, y)\, dx\, dy$$

is finite. Hence, by Theorem 4.1, $E(aX + bY)$ is defined and

$$\begin{aligned}
E(aX + bY) &= \int_{-\infty}^{\infty}\int_{-\infty}^{\infty} g(x, y) f_{X,Y}(x, y)\, dx\, dy \\
&= \int_{-\infty}^{\infty}\int_{-\infty}^{\infty} (ax + by) f_{X,Y}(x, y)\, dx\, dy \\
&= \int_{-\infty}^{\infty}\int_{-\infty}^{\infty} ax f_{X,Y}(x, y)\, dx\, dy + \int_{-\infty}^{\infty}\int_{-\infty}^{\infty} by f_{X,Y}(x, y)\, dx\, dy \\
&= a\int_{-\infty}^{\infty} x\left[\int_{-\infty}^{\infty} f_{X,Y}(x, y)\, dy\right] dx + b\int_{-\infty}^{\infty} y\left[\int_{-\infty}^{\infty} f_{X,Y}(x, y)\, dx\right] dy \\
&= a\int_{-\infty}^{\infty} x f_X(x)\, dx + b\int_{-\infty}^{\infty} y f_Y(y)\, dy \quad \text{by 3.2, Chapter 5} \\
&= aE(X) + bE(Y),
\end{aligned}$$

concluding the proof.

Notice that nothing is required of $X$ and $Y$ other than that they possess finite expectations in order that $E(aX + bY)$ exist and equal $aE(X) + bE(Y)$. This is actually quite reasonable in view of the intuitive discussion preceding the proof.

By induction, we have the following corollary.

**4.3 *Corollary:*** $E(a_1X_1 + \cdots + a_kX_k)$, $E\left(\frac{1}{k}\sum_{i=1}^{k} X_i\right)$

If the random variables $X_1, \ldots, X_k$ defined on the same probability space all have expectations, then so does $a_1X_1 + \cdots + a_kX_k$, and

$$E(a_1X_1 + \cdots + a_kX_k) = a_1E(X_1) + \cdots + a_kE(X_k).$$

If $E(X_i) = \mu$ for all $i$, then by taking $a_i = 1/k$, we have

$$E\left(\frac{1}{k}\sum_{i=1}^{k} X_i\right) = \mu.$$

**4.4 *Example:* E(2X − Y − 3)**

Suppose $E(X)$ and $E(Y)$ exist. Then

$$E(2X - Y - 3) = 2E(X) - E(Y) - 3,$$

since $-3$ can be considered a constant random variable whose expectation is $-3$. △

### 4.5 EXERCISES

We assume that the random variables $X_i$ have expectation $E(X_i) = \mu$, $\text{Var } X_i = \sigma^2$. Find:

1. $E(X_1 + X_2)$.
2.° $E(5X_1 + 7X_2)$.
3. $E(\tfrac{1}{2}[X_1 + X_2])$.
4.° $E(\tfrac{1}{2}[X_1 + X_2] - \mu)$.
5.° $E\left(\frac{1}{\sigma}\{\tfrac{1}{2}[X_1 + X_2] - \mu\}\right)$.
6.° $E\left[\frac{1}{n}\sum_{i=1}^{n} X_i\right]$.
7. $E\left[\frac{1}{n}\sum_{i=1}^{n} (X_i - \mu)\right]$.
8. $E\left[\left(\frac{1}{n}\sum_{i=1}^{n} X_i\right) - \mu\right]$.
9.° $E\left[\frac{1}{\sqrt{n}\sigma}\sum_{i=1}^{n} (X_i - \mu)\right]$.
10.° $E\left[\frac{1}{\sqrt{n}}\sum_{i=1}^{n} \frac{X_i - \mu}{\sigma}\right]$
11. $E\left[\frac{1}{n}\sum_{i=1}^{n} \frac{X_i - \mu}{\sigma}\right]$.

Let us now examine the result of Problem 3.13(4), comparing it with the answer to 3.12(7). To do this, put $N = M_1 + M_2 + \cdots + M_k$, where $M_1, \ldots, M_k$ are independent random variables representing the number of trials until a success. We notice that

$$\text{Var}(M_1 + \cdots + M_k) = \text{Var } M_1 + \cdots + \text{Var } M_k.$$

This suggests that a result similar to Corollary 4.3 might hold for variances. However, the variance is never negative. Hence, if $\text{Var } X \neq 0$, we know that

$$0 = \text{Var}(X - X) \neq \text{Var } X + \text{Var}(-X).$$

Therefore, any relation similar to Corollary 4.3 will only hold under certain restrictions. As we investigate this problem, keep in mind the importance of the variance as indicated by Chebyshev's inequality (3.14). In order to simplify our investigation, we anticipate some of the computations and do some necessary preliminaries.

**4.6 Definition: Covariance.** Let $X$ and $Y$ be random variables defined on the same probability space, and having expectations $\mu_x$ and $\mu_y$, respectively. If the quantity $E[(X - \mu_x)(Y - \mu_y)]$ exists, it is called the *covariance of $X$ and $Y$*, and is denoted by the symbol $\text{cov}(X, Y)$. We shall see that $\text{cov}(X, Y)$ arises in attempting to compute $\text{Var}(X + Y)$.

### *4.7 Theorem: Existence of Covariance*

If $X$ and $Y$ are random variables on the same probability space, both having finite variances, then $\text{cov}(X, Y)$ exists.

The proof is left to Problem 4.15(1).

### *4.8 Theorem: Variance of Linear Combination*

Suppose $X$ and $Y$ are random variables defined on the same probability space, each possessing a finite variance. Then for any numbers $a$ and $b$, the random variable $aX + bY$ has a finite variance given by

$$\text{Var}[aX + bY] = a^2 \text{ Var } X + b^2 \text{ Var } Y + 2ab \text{ cov}(X, Y).$$

*Remark.* Here we make frequent use of Theorem 4.1 on evaluating $E[g(X)]$.

*Proof:* We can show that $\text{Var}[aX + bY]$ is finite as we proved Theorem 4.2. We will omit the tedious but straightforward details, and give the remainder of the proof in the density case only. The discrete case is similar. By Theorem 4.2, $E(aX + bY) = aE(X) + bE(Y)$. Hence

$$\begin{aligned}\text{Var}(aX + bY) &= E[\{aX + bY - E(aX + bY)\}^2] \\ &= E[\{aX + bY - aE(X) - bE(Y)\}^2] \\ &= E[\{a(X - E(X)) + b(Y - E(Y))\}^2] \\ &= \int_{-\infty}^{\infty}\int_{-\infty}^{\infty} \{a(x - E(X)) + b(y - E(Y))\}^2 f_{X,Y}(x, y)\, dx\, dy \qquad \text{by Theorem 4.1} \\ &= a^2 \int_{-\infty}^{\infty}\int_{-\infty}^{\infty} [x - E(X)]^2 f_{X,Y}(x, y)\, dx\, dy \\ &\quad + b^2 \int_{-\infty}^{\infty}\int_{-\infty}^{\infty} [y - E(Y)]^2 f_{X,Y}(x, y)\, dx\, dy \\ &\quad + 2ab \int_{-\infty}^{\infty}\int_{-\infty}^{\infty} [x - E(X)][y - E(Y)] f_{X,Y}(x, y)\, dx\, dy.\end{aligned}$$

Now we again use Theorem 4.1 on each of the three integrals, as follows: In the first one, $g(x, y) = (x - E(X))^2$. Hence, the first integral is $E[g(X, Y)] = E[(X - E(X))^2] = \text{Var } X$. The second integral is found in the same way to be Var $Y$. In the third integral, we have $g(x, y) = (x - E(X))(y - E(Y))$. Hence, the third integral is

$$E[g(X, Y)] = E[(X - E(X))(Y - E(Y))] = \text{cov}(X, Y).$$

Substituting for these three integrals yields

$$\text{Var}[aX + bY] = a^2 \text{ Var } X + b^2 \text{ Var } Y + 2ab \text{ cov}(X, Y),$$

as asserted.

In order to establish the basic result on the variance of a sum of independent random variables, we prove the following lemma.

### 4.9 *Lemma: E(WZ) = E(W)E(Z) for W and Z Independent*

Suppose $W$ and $Z$ are independent random variables, each of which has an expectation. Then $E[WZ]$ exists and

$$E[WZ] = E(W)E(Z).$$

*Proof:* Under the given hypotheses, it is not difficult to see that

$$\int_{-\infty}^{\infty} \int_{-\infty}^{\infty} |wz| f_W(w) f_Z(z) \, dw \, dz$$

exists. Hence, by Theorem 4.1, $E[WZ]$ exists and

$$\begin{aligned} E[WZ] &= \int_{-\infty}^{\infty} \int_{-\infty}^{\infty} wz f_W(w) f_Z(z) \, dw \, dz \\ &= \left( \int_{-\infty}^{\infty} w f_W(w) \, dw \right) \left( \int_{-\infty}^{\infty} z f_Z(z) \, dz \right) \\ &= E(W)E(Z). \end{aligned}$$

There is a simple corollary.

### 4.10 *Corollary: W and Z Independent ⇒ cov(W, Z) = 0*

If $W$ and $Z$ are random variables defined on the same probability space and having finite expectations, then

$$\text{cov}(W, Z) = 0.$$

The proof follows from the fact that if $W$ and $Z$ are independent, then so are $W - E(W)$ and $Z - E(Z)$. Hence

$$\begin{aligned}\operatorname{cov}(W, Z) &= E[(W - E(W))(Z - E(Z))] \\ &= (E[W - E(W)])(E[Z - E(Z)]) \\ &= 0 \cdot 0.\end{aligned}$$

The main result on the variance of a sum of independent variables now follows.

### 4.11 *Theorem: Variance of Linear Combination of Independent Random Variables*

If $X_1, \ldots, X_n$ are independent random variables with finite variances and $a_1, \ldots, a_n$ are real numbers, then $a_1X_1 + \cdots + a_nX_n$ has a variance and

$$\operatorname{Var}(a_1X_1 + \cdots + a_nX_n) = a_1^2 \operatorname{Var} X_1 + \cdots + a_n^2 \operatorname{Var} X_n.$$

The proof that $\operatorname{Var}(a_1X_1 + a_2X_2) = a_1^2 \operatorname{Var} X_1 + a_2^2 \operatorname{Var} X_2$ follows from Theorem 4.8 and Corollary 4.10. The generalization to $n$ variables follows by induction.

### 4.12 *Corollary:* $\operatorname{Var} \frac{1}{n} \sum_{i=1}^{n} X_i$

If $X_1, \ldots, X_n$ are independent random variables with common variance $\sigma^2$, then it follows from Theorem 4.11 that

$$\operatorname{Var} \frac{1}{n} \sum_{i=1}^{n} X_i = \frac{\sigma^2}{n}$$

The significant results that arise from combining this corollary with Chebyshev's inequality (3.14) will be explored in Section 6.

### 4.13 *Example: Variance of Binomial Random Variable*

We know that if $X$ is the number of successes on $n$ independent trials, each with probability $p$ of success, then $X = X_1 + X_2 + \cdots + X_n$, where $X_1, \ldots, X_n$ are independent, $X_i$ being the number of successes (0 or 1) on the $i$th trial. But in Example 3.9, we showed that $\operatorname{Var} X_i = p(1 - p)$. Hence, by using 4.11 (variance of linear combination), we have $\operatorname{Var} X = np(1 - p)$. △

## *4.14 EXERCISES*

1. Suppose that $X_1, \ldots, X_n$ are independent with expectation $\mu$ and variance $\sigma^2$. Find:

(a)° $\operatorname{Var}(X_1 + X_2)$.

(b) $\operatorname{Var}(5X_1 + 7X_2)$.

(c)° $\operatorname{Var}(\tfrac{1}{2}[X_1 + X_2])$.

(d) $\operatorname{Var}(\tfrac{1}{2}[X_1 + X_2] - \mu)$.

(e) $\operatorname{Var}\left(\frac{1}{\sigma}\{\tfrac{1}{2}[X_1 + X_2] - \mu\}\right)$.

(f) $\operatorname{Var}\left[\frac{1}{n}\sum_{i=1}^{n} X_i\right]$.

(g)° $\operatorname{Var}\left[\frac{1}{n}\sum_{i=1}^{n} (X_i - \mu)\right]$.

(h)° $\operatorname{Var}\left[\left(\frac{1}{n}\sum_{i=1}^{n} X_i\right) - \mu\right]$.

(i) $\operatorname{Var}\left[\frac{1}{n\sigma}\sum_{i=1}^{n} (X_i - \mu)\right]$.

(j) $\operatorname{Var}\left[\frac{1}{n}\sum_{i=1}^{n} \left(\frac{X_i - \mu}{\sigma}\right)\right]$.

(k)° $\operatorname{Var}\left(\frac{1}{\sqrt{n}}\sum_{i=1}^{n} X_i\right)$.

(l)° $\operatorname{Var}\left(\frac{1}{\sqrt{n}}\sum_{i=1}^{n} \left(\frac{X_i - \mu}{\sigma}\right)\right)$.

2. Suppose that $X$ and $Y$ are random variables with $\operatorname{Var} X = 4$, $\operatorname{Var} Y = 9$, $\operatorname{cov}(X, Y) = 1$. What is:

(a)° $\operatorname{Var}(X + Y)$?

(b) $\operatorname{Var}(X - Y)$?

(c)° $\operatorname{Var}(2X - 3Y)$?

(d) $\operatorname{Var}(2X + 7Y)$?

3. If $X$ and $Y$ are independent with $E(X) = 0 = E(Y)$, $\operatorname{Var} X = 1 = \operatorname{Var} Y$:

(a)° Find $\operatorname{cov}(X, X + Y)$.

(b) Find $\operatorname{cov}(X - Y, X + Y)$.

4.° Is it possible to have $\operatorname{Var} X = 1 = \operatorname{Var} Y$ and $\operatorname{cov}(X, Y) = -3$? Justify. Hint: Look at what the variance of $X + Y$ would be.

5. Apply Theorem 4.11 to show that $\operatorname{Var}(aX + b) = a^2 \operatorname{Var} X$.

## *4.15 PROBLEMS*

1. Let $X$ and $Y$ be discrete random variables defined on the same probability space and having finite variances. Show that $\operatorname{cov}(X, Y)$ is well defined.

**Outline.** We first prove that if $E(X^2)$ and $E(Y^2)$ are finite, then so is $E(XY)$, and

$$(*) \qquad E(|XY|) \leq \sqrt{E(X^2)E(Y^2)}.$$

To do this we lose no generality in assuming $E(Y^2) > 0$, since when $E(Y^2) = 0$ we have $Y^2 = 0$; in which case the result is surely true.

(a) First prove $(*)$ for the case in which $X$ and $Y$ can only assume a finite set of values by looking at

$$0 \leq E[(|X| + \alpha\,|Y|)^2] = E(X^2) + 2\alpha E(|XY|) + \alpha^2 E(Y^2) \equiv g(\alpha).$$

So $\min_\alpha g(\alpha) \geq 0$. Use calculus techniques to find $\min_\alpha g(\alpha)$, and show that this yields $(*)$.

(b) To prove $(*)$ in general, let

$$X_n = \begin{cases} X & \text{if } |X| \leq n \\ 0 & \text{otherwise,} \end{cases}$$

$$Y_n = \begin{cases} Y & \text{if } |Y| \leq n \\ 0 & \text{otherwise.} \end{cases}$$

By (a),

$$E(|X_nY_n|) \leq \sqrt{E(X_n^2)E(Y_n^2)}.$$

Now let $n$ approach $\infty$ to obtain $(*)$ for the general discrete case.

Now use the fact that since the variances are finite, both $E(X)$ and $E(Y)$ exist (see Remark 3.6). Hence, if $E(|XY|)$ exists, so does $\text{cov}(X, Y)$ and

$$\text{cov}(X, Y) = E[(X - E(X))(Y - E(Y))] = E(XY) - E(X)E(Y).$$

2.° A list of $n$ names is initially in random order. Interchanges are made between adjacent elements if they are not in correct alphabetical order, until the list is completely alphabetized. What is the expected number of interchanges?

This problem is of importance in data processing, since it leads to an estimate of the computer time needed to alphabetize a list using an "interchange sorting scheme." Hint: Let $N_1$ be the number of interchanges required to move the name that is first in alphabetical order to its proper place. Let $N_2$ be the number of interchanges that *then* must be done on the name that comes second in alphabetical order to put it into its proper position. Generally, let $N_i$ be the number of interchanges that must be made on the name $i$th in alphabetical order in order to get it into position $i$, *after* the names that are in alphabetical positions 1 to $i - 1$ have been placed in their proper positions. The total number of interchanges is simply $\sum_{i=1}^{n} N_i$. But (once the first $i - 1$ names in the alphabet have been placed in their proper positions) the number of interchanges needed by the $i$th name in alphabetical order satisfies

$$P\{N_i = j\} = \frac{1}{n - i + 1}$$

for $j = 0, 1, \ldots, n - i$. Next, compute $E(N_i)$, using the formula

$$\sum_{j=1}^{k} j = \frac{k(k+1)}{2}$$

(which can be verified by induction), and then find $E\left(\sum_{i=1}^{n} N_i\right)$ using Corollary 4.3 on the expectation of a linear combination and the given formula for $\sum_{j=1}^{k} j$.

3.° Refer to the previous problem. Let $I(n) = \sum_{i=1}^{n} N_i$ be the number of interchanges required to put the list in order.

(a) Prove that

$$\text{Var}(I(n)) = \frac{(2n + 5)n(n - 1)}{72}.$$

Hints: Show that $N_1, \ldots, N_n$ are independent random variables. Use the formulae

$$\sum_{j=1}^{k} j = \frac{k(k + 1)}{2}, \qquad \sum_{j=1}^{k} j^2 = \frac{k(k + 1)(2k + 1)}{6}.$$

(b) If each interchange takes 1/10,000 of a second on a computer, furnish a pair of values $L(n)$ and $U(n)$ that, with probability at least 8/9, are lower and upper bounds respectively for the total time it will take the computer to accomplish the alphabetization.
Hint: Use the Chebyshev inequality (3.14).

(c) Those with access to a digital computer might actually want to write a program to alphabetize by interchanges. Knowledge of the speed of the machine can then be used to compare the predictions of the type given in (b) with reality. (This problem arose from such a practical case. Its solution proved extremely useful in deciding whether such a sorting procedure was feasible. Furthermore, the solution was quite accurate in predicting the actual time spent by the computer.)

## SECTION 5
## Correlation

We showed in Corollary 4.10 that if two *independent* random variables $X$ and $Y$ both have expectations, then $\text{cov}(X, Y) = 0$. This would lead us to believe that $\text{cov}(X, Y)$ measures the dependence of $X$ and $Y$ in some sense, and thus is worth investigating. To do so, we introduce here a "standardized" covariance called the correlation coefficient and study some of its properties.

### 5.1 *Definition: Correlation Coefficient*

Let $X$ and $Y$ be two random variables both defined on the same probability space and having finite non-zero variances $\sigma_x$ and $\sigma_y$. The quantity

$$\rho_{X,Y} = \text{cov}\left(\frac{X - E(X)}{\sigma_x}, \frac{Y - E(Y)}{\sigma_y}\right)$$

is called the *correlation coefficient*, or more properly, the *coefficient of linear correlation between X and Y.*

Note that

$$\frac{X - E(X)}{\sigma_x}$$

and

$$\frac{Y - E(Y)}{\sigma_y}$$

are the standardized versions of $X$ and $Y$, having expectation 0 and variance 1 (see Lemma 3.11). From this we see that

**5.2**
$$\rho_{X,Y} = \frac{E[(X - E(X))(Y - E(Y))]}{\sigma_x \sigma_y} = \frac{\text{cov}(X, Y)}{\sqrt{[\text{Var } X][\text{Var } Y]}}$$

The correlation coefficient is of most important practical use when we assume that the random vector $(X, Y)$ has a bivariate normal distribution (see Equation 3.5 in Chapter 5). We will investigate the theoretical basis for this important model in Theorem 3.21 of Chapter 7. The model is frequently used when $X$ and $Y$ represent related measurements on a given individual, such as height and weight, high school grades and college grades, or blood pressure and arterial elasticity.

### 5.3 *Theorem: Meaning of Bivariate Normal Parameters*

Suppose that $X$ and $Y$ have a bivariate normal distribution

$$f_{X,Y}(x, y) = \frac{1}{2\pi\sigma_x\sigma_y\sqrt{1 - \rho^2}} \exp \frac{-1}{2(1 - \rho^2)}\left[\left(\frac{x - \mu_x}{\sigma_x}\right)^2 - 2\rho\left(\frac{x - \mu_x}{\sigma_x}\right)\left(\frac{y - \mu_y}{\sigma_y}\right) + \left(\frac{y - \mu_y}{\sigma_y}\right)^2\right].$$

Then $\text{Var } X = \sigma_x^2$, $\text{Var } Y = \sigma_y^2$, $E(X) = \mu_x$, $E(Y) = \mu_y$, and the coefficient of linear correlation between $X$ and $Y$, $\rho_{X,Y} = \rho$. Hence, in the bivariate normal case, $\rho_{X,Y} = 0$ if and only if $X$ and $Y$ are independent. The student is asked for the proof in Problem 5.8(1).

In Example 7.9 of Chapter 5, we saw that the correlation coefficient helped us use the observed value of $X$ to improve our prediction for $Y$ in the bivariate normal case. We will now investigate the meaning of the correlation coefficient in a more general setting.

We have implied that $\rho_{X,Y}$ is a measure of the dependence between $X$ and $Y$, since $\rho_{X,Y} = 0$ when $X$ and $Y$ are independent. It would appear from the bivariate normal case that the further $\rho_{X,Y}$ departs from 0, the more dependent $X$ and $Y$ become. Actually, the matter is not nearly this simple, since it is possible to have $\rho_{X,Y} = 0$ even when $Y$ is completely determined by $X$ (see Exercise 5.7(1)). You will be better able to understand this apparent paradox when we treat the vector Central Limit Theorem in the next chapter. However, some of the universal meaning of $\rho_{X,Y}$ is furnished by the theorems that follow.

### 5.4 *Theorem: |Correlation Coefficient| ≤ 1*

If $\rho$ is the coefficient of linear correlation between the random variables $X$ and $Y$, then $|\rho| \leq 1$. Furthermore, $|\rho| = 1$ if and only if for some real numbers $\alpha \neq 0$ and $\beta$ we have $X + \alpha Y = \beta$. (This explains the reason for calling $\rho$ the coefficient of *linear* correlation.)

*Proof*: For each real number $\alpha$, we know that $\text{Var}(Y + \alpha X) \geq 0$. Using Theorem 4.8 on the variance of a linear combination, we see that this inequality is equivalent to

$$(\text{Var } X)\alpha^2 + [2 \text{ cov}(X, Y)]\alpha + \text{Var } Y \geq 0,$$

that is,

$$\alpha^2 + 2\frac{\text{cov}(X, Y)}{\text{Var } X}\alpha + \frac{\text{Var } Y}{\text{Var } X} \geq 0$$

(Var $X$ is non-zero from the definition of correlation coefficient). Using completion of the square, we write

$$\alpha^2 + 2\frac{\text{cov}(X, Y)}{\text{Var } X}\alpha + \left[\frac{\text{cov}(X, Y)}{\text{Var } X}\right]^2 + \frac{\text{Var } Y}{\text{Var } X} - \left[\frac{\text{cov}(X, Y)}{\text{Var } X}\right]^2 \geq 0$$

or

$$\left[\alpha + \frac{\text{cov}(X, Y)}{\text{Var } X}\right]^2 + \frac{\text{Var } Y}{\text{Var } X} - \left[\frac{\text{cov}(X, Y)}{\text{Var } X}\right]^2 \geq 0$$

for all real $\alpha$. From this it follows that the sum of the last terms must be non-negative, since we are allowed to choose

$$\alpha = \frac{-\text{cov}(X, Y)}{\text{Var } X}.$$

Thus

$$\frac{\text{Var } Y}{\text{Var } X} - \left[\frac{\text{cov}(X, Y)}{\text{Var } X}\right]^2 \geq 0.$$

Dividing by

$$\frac{\text{Var } Y}{\text{Var } X}$$

and rearranging yields

$$\left[\frac{\text{cov}(X, Y)}{\sqrt{(\text{Var } X)(\text{Var } Y)}}\right]^2 \leq 1,$$

from which it follows that $|\rho| \leq 1$. Now $\text{Var}(Y + \alpha X) = 0$ if and only if $Y + \alpha X$ is constant, that is, for some $\beta$, $Y + \alpha X = \beta$. Also, $\alpha$ cannot be 0 in this case, since this would imply that Var $Y = 0$. But if we examine the earlier part of the proof, we also see that $\text{Var}(Y + \alpha X) = 0$ if and only if $|\rho| = 1$. Thus, $|\rho| = 1$ if and only if $Y + \alpha X = \beta$ for some $\beta$ and $\alpha \neq 0$, concluding the proof.

In view of this result, we might expect $X$ and $Y$ to be "almost" linearly related if $|\rho|$ is close to 1. We will explore this in Problem 5.8(2). This will give an operational meaning for values of $|\rho|$ near 1.

The following theorem helps clarify the role of $\rho_{X,Y}$ as a measure of dependence.

### 5.5 *Theorem: Proportion of Variability of X "due" to Y*

Suppose the coefficient of linear correlation between $X$ and $Y$ is $\rho$. Then

$$\min_{\alpha} \text{Var}(Y + \alpha X) = (1 - \rho^2)\text{Var } X$$

where $\min_{\alpha} \text{Var}(Y + \alpha X)$ is the smallest value of $g(\alpha) = \text{Var}(Y + \alpha X)$.

*Proof:*

$$\begin{aligned}
\operatorname{Var}(Y+\alpha X) &= \operatorname{Var} Y + 2\alpha \operatorname{cov}(X, Y) + \alpha^2 \operatorname{Var} X \\
&= (\operatorname{Var} X)\left[\frac{\operatorname{Var} Y}{\operatorname{Var} X} + 2\,\frac{\operatorname{cov}(X, Y)}{\operatorname{Var} X}\,\alpha + \alpha^2\right] \\
&= (\operatorname{Var} X)\left[\alpha^2 + 2\,\frac{\operatorname{cov}(X, Y)}{\operatorname{Var} X}\,\alpha + \left(\frac{\operatorname{cov}(X, Y)}{\operatorname{Var} X}\right)^2 \right. \\
&\qquad \left. + \frac{\operatorname{Var} Y}{\operatorname{Var} X} - \left(\frac{\operatorname{cov}(X, Y)}{\operatorname{Var} X}\right)^2\right]
\end{aligned}$$

completing the square

$$= (\operatorname{Var} X)\left[\left(\alpha + \frac{\operatorname{cov}(X, Y)}{\operatorname{Var} X}\right)^2 + \frac{\operatorname{Var} Y}{\operatorname{Var} X} - \left(\frac{\operatorname{cov}(X, Y)}{\operatorname{Var} X}\right)^2\right].$$

Letting

$$\alpha = \frac{-\operatorname{cov}(X, Y)}{\operatorname{Var} Y} = -\rho\sqrt{\frac{\operatorname{Var} Y}{\operatorname{Var} X}},$$

we have

$$\begin{aligned}
\min_{\alpha} \operatorname{Var}(Y+\alpha X) &= \operatorname{Var} Y - \frac{[\operatorname{cov}(X, Y)]^2}{\operatorname{Var} X} \\
&= \operatorname{Var} Y - \frac{[\rho^2(\operatorname{Var} X)(\operatorname{Var} Y)]}{\operatorname{Var} X} \\
&= (1-\rho^2)\operatorname{Var} Y,
\end{aligned}$$

concluding the proof.

### 5.6 *Remarks on Variance Reduction via Correlation*

We may think of $X$ as accounting for a proportion $\rho^2$ of the variance of $Y$, or being usable to reduce the variance of $Y$ by a factor of $1 - \rho^2$. Such a variance reduction might be useful in the following situation. Suppose $Y$ represents actual error from the desired impact point of a space vehicle to the moon, assuming no correction is made on the initial trajectory of the vehicle. Unfortunately, at best $Y$ is not observable for quite a while. (We consider $Y$ to be random because of such factors as uneven burning of fuel, obstacles, imperfect settings of controls, and so on.) However, an early measurement of $X$ "on" $Y$ is available. We assume that $E(Y) = 0$ (that is, on the average you'll land where you want) and that $E(X) = 0$ (that is, our measurement of impact error is not biased in either the positive or negative direction). If you are able to apply a specified "correction" to the final impact point, then it seems reasonable to apply a correction of $\alpha X$ (where $\alpha = -\rho\sqrt{\operatorname{Var} Y/\operatorname{Var} X}$) so that the "corrected" impact point error is now $Y + \alpha X$. Since $\operatorname{Var}(Y + \alpha X) = (1 - \rho^2)\operatorname{Var} Y$, the error variance is now reduced by a factor of $1 - \rho^2$, giving a less variable trajectory that still is "on the average" on course because $E(Y + \alpha X) = 0$. The Chebyshev inequality (3.14) shows that such variance reduction is indeed operationally useful rather than being simply intuitively appealing.

## 5.7 EXERCISES

1. Show that the random variable $Y$ may actually be a function of $X$, and yet their coefficient $\rho_{X,Y}$ of linear correlation may be 0.
Hint: Look at $\text{cov}(X, Y)$ where $Y = X^2$ and $X$ has a standard normal density ($E(X) = 0$, $\text{Var } X = 1$).

2.° Let $X$ be uniformly distributed on $[-.5, .5]$, and let $Y = X^3$. Find the correlation coefficient between $X$ and $Y$.

3. Suppose that $X$ and $Y$ are independent, $E(X) = 0 = E(Y)$, $\text{Var } X = 1 = \text{Var } Y$. Find $\rho_{X,X+Y}$.

## 5.8 PROBLEMS

1. Prove Theorem 5.3 (the meaning of the bivariate normal parameters).

2. Show that if $|\rho_{X,Y}|$ is near 1, then $X$ and $Y$ very likely are "close" to being linearly related. More precisely, prove that if $E(X) = \mu_x$, $E(Y) = \mu_y$, $\text{Var } X = \sigma_x^2$, $\text{Var } Y = \sigma_y^2$, then for any given (large) $t > 0$, for $|\rho_{X,Y}|$ sufficiently close to 1, say $\rho_{X,Y}^2 = 1 - \varepsilon^2$, $\varepsilon > 0$ close to 0,

$$P\{(X, Y) \in B\} \geq 1 - 1/t^2$$

where

$$B = \left\{(x, y) \in \mathscr{R}_2 : \left| y - \rho_{X,Y} \frac{\sigma_y}{\sigma_x} x - \mu_y + \rho_{X,Y} \mu_x \frac{\sigma_y}{\sigma_x} \right| \leq t \, \varepsilon \, \sigma_y \right\}.$$

In our sketch, $B$ is the shaded region. For fixed $t$, $\mu_x$, $\mu_y$, $\sigma_x$, $\sigma_y$ as $\varepsilon$ goes to 0, the region $B$ squeezes down to the "line"

$$y = \frac{\sigma_y}{\sigma_x} x + \mu_y - \mu_x \frac{\sigma_y}{\sigma_x}.$$

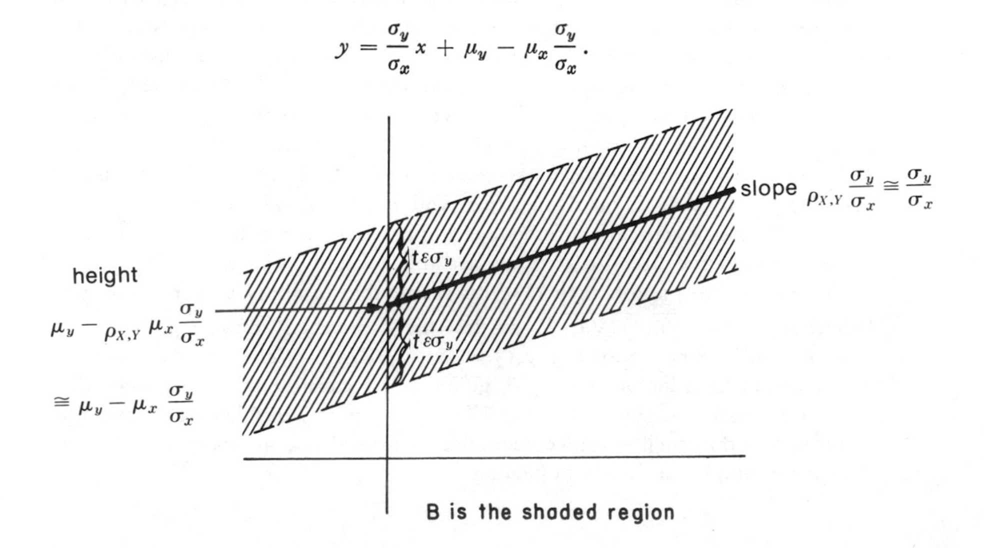

B is the shaded region

So for $\mu_x, \mu_y, \sigma_x, \sigma_y$ fixed and $t$ fixed and large for $|\rho_{X,Y}|$ close to 1, we see that $(X, Y)$ has a high probability of lying in $B$, that is, near the line

$$y = \rho_{X,Y} \frac{\sigma_y}{\sigma_x} x + \mu_y - \rho_{X,Y} \mu_x \frac{\sigma_y}{\sigma_x}.$$

**Outline.** First note that $\rho_{X,Y} = \rho_{X^*,Y^*}$ where

$$X^* = \frac{X - E(Y)}{\sigma_x}$$

and

$$Y^* = \frac{Y - E(Y)}{\sigma_y}.$$

So let us first work with $X^*$ and $Y^*$, noting that these variables have mean 0 and unit variance. Now suppose that $\rho^2_{X^*,Y^*} = \rho^2 = 1 - \varepsilon^2$ where $\varepsilon^2$ is small. We know from the proof of Theorem 5.5 that

$$\mathrm{Var}(Y^* - \rho X^*) = 1 - \rho^2 = \varepsilon^2.$$

Now apply the Chebyshev inequality (3.14) to $Y^* - \rho X^*$ to obtain

$$P\{|Y^* - \rho X^*| \le t\varepsilon\} \ge 1 - \frac{1}{t^2},$$

that is,

$$P\left\{\left|\frac{Y - \mu_y}{\sigma_y} - \rho \frac{X - \mu_x}{\sigma_x}\right| \le t\varepsilon\right\} \ge 1 - \frac{1}{t^2}$$

or

$$P\{(X, Y) \subset B\} \ge 1 - \frac{1}{t^2}$$

where

$$B = \left\{(x, y) \subset \mathscr{R}_2 : \left|\frac{y - \mu_y}{\sigma_y} - \rho \frac{x - \mu_x}{\sigma_x}\right| \le t\varepsilon\right\}.$$

## SECTION 6
## Error Reduction, the Weak Law

In this section, we derive some fundamental results about the behavior of sums and averages of independent random variables. Our derivations depend mainly on the Chebyshev inequality and the laws governing expectations and variances of sums of random variables.

### 6.1 *Theorem: Deviation of* $\frac{1}{n}\sum_{i=1}^{n} X_i$ *from* $\mu$

If $X_1, X_2, \ldots, X_n$ are independent random variables with $E(X_i) = \mu$ and $\mathrm{Var}\, X_i = \sigma^2$ for all $i$, then

$$P\left\{\left|\frac{1}{n}\sum_{i=1}^{n} X_i - \mu\right| \ge t\frac{\sigma}{\sqrt{n}}\right\} \le \frac{1}{t^2}.$$

*Proof:* The fact that

$$E\left[\frac{1}{n}\sum_{i=1}^{n} X_i\right] = \mu$$

was proved in Corollary 4.3. In Corollary 4.12, we saw that

$$\operatorname{Var}\left(\frac{1}{n}\sum_{i=1}^{n} X_i\right) = \frac{\sigma^2}{n}.$$

Applying the Chebyshev inequality (3.14) to the event

$$\left\{\left|\frac{1}{n}\sum_{i=1}^{n} X_i - \mu\right| \geq t\frac{\sigma}{\sqrt{n}}\right\}$$

yields our desired result.

*We see that in a sense the average*

$$\frac{1}{n}\sum_{i=1}^{n} X_i$$

*approaches its expectation at a rate of at least the order of magnitude of* $1/\sqrt{n}$.

## 6.2 *Example: Comparison of Growth of Systematic and Random Error*

If we have a scale for weighing objects, then the reading of this scale is the value of a random variable $X$. We assume that if the object being weighed really has weight $\mu$, then $E(X) = \mu + \delta$ and $\operatorname{Var} X = \sigma^2$. We refer to $\delta$ as the *bias* or *systematic error* of the scale. Now suppose that we independently weigh $n$ objects, having true weights $\mu_1, \ldots, \mu_n$ respectively. Denote our measurements by $X_1, \ldots, X_n$. The error on the $i$th measurement is $X_i - \mu_i$. Let $e_n$ be the total error accumulated in the $n$ weighings. Then $e_n$ is given by

$$e_n = (X_1 - \mu_1) + (X_2 - \mu_2) + \cdots + (X_n - \mu_n).$$

From Corollary 4.3 and Theorem 4.11, we find that

$$E(e_n) = n\delta, \qquad \operatorname{Var}(e_n) = n\sigma^2.$$

If $\delta = 0$, we say that the scale is *unbiased*, and refer to the total accumulated error $e_n$ as *purely random*. In the case of an unbiased scale, we see from the Chebyshev inequality that

$$P\{|e_n| \geq t\sqrt{n}\,\sigma\} \leq 1/t^2.$$

Thus, in a sense, the purely random accumulated error tends to grow no faster with $n$ than a multiple of $\sqrt{n}$.

On the other hand, if the systematic error $\delta$ is not zero, then the accumulated error $e_n$ would be about $n\delta$. Notice that the total accumulated error does not seem to grow nearly as fast on an unbiased scale as it does when there is a non-zero systematic error present. △

Here is a practical use of Chebyshev's inequality.

### 6.3 *Example: Effect of Roundoff*

A large business firm wants a quick estimate of total dollar sales in a day. To save time, the amount of each sale is rounded to the nearest dollar. The model chosen to describe this situation is that costs from sale to sale are independent and (for simplicity) that the roundoff error has a uniform density on [—.5, .5] (dollars). Let us determine an upper bound for the probability that the error in totaling 10,000 sales exceeds \$500.00. Note that the maximum possible error is \$5000.00. We are investigating

$$(*) \qquad P\left\{\left|\sum_{i=1}^{10{,}000} X_i\right| \geq 500\right\}$$

where $X_i$ is the roundoff error associated with the $i$th sale. We see that

$$E(X_i) = \int_{-.5}^{.5} x\,dx = 0 \quad \text{and} \quad \operatorname{Var} X_i = E(X_i^2) = \int_{-.5}^{.5} x^2\,dx = \left.\frac{x^3}{3}\right|_{-.5}^{.5} = \frac{1}{12}.$$

Hence, from Corollary 4.3 we have $E\left(\sum_{i=1}^{10{,}000} X_i\right) = 0$, and from Theorem 4.11 we have

$$\operatorname{Var}\left(\sum_{i=1}^{10{,}000} X_i\right) = \frac{10{,}000}{12}.$$

In order to use the Chebyshev inequality (3.14) on (*), we identify

$$500 \text{ with } t\sigma \text{ and } \sqrt{\frac{10{,}000}{12}} \text{ with } \sigma,$$

so that

$$500 = t\sqrt{\frac{10{,}000}{12}} = t\sigma.$$

We find that $t = 5\sqrt{12}$. Hence

$$P\left\{\left|\sum_{i=1}^{10{,}000} X_i\right| \geq 500\right\} \leq \frac{1}{t^2} = \frac{1}{300}.$$

That is, there is less than one chance in 300 that the total error (which could conceivably be \$5000) will exceed \$500.00. Although this result may surprise some, it is actually very conservative. The true probability is more surprising, as we shall see in the next chapter. △

Note that if $E(X_i) = 0$, $\text{Var } X_i = \sigma^2$ for $i = 1, 2, \ldots$ then for large $n$, $\left|\sum_{i=1}^{n} X_i\right|$ is not likely to exceed some multiple of $\sqrt{n}$. This follows from the Chebyshev inequality, since

$$P\left\{\left|\sum_{i=1}^{n} X_i\right| \leq t\sqrt{n}\,\sigma\right\} \geq 1 - \frac{1}{t^2}.$$

## 6.4 *Example: Approach of Proportion to Probability*

A fair coin is tossed independently $n$ times. If we want 95 per cent assurance that the proportion of heads will be within .01 of the probability (.5) of heads, how large should $n$ be? Here we let $Y_n$ be the proportion of heads in $n$ trials;

$$Y_n = \frac{1}{n}\sum_{i=1}^{n} X_i$$

where $X_i$ is the number of heads on the $i$th trial, $P\{X_i = 1\} = 1/2 = P\{X_i = 0\}$. Then we know from 2.5 and 3.9 that $E(X_i) = 1/2$, $\text{Var } X_i = 1/4$. Therefore, by Corollary 4.3 and Theorem 4.11,

$$E(Y_n) = \frac{1}{2}, \quad \text{Var } Y_n = \frac{1}{4n},$$

and thus $\sqrt{\text{Var } Y_n} = 1/(2\sqrt{n})$. We are asked to determine $n$ so that $P\{|Y_n - .5| \leq .01\} \geq .95$, that is, so that $P\{|Y_n - .5| > .01\} \leq .05$. If we now identify .05 with $1/t^2$ of the Chebyshev inequality 3.14 and .01 with $t\sigma = t/(2\sqrt{n})$ of 3.14, we find $t^2 = 20$ and $\sqrt{20}/(2\sqrt{n}) = .01$. Solving this equation, we find $n = 50{,}000$. That is, *if we toss a fair coin* 50,000 *times, then the probability is at least* .95 *that the proportion of heads will be within* .01 *of* 1/2.

In general, if $Y_n$ is the proportion of occurrences of some given event $A$ in $n$ independent repetitions of some experiment (that is, the proportion of experiments in which the outcome belongs to $A$) and $P(A) = p$, then by arguments similar to those preceding (where $X_i$ is the number of occurrences of $A$ on the $i$th trial), we find that

$$P\{|Y_n - p| \leq t\sqrt{p(1-p)/n}\} \geq 1 - \frac{1}{t^2}.$$

In essence, this says that *in some sense the proportion of occurrences of an event $A$ in $n$ independent trials approaches its probability $p = P(A)$* at about a rate of at least $1/\sqrt{n}$. △

We conclude this section with two important theoretical consequences of the Chebyshev inequality.

### 6.5 *Theorem: Weak Law of Large Numbers*

Let $X_1, X_2, \ldots$ be a sequence of independent random variables with $E(X_i) = \mu$ and $\operatorname{Var} X_i = \sigma^2$. Then for each $\varepsilon > 0$,

$$\lim_{n\to\infty} P\left\{\left|\frac{1}{n}\sum_{i=1}^{n} X_i - \mu\right| < \varepsilon\right\} = 1.$$

*Proof:* If we set $t\sigma/\sqrt{n} = \varepsilon$, or equivalently $1/t^2 = \sigma^2/(\varepsilon^2 n)$, in Theorem 6.1, then we have

$$P\left\{\left|\frac{1}{n}\sum_{i=1}^{n} X_i - \mu\right| \geq \varepsilon\right\} \leq \frac{\sigma^2}{\varepsilon^2 n}$$

or

$$P\left\{\left|\frac{1}{n}\sum_{i=1}^{n} X_i - \mu\right| < \varepsilon\right\} \geq 1 - \frac{\sigma^2}{\varepsilon^2 n}.$$

The result now follows by letting $n \to \infty$.

The conclusion of this theorem can be shown, by advanced methods, to hold if the existence of $\sigma^2$ is replaced by the condition that all $X_i$ have the same distribution with finite expectation.

**6.6 Definition: Sample Moments.** If $X_1, X_2, \ldots, X_n$ are random variables with a common distribution, then

$$\frac{1}{n}\sum_{i=1}^{n} X_i^k$$

is called the k*th sample moment* based on $X_1, \ldots, X_n$.

### 6.7 *Corollary to 6.5: Convergence of Sample Moments*

If $Y_1, Y_2, \ldots$ are independent random variables with a common distribution and for which $\nu_k = E(Y_i^k)$ and $\nu_{2k} = E(Y_i^{2k})$ are finite, then

$$\lim_{n\to\infty} P\left\{\left|\frac{1}{n}\sum_{i=1}^{n} Y_i^k - \nu_k\right| < \varepsilon\right\} = 1 \qquad \text{for each} \quad \varepsilon > 0.$$

*Proof:* Let $X_i = Y_i^k$ in Theorem 6.5 and notice that

$$\begin{aligned}\operatorname{Var} X_i &= E(X_i^2) - (E[X_i])^2 \qquad \text{by 3.8}\\ &= \nu_{2k} - \nu_k^2.\end{aligned}$$

## 6.8 EXERCISES

1. Let $T_i$, $1 \leq i \leq 9$, be independent random variables representing the lifetimes of high performance transistors having an expected lifetime of 5 years and a standard deviation of .4 years.

(a)° Find a lower bound for the probability that their average observed lifetime

$\frac{1}{9}\sum_{i=1} T_i$ lies in the interval [3, 7].

(b) If $T_1, T_2, \ldots$ are the lifetimes of independent transistors, find a lower bound for

$$P\left\{\frac{1}{n}\sum_{i=1}^{n} T_i \in [3, 7]\right\}.$$

2. Suppose that your electric bills $B_1, B_2, \ldots$ for months $1, 2, \ldots$ are independent random variables with $E(B_i) = \$16$, and $\sqrt{\text{Var } B_i} = \$1$. You want to plan ahead for the next year. To do this, find an amount $a$ dollars such that

(a)° $P\left\{\sum_{i=1}^{12} B_i \leq a\right\} \geq .9$ (b) $P\left\{\sum_{i=1}^{12} B_i \leq a\right\} \geq .99.$

3. Find a simple lower bound for the probability:

(a)° that the number of even outcomes in 10,000 tosses of a pair of dice lies in the interval [4600, 5400].

(b) that the number of snake-eyes—(1, 1)—in 36,000 tosses of a pair of dice lies in the interval [950, 1050].

4.° Suppose your watch gains or loses at most 30 seconds per day, but is properly calibrated (that has no average bias or systematic error; see Example 6.2). Make some reasonable assumptions to determine an upper bound for the probability that in a one-year period your watch is off by more than half an hour.

## 6.9 PROBLEM

If in Example 6.3 we had assumed a discrete uniform distribution on the 100 points $-.495, -.485, \ldots, -.005, .005, \ldots, .485, .495$, by how much could our result have been changed? Here are three methods you should try on this problem.

(a)° Do the computation directly, using the formula

$$\sum_{j=1}^{n} j^2 = \frac{n(n+1)(2n+1)}{6}.$$

(b)° Use the fact that if $X^*$ is the random variable obtained from $X$ by roundoff to the nearest discrete point of the sample space of the discrete uniform model, then $X^*$ is governed by this discrete uniform model and

$$(*) \qquad \left|\sum_{i=1}^{10,000} X_i = \sum_{i=1}^{10,000} X_i^*\right| \leq 10,000 \times .005 = 50.$$

Compute an upper bound for $P\left\{\left|\sum_{i=1}^{10,000} X_i\right| \geq 450\right\}$ (as we did for $P\{|\sum X_i| \geq 500\}$) and use it with (*) to compute the desired upper bound for $P\{|\sum X_i^*| \geq 500\}$.

(c) Alternatively, defining $X_i^*$ as in (b), use the fact that

$$\operatorname{Var} \sum_{i=1}^{10,000} (X_i - X_*^i) \leq 10,000 \times .000025$$

and hence that $\sqrt{\operatorname{Var} \sum_{i=1}^{10,000} (X_i - X_i^*)} \leq 100 \times .005 = .5$ to compute, via the Chebyshev inequality, an upper bound for

$$(**) \quad P\left\{\left|\sum_{i=1}^{10,000} X_i - \sum_{i=1}^{10,000} X_i^*\right| \geq 25\right\} = P\left\{\left|\sum_{i=1}^{10,000} (X_i - X_i^*)\right| \geq 25\right\}$$

(the choice of 25 is arbitrary). Now compute an upper bound for $P\left\{\left|\sum_{i=1}^{10,000} X_i\right| \geq 475\right\}$ (as we did for $P\{|\sum X_i| \geq 500\}$) and combine it with the upper bound for (**) to obtain an upper bound for $P\{|\sum X_i^*| \geq 500\}$.

## SECTION 7
## Moment Generating Functions

The problem of computing moments comes up often enough to make it worthwhile to find efficient methods of doing this. The tool that we introduce at this point has extensive applications in addition to facilitating moment computations.

**7.1 Definition: Moment Generating Function.** The function $\varphi_X$ defined for all $t$ at which

$$\varphi_X(t) \equiv E(e^{tX})$$

exists is called the *moment generating function of the random variable X*. We shall abbreviate this to m.g.f.

The proofs of many of the important properties of moment generating functions are beyond the level of this text. Nonetheless, we shall make use of all such needed results after stating them carefully and presenting arguments to show their plausibility. Before investigating the reason for calling $\varphi_X$ the moment generating function and investigating its practical uses, we give a few examples.

### 7.2 *Example: Poisson m.g.f.*

If

$$P\{X = k\} = \frac{e^{-\lambda}\lambda^k}{k!} \qquad \text{for} \qquad k = 0, 1, 2, \ldots$$

then (by Theorem 3.2 on evaluating $E[g(X)]$, with $g(X) = e^{Xt}$), we see that

$$\begin{aligned}
\varphi_X(t) = E(e^{Xt}) &= \sum_{k=0}^{\infty} e^{kt} e^{-\lambda} \frac{\lambda^k}{k!} \\
&= e^{-\lambda} \sum_{k=0}^{\infty} (\lambda e^t)^k / k! \\
&= e^{-\lambda} e^{\lambda e^t} \qquad \left(\text{using the expansion } e^y = \sum_{k=0}^{\infty} \frac{y^k}{k!}\right) \\
&= e^{\lambda(e^t - 1)}.
\end{aligned}$$

*That is, if $X$ is Poisson with $E(X) = \lambda$, then $\varphi_X(t) = e^{\lambda(e^t-1)}$ for all $t$.* △

### 7.3 *Example: Uniform m.g.f.*

$$\text{If } f_X(x) = \begin{cases} 1 & \text{for} \quad -1/2 \le x \le 1/2 \\ 0 & \text{otherwise,} \end{cases}$$

then for $t \neq 0$

$$\varphi_X(t) = E(e^{tX}) = \int_{-1/2}^{1/2} e^{tx}\, dx = \frac{1}{t}\,[e^{t/2} - e^{-t/2}]$$

$$\varphi_X(0) = E(e^0) = 1.$$

Hence

$$\varphi_X(t) = \begin{cases} \dfrac{1}{t}\,[e^{t/2} - e^{-t/2}] & \text{for} \quad t \neq 0 \\ 1 & \text{for} \quad t = 0. \end{cases}$$ △

Now let us see how $\varphi_X$ can be used to generate the moments of $X$. Using the well-known expansion

$$e^y = 1 + y + \frac{y^2}{2!} + \frac{y^3}{3!} + \cdots$$

we have

$$\varphi_X(t) = E(e^{tX}) = E\left(1 + tX + \frac{t^2X^2}{2!} + \frac{t^3X^3}{3!} + \cdots\right).$$

Assuming a generalized (infinite series) version of $E(aY + bZ) = aE(Y) + bE(Z)$, we may write

$$\varphi_X(t) = 1 + E(X)t + \frac{E(X^2)}{2!}\,t^2 + \frac{E(X^3)}{3!}\,t^3 + \cdots.$$

From this it would follow that if we could find a power series representation of $\varphi_X$, namely

$$\varphi_X(t) = a_0 + a_1 t + a_2 t^2 + \cdots,$$

then

$$\begin{aligned} a_0 &= 1 \\ a_1 &= E(X) \\ 2!\, a_2 &= E(X^2) \\ &\cdot \\ &\cdot \\ &\cdot \end{aligned}$$

In fact, assuming that the derivative of the sum is the sum of the derivatives, then

$$\begin{aligned} \varphi_X(0) &= a_0 \\ \varphi'_X(0) &= a_1 \\ \varphi''_X(0) &= 2a_2, \text{ and so on,} \end{aligned}$$

so that

$$\begin{aligned} E(X) &= \varphi'_X(0) \\ E(X^2) &= \varphi''_X(0) \\ E(X^3) &= \varphi'''_X(0), \text{ and so on.} \end{aligned}$$

These results actually hold quite generally. Although we are unable to prove them, we state them in the following theorem.

### 7.4 *Theorem: Moments and the Moment Generating Function*

Suppose that $X$ is a random variable whose moment generating function $\varphi_X$ is defined in some interval $|t| \leq t_0$ with $t_0 > 0$. Then

(i) All moments of $X$ exist.

(ii) $\varphi_X(t)$ has a unique power series representation in $(-t_0, t_0)$ given by

$$\varphi_X(t) = 1 + E(X)t + \frac{E(X^2)}{2!}t^2 + \frac{E(X^3)}{3!}t^3 + \cdots.$$

(iii) For each counting number $k$,

$$E(X^k) = \varphi_X^{(k)}(0).$$

### 7.5 *Example: Moments of a Poisson Variable*

From Example 7.2, we have found that when

$$P\{X = k\} = e^{-\lambda}\frac{\lambda^k}{k!}, \qquad k = 0, 1, 2, \ldots,$$

then $\varphi_X(t) = e^{\lambda(e^t-1)}$. Then

$$\varphi'_X(t) = \lambda e^t e^{\lambda(e^t-1)}, \qquad \varphi'_X(0) = \lambda$$

$$\varphi''_X(t) = \lambda e^t e^{\lambda(e^t-1)} + \lambda^2 e^{2t} e^{\lambda(e^t-1)}.$$

Hence $\varphi''_x(0) = \lambda + \lambda^2$. Thus, by Theorem 7.4,

$$E(X) = \lambda, \qquad E(X^2) = \lambda + \lambda^2,$$

and hence (by Lemma 3.8) Var $X = \lambda$. △

### 7.6 *Example: Moments of a Standard Normal Variable*

If

$$f_Y(y) = \frac{1}{\sqrt{2\pi}}e^{-y^2/2},$$

then

$$\varphi_Y(t) = \int_{-\infty}^{\infty} e^{ty}\frac{1}{\sqrt{2\pi}}e^{-y^2/2}\,dy$$

$$= \int_{-\infty}^{\infty}\frac{1}{\sqrt{2\pi}}e^{-1/2[y^2-2ty+t^2-t^2]}\,dy$$

$$= \int_{-\infty}^{\infty}\frac{1}{\sqrt{2\pi}}e^{-1/2(y-t)^2}e^{t^2/2}\,dy$$

$$= e^{t^2/2}\int_{-\infty}^{\infty}\frac{1}{\sqrt{2\pi}}e^{-1/2(y-t)^2}\,dy$$

$$= e^{t^2/2},$$

since the integrand is the normal density with mean $t$ and variance 1 (see Example 2.18, Chapter 4). Thus

**7.7** $$\varphi_Y(t) = e^{t^2/2}.$$

Hence

$$\begin{aligned}\varphi_Y(t) &= 1 + \frac{t^2}{2} = \frac{(t^2/2)^2}{2!} + \frac{(t^2/2)^3}{3!} + \cdots \\ &= 1 + \frac{1}{2}t^2 + \frac{1}{2^2 \times 2!}(t^2)^2 + \frac{1}{2^3 \times 3!}(t^2)^3 + \cdots \\ &= \sum_{n=0}^{\infty} a_n t^n\end{aligned}$$

where

$$a_n = \begin{cases} 0 & \text{if } n \text{ is odd} \\ \dfrac{1}{2^{n/2}\left(\dfrac{n}{2}\right)!} & \text{if } n \text{ is even.}\end{cases}$$

But, from Theorem 7.4,

$$\varphi_Y(t) = \sum_{n=0}^{\infty} \frac{E(Y^n)}{n!} t^n.$$

Hence

$$\frac{E(Y^n)}{n!} = \begin{cases} 0 & \text{if } n \text{ is odd} \\ \dfrac{1}{2^{n/2}\left(\dfrac{n}{2}\right)!} & \text{if } n \text{ is even.}\end{cases}$$

Thus, we find that if $Y$ is a standard normal random variable, then

$$E(Y^n) = \begin{cases} 0 & \text{if } n \text{ is odd} \\ \dfrac{n!}{2^{n/2}\left(\dfrac{n}{2}\right)!} & \text{if } n \text{ is even.}\end{cases}$$

Note that it would be difficult to determine $E(Y^n)$ by direct computation of $\varphi_Y^{(n)}(0)$. △

The following theorem is an extremely important result that is plausible (although we cannot prove it here), in view of Remark 3.6, about moments determining distributions.

### *7.8 Theorem: m.g.f. and Distribution*

If $X$ has moment generating function $\varphi_X$ which is defined for all $|t| \leq t_0$, $t_0 > 0$, then $\varphi_X$ completely determines the distribution $P_X$.

In particular, if we recognize from a previous computation that

$$\varphi_X(t) = \int_{-\infty}^{\infty} e^{tx} f(x)\, dx$$

for $|t| \leq t_0$, then $X$ must have density $f$. If $\varphi_X(t) = \sum_k e^{tk} p_k$ for $|t| \leq t_0$, then $X$ must have discrete probability function $\{p_k\}$.

The significance of 7.8 lies in the fact that certain results concerning probability distributions are much easier to establish by means of their generating functions. This is especially true in regard to sums of independent random variables, which we will discuss in the next chapter.

### 7.9 *Example: Poisson m.g.f.*

If the m.g.f. $\varphi_X(t) = e^{\lambda(e^t - 1)}$, then from Example 7.2 and Theorem 7.8 we know that $X$ is Poisson with mean $\lambda$, that is,

$$P\{X = k\} = e^{-\lambda} \frac{\lambda^k}{k!} \qquad \text{for} \quad k = 0, 1, 2, \ldots. \qquad \triangle$$

---

### *7.10 EXERCISES*

1.° Find the m.g.f. $\varphi_X$ when $P\{X = 1\} = p = 1 - P\{X = 0\}$.

2. (a) Find the m.g.f. $\varphi_X$ when $P\{X = k\} = \binom{n}{k} p^k (1 - p)^{n-k}$ for $k = 0, 1, \ldots, n$.

Hint: Use the binomial theorem.

(b) Find $E(X)$ and Var $X$.

3. (a) Find the m.g.f. $\varphi_X$ when $X$ is uniformly distributed on $[a, b]$.
   (b) Find $E(X^n)$ from $\varphi_X$ using Theorem 7.4(ii).

4.° (a) Find the value $\varphi_X(s)$ of the m.g.f. when

$$f_X(t) = \begin{cases} \dfrac{1}{\theta} e^{-t/\theta} & \text{for} \quad t \geq 0 \\ 0 & \text{otherwise.} \end{cases}$$

(b) Using Theorem 7.4(ii), find $E(X^n)$.

We asserted that m.g.f.'s are of great help in studying sums of independent random variables. The following theorem supports this assertion.

### 7.11 *Theorem: m.g.f. of Linear Combinations of Independent Random Variables*

Suppose that $X$ and $Y$ are independent random variables with m.g.f.'s $\varphi_X$ and $\varphi_Y$, respectively. Then for all real numbers $a$, $b$, and $c$, the random variable $aX + bY + c$ has a moment generating function $\varphi_{aX+bY+c}$ given by the relation

$$\varphi_{aX+bY+c}(t) = e^{ct} \varphi_X(at) \varphi_Y(bt).$$

*Proof:* We will prove only the density case. Here

$$\varphi_{aX+bY+c}(t) = E[e^{(aX+bY+c)t}]$$

$$= \int_{-\infty}^{\infty}\int_{-\infty}^{\infty} e^{(ax+by+c)t} f_{X,Y}(x, y)\, dx\, dy$$

(by Theorem 4.1)

$$= \int_{-\infty}^{\infty}\int_{-\infty}^{\infty} e^{(ax+by+c)t} f_X(x) f_Y(y)\, dx\, dy$$

(using the independence of $X$ and $Y$)

$$= e^{ct}\left[\int_{-\infty}^{\infty} e^{(at)x} f_X(x)\, dx\right]\left[\int_{-\infty}^{\infty} e^{(bt)y} f_Y(y)\, dy\right]$$

$$= e^{ct}\varphi_X(at)\varphi_Y(bt),$$

concluding the proof.

### 7.12 *Example: Sum of Independent Poisson Variables*

Recall that in Example 4.5 of Chapter 5 we showed directly that if $X$ and $Y$ are independent with

$$P\{X = k\} = \frac{e^{-\mu}\mu^k}{k!}, \qquad P\{Y = k\} = \frac{e^{-\nu}\nu^k}{k!}, \qquad k = 0, 1, \ldots,$$

then

$$P\{X + Y = k\} = \frac{e^{-(\mu+\nu)}(\mu + \nu)^k}{k!}.$$

We can prove this result with much less effort, using m.g.f.'s. By Example 7.2, $\varphi_X(t) = e^{\mu(e^t-1)}$, $\varphi_Y(t) = e^{\nu(e^t-1)}$. Hence, by Theorem 7.11

$$\varphi_{X+Y}(t) = e^{\mu(e^t-1)}e^{\nu(e^t-1)} = e^{(\mu+\nu)(e^t-1)}.$$

Again, using Example 7.2 we recognize this as the m.g.f. of a Poisson variable with mean $\mu + \nu$. Thus, by Theorem 7.8 (unique determination of a distribution by its m.g.f.), $X + Y$ must be Poisson with mean $\mu + \nu$. △

---

### *7.13 EXERCISES*

1.° Recall the result of Exercise 7.10(1), on the m.g.f. of a Bernoulli random variable, and the fact that if $X$ is a binomial random variable with parameters $n$ and $p$ (number of successes in $n$ independent trials, each with probability $p$ of success), then $X = Y_1 + \cdots + Y_n$ where $Y_1, \ldots, Y_n$ are independent with $P\{Y_i = 1\} = p = 1 - P\{Y_i = 0\}$ ($Y_i$ represents the number of successes on trial $i$). From these results, find the m.g.f. of $X$.

2.° Let

$$f_X(x) = \frac{1}{\sqrt{2\pi\sigma^2}} e^{-(x-\mu)^2/(2\sigma)^2}.$$

Find the m.g.f. of $X$.
Hint: In Example 2.25, Chapter 4, it was shown that $Y = (X - \mu)/\sigma$ has a standard normal density, but the m.g.f. of $Y$ was calculated in Example 7.6 (see 7.7). Now use the fact that $X = \sigma Y + \mu$ and Theorem 7.11 with $a = \sigma$, $b = 0$, $c = \mu$.

3. Let $f_X(x) = \dfrac{1}{\sqrt{2\pi}} e^{-x^2/2}$.

(a) Find $\varphi_{X^2}(t)$.
Hint: Use Theorem 3.2 on the evaluation of $E[g(X)]$.

(b) Find $E(X^4)$ and $E(X^6)$ from $\varphi_{X^2}$ and compare them to the answers found in Example 7.6.

(c) Let $X_1, X_2, \ldots, X_n$ be independent, all with density $f_X$ given previously. Find the m.g.f. of $X_1^2 + X_2^2 + \cdots + X_n^2$.

4.° Show that if $X$ and $Y$ are independent normal random variables with $E(X) = \mu_x$, $\operatorname{Var} X = \sigma_x^2$, $E(Y) = \mu_y$, and $\operatorname{Var} Y = \sigma_y^2$, then $Z = X + Y$ is normal with $E(Z) = \mu_x + \mu_y$, $\operatorname{Var} Z = \sigma_x^2 + \sigma_y^2$.
Hint: Use the result of Exercise 2. (This result was proved earlier in Example 4.6, Chapter 5.)

5. Let

$$f(z) = \begin{cases} z & \text{for} \quad 0 \le z \le 1 \\ 2 - z & \text{for} \quad 1 \le z \le 2 \\ 0 & \text{otherwise.} \end{cases}$$

Show that $f$ is the density of the sum of two independent random variables, each of which is uniformly distributed on $[0, 1]$.

6. Find a random variable $X$ such that $\varphi_X$ fails to exist except at $t = 0$.
Hint: If $\varphi_X$ exists in an interval $|t| \le t_0$ with $t_0 > 0$, then $X$ possesses all moments.

---

### *7.14 PROBLEM*

Let $X_1, \ldots, X_n, \ldots$ be independent and uniformly distributed on $[-\sqrt{3}, \sqrt{3}]$. Let

$$A_n = \frac{1}{\sqrt{n}}(X_1 + X_2 + \cdots + X_n).$$

Try to find $\lim_{n\to\infty} \varphi_{A_n}(t)$ by using the approximation

$$g(t) \simeq g(0) + g'(0)t + g''(0)\frac{t^2}{2} + g'''(0)\frac{t^3}{3!}$$

(the right side is a polynomial in $t$ whose first, second, and third derivatives at $t = 0$ agree with those of $g$), and the formula

$$\lim_{n\to\infty}\left(1+\frac{y}{n}\right)^n = e^y.$$

Does your result suggest anything interesting? (See 7.7.)

---

*7.15 PROBLEM*

Using the results of Exercise 7.13(2), prove the following result, which is important for reference purposes.

### 7.16 *Theorem: Distribution of Linear Combinations of Independent Normal Variables*

Let $X_1, \ldots, X_n$ be independent normal random variables with $E(X_i) = \mu$, $\operatorname{Var} X_i = \sigma^2$. Then:

(a) $\bar{X}_{(n)} \equiv \frac{1}{n}\sum_{i=1}^{n} X_i$

is normal with mean $\mu$ and variance $\sigma^2/n$.

(b) $P\{|\bar{X}_{(n)} - \mu| \le \delta\} = \int_{-\sqrt{n}\delta/\sigma}^{\sqrt{n}\delta/\sigma} \frac{1}{\sqrt{2\pi}} e^{-t^2/2}\, dt.$

(c) $\frac{1}{\sqrt{n}}\sum_{i=1}^{n} \frac{X_i - \mu}{\sigma}$

has a standard normal distribution.

(d) $\sum_{i=1}^{n} X_i$ is normal with mean $n\mu$ and variance $n\sigma^2$.

## SECTION 8
## Conditional Expectation*

The concept of conditional expectation bears the same relation to that of expectation as conditional probability does to probability. Namely, we may think of conditional expectation as expectation in a subpopulation. Hence, if $X$ is a random variable, and if information becomes available during the experiment

---

* This section may be regarded as optional.

that causes us to change the distribution of $X$ to a conditional one, this will also cause a change in the expectation of $X$. We give the definitions only for discrete and density cases.

**8.1 Definition:** $E(X|Y = y)$, $E(X|A)$. Let $X$ and $Y$ be random variables defined on the same probability space. If $X$ and $Y$ are discrete random variables, recall from Definition 6.1, Chapter 5 that

$$P\{X = x \mid Y = y\} = \frac{P\{X = x, Y = y\}}{P\{Y = y\}}$$

whenever the denominator is not zero. In this case, we define the *conditional expectation of X given Y* $= y$, written $E(X \mid Y = y)$, by

$$E(X \mid Y = y) = \sum_{\substack{\uparrow \\ \text{all values} \\ x \text{ of } X}} xP\{X = x \mid Y = y\}$$

provided $\sum_{\substack{\uparrow \\ \text{all values} \\ x \text{ of } X}} |x|\, P\{X = x \mid Y = y\}$ is finite. If $X$ and $Y$ have joint density $f_{X,Y}$, recall that when $f_Y(y) \neq 0$, the conditional density $f_{X.Y}(x \mid y)$ is given by

$$f_{X|Y}(x \mid y) = \frac{f_{X.Y}(x, y)}{f_Y(y)}$$

(see Definition 7.1, Chapter 5). In this case we define $E(X \mid Y = y)$ by

$$E(X \mid Y = y) = \int_{-\infty}^{\infty} x f_{X|Y}(x \mid y)\, dx$$

provided $\int_{-\infty}^{\infty} |x| f_{X|Y}(x \mid y)\, dx$ is finite.

The *conditional variance* of $X$ given $Y = y$ is defined as

$$\text{Var}(X \mid Y = y) = E[(X - E(X \mid Y = y))^2 \mid Y = y].$$

Similarly, let $A$ be any event of positive probability. We define

$$E(X \mid A) = \sum xP\{X = x \mid A\}$$

for the discrete case, provided $\sum |x|\, P\{X = x \mid A\}$ is finite.

For the density case,

$$E(X \mid A) = \int_{-\infty}^{\infty} x f_{X|A}(x \mid A)\, dx$$

where

$$f_{X|A}(x \mid A) = \frac{d}{dx} P\{X \leq x \mid A\},$$

provided $\int_{-\infty}^{\infty} |x| f_{X}|_{A}(x \mid A)\, dx$ is finite.

### *8.2 Example: Dice*

Suppose a pair of dice are tossed and $X_i$ is a random variable whose value is the result on die $i$, $i = 1, 2$, and $S$ is a random variable whose value is the sum on the dice ($S = X_1 + X_2$). We find $E(X_1 \mid S = 6)$:

$$P\{X_1 = j \mid S = 6\} = \frac{P\{X_1 = j, S = 6\}}{P\{S = 6\}}$$

$$= \frac{P\{X_1 = j, X_2 = 6 - j\}}{P\{S = 6\}}$$

$$= \frac{P\{X_1 = j\}P\{X_2 = 6 - j\}}{P\{S = 6\}}$$

$$= \begin{cases} \dfrac{1/36}{5/36} & \text{for } j = 1, 2, 3, 4. 5 \\ 0 & \text{otherwise.} \end{cases}$$

Thus $E(X_1 \mid S = 6\} = \sum jP\{X_1 = j \mid S = 6\}$

$$= (1 + 2 + 3 + 4 + 5) \cdot 1/5$$

$$= 3.$$

△

### 8.3 *Example: Coins*

Suppose a fair coin is tossed independently 10 times. Let $X_i$ be the number of heads on the $i$th toss and let $T$ be the total number of heads. Find $E(X_1 \mid T = 3)$.

$$P\{X_1 = j \mid T = 3\} = \frac{P\{X_1 = j, T = 3\}}{P\{T = 3\}}$$

$$= \frac{P\{X_1 = j, X_2 + \cdots + X_{10} = 3 - j\}}{P\{T = 3\}}$$

$$= \frac{P\{X_1 = j\}P\{X_2 + \cdots + X_{10} = 3 - j\}}{P\{T = 3\}}$$

$$= \begin{cases} \dfrac{\frac{1}{2} \cdot \binom{9}{3}\left(\frac{1}{2}\right)^9}{\binom{10}{3}\left(\frac{1}{2}\right)^{10}} & \text{for } j = 0 \\ \dfrac{\frac{1}{2} \cdot \binom{9}{2}\left(\frac{1}{2}\right)^9}{\binom{10}{3}\left(\frac{1}{2}\right)^{10}} & \text{for } j = 1 \\ 0 & \text{otherwise} \end{cases}$$

$$= \begin{cases} 7/10 & \text{for } j = 0 \\ 3/10 & \text{for } j = 1 \\ 0 & \text{otherwise.} \end{cases}$$

Hence, $E(X_1 \mid T = 3) = 0 \cdot 7/10 + 1 \cdot 3/10 = 3/10$. This result should not be surprising. △

### 8.4 *Example: Bivariate Normal Conditional Expectations*

Suppose that

$$f_{X,Y}(x, y) = \frac{1}{2\pi\sqrt{1-\rho^2}} e^{-(1/2\rho^2)[x^2-2\rho xy+y^2]}.$$

Then we know from Example 7.7 in Chapter 5 that

$$f_{X|Y}(x \mid y) = \frac{1}{\sqrt{2\pi(1-\rho^2)}} e^{-[1/2(1-\rho^2)][x-\rho y]^2}.$$

Thus, the conditional distribution of $X$ given $Y = y$ is normal, with $E(X \mid Y = y) = \rho y$, $\text{Var}\,(X \mid Y = y) = 1 - \rho^2$. △

The following result, which intuitively holds because $E(X \mid Y = y)$ is the expectation in a subpopulation, simplifies many manipulations.

### 8.5 *Theorem: Properties of Conditional Expectation*

Let $a$, $b$, and $z$ be real numbers and suppose that $E(X \mid Z = z)$ and $E(Y \mid Z = z)$ exist. Then $E(aX + bY \mid Z = z)$ exists and

$$E(aX + bY \mid Z = z) = aE(X \mid Z = z) + bE(Y \mid Z = z).$$

Furthermore, for each real valued function $g$, the value $E(Xg(Z) \mid Z = z)$ exists and

$$E(Xg(Z) \mid Z = z) = g(z)E(X \mid Z = z).$$

The proof is left to the student.

---

### 8.6 EXERCISES

1. (a) Suppose that a fair coin is tossed independently 10 times. Let $Y$ be the total number of heads on the first *and* second tosses and $T$ be the total number of heads on the 10 tosses. Find $E(Y \mid T = 3)$.

(b)° Same as (a) for a coin whose probability of heads is $p$.

2. In the toss of three fair dice, find the conditional expectation of the value on the first die, given:

(a) that the sum is 4.
(b) that the sum is 5.
(c)° that the sum is in {4, 5}.

3.° Let $f_{X,Y}(x,y) = \begin{cases} 1/\pi & \text{for} \quad x^2 + y^2 \le 1 \\ 0 & \text{otherwise.} \end{cases}$

(a) Find $E(X \mid Y = y)$.
(b) Find Var $(X \mid Y = y)$.

4. Let $f_{X,Y}(x,y) = \begin{cases} x + y & \text{for} \quad 0 \le x \le 1 \text{ and } 0 \le y \le 1 \\ 0 & \text{otherwise.} \end{cases}$

(a) Find $E(X \mid Y = y)$.
(b) Find Var $(X \mid Y = y)$.

5. Let

$$f_{X^*,Y^*}(x^*,y^*) = \frac{1}{2\pi\sigma_x\sigma_y\sqrt{1-\rho^2}} \exp \frac{-1}{2(1-\rho^2)}\left[\left(\frac{x^* - \mu_x}{\sigma_x}\right)^2 - 2\rho\left(\frac{x^* - \mu_x}{\sigma_x}\right)\left(\frac{y^* - \mu_y}{\sigma_y}\right) + \left(\frac{y^* - \mu_y}{\sigma_y}\right)^2\right].$$

(a) Show that if

$$X = \frac{X^* - \mu_x}{\sigma_x}, \qquad Y = \frac{Y^* - \mu_y}{\sigma_y},$$

then

$$f_{X,Y}(x,y) = \frac{1}{2\pi\sqrt{1-\rho^2}} \exp \frac{-1}{2(1-\rho^2)} [x^2 - 2\rho xy + y^2].$$

Hint: Compute an expression for $F_{X,Y}(x,y) = P\{X \le x, Y \le y\}$ first.

(b) Show that

$$E(X^* \mid Y^* = y^*) = \mu_x + \rho(y^* - \mu_y)\frac{\sigma_x}{\sigma_y}.$$

Hint: Use the result of (a) and Example 8.4.

(c) Show that Var $(X^* \mid Y^* = y^*) = (1 - \rho^2)\sigma_x^2$.

Hint: Use the result of (a) and Example 8.4.

6. Prove Theorem 8.5 in the discrete case and in the density case.

In many applications we do not know which value of $E(X \mid Y = y)$ to use until $Y$ has been observed. Suppose we expect to be given information as to which value $Y$ assumes. For prediction of the value of $X$, we would wait until $Y$ has been observed, and if $Y$ assumed the value $y$ we would use $E(X \mid Y = y)$. Hence we may think of $E(X \mid Y = y)$ as the value of a random variable. We make this precise in the following definition.

**8.7 Definition: $E(X \mid Y)$.** The *conditional expectation* $E(X \mid Y)$ is a random variable $g(Y)$ whose value $g(y)$ when $Y$ assumes the value $y$ is given by $E(X \mid Y = y)$.

### 8.8 *Example: Conditional Normal Expectation*

Recall that in Example 8.4 we found that $E(X \mid Y = y) = \rho y$. Hence, when

$$f_{X,Y}(x, y) = \frac{1}{2\pi(1-\rho^2)} \exp \frac{-1}{2(1-\rho^2)} [x^2 - 2\rho xy + y^2],$$

we see that $E(X \mid Y) = \rho Y$. △

Notice that $E(X \mid Y)$ is a random variable whose value is determined by the value which $Y$ assumes. ***The value assumed by $X$ plays no role at all in the determination of the value assumed by $E(X \mid Y)$.***

One of the main powerful uses of $E(X \mid Y)$ comes about from the following theorem.

### 8.9 *Theorem: Expectation of Conditional Expectation*

$$E[E(X \mid Y)] = E(X).$$

*Remark.* This result is merely an extension of the stratified sampling theorem (4.11 in Chapter 3). It states, in essence, that the population average may be obtained by suitably averaging the subpopulation averages.

*Proof:* We give the proof for the discrete case only; that of the density case is similar.

Let $g(Y)$ denote $E(X \mid Y)$. Then by Theorem 3.2,

$$E[g(Y)] = \sum_{\substack{\text{all values } y \\ \text{for which} \\ P\{Y=y\} \neq 0}} g(y) P\{Y = y\}$$

$$= \sum_{\substack{\text{all values } y \\ \text{with } P\{Y=y\} \neq 0}} \left[ \sum_{\substack{\text{all values} \\ x \text{ of } X}} x P\{X = x \mid Y = y\} \right] P\{Y = y\}$$

$$= \sum_{\substack{\text{all values } y \\ \text{with } P\{Y=y\} \neq 0}} \left[ \sum_{\substack{\text{all values} \\ x \text{ of } X}} x \frac{P\{X = x, Y = y\}}{P\{Y = y\}} \right] P\{Y = y\}$$

$$= \sum_{\substack{\text{all values } y \\ \text{with } P\{Y=y\} \neq 0}} \sum_{\substack{\text{all values} \\ x \text{ of } X}} x P\{X = x, Y = y\}$$

$$= \sum_{\substack{\text{all values} \\ (x,y) \text{ of } (X,Y)}} x P\{X = x, Y = y\}$$

$$= E(X) \qquad \text{using Theorem 4.1 with } g(X, Y) = X,$$

concluding the proof.

### *8.10 EXERCISE*

Prove Theorem 8.9 in the density case.

### 8.11 *Example:* $E[E(X \mid Y)]$ *in the Bivariate Normal Case*

Referring to Example 8.8, where

$$f_{X,Y}(x, y) = \frac{1}{2\pi(1-\rho^2)} \exp \frac{-1}{2(1-\rho^2)} (x^2 - 2\rho xy + y^2),$$

we see that

$$E(X) = E[E(X \mid Y)] = E[\rho Y] = 0.$$

This is consistent with the result that

$$E(X) = \int_{-\infty}^{\infty} x f_X(x)\, dx \qquad \text{where} \qquad f_X(x) = \frac{1}{\sqrt{2\pi}} e^{-x^2/2},$$

as found in Example 3.3, Chapter 5. △

### 8.12 *Example: Escape Problem*

A prisoner is in a cell with three doors, as illustrated in Figure 6.2. The prisoner chooses one of the doors ($D_1$, $D_2$, or $D_3$) in attempting to escape. Unfortunately, doors $D_1$ and $D_2$ are traps. If he chooses door $D_1$, the prisoner will wander in a tunnel for 13 days and return to his cell with a complete loss of memory; if he chooses door $D_2$, he will be in a tunnel for eight days. Door $D_3$ leads to immediate freedom.

Let $T$ be a random variable whose value is the number of days required until freedom is attained. We want to compute $E(T)$, assuming that the doors are

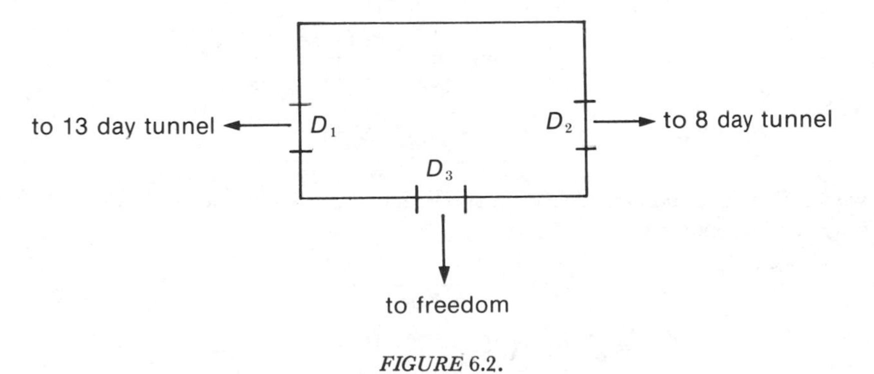


*FIGURE* 6.2.

chosen with equal probability, and independence of trials (loss of memory). Normally, computation of $E(T)$ would require determination of the probability distribution of $T$, a very difficult job indeed. Use of Theorem 8.9 greatly simplifies matters. Let $N$ be the random variable denoting the number on the door chosen on the first trial. Now

$$(*) \qquad E(T) = E[E(T \mid N)] = \sum_{i=1}^{3} E(T \mid N = i) P\{N = i\},$$

and $P\{N = i\} = 1/3$ for $i = 1, 2, 3$. Intuitively, $E(T \mid N = 1) = 13 + E(T)$, since, if the prisoner chooses $D_1$, he wanders for 13 days, returns to his cell, and must start again. Similarly, $E(T \mid N = 2) = 8 + E(T)$, but $E(T \mid N = 3) = 0$. Using this information in (*) yields

$$E(T) = \tfrac{1}{3}[(13 + E(T)) + (8 + E(T)) + 0],$$

from which we find

$$\tfrac{1}{3}E(T) = \tfrac{21}{3}$$

so that

$$E(T) = 21.$$

A practical use of this situation would be if we wanted to find the expected survival time of an insect in an insect trap.

The preceding intuitive arguments concerning $E(T \mid N = i)$ were reasonable, but were not precise enough to give us the ability to compute, say, $E(T^2 \mid N = i)$ with any degree of certainty. In the following paragraph, we shall show that if $H$ is any real valued function of a real variable, then

$$(\substack{*\\*}) \qquad E(H(T) \mid N = i) = \begin{cases} E[H(T + d_i)] & \text{for} \quad i = 1, 2 \\ H(0) & \text{for} \quad i = 3 \end{cases}$$

where $d_1 = 13$ and $d_2 = 8$.

In order to present a convincing proof, we must build a model that embodies the stated assumptions. As a sample space, we take the set of all sequences $(a_i, a_2, \ldots)$ where each $a_i \in \{1, 2, 3\}$; here $a_i$ represents the door chosen on trial $i$, where the trials are independent with a probability of 1/3 for each of the three possibilities on every trial. Now, for $i = 1, 2$,

$$\begin{aligned} E[H(T) \mid N = i] &= \sum_j H(j) P\{T = j \mid N = i\} \\ &= \sum_j H(j) \frac{P\{T = j, N = i\}}{P\{N = i\}} \\ &= \sum_j H(j) \frac{P\{T_2 = j - d_i, N = i\}}{P\{N = i\}} \end{aligned}$$

where $T_2$ is the time spent after the first return to the cell ($T_2$ is determined only by trials after the first one)

$$\begin{aligned} &= \sum_j H(j) \frac{P\{T_2 = j - d_i\} P\{N = i\}}{P\{N = i\}} \\ &= \sum_j H(j) P\{T_2 = j - d_i\}. \end{aligned}$$

But $T_2$ has the same distribution as $T$, since the trials are independent repetitions of the same experiment. Hence this last expression is equal to

$$\sum_k H(j)P\{T = j - d_i\}$$

$$= \sum_k H(k + d_i)P\{T = k\} \qquad (\text{letting } k = j - d_i)$$

$$= E(H[T + d_i]),$$

as asserted. For $i = 3$, the result is evident.

To avoid unnecessary care in the bookkeeping, we may take the various sums over all integers. Then the unwanted terms, such as $P\{T = -13\}$, will automatically drop out. △

The following theorem is important in both theoretical and practical work involving conditional expectation.

### *8.13 Theorem: Properties of Conditional Expectation*

Let $a$ and $b$ be real numbers and suppose that $E(X \mid Z)$ and $E(Y \mid Z)$ exist. Then $E(aX + bY \mid Z)$ exists and

$$E(aX + bY \mid Z) = aE(X \mid Z) + bE(Y \mid Z).$$

Furthermore, for each real valued function $g$, the expectation $E(Xg(Z) \mid Z)$ exists and

$$E(Xg(Z) \mid Z) = g(Z)E(X \mid Z).$$

This theorem is just the natural extension of Theorem 8.5 to conditional expectations considered as random variables.

---

### *8.14 PROBLEMS*

1. Using Theorem 8.9 on the expectation of the conditional expectation, compute the expected number of tosses of a die until a 5 comes up.

2. Solve the three doors problem (8.12) for the case of arbitrary probabilities $p_1, p_2, p_3$ where $p_i = P\{N = i\}$ and $p_1 + p_2 + p_3 = 1$.

3.° Find the variance of the time to freedom of Example 8.12 and use it to compute an upper bound for the probability that it takes over 150 days to attain freedom.

4.° Using Theorem 8.9, on the expectation of the conditional expectation, twice, determine the expected number of tosses of a fair die until 4 and 3 turn up in succession.

## SECTION 9
## Quantiles

Earlier, we introduced the moments of a distribution function $F_X$, which in some sense describe the distribution whenever they exist. For example, the first moment $E(X)$ may be thought of as a useful "center" for the distribution from which to measure deviations. Similarly, the standard deviation, $\sigma_x$, provides a useful scale for such measurements. Of course, as we have seen, such moments may fail to exist (see Counterexample 2.10). In this case, the theory we developed concerning expectations is of no value. However, there are other numerical quantities, called *quantiles*, that play a similar descriptive role for the distribution and have the added advantage that they always exist. In particular, the *median* is a quantile that, like $E(X)$, acts as a "center" for the distribution. The *interquartile range*, which is the difference between two special quantiles, acts like the standard deviation $\sigma$, as a reasonable scale for measurement.

Normally, we would like to define the $p$ quantile, $\theta_p$, of the distribution of $X$ as the value such that

$$P\{X \leq \theta_p\} = p.$$

Since there may not be any such value, we are forced to define it as follows.

### 9.1 *Defiinition: p Quantile, Interquartile Range*

A *p-quantile* of the distribution $P_X$ (or the random variable $X$) is any number $\theta_p$ such that

$$P\{X < \theta_p\} \leq p \leq P\{X \leq \theta_p\}.$$

The value $\theta_{1/2}$ is called a *median* of $P_X$, and $\theta_{3/4} - \theta_{1/4}$ is called the *interquartile range*.

In the case of a density, Figure 6.3 illustrates $\theta_p$.

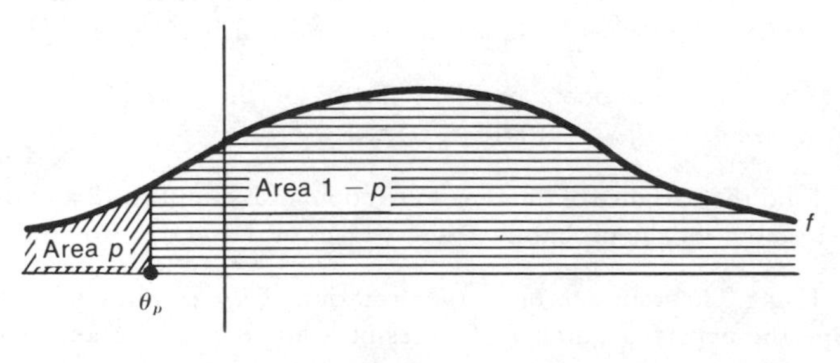


*FIGURE* 6.3.

### 9.2 *Example: Uniform Distribution Quantiles*

Since the uniform distribution function on [0, 1] is given by

$$F(x) = \begin{cases} 0 & \text{if} \quad x < 0 \\ x & \text{if} \quad 0 \le x \le 1 \\ 1 & \text{if} \quad x > 1, \end{cases}$$

we see that here $\theta_p = p$. △

### 9.3 *Example: Normal Distribution Quantiles*

If

$$F(x) = \int_{-\infty}^{x} \frac{1}{\sqrt{2\pi}} e^{-t^2/2}\, dt,$$

then it is easy to see by symmetry that $F(0) = 1/2$, and hence here $\theta_{1/2} = 0$. To find $\theta_p$ for any other $p$ requires numerical solution for $x$ of the equation

$$\int_{-\infty}^{x} \frac{1}{\sqrt{2\pi}} e^{-t^2/2}\, dt = p.$$

A glance at the tables of the standard normal distribution function in the Appendix shows that we have the approximations

$$\theta_{1/4} \cong -.676, \qquad \theta_{3/4} \cong .676, \qquad \theta_{1/10} \cong -1.285, \qquad \theta_{9/10} \cong 1.285.$$ △

### 9.4 *Example: Poisson Distribution Quantiles*

From the tables of the Poisson distributions, we find for $E(X) = 1$,

$$\theta_{.3} = 0, \qquad \theta_{.4} = 1, \qquad \theta_{.5} = 1, \qquad \theta_{.8} = 2.$$

For $E(X) = 3$,

$$\theta_{.3} = 1, \qquad \theta_{.4} = 2, \qquad \theta_{.5} = 2, \qquad \theta_{.8} = 3.$$ △

---

## 9.5 *EXERCISES*

1. From tables of the binomial distribution with $n = 10$ and $p = .4$, find:

   (a)° $\theta_{.3}$.
   (b) $\theta_{.5}$.
   (c) $\theta_{.8}$.

2. Find the following quantiles when $X$ is normal with $E(X) = 2$, Var $X = 9$. Hint: Use Example 2.25 in Chapter 4.

(a)° $\theta_{1/4}$.
(b) $\theta_{3/4}$.
(c) $\theta_{1/2}$.
(d) $\theta_{1/10}$.

3. If

$$f_X(x) = \begin{cases} \frac{1}{2}e^{-x/2} & \text{for} \quad x > 0 \\ 0 & \text{otherwise,} \end{cases}$$

find:

(a)° the median of the distribution of $X$.
(b) the interquartile range of the distribution of $X$.

4. Let

$$f_X(x) = \frac{1}{\pi(1 + [x - \theta]^2)}.$$

Find the median of the distribution of $X$.

If $X_1, \ldots, X_{2n+1}$ are independent random variables, each with the same distribution function, $F$, then if you observe one value of each it is natural to arrange these observed values in order of magnitude, denoting by $X_{[1]}$ the random variable whose value is the smallest observed value, by $X_{[2]}$ the one with the next largest value, and so on. It is also reasonable, then, to expect that there is a high probability that the observed value of $X_{[n+1]}$ will be "close" to the median of $F$. This is in fact the case. If we believe in some given situation that the median of $F$ is a more useful measure than its expectation of the "center" of $F$, then it would appear reasonable to use the value of $X_{[n+1]}$ rather than the value of

$$\frac{1}{2n+1}\sum_{i=1}^{2n+1} X_i$$

as an empirical measure of the "center." The question just raised actually has no simple answer in general, since the answer varies with the distribution being considered.

Intuitively, we might believe that the median would be preferable to the expectation in describing a distribution if extremely high or low outcomes do not (or should not) have an overpowering effect, as in the case of one low test score. On the other hand, if a sufficiently bad component can cause failure of a whole structure, then possibly the expectation is more useful.

We conclude this section by defining the sample quantities related to the quantiles and giving a statement of one of their important properties.

**9.6 Definition: Order Statistics, Sample Quantiles.** Let $X_1, X_2, \ldots, X_n$ be random variables all defined on the same probability space. The *order statistics* $X_{[1],n}, X_{[2],n}, \ldots, X_{[n],n}$ are defined as follows.

The sequence $(X_{[1],n}(s), X_{[2],n}(s), \ldots, X_{[n],n}(s))$ is the rearrangement of the sequence $(X_1(s), X_2(s), \ldots, X_n(s))$ of observed values according to order of magnitude, that is, $X_{[1],n}(s) \leq X_{[2],n}(s) \leq \cdots \leq X_{[n],n}(s)$. Hence, $X_{[1],n}(s)$ is the smallest observed value in the sample, $X_{[2],n}(s)$ is next smallest, and so on.

The *sample p-quantile* where $0 < p < 1$ is that $X_{[j],n}$ where $j$ is the smallest integer greater than or equal to $np$; we use the symbol $X_{(p),n}$ to denote the sample $p$-quantile. Note that *roughly speaking*, $X_{(p),n}$ *exceeds a proportion p of the observations in the sample* $X_1, \ldots, X_n$. The sample 1/2-quantile $X_{(1/2),n}$ is called the *sample median.*

### 9.7 *Example: Data*

Suppose that the observed sample values are $X_1(s) = 2$, $X_2(s) = 1.5$, $X_3(s) = 1.4$, and $X_4(s) = 6$. Then

$$X_{[1],4}(s) = 1.4, \qquad X_{[2],4}(s) = 1.5, \qquad X_{[3],4}(s) = 3, \qquad X_{[4],4}(s) = 6$$

$$X_{(.1),4}(s) = X_{[1],4}(s) = 1.4$$

$$X_{(.75),4}(s) = X_{(.71),4}(s) = X_{[3],4}(s) = 3, \qquad \text{and so on.} \qquad \triangle$$

We ask for the proof of the following important theoretical result in Problem 9.9(1).

### 9.8 *Theorem: Convergence of Sample Quantiles*

Suppose that $X_1, X_2, \ldots$ are independent random variables, all with the same distribution function $F$, where $F$ has a unique $p$-quantile $\theta_p$. Then for each $\varepsilon > 0$,

$$\lim_{n\to\infty} P\{|X_{(p),n} - \theta_p| \leq \varepsilon\} = 1.$$

That is, when $n$ is large, it is highly probable that the sample $p$-quantile $X_{(p),n}$ is close to the population $p$-quantile $\theta_p$.

---

## 9.9 *PROBLEMS*

1. Use the following outline to prove Theorem 9.8. Let $[np]$ represent the smallest integer $\geq np$. The event $\{X_{(p),n} \leq \theta_p + \varepsilon\}$, that is, that $[np]$ *or more* observations are to the left of $\theta_p + \varepsilon$, is the same as the event $\{S_n \geq [np]\}$, where $S_n$ is the number of observations $X_i$ not exceeding $\theta + \varepsilon$; here $S_n$ is binomial with parameters $n$ and $p + \delta$ (where $P\{X_i \leq \theta_p + \varepsilon\} = p + \delta > p$). Now apply the Chebyshev inequality to the event

$$\{S_n \geq [np]\} = \{S_n - n(p + \delta) \geq [np] - n(p + \delta)\}$$

and then let $n \to \infty$. This shows that $\lim_{n\to\infty} P\{X_{(p),n} \leq \theta_p + \varepsilon\} = 1$. Similarly, it can be established that $\lim_{n\to\infty} P\{X_{(p),n} \geq \theta_p - \varepsilon\} = 1$, from which the desired result follows.

2. Suppose we are given independent random variables $X_1, \ldots, X_{2n-1}$ with common density $f$ possessing a unique median. Find the density of the sample median $X_{[n],\,2n-1}$.
Hints: In order for $X_{[n],2n-1}$ to lie in $[x, x + \Delta x]$ we must have (i) exactly $n - 1$ of the $X_i$ to the left of $[x, x + \Delta x]$, (ii) precisely 1 observation in $[x, x + \Delta x]$, and (iii) the $n - 1$ remaining observations to the right of $[x, x + \Delta x]$ for $\Delta x$ sufficiently small. Compute this probability from the multinomial probability distribution (Example 2.6, Chapter 5) with three possible outcomes (corresponding to the three cases above), with

$$p_1 = P\{X_i < x\} = \int_{-\infty}^{x} f(t)\, dt,$$

$$p_2 = P\{x \le X_i \le x + \Delta x\} = \int_{x}^{x+\Delta x} f(t)\, dt \cong f(x)\, \Delta x,$$

$$p_3 = 1 - p_1 - p_2,$$

$$j_1 = n - 1,$$

$$j_2 = 1,$$

$$j_3 = n - 1.$$

3. Let

$$f(x) = \frac{1}{\pi(1 + x^2)}.$$

(a) Show that $X$ does not have an expectation if $f_X = f$.
(b) Find the density of

$$\frac{X + Y}{2}$$

when $X$ and $Y$ are independent each with density $f$ above.
Hints: Let $Z = X + Y$. Then

$$F_Z(z) = P\{X + Y \le z\} = \iint_A f_X(x) f_Y(y)\, dx\, dy \quad \text{where} \quad A = \{(x, y) \in \mathscr{R}_2: x + y \le z\}$$

$$= \int_{-\infty}^{\infty} \int_{-\infty}^{z-y} f_X(x) f_Y(y)\, dx\, dy.$$

Hence, differentiating under the integral and using the Fundamental Theorem of Calculus (see the Appendix),

$$f_Z(z) = F_Z'(z) = \int_{-\infty}^{\infty} f_X(z - y) f_Y(y)\, dy.$$

Now note that

$$\frac{1}{\pi^2} \frac{1}{1 + y^2} \frac{1}{1 + [z - y]^2} = \frac{a + by}{1 + y^2} + \frac{c + dy}{1 + [z - y]^2}$$

where

$$a = \frac{1}{\pi^2(z^2 + 4)}, \qquad c = \frac{3}{\pi^2(z^2 + 4)}.$$

$$b = \frac{2}{\pi^2 z(z^2 + 4)}, \qquad d = \frac{-2}{\pi^2 z(z^2 + 4)}.$$

From this we can find a formula for $f_Z(z)$. To find the desired density of $U = Z/2$ is now routine.

(c) Having shown that

$$\frac{X + Y}{2}$$

also has density $f$, show that the averages

$$\frac{1}{n}\sum_{i=1}^{n} X,$$

where $X_i$ are independent with density $f$, do not settle down.

Hints: (a) All averages

$$\frac{1}{n}\sum_{1=i}^{n} X_i \qquad \text{where} \qquad n = 2^m,\ m = 1, 2, \ldots,$$

have density $f$.

(b)
$$\frac{1}{n}\sum_{1=i}^{n} X_i = \frac{k}{n}\left(\frac{1}{k}\sum_{i=1}^{k} X_i\right) + \frac{n-k}{n}\left(\frac{1}{n-k}\sum_{i=k+1}^{n} X_i\right).$$

We can choose both $k$ and $n$ to be powers of 2 with $k$ large, but with $n$ much larger than $k$. From this we see that

$$\frac{1}{k}\sum_{i=1}^{k} X_i \qquad \text{and} \qquad \frac{1}{n}\sum_{i=1}^{n} X_i$$

are effectively independent, each with density $f$.

# 7

# Approximations to Distributions

## SECTION 1
## Introduction

It is frequently true that the formula for the probability of some event is easy to express, but useless from a computational viewpoint. Furthermore, it may not be very enlightening in any sense as it stands. For example, suppose $10^{15}$ independent experiments are performed, each with probability $10^{-14}$ of success. Such a situation would arise if we were examining the behavior of a computer over a three-year period. In this case a "success" is an error in computation, and our computer performs about one computation every .1 microsecond. The probability of not more than 18 successes (that is, of an observed average of not more than one error every other month) is

$$\sum_{k=0}^{18} \binom{10^{15}}{k} 10^{-14k}(1 - 10^{-14})^{10^{15}-k} .$$

*As it stands*, this formula is of no practical use.

A similar situation often occurs when we try to compute the distribution of a function $g(X_1, \ldots, X_n)$ where $f_{X_1,\ldots,X_n}$ is known. For example, if $X_1, \ldots, X_n$ are independent random variables with a common univariate density $f$, then

$$P\{X_1 + X_2 + \cdots + X_n \leq a\} = \int_A \cdots \int f(x_1) \cdots f(x_n)\, dx_1 \cdots dx_n,$$

where $A$ is the set of points $(x_1, \ldots, x_n)$ in $n$-dimensional space such that $x_1 + x_2 + \cdots + x_n \leq a$. Here we can actually derive by induction the formula

$$f_{X_1+\cdots+X_n}(t) = \int_{-\infty}^{\infty} \cdots \int_{-\infty}^{\infty} f(t_1)f(t_2 - y_1)f(t_3 - y_2) \cdots f(t - y_{n-1})\, dy_1 \cdots dy_{n-1}$$

for the density of $X_1 + \cdots + X_n$. However, there are very few cases in which we can make immediate use of this formula to compute probabilities.

Because we encounter such situations often enough, it is worthwhile to find approximations that are sufficiently accurate for practical purposes. In this chapter we shall describe some of the tools that enable us to derive such approximations in many important cases.

## SECTION 2
## The Poisson Approximation to the Binomial

The first approximation we present concerns the computation of binomial probabilities when $n$ is large, and $p$ is small, but $np$ is moderate. Such a situation arises, for example, when computing the probability that exactly $k$ atoms will split in a given time interval. Here there are many atoms, on the order of magnitude of $10^{23}$, but each atom has very small probability of splitting, say on the order of magnitude of $10^{-23}$. The main result of this section is embodied in the following theorem.

### *2.1 Theorem: Rare Events*

Suppose that for each $n$, the random variable $X_n$ has a binomial distribution with parameters $n$ and $p_n$, that is, $X_n$ is the number of successes on $n$ experiments, each having probability $p_n$ of success. If $\lim_{n\to\infty} np_n = \lambda$ (recall that $E(X_n) = np_n$), then for each fixed non-negative integer $k$,

$$\lim_{n\to\infty} P\{X_n = k\} = e^{-\lambda}\frac{\lambda^k}{k!}.$$

*Proof:*

$$P\{X_n = k\} = \binom{n}{k} p_n^k (1 - p_n)^{n-k}$$

$$= \frac{n\cdot(n-1)\cdots(n-k+1)}{k!} p_n^k (1 - p_n)^n (1 - p_n)^{-k}.$$

Let $\lambda_n = np_n$; we have $\lim_{n\to\infty} \lambda_n = \lambda$. We see that

$$P\{X_n = k\} = \frac{1}{k!}\frac{n\cdot(n-1)\cdots(n-k+1)}{\underbrace{n\cdot n\cdots n}_{k \text{ factors}}}\lambda_n^k\left(1 - \frac{\lambda_n}{n}\right)^n (1 - p_n)^{-k}$$

$$= \underbrace{\frac{1}{k!}\cdot 1\cdot\left(1 - \frac{1}{n}\right)\cdots\left(1 - \frac{k-1}{n}\right)}_{(1)}\underbrace{\lambda_n^k}_{(2)}\underbrace{\left(1 - \frac{\lambda_n}{n}\right)^n}_{(3)}\underbrace{(1 - p_n)^{-k}}_{(4)}.$$

As $n$ becomes large, each of the $k$ factors above brace (1) approaches 1. Since $k$ is fixed, the factor above brace (1) approaches 1 as $n \to \infty$. Since $\lambda_n \to \lambda$, the factor above brace (2) approaches $\lambda^k$. For the factor above brace (3), we make use of the well-known fact that

$$\lim_{n\to\infty}\left(1+\frac{x}{n}\right)^n = e^x,$$

which, together with the fact that $\lambda_n$ is approaching $\lambda$, yields

$$\lim_{n\to\infty}\left(1-\frac{\lambda_n}{n}\right)^n = e^{-\lambda}.$$

The factor above brace (4) approaches 1 because $k$ is fixed and $p_n$ is approaching 0. Combining these results yields

$$\lim_{n\to\infty} P\{X_n = k\} = \frac{1}{k!}\cdot 1\cdot \lambda^k\cdot e^{-\lambda}\cdot 1 = e^{-\lambda}\frac{\lambda^k}{k!}.$$

It is common practice to use this approximation whenever $n$ is large and $p$ is small but $np$ is moderate. In this regard, a useful guide is the following result, derived by Uspensky [20].

### 2.2 *Theorem: Error in Poisson Approximation*

Let $S$ be a random variable having a binomial distribution with parameters $n$ and $p$; let $\lambda = np$. Then

$$\left| P\{S \le m\} - \sum_{k=0}^{m} e^{-\lambda}\frac{\lambda^k}{k!}\right| \le \exp\frac{\lambda + 1/4 + \lambda^3/n}{2(n-\lambda)} - 1.$$

Note that the sum is likely to be rather difficult to compute when $m$ is large.

### 2.3 *Example: Size of Error*

Since $e^x$ is close to 1 when $x$ is close to 0—in fact, $e^x \cong 1 + x$—the error in Theorem 2.2 will be small when

$$\frac{\lambda + 1/4 + \lambda^3/n}{2(n-\lambda)}$$

is close to 0. Since

$$\frac{\lambda + 1/4 + \lambda^3/n}{2(n-\lambda)} = \frac{p}{2(1-p)} + \frac{1}{8n(1-p)} + \frac{np^3}{2(1-p)},$$

we see that in order for the error to be near 0 we must have $p$ near 0 and $np^3$ near 0. For example, if $p = .01$ and $n = 100$, then the error is about .005. △

### 2.4 *Example: Mutations*

A collection of 3200 fruit flies are subjected to rather large doses of radiation in an attempt to study mutations. If the probability of mutation in any fruit fly is .0007, what is the probability of more than three mutations? Our approximation to this probability is

$$1 - \sum_{k=0}^{3} \frac{e^{-\lambda}\lambda^k}{k!} \quad \text{where} \quad \lambda = np = 2.24.$$

This yields

$$1 - .106\left(1 + 2.24 + \frac{2.24^2}{2} + \frac{2.24^3}{6}\right) \cong .19.$$

The error in this computation does not exceed

$$\exp\left(\frac{2.24 + .25 + .0035}{2(3200 - 2.24)}\right) - 1 \cong e^{.00039} - 1 \cong .00039$$

(using the differential approximation $e^x \cong 1 + x$, valid for $x$ near 0). △

### 2.5 *Example: Atomic Splittings*

In a batch of $10^{23}$ atoms, the probability of any given atom splitting in a 24-hour period is $10^{-23}$. Find the probability that fewer than five atoms will split. This probability is approximately

$$\sum_{k=0}^{4} \frac{e^{-1}}{k!} \cong \frac{1}{2.718}[1 + 1 + 1/2 + 1/6 + 1/24]$$

$$\cong \frac{2.708}{2.718} = 1 - \frac{.01}{2.718} \cong 1 - .00368 = .99632.$$

The error in our approximation is, by Theorem 2.2, less than

$$\exp\frac{1 + 1/4 + 10^{-23}}{2(10^{23} - 1)} - 1 \cong e^{1.25/2\cdot 10^{23}-1} \cong .625 \cdot 10^{-23}.$$ △

---

## 2.6 *EXERCISES*

1.° In a book with 200,000 words, suppose that the probability that any given word is set in type incorrectly is 1/50,000.

(a) Approximately what are the chances that every word is set correctly? Provide decimal upper and lower bounds for the desired probability.

(b) Provide decimal upper and lower bounds for the probability that more than six words will be set incorrectly.

2. An experiment with probability of success .003 is performed 500 times. Furnish decimal upper and lower bounds for the probability of fewer than four successes.

3. In a bacteria culture, it is determined that in any given square centimeter the probability of at least one bacterium is .001. Suppose we have a 200 cm. × 50 cm. culture. What is the probability of finding more than 14 bacteria? (Assume that if $A_1, \ldots, A_j$ are any disjoint areas, then the numbers $X_1, \ldots, X_j$ of bacteria on each of these areas are independent random variables. This is reasonable if the bacteria fall from the air at random on the culture.)

4.° It is claimed that if an experiment with probability of success .001 is repeated 3000 times, then the probability of at least two successes is *very high*. Is this true?

5.° Suppose $n$ experiments are performed, each with probability .01 of success. Approximately how large should $n$ be in order that the probability of at least three successes is at least .9? Using the error bound given in Theorem 2.2, furnish an $n$ that guarantees what is desired.

6. Repeat Exercise 5, but now we only desire to guarantee a probability of .8 of at least three successes.

## 2.7 The Poisson Process

We have mentioned many times that the Poisson distribution is used in models for phone traffic, arrivals at a service counter, and so forth. In this section we shall show how this model arises from elementary considerations, with the use of Theorem 2.1.

Suppose that as time passes there occurs a sequence of events (such as initiation of phone calls, or atomic disintegrations, or accidents) with the following probabilistic properties: (a) The number of occurrences $n_1, n_2, \ldots$ that happen in any set of given disjoint intervals $I_1, I_2, \ldots$ are the values of independent random variables $N_1, N_2, \ldots$. (b) The probability of exactly one such occurrence in any interval $[t, t + h]$ is $\lambda h + \theta(h) \cdot h$ where $\theta$ is a function not dependent on $t$, satisfying $\lim_{h \to 0} \theta(h) = 0$ and $\lambda$ is some positive constant. (c) The probability of more than one such occurrence in $[t, t + h]$ is $\mathscr{L}(h) \cdot h$ (independent of $t$) where $\mathscr{L}$ is a function satisfying $\lim_{h \to 0} \mathscr{L}(h) = 0$.

That is, *for small intervals* the probability of exactly one occurrence is, for practical purposes, proportional to the length of the interval. We can neglect the possibility of more than one occurrence.

*We often refer to assumptions* (a), (b) *and* (c) *by saying that these events occur independently and at random.*

Now we ask for the probability of exactly $k$ occurrences of the events in question in any given interval $[t_0, t_0 + t]$. To find this probability, we break up this interval into $n$ parts, each of length $t/n$. We see that the $n$ intervals lead to $n$ independent trials, each with probability

$$p_n = \lambda \frac{t}{n} + \theta\left(\frac{t}{n}\right)\frac{t}{n} + \mathscr{L}\left(\frac{t}{n}\right)\frac{t}{n}$$

of success, where *success in the* j*th interval means at least one occurrence in this interval.* Thus since $np_n \to \lambda t$, we can apply Theorem 2.1 to conclude that the probability of $k$ *successes* is

$$e^{-\lambda t}\frac{(\lambda t)^k}{k!}.$$

But if we let $D_n$ be the event that more than one occurrence happens in even one of the $n$ equal length subintervals of $[t_0, t_0 + t]$, then, using assumption (c) and $P(\bigcup_k A_k) \leq \sum_k P(A_k)$ [Theorem 3.5(d), Chapter 3], we see that $\lim P(D_n) = 0$. Hence, we see that the probability of exactly $k$ occurrences in $[t_0, t_0 + t]$ is

$$e^{-\lambda t}\frac{(\lambda t)^k}{k!}.$$

So, the assumption of events occurring independently and at random leads to the following model: If $N_1, N_2, \ldots$ are random variables that represent the number of occurrences of some phenomenon in disjoint time intervals $I_1, I_2, \ldots$, then

(a) $N_1, N_2, \ldots$ are independent and

(b) $P\{N_j = k\} = e^{-\lambda t_j}\dfrac{(\lambda t_j)^k}{k!}$

where $t_j$ is the length of $I_j$. This model is called the *Poisson Process.*

---

## 2.8 PROBLEMS

1. Find the density of $X_{[1],n}, \ldots, X_{[n],n}$ where $X_1, \ldots, X_n$ are independent and uniformly distributed on $[t_0, t_0 + t]$.
Hint: $X_{[1],n} \in I_1$ and $X_{[2],n} \in I_2$ and $\cdots$ and $X_{[n],n} \in I_n$, where $I_j$ is to the left of $I_{j+1}$ for $j = 1, \ldots, n - 1$ if and only if for some permutation $i_1, \ldots, i_n$ of $1, 2, \ldots, n$ we have $X_{i_1} \in I_1$ and $X_{i_2} \in I_2$ and $\cdots$ and $X_{i_n} \in I_n$.

2. Show that the conditional distribution of the times $T_i$, $i = 1, \ldots, n$, $T_i < T_{i+1}$, at which the occurrences of a Poisson process happen, given $n$ such occurrences in the time interval $[t_0, t_0 + t]$, is that of $X_{[1],n}, \ldots, X_{[n],n}$, where $X_1, \ldots, X_n$ are independent with the uniform distribution on $[t_0, t_0 + t]$.
Hint: Let $\Delta t_0 = 0$, $t_0 + t = t_{n+1}$ and let
$A_i$ be the event: no occurrence in $[t_{i-1} + \Delta t_{i-1}, t_i]$ for $i = 1, \ldots, n + 1$
$B_j$ be the event: one occurrence in $[t_j, t_j + \Delta t_j]$ for $j = 1, \ldots, n$.
Then for $t_j + \Delta t_j \leq t_{j+1}$ for all $j$,

$$f_{T_1, \ldots, T_n \mid n \text{ occurrences in } [t_0; t_0+t]}(t_1, \ldots, t_n)\,\Delta t_1 \cdots \Delta t_n \cong \frac{P\{A_1 \cap B_1 \cap A_2 \cap \cdots \cap A_n \cap B_n \cap A_{n+1}\}}{P\{n \text{ occurrences in } [t_0, t_0 + t]\}}.$$

(Here we are assuming that a conditional density exists, the proof of which is non-trivial.)

This result explains why assumptions (a), (b) and (c) of 2.7 are referred to as "events occurring independently and at random" (that is, uniformly in the foregoing sense).

## SECTION 3
## The Central Limit Theorem

The most frequently encountered function of random variables is given by $g(X_1, \ldots, X_n) = X_1 + X_2 + \cdots + X_n$. It arises when we want to compute the total accumulated error in timing, or in fitting together various components to form a single unit. It also arises when we want to compute the total number of successes in $n$ repetitions of an experiment or the total roundoff error in a series of additions. This section is therefore devoted to the problem of approximating the distribution of sums of random variables. Our main theorem concerns sums of independent identically distributed random variables with mean $\mu$ and variance $\sigma^2$. It is remarkable for its universal applicability. The first version of this theorem was discovered by direct numerical computations on the binomial distribution by De Moivre in 1793. The first general proof was given in 1901 by Liapunov; the technique of proof given here was essentially introduced by Lévy in 1935. We shall first state a general version of the Central Limit Theorem proved by Berry in 1941.

### 3.1 *Central Limit Theorem with Error Term (Berry)*

Suppose that $X_1, X_2, \ldots$ are independent random variables with a common distribution function $F_{X_i} = F$, finite expectation $E(X_i) = \mu$, and a finite variance $E[(X_i - \mu)^2] = \sigma^2$. Let

$$\Phi(t) = \int_{-\infty}^{t} \frac{1}{\sqrt{2\pi}} e^{-u^2/2}\, du$$

be the standard normal distribution function. Then for each real $t$,

**3.2**
$$\lim_{n\to\infty} P\left\{\frac{1}{\sqrt{n}} \sum_{i=1}^{n} \frac{X_i - \mu}{\sigma} \le t\right\} = \Phi(t).\dagger$$

Moreover, if

$$\Gamma = E\left[\left|\frac{X_i - \mu}{\sigma}\right|^3\right] = \frac{E[|X_i - \mu|^3]}{\sigma^3}$$

is finite, then‡

**3.3**
$$\left| P\left\{\frac{1}{\sqrt{n}} \sum_{i=1}^{n} \frac{X_i - \mu}{\sigma} \le t\right\} - \Phi(t) \right| \le \frac{2\Gamma}{\sqrt{n}}.$$

---

† Note that $E\left[\frac{1}{\sqrt{n}} \sum_{i=1}^{n} \frac{X_i - \mu}{\sigma}\right] = 0$ and $\mathrm{Var}\left[\frac{1}{\sqrt{n}} \sum_{i=1}^{n} \frac{X_i - \mu}{\sigma}\right] = 1$.

‡ Berry's paper contained a minor numerical error. The result given here has been verified by the authors.

Before giving a sketch of the proof of Equation 3.2, we will make some remarks and present an example of its use. We may think of *the Central Limit Theorem as a refinement of the Chebyshev inequality*: Recall that for the case of independent random variables $X_1, \ldots, X_n$ with $E(X_i) = \mu$, Var $X_i = \sigma^2$, the Chebyshev inequality takes the form

**3.4** $$P\left\{\left|\frac{1}{n}\sum_{i=1}^{n} X_i - \mu\right| \leq t\frac{\sigma}{\sqrt{n}}\right\} \geq 1 - \frac{1}{t^2}.$$

We may also write this as

$$P\left\{\left|\frac{1}{\sqrt{n}}\sum_{i=1}^{n} \frac{X_i - \mu}{\sigma}\right| \leq t\right\} \geq 1 - \frac{1}{t^2}.$$

That is, ***we may conclude from the Chebyshev inequality that***

**3.5** $$P\left\{-t \leq \frac{1}{\sqrt{n}}\sum_{i=1}^{n} \frac{X_i - \mu}{\sigma} \leq t\right\} \geq 1 - \frac{1}{t^2}.$$

On the other hand, note that ***we may conclude from the Central Limit Theorem that***

**3.6** $$P\left\{-t \leq \frac{1}{\sqrt{n}}\sum_{i=1}^{n} \frac{X_i - \mu}{\sigma} \leq t\right\} \cong \int_{-t}^{t} \frac{1}{\sqrt{2\pi}} e^{-u^2/2}\, du, \text{ for large } n.$$

Now, recall that the standard deviation (s.d.) is the non-negative square root of the variance. From Equation 3.4, we see that the Chebyshev inequality tells us that *the probability that*

$$\frac{1}{n}\sum_{i=1}^{n} X_i \textit{ will be within } t \textit{ of its s.d.'s of its mean } \mu \textit{ exceeds } 1 - \frac{1}{t^2},$$

whereas from 3.4, 3.5, and 3.6 we see that the Central Limit Theorem tells us that *the probability that*

$$\frac{1}{n}\sum_{i=1}^{n} X_i \textit{ will be within } t \textit{ of its s.d.'s of its mean } \mu \textit{ is approximately } \int_{-t}^{t} \frac{1}{\sqrt{2\pi}} e^{-u^2/2}\, du,$$

which is the probability mass within $t$ standard deviations of the mean of the standard normal distribution.

In the case in which $X_1, \ldots, X_n$ are independent normal random variables with mean $\mu$ and variance $\sigma^2$, it follows from Example 4.6 of Chapter 5 that the

Central Limit Theorem is exact, that is,

$$P\left\{\frac{1}{\sqrt{n}}\sum_{i=1}^{n}\frac{X_i-\mu}{\sigma}\le t\right\}=\int_{-\infty}^{t}\frac{1}{\sqrt{2\pi}}e^{-u^2/2}\,du,$$

whereas Berry has shown that in certain cases the error is of the order of magnitude given in Equation 3.3.

***It is traditional to use the Central Limit Theorem to compute a practical approximation to***

$$P\left\{\frac{1}{\sqrt{n}}\sum_{i=1}^{n}\frac{X_i-\mu}{\sigma}\le t\right\},$$

***that is, to use the approximation***

**3.7**
$$P\left\{\frac{1}{\sqrt{n}}\sum_{i=1}^{n}\frac{X_i-\mu}{\sigma}\le t\right\}\cong\int_{-\infty}^{t}\frac{1}{\sqrt{2\pi}}e^{-u^2/2}\,du.$$

It is wise to use caution in following this tradition; in cases in which an error in the "wrong direction" can be critical, it may be wiser to use the conservative Chebyshev inequality, or to make use of the error term. The error term is sometimes useful in showing that the actual error in using Equation 3.7 is not too great.

### *3.8 Example: Accuracy of a Wristwatch*

A wristwatch is known to make an error of at most half a minute per day, but on the average to keep good time. Hence, as a model for accuracy, we let $X_i$ be the error actually made on the $i$th day and assume that

(a) $X_1, X_2, \ldots$ are independent (this might be justified if the error is dependent on the wearer's physical activity and this activity is independent from day to day) and

(b) $X_i$ has a uniform density on $[-1/2, 1/2]$.

The total error made in a year is $\sum_{i=1}^{365} X_i$. We want to determine the probability that the watch is in error after a year by, at most, half an hour. Note that the maximum possible error is about three hours. To solve this problem, we wish to compute

$$P\left\{\left|\sum_{i=1}^{365} X_i\right|\le 30\right\}.$$

The way to use the Central Limit Theorem is to rewrite the event $\left\{\left|\sum_{i=1}^{365} X_i\right|\le 30\right\}$ so that it resembles the one in Equation 3.7. That is, we must convert

$$\sum_{i=1}^{365} X_i \quad\text{to}\quad \frac{1}{\sqrt{365}}\sum_{i=1}^{365}\frac{X_i-\mu}{\sigma}.$$

We proceed as follows: Note that $\mu = E(X_i) = 0$. Thus, Var $X_i = \int_{-1/2}^{1/2} x^2\,dx = 1/12$, and

$$\left\{\left|\sum_{i=1}^{365} X_i\right| \le 30\right\} = \left\{-30 \le \sum_{i=1}^{365} X_i \le 30\right\}$$

$$= \left\{\frac{-30}{\left(\frac{1}{\sqrt{12}}\right)} \le \sum_{i=1}^{365} \frac{X_i - \mu}{\sigma} \le \frac{30}{\left(\frac{1}{\sqrt{12}}\right)}\right\}$$

$$= \left\{-\frac{30\sqrt{12}}{\sqrt{365}} \le \frac{1}{\sqrt{365}} \sum_{i=1}^{365} \frac{X_i - \mu}{\sigma} \le \frac{30\sqrt{12}}{\sqrt{365}}\right\}.$$

That is,

$$\left\{\left|\sum_{i=1}^{365} X_i\right| \le 30\right\} = \left\{-\frac{30\sqrt{12}}{\sqrt{365}} \le \frac{1}{\sqrt{365}} \sum_{i=1}^{365} \frac{X_i - \mu}{\sigma} \le \frac{30\sqrt{12}}{\sqrt{365}}\right\}.$$

Thus, from Equation 3.2,

$$P\left\{\left|\sum_{i=1}^{365} x_i\right| \le 30\right\} \cong \int_{-30\sqrt{12}/\sqrt{365}}^{30\sqrt{12}/\sqrt{365}} \frac{1}{\sqrt{2\pi}} e^{-u^2/2}\,du \cong 1,$$

as can be seen from tables of the normal distribution function. That is, knowing that the daily error is uniform on $[-1/2, 1/2]$, and that daily errors are independent, it is almost certain that the yearly error does not exceed 30 minutes, subject only to the *error of our central limit theorem approximation. It can be shown that in the case of uniform random variables on* $[a, b]$, *the error in the Central Limit Theorem approximation does not exceed*

**3.9**
$$\frac{1}{\pi n}\left[\frac{2}{7} + 3.7e^{-n/8} + \frac{1}{3^{n/2}}\right]$$

where $n$ is the number of terms being added.† In the case considered here, the error is less than 1/4000. The upper bound for the error given by Equation 3.3 is about $2.598/\sqrt{n}$ for uniformly distributed $X_i$. In the case just treated, this bound is .136, which is quite inferior to the one given in 3.9. △

We now give an outline of a proof of the Central Limit Theorem in the case in which $X_i$ has a moment generating function.

Our arguments depend on the following "not unreasonable" result, which we state without proof (see Cramér [6]).

---

† This result, established by Rosenblatt, is unpublished.

### 3.10 Continuity Theorem

If $Y_1, Y_2, \ldots ; Y$ are random variables with m.g.f.'s $\varphi_{Y_1}, \varphi_{Y_2}, \ldots ; \varphi_Y$ defined for $|t| < t_0$ with $t_0 > 0$, and $\lim_{n\to\infty} \varphi_{Y_n}(t) = \varphi_Y(t)$ for each real $t$ in $(-t_0, t_0)$, then $\lim_{n\to\infty} F_{Y_n}(y) = F_Y(y)$ for each real $y$ at which $F_Y$ is continuous. In the theorem we are attempting to prove, we have

$$Y_n = \frac{1}{\sqrt{n}} \sum_{i=1}^{n} \frac{X_i - \mu}{\sigma} = \frac{1}{\sqrt{n}} \sum_{i=1}^{n} Z_i.$$

where $E(Z_i) = 0$ and Var $Z_i = 1$, by Lemma 3.11, Chapter 6. Since the $Z_i$ all have the same distribution, let us denote the common m.g.f. of $\varphi_{Z_i}$ of the $Z_i$ by $\varphi$. Then, applying Theorem 7.11, Chapter 6, sufficiently often, we have

$$\varphi_{Y_n}(t) = \left[\varphi\left(\frac{t}{\sqrt{n}}\right)\right]^n.$$

Now $\varphi(0) = E(e^{0\cdot Z}) = 1$, since $\varphi_Z(t) = E(e^{tZ})$. Furthermore, by Theorem 7.4 of Chapter 6, $\varphi'(0) = E(Z) = 0$ and $\varphi''(0) = E(Z^2) = \text{Var } Z = 1$. This suggests a Taylor expansion (see the Appendix)

$$\varphi(\tau) = \varphi(0) + \varphi'(0)\tau + \varphi''(0)\frac{\tau^2}{2} + \theta(\tau)\frac{\tau^2}{2}$$

where $\theta$ is a function such that $\lim_{\tau\to 0} \theta(\tau) = 0$. Hence

$$\varphi\left(\frac{t}{\sqrt{n}}\right) = \varphi(0) + \varphi'(0)\frac{t}{\sqrt{n}} + \varphi''(0)\frac{t^2}{2n} + \theta\left(\frac{t}{\sqrt{n}}\right)\frac{t^2}{2n}$$

$$\varphi_{Y_n}(t) = \left[1 + \frac{t^2}{2n} + \theta\left(\frac{t}{\sqrt{n}}\right)\frac{t^2}{2n}\right]^n = \left[1 + \frac{(t^2/2)(1 + \theta(t/\sqrt{n}))}{n}\right]^n.$$

We now use the fact that

$$\lim_{n\to\infty}\left(1 + \frac{x}{n}\right)^n = e^x$$

to see that for

$$Y_n = \frac{1}{\sqrt{n}} \sum_{i=1}^{n} \frac{X_i - \mu}{\sigma}$$

we have $\lim_{n\to\infty} \varphi_{Y_n}(t) = e^{t^2/2}$. But $e^{t^2/2}$ is the value, $\varphi_Y(t)$, of the m.g.f. of a standard normal random variable $Y$ (see Equation 7.7 in Chapter 6). Thus, from the

Continuity Theorem (3.10), we have

$$F_{Y_n}(y) = P\left\{\frac{1}{\sqrt{n}}\sum_{i=1}^{n}\frac{X_i-\mu}{\sigma}\le y\right\}\to\int_{-\infty}^{y}\frac{1}{\sqrt{2\pi}}e^{-u^2/2}\,du,$$

which concludes the proof of the Central Limit Theorem.

### 3.11 *Example: Repairs*

Suppose that a given type of electronic calculator contains four printed circuit boards. Let $p_i$ be the probability that a unit sent for repair needs $i$ of these boards replaced, $i = 1, 2, 3, 4$, where $p_1 = 1/2$, $p_2 = 1/4$, and $p_3 = 1/8 = p_4$. In 10,000 units sent for repair, what is the probability that fewer than 18,875 circuit boards will be needed?

We let $X_j$ be the number of circuit boards needing replacement on the $j$th calculator. The total number of circuit boards needed is $\sum_{j=1}^{10,000} X_j$. We assume $X_1, \ldots, X_{10,000}$ to be independent. From our assumptions

$$E(X_j) = \sum_{i=1}^{4} ip_i = 1\cdot\frac{1}{2}+2\cdot\frac{1}{4}+3\cdot\frac{1}{8}+4\cdot\frac{1}{8}=\frac{15}{8},$$

$$E(X_j^2) = \sum_{i=1}^{4} i^2p_i = \frac{37}{8}.$$

Since $\operatorname{Var} X_j = E(X_j^2) - [E(X_j)]^2$, by Lemma 3.8, Chapter 6 we have

$$\operatorname{Var} X_j = 71/64 \cong (1.054)^2.$$

The event of interest is $\left\{\sum_{j=1}^{10,000} X_j \le 18,875\right\}$. We want to rewrite this event in the form suitable to apply the approximation 3.7 suggested by the Central Limit Theorem. But

$$\begin{aligned}
P\left\{\sum_{j=1}^{10,000} X_j \le 18,875\right\} &= P\left\{\sum_{j=1}^{10,000}(X_j - 15/8) \le 18,875 - \frac{150,000}{8}\right\}\\
&= P\left\{\sum_{j=1}^{10,000}(X_j - 15/8) \le \frac{1,000}{8}\right\}\\
&\cong P\left\{1/100 \sum_{j=1}^{10,000}\frac{X_j - 15/8}{1.054} \le \frac{1,000}{843.2}\right\}\\
&\cong P\left\{\frac{1}{\sqrt{10,000}}\sum_{j=1}^{10,000}\frac{X_j - E(X_j)}{\sigma_{X_j}} \le 1.185\right\}\\
&\cong .882
\end{aligned}$$

by Equation 3.7. Thus, the chances of needing fewer than 18,875 circuit boards are about .882. Note that the expected number of circuit boards needed is $n\mu = 10{,}000 \cdot 15/8 = 18{,}750$. We shall see in Exercise 3.12(9) that the upper bound for the error in this computation is .0293. △

### 3.12 *Example: Electron Arrivals*

Suppose that in any given microsecond (one millionth of a second), the probability that $k$ electrons will arrive at the plate of a vacuum tube is

$$e^{-.1}\frac{(.1)^k}{k!}.$$

What is the probability that the number of arrivals in one second lies between 99,500 and 101,000?

Here we assume that if $N_1, N_2, \ldots, N_k$ are the arrival numbers in disjoint time periods, then $N_1, \ldots, N_k$ are independent random variables. Now we know that the assumptions here lead to the conclusion that the desired probability is

$$\sum_{j=99{,}500}^{101{,}000} e^{-100{,}000}\frac{(100{,}000)^j}{j!},$$

using the fact that the sum of independent Poisson variables is Poisson (see Example 4.5, Chapter 5). Since this expression is of very little computational use to us, we turn to the Central Limit Theorem. Let $X_i$ be the random variable whose value is the number of electron arrivals in the $i$th microsecond. We want to compute

$$P\left\{99{,}500 \le \sum_{i=1}^{10^6} X_i \le 101{,}000\right\}.$$

As in the previous example, we rewrite the event of interest in a form resembling 3.7. To do this, we note that $\mu = E(X_i) = .1$, $\sigma = \sqrt{\text{Var } X_i} = \sqrt{.1}$.

Then

$$\begin{aligned}
&\left\{99{,}500 \le \sum_{i=1}^{10^6} X_i \le 101{,}000\right\} \\
&\quad = \left\{99{,}500 - 100{,}000 \le \sum_{i=1}^{10^6} (X_i - \mu) \le 101{,}000 - 100{,}000\right\} \\
&\quad = \left\{\frac{-500}{10^3\sqrt{.1}} \le \frac{1}{\sqrt{n}}\sum_{i=1}^{n}\frac{X_i - \mu}{\sigma} \le \frac{1{,}000}{10^3\sqrt{.1}}\right\} \\
&\quad \cong \left\{-1.58 \le \frac{1}{\sqrt{n}}\sum_{i=1}^{n}\frac{X_i - \mu}{\sigma} \le 3.16\right\}.
\end{aligned}$$

Thus, by the Central Limit Theorem, the desired probability is approximately

$$\int_{-1.58}^{3.16} \frac{1}{\sqrt{2\pi}} e^{-u^2/2}\, du \simeq .942.$$

We can use Equation 3.3 to determine the accuracy of this approximation.

To do this, we note that

$$\begin{aligned}\Gamma = \frac{E[|X_i - \mu|^3]}{\sigma^3} &= \frac{E[|X_i^3 - 3\mu X_i^2 + 3\mu^2 X_i - \mu^3|]}{\sigma^3} \\ &= \frac{E[|X_i^3 - .3X_i^2 + .03X_i - .001|]}{\sigma^3} \\ &\leq \frac{E\,|X_i^3| + .3E\,|X_i^2| + .03E\,|X_i| + .001}{.1^{3/2}}\end{aligned}$$

Using the m.g.f. of a Poisson random variable or direct computation, we obtain

$$\begin{aligned}E(X_i) &= \lambda = .1 \\ E(X_i^2) &= \lambda^2 + \lambda = .11 \\ E(X_i^3) &= \lambda^3 + 3\lambda^2 + \lambda = .131.\end{aligned}$$

Thus

$$\Gamma \leq \frac{.168}{.1^{3/2}} \simeq \frac{.168}{.0316} \simeq 5.34.$$

From Equation 3.3, we see that $2\Gamma/\sqrt{n} \leq 10.68/10^3 = .01068$, so that the error in our approximation does not exceed .02136. △

---

## 3.13 EXERCISES

1. Let $X_1, \ldots, X_{100}$ be uniformly distributed on [0, 1].

(a) Find an approximation to $P\left\{\sum_{i=1}^{100} X_i \geq 60\right\}$.

(b) Using Equation 3.9, determine an upper bound for the error in part (a).

2.° Let $S_{200}$ be the number of successes in 200 independent experiments, each of which has probability .75 of success. (Note that $S_{200} = \sum_{i=1}^{200} X_i$ where the $X_i$ are independent with $P\{X_i = 1\} = p = 1 - P\{X_i = 0\}$.)

(a) Find an approximation to the probability of at least 155 successes.

(b) Using Equation 3.3, find an upper bound for the error in your approximation. (A better upper bound will be presented in the optional part of this section.)

3. In mathematical computations on a large digital computer, numbers are "chopped" after eight decimal digits. Assuming, therefore, that the error is uniform on $[0, 10^{-8}]$:

(a) approximately what is the probability that in 1,000,000 additions the error exceeds $(1/2) \cdot 10^{-2} + 10^{-5}$?

(b) Use Equation 3.9 to find an upper bound for the percentage error in your computation of the probability of part (a).

4.° Forty-nine pieces of material are to be fitted together to form one large section. If the error made in each piece is uniform on $[-1/8, 1/8]$ and the errors are independent:

(a) Approximately what is the probability that the magnitude of the error exceeds 1/4?

(b) Use Equation 3.9 to determine an upper bound for the error in your approximation. (This problem arose from the actual attempt to sew pleats that would fit into the sleeve of an academic gown. The answer was quite realistic in implying a lack of success in this endeavor.)

5. Suppose that lightbulbs have a lifetime governed by the exponential density, and that if $T_i$ is the lifetime of the $i$th bulb, then

$$f_{T_i}(t) = \begin{cases} \dfrac{1}{200} e^{-t/200} & \text{for} \quad t \geq 0 \\ 0 & \text{otherwise.} \end{cases}$$

Using a Central Limit Theorem approximation, determine how many bulbs should be purchased so that with probability at least .95 the supply should last for at least six months in a store that uses a total of 150 bulbs, each for 8 hours per day, 5 days per week.

6. Suppose you know that the number of gallons of gasoline used by your car per mile is between .0833 and .1 (10 to 12 miles per gallon). You have a 3300-mile trip to make. Find numbers $a$ and $b$ such that the probability that the number of gallons needed lies between $a$ and $b$ is approximately .95. Assume that the number of gallons per mile has a uniform distribution and that the numbers of gallons needed for disjoint mile-length intervals are independent.

7.° Lots of 100 units, of which 90 are good and 10 are defective, are sampled 100 times in one day. If each sample consists of 5 units, what is the approximate probability that the total number of defectives observed is between 40 and 55?
Hint: If you find computation of the mean and variance for the hypergeometric distribution too unpleasant, assume that the sampling is done with replacement.

8. Suppose that for each $\ell$ the number of defects on a magnetic tape of length $\ell$ feet is a Poisson random variable with mean $\ell/600$. Approximately what is the distribution of the total number of defects in a lot of 100 reels, each 1800 feet long?

9.° Referring to Example 3.11, show that the error in the computation there does not exceed .0293.

---

### *3.14 PROBLEM*

In the game of bridge we assign "points" to the cards as follows:

Ace, 4; King, 3; Queen, 2; Jack, 1.

All other cards are assigned 0 points. In a bridge hand of 13 cards, find:

(a) the expected number of points.

(b) the variance of the number of points.

Hint: Let $X_i$ be the number of points on the $i$th card dealt to the hand. $E(X_i)$ is easy to find. Hence, so is $E\left(\sum_{i=1}^{13} X_i\right)$, the desired expectation. To find $\text{Var} \sum_{i=1}^{13} X_i$ you must determine $E(X_iX_j)$. For this, you must determine

$$P\{X_i = a, X_j = b\} = \begin{cases} P\{X_1 = a, X_2 = b\} & \text{if} \quad i \neq j \\ P\{X_1 = a\} & \text{if} \quad i = j \text{ and } a = b. \end{cases}$$

But $P\{X_1 = a, X_2 = b\} = P\{X_1 = a\}P\{X_2 = b \mid X_1 = a\}$.

(c) Find an approximation to the probability that in 100 hands the total number of points fails to exceed 900.

We now focus our attention on a particularly important special case of the Central Limit Theorem.

### 3.15 The Normal Approximation to the Binomial Distribution

Suppose that $S_n$ is a binomial random variable with parameters $n$ and $p$. Then $S_n$ represents the number of successes in $n$ independent repetitions of an experiment whose probability of success is $p$. Thus, we may write

$$S_n = \sum_{i=1}^{n} X_i$$

where $X_1, \ldots, X_n$ are independent with $P\{X_i = 1\} = p = 1 - P\{X_i = 0\}$. Here $X_i$ represents the number of successes on the $i$th repetition of the experiment.

The Central Limit Theorem is certainly applicable to the approximate computation of $P\{a \leq S_n \leq b\}$. In fact, since $E(X_i) = p$, $\text{Var } X_i = p(1 - p)$, we see that

$$\begin{aligned} P\{a \leq S_n \leq b\} &= P\left\{a \leq \sum_{i=1}^{n} X_i \leq b\right\} \\ &= P\left\{a - np \leq \sum_{i=1}^{n} (X_i - p) \leq b - np\right\} \\ &= P\left\{\frac{a - np}{\sqrt{n}\sqrt{p(1-p)}} \leq \frac{1}{\sqrt{n}} \sum_{i=1}^{n} \frac{X_i - p}{\sqrt{p(1-p)}} \leq \frac{b - np}{\sqrt{n}\sqrt{p(1-p)}}\right\} \\ &= P\left\{\frac{a - np}{\sqrt{np(1-p)}} \leq \frac{1}{\sqrt{n}} \sum_{i=1}^{n} \frac{X_i - \mu}{\sigma} \leq \frac{b - np}{\sqrt{np(1-p)}}\right\} \\ &\cong \int_{(a-np)/\sqrt{np(1-p)}}^{(b-np)/\sqrt{np(1-p)}} \frac{1}{\sqrt{2\pi}} e^{-u^2/2}\, du. \end{aligned}$$

The contents of the foregoing discussion may be summarized as follows: If $S_n$ is a binomial random variable with parameters $n$ and $p$, then for fixed $p$ and large

$n$, we have the approximation

$$P\{a \leq S_n \leq b\} = \sum_{k=a}^{b} \binom{n}{k} p^k(1-p)^{n-k} \cong \int_{(a-np)/\sqrt{np(1-p)}}^{(b-np)/\sqrt{np(1-p)}} \frac{1}{\sqrt{2\pi}} e^{-u^2/2}\, du.$$

We now clarify a confusing point regarding the existence of two approximations to the binomial distribution, the Poisson approximation developed in Section 2 and the normal approximation. The important point to bear in mind is that ***the Poisson approximation is useful when $n$ is large, $p$ is small, and $np$ is moderate*** (more precisely, when $p$ is small and $np^3$ is small). ***The normal approximation is more useful when $n$ is large but $p$ is near*** $1/2$ (or for fixed $p$ as $n$ grows large).

In the optional part to follow, we state and apply a very useful result that Uspensky developed [20] concerning the normal approximation to the binomial distribution.

### *3.16 Theorem: Normal Approximation to Binomial [Uspensky]*

Let $Z_1, \ldots, Z_n$ be independent, with $P\{Z_1 = 1\} = p = 1 - P\{Z_i = 0\}$. Then, if $b$ is a non-negative integer,

$$P\left\{\sum_{i=1}^{n} Z_i \leq b\right\} = \frac{1}{\sqrt{2\pi}} \int_{-\infty}^{w} e^{-u^2/2}\, du + \frac{1-2p}{6\sqrt{2\pi np(1-p)}}(1-w^2)e^{-w^2/2} + \varepsilon$$

where

$$w = \frac{b - np + \frac{1}{2}}{\sqrt{np(1-p)}} \qquad \text{and} \qquad |\varepsilon| \leq \frac{.065 + .09\,|1-2p|}{np(1-p)} + \tfrac{1}{2}e^{-1.5\sqrt{np(1-p)}},$$

provided $np(1-p) \geq 25$.

### *3.17 Example:* 100 *Tosses of a Fair Coin*

Find upper and lower bounds for the probability that the number $S_{100}$ of heads of 100 tosses of a fair coin satisfies $40 \leq S_{100} \leq 60$.

Here $S_{100} = \sum_{i=1}^{100} Z_i$ where $Z_i$ is the number of heads on the $i$th toss. For this, we shall find bounds for

$$P\{S_{100} \leq 60\} \qquad \text{and} \qquad P\{S_{100} \leq 39\},$$

using Uspensky's approximation. Now,

$$P\{S_{100} \leq 60\} = \frac{1}{\sqrt{2\pi}} \int_{-\infty}^{10.5/5} e^{-u^2/2}\, du + \varepsilon_{60}$$

where

$$|\varepsilon_{60}| \leq \frac{.065}{25} + \frac{e^{-7.5}}{2} \cong .0029.$$

Furthermore,

$$P\{S_{100} \leq 39\} = \frac{1}{\sqrt{2\pi}} \int_{-\infty}^{-10.5/5} e^{-u^2/2}\, du + \varepsilon_{39}$$

where

$$|\varepsilon_{39}| \leq .0029.$$

Thus

$$P\{40 \leq S_{100} \leq 60\} \cong \frac{1}{\sqrt{2\pi}} \int_{-10.5/5}^{10.5/5} e^{-u^2/2}\, du \cong .9642$$

with an error of at most .0058. That is,

$$.9584 \leq P\{40 \leq S_{100} \leq 60\} \leq .9700.$$

Note that the computed error bound $|\varepsilon| \leq .0029$ is valid for all computations concerning 100 tosses of a fair coin.

By way of comparison with the general Central Limit Theorem, the Berry result (3.3) yields the following:

$$P\{40 \leq S_{100} \leq 60\} \cong \int_{-2}^{2} \frac{1}{\sqrt{2\pi}} e^{-u^2/2}\, du \cong .9545,$$

with an error of at most .4 (since

$$\frac{2\Gamma}{\sqrt{n}} = \frac{1}{5} \frac{E\,|X_i - .5|^3}{.5^3} = \frac{1}{5},$$

but the error bound must be doubled because we are computing a "two-sided" inequality). This is a much less precise result. △

### *3.18 Example: Dice*

A die is tossed 6000 times. Obtain upper and lower bounds for $P\{950 \leq N_1 \leq 1030\}$, where $N_1$ is the total number of 1's that turn up. As before, we first compute upper and lower bounds for

$$P\{N_1 \leq 949\} \quad \text{and} \quad P\{N_1 \leq 1030\}.$$

First we find

$$w_{949} = \frac{949 - 1000 + \tfrac{1}{2}}{\sqrt{5000/6}} \cong -1.75 \quad \text{and} \quad w_{1030} = \frac{1030 - 1000 + \tfrac{1}{2}}{\sqrt{5000/6}} \cong 1.057;$$

$$\begin{aligned} P\{N_1 \leq 949\} &= \int_{-\infty}^{-1.75} \frac{1}{\sqrt{2\pi}} e^{-u^2/2}\, du + \frac{\tfrac{2}{3}}{6\sqrt{\pi \cdot 10000/6}} (1 - 1.75^2) e^{-17.5^2/2} + \varepsilon_{949} \\ &\cong .0401 - .0007 + \varepsilon_{949} \\ &= .0394 + \varepsilon_{949} \end{aligned}$$

where

$$|\varepsilon_{949}| \leq \frac{.065 + .06}{5000/6} + e^{-1.5\sqrt{5000/6}}/2 \cong .00015.$$

$$\begin{aligned} P\{N_1 \leq 1030\} &= \int_{-\infty}^{1.057} \frac{1}{\sqrt{2\pi}} e^{-u^2/2}\, du + \frac{2/3}{6\sqrt{\pi \cdot 10000/6}} (1 - 1.057^2)e^{-1.057^2/2} + \varepsilon_{1030} \\ &= .8547 + .00005 + \varepsilon_{1030} \\ &\cong .85475 + \varepsilon_{1030}. \end{aligned}$$

Here the upper bound for $\varepsilon_{1030}$ is the same as that for $\varepsilon_{949}$, that is, $|\varepsilon_{1030}| \leq .00015$. Thus we find

$$P\{950 \leq N_1 \leq 1030\} = .81535 + \varepsilon$$

where $|\varepsilon| \leq .0003$. That is,

$$.81505 \leq P\{950 \leq N_1 \leq 1030\} \leq .81565.$$

We again compare this result with that obtained using Berry's result (3.3), which yields

$$P\{950 \leq N_1 \leq 1030\} \cong \int_{-1.73}^{1.04} \frac{1}{\sqrt{2\pi}} e^{u^2/2}\, du \cong .809$$

with an error of at most .1 (since here $2\Gamma \cong 3.86$). Again, this is not nearly as good a result as that furnished by Uspensky's theorem. △

---

## 3.19 EXERCISES

1. An experiment has probability .75 of success. Find upper and lower bounds for the probability of at least 155 successes in 200 repetitions of this experiment.

2.° Consider an experiment with probability .7 of success. How many times should the experiment be repeated so that the probability of at least 80 successes is greater than or equal to .9?

3. Suppose that injected penicillin causes an allergic reaction in 10 per cent of people. Find an upper bound for the probability of fewer than 120 reactions among 1000 people injected with penicillin.

The final topic in this section is the Central Limit Theorem for random vectors. We are including this topic mainly to help the reader understand why the bivariate normal density is used as a model for a pair of measurements (for example, a student's high school and college performance). We first must generalize the concept of independent random variables.

### 3.20 *Definition: Independent Random Vectors*

Let $X_1, Y_1, X_2, Y_2, \ldots, X_n, Y_n$ be random variables with joint distribution function $F_{X_1,Y_1,\ldots,X_n,Y_n}$. We say that the random vectors $(X_1, Y_1), \ldots, (X_n, Y_n)$ are mutually independent if for all sequences $x_1, y_1, \ldots, x_n, y_n$ of reals we have $F_{X_1,Y_1,\ldots,X_n,Y_n}(x_1, y_1, \ldots, x_n, y_n) = F_{X_1,Y_1}(x_1, y_1) \cdots F_{X_n,Y_n}(x_n, y_n)$. Note that $X_i$ and $Y_i$ may be dependent.

### 3.21 *Vector Central Limit Theorem (Bivariate)*

Let $(X_1, Y_1), \ldots, (X_n, Y_n), \ldots$ be mutually independent random vectors, each with the same joint distribution function $F_{X_i,Y_i} = F$, and suppose $E(X_i) = \mu_x$, $\operatorname{Var} X_i = \sigma_x^2 > 0$, $E(Y_i) = \mu_y$, $\operatorname{Var} Y_i = \sigma_y^2 > 0$, and the correlation coefficient $\rho_{X_i,Y_i} = \rho$. Let

$$V_n = \frac{1}{\sqrt{n}} \sum_{i=1}^{n} \frac{X_i - \mu_x}{\sigma_x}, \qquad W_n = \frac{1}{\sqrt{n}} \sum_{i=1}^{n} \frac{Y_i - \mu_y}{\sigma_y}.$$

Then

$$\lim_{n\to\infty} F_{V_n,W_n}(v, w) = \lim_{n\to\infty} P\{V_n \le v, W_n \le w\}$$

$$= \int_{-\infty}^{v} \int_{-\infty}^{w} \frac{1}{2\pi\sqrt{1-\rho^2}} \exp \frac{-1}{2(1-\rho^2)} [x^2 - 2\rho xy + y^2]\, dy\, dx.$$

This theorem can be proved by an extension of the method previously used, defining the bivariate m.g.f. $\varphi_{X,Y}(s, t) = E[e^{sX+tY}]$.

In practical applications leading to random variables $V_n$ and $W_n$, we approximate $F_{V_n,W_n}$ by the bivariate normal distribution with both means 0, unit variances and correlation coefficient $\rho$.

The bivariate normal model is used extensively to describe the behavior of pairs of measurements $(V, W)$ when it is plausible to think of $V$ and $W$ as sums similar to $V_n$ and $W_n$. For example, if $V$ is high school performance and $W$ is college performance, we may think of $V$ as a sum $\sum_i X_i$ and of $W$ as a sum $\sum_i Y_i$ where $X_i$ is the contribution of a given factor (such as physical endurance) to high school performance, $Y_i$ is the contribution to college performance of this factor.

Note that we may have $Y_i$ actually a function of $X_i$ but $\rho = 0$ (see Exercise 5.7(1), Chapter 6). However, the vector Central Limit Theorem shows that when $\rho = 0$, the random variables $\sum_{i=1}^{n} X_i$ and $\sum_{i=1}^{n} Y_i$ tend to "become" independent as $n$ grows large. Thus we see from the vector Central Limit Theorem that $\rho$ is, in a sense, a measure of the dependence of $\sum_{i=1}^{n} X_i$ and $\sum_{i=1}^{n} Y_i$.

*Final Remarks on the Central Limit Theorem.* It should be mentioned that the conclusion of the Central Limit Theorem holds under more general conditions than the ones given here. If the given random variables are dependent, but not "too" dependent, or have different distributions that are, however, not "too" different, then the limiting normality still holds. However, the reader is cautioned that a very large $n$ may be required before the approximation is useful. For example, if $X_i$ are independent Poisson with mean .1, then from tables of the Poisson distribution we find that

$$P\left\{\sum_{i=1}^{200} X_i \leq 16\right\} = .2211.$$

The Central Limit Theorem approximation yields

$$P\left\{\sum_{i=1}^{200} X_i \leq 16\right\} \cong .1854,$$

which is in error by a substantial percentage.

## SECTION 4
## Limiting Distribution of Sample Quantiles

Recall that we proposed the quantiles as an alternative to using moments to describe a distribution. In the last chapter, we showed that for large $n$, it is highly probable that the sample quantile $X_{(p),n}$ is close to the population quantile $\theta_p$ (see Theorem 9.8 and Problem 9.9(1) in Chapter 6). In this section we shall derive and apply a general theorem concerning the large-sample distribution of the sample quantiles.

### 4.1 *Theorem: Limiting Distribution of Sample Quantiles*

Suppose $X_1, X_2, \ldots, X_n$ are independent random variables having a density $f > 0$. If $f$ has a derivative that is continuous on some non-zero length interval centered around the $p$-quantile, $\theta_p$, then

**4.2**
$$\lim_{n\to\infty} P\left\{\sqrt{\frac{n}{p(1-p)}}\, f(\theta_p)[X_{(p),n} - \theta_p] \leq y\right\} = \int_{-\infty}^{y} \frac{e^{-u^2/2}}{\sqrt{2\pi}}\, du.$$

*Remarks.* Actually, it is possible to prove a much stronger theorem under fewer restrictions on $f$, namely that if $f > 0$ is continuous in a non-zero length interval centered at $\theta_p$, then

$$\lim_{n\to\infty} P\left\{\sqrt{\frac{n}{p(1-p)}}\, f(\theta_p)[X_{(p),n} - \theta_p] \in A\right\} = \int_{A} \frac{e^{-u^2/2}}{\sqrt{2\pi}}\, du.$$

Derivation of this latter result is too tedious to include here. In most "practical" applications, the right side of Equation 4.2 is used to approximate

$$P\left\{\sqrt{\frac{n}{p(1-p)}}\, f(\theta_p)[X_{(p),n} - \theta_p] \leq y\right\}$$

for "reasonably large $n$." We defer the proof of Theorem 4.1 until the end of this section.

### 4.3 *Example: Sample Median*

Suppose $X_1, \ldots, X_{25}$ are independent random variables with a Cauchy density

$$f(x) = \frac{1}{\pi(1 + x^2)}.$$

Recall how these random variables arose from a spinning pointer (Problem 3.5(3), Chapter 4). Also recall that these random variables do not have an expectation. We ask for the probability that the sample median of the $X_j$ lies in $[-.5, .5]$, that is,

$$P\{-.5 \leq X_{[13],25} \leq .5\}.$$

Now $\theta_{.5} = 0$, $f(\theta_{.5}) = 1/\pi$. Thus

$$\begin{aligned} P\{-.5 \leq X_{[13],25} \leq .5\} &= P\left\{-\frac{5}{\pi} \leq 2\sqrt{n}\, f(\theta_{.5})[X_{(.5),n} - \theta_{.5}] \leq \frac{5}{\pi}\right\} \\ &= P\left\{2\sqrt{n}\, f(\theta_{.5})[X_{(.5),n} - \theta_{.5}] \in \left[-\frac{5}{\pi}, \frac{5}{\pi}\right]\right\} \\ &= \int_{-5/\pi}^{5/\pi} \frac{1}{\sqrt{2\pi}} e^{-u^2/2}\, du \\ &\cong .888. \end{aligned}$$

Note that in the case of a single observation, $X_i$,

$$P\{-.5 \leq X_i \leq .5\} = \tfrac{1}{3}[\arctan .5 - \arctan(-.5)] \cong \frac{2}{\pi} \times .465 \cong .296.$$

It follows from Theorem 4.1 that in a sense the sample median settles down to the true median at a rate of about $1/\sqrt{n}$. It can also be shown that the sample average

$$\frac{1}{n} \sum_{i=1}^{n} X_i$$

still has a Cauchy density and does not settle down at all as $n$ grows large (see Problem 9.9(3), Chapter 6). △

## *4.4 EXERCISES*

1.° Suppose $X_1, \ldots, X_{49}$ are independent random variables with a Cauchy density

$$f(x) = \frac{1}{\pi(1 + x^2)}.$$

Find approximately $P\{-1.25 \leq X_{(1/4),49} \leq -.75\}$.

2. Suppose $X_1, \ldots, X_n$ are distributed as in Exercise 1. How large should $n$ be so that $P\{-1.1 \leq X_{(1/4),n} \leq -.9\} \cong .95$?

3.° For $X_1, \ldots, X_n$ independent and normal with $E(X_i) = 0$ and $\operatorname{Var} X_i = 1$, how large should $n$ be so that $P\{-.05 \leq X_{(.5),n} \leq .05\} \geq .9$? Compare this with the sample size $n$ required in order that

$$P\left\{-.05 \leq \frac{1}{n}\sum_{i=1}^{n} X_i \leq .05\right\} \geq .9.$$

4. Suppose $X_1, \ldots, X_{100}$ are uniformly distributed on $[-1/2, 1/2]$. Find approximately:

(a) $P\{X_{(.5),100} \in [-.1; .1]\}$.
(b) $P\{X_{(.25),100} \leq -.33\}$.

*Proof of Theorem 4.1* (suggested by Professor David Moore of Purdue University). We note that

**4.5**
$$\sqrt{\frac{n}{p(1-p)}}\, f(\theta_p)[X_{(p),n} - \theta_p] \leq y$$

if and only if

$$X_{(p),n} \leq \theta_p + \sqrt{\frac{p(1-p)}{n}}\,\frac{y}{f(\theta_p)}.$$

Define $\delta_n$ by the equation

**4.6**
$$\theta_{p+\delta_n} = \theta_p + \sqrt{\frac{p(1-p)}{n}}\,\frac{y}{f(\theta_p)}.$$

Then we see that 4.5 holds precisely when

$$X_{(p),n} \leq \theta_{p+\delta_n}.$$

But $X_{(p),n} \leq \theta_{p+\delta_n}$ if and only if a proportion $p$ of the variables $X_1, \ldots, X_n$ assume a value less than or equal to $\theta_{p+\delta_n}$. We let

$$Z_i = \begin{cases} 1 & \text{if} \quad X_i \leq \theta_{p+\delta_n} \\ 0 & \text{otherwise,} \end{cases}$$

and note that

**4.7**
$$P\{Z_i = 1\} = p + \delta_n.$$

Furthermore,

$$X_{(p),n} \leq \theta_{p+\delta_n} \qquad \text{if and only if} \qquad \sum_{i=1}^{n} Z_i \geq [np]$$

where $[np]$ is the smallest integer greater than or equal to $np$. Thus

$$P\left\{\sqrt{\frac{n}{p(1-p)}} f(\theta_p)[X_{(p),n} - \theta_p] \leq y\right\} = P\left\{\sum_{i=1}^{n} Z_i \geq [np]\right\}$$

$$= 1 - P\left\{\sum_{i=1}^{n} Z_i \leq [np] - 1\right\}.$$

Now, using Uspensky's result (3.16), we identify $b$ with $[np] - 1$ and $p$ with $p + \delta_n$ (in light of 4.7) to see that this last expression equals

**4.8** $$1 - \frac{1}{\sqrt{2\pi}}\int_{-\infty}^{w_n} e^{-u^2/2}\,du - \frac{1 - 2(p+\delta_n)}{6\sqrt{2\pi n(p+\delta_n)(1-p-\delta_n)}}(1 - w_n^2)e^{-w_n^2/2} + \varepsilon_n$$

$$= P_n \qquad + \qquad c_n \qquad + \varepsilon_n$$

where

**4.9** $$w_n = \frac{[np] - 1 - n(p+\delta_n) + \tfrac{1}{2}}{\sqrt{n(p+\delta_n)(1-p-\delta_n)}}$$

and

$$|\varepsilon_n| \leq \frac{.065 + .09\,|1 - 2(p+\delta_n)|}{n(p+\delta_n)(1-p-\delta_n)} + \tfrac{1}{2}e^{-1.5\sqrt{n(p+\delta_n)(1-p-\delta_n)}}$$

and we must have

$$n \geq \frac{25}{(p+\delta_n)(1-p-\delta_n)}.$$

In order to prove the theorem, we shall show that $c_n$ and $\varepsilon_n$ approach 0 as $n$ grows large, and $\lim_{n\to\infty} w_n = -y$, so that

$$1 - \frac{1}{\sqrt{2\pi}}\int_{-\infty}^{w_n} e^{-u^2/2}\,du \to 1 - \frac{1}{\sqrt{2\pi}}\int_{-\infty}^{-y} e^{-u^2/2}\,du = \frac{1}{\sqrt{2\pi}}\int_{-\infty}^{y} e^{-u^2/2}\,du.$$

The assertion concerning $c_n$ follows from the fact that $g(w) = (1 - w^2)e^{-w^2/2}$ represents a bounded function, and $\lim_{n\to\infty} \delta_n = 0$ due to the existence of a positive continuous density.

Even simpler arguments handle $\varepsilon_n$.

To handle $w_n$, we note that applying the distribution function $F$ to both sides of Equation 4.6, we obtain

$$p + \delta_n = F\left(\theta_p + \sqrt{\frac{p(1-p)}{n}}\,\frac{y}{f(\theta_p)}\right)$$

or

$$\delta_n = F\left(\theta_p + \sqrt{\frac{p(1-p)}{n}}\,\frac{y}{f(\theta_p)}\right) - p.$$

Now applying Taylor's Theorem (Appendix A1.5) to $F$, we find that

$$\delta_n = F(\theta_p) + F'(\theta_p)\sqrt{\frac{p(1-p)}{n}}\,\frac{y}{f(\theta_p)} + F''(c)\frac{p(1-p)}{2n}\,\frac{y^2}{f^2(\theta_p)} - p.$$

That is,

$$\delta_n = p + y\sqrt{\frac{p(1-p)}{n}} + f'(c)\,\frac{p(1-p)}{2n}\,\frac{y^2}{f^2(\theta_p)} - p.$$

Hence, selective substitution in Equation 4.9 yields

$$w_n = \frac{[[np]] - np - \tfrac{1}{2}]}{\sqrt{n(1+\delta_n)(1-p-\delta_n)}} - \frac{y\sqrt{p(1-p)} + f'(c)\,p(1-p)\,y^2/[2\sqrt{n}\,f^2(\theta_p)]}{\sqrt{(1+\delta_n)(1-p-\delta_n)}}.$$

Now, using the fact that $\lim_{n\to\infty} \delta_n = 0$, it follows that $\lim_{n\to\infty} w_n = -y$, proving the theorem.

We went through all this detail because these computations can actually yield very good approximations for

$$P\left\{\sqrt{\frac{n}{p(1-p)}}\,f(\theta_p)[X_{(p),n} - \theta_p] \le y\right\}.$$

Knowing $F$, we can evaluate the terms in Equation 4.8, using $P_n + c_n$ as the approximation, with an error smaller than the bound on $w_n$.

---

*4.10 PROBLEM°*

Using the methods of the proof just given, find upper and lower bounds for $P\{X_{[201],401} \le \theta_{1/2} + .05\}$ when $X_1, \ldots, X_{401}$ are independent with density

$$f(x) = \frac{1}{\pi[1 + (x-\theta)^2]}.$$

## SECTION 5
## Final Remarks

This chapter constitutes only an introduction to the subject of approximations in probability. We have omitted some topics, such as the approximation of the hypergeometric distribution (sampling without replacement) by the binomial, and the important limit theorems connected with $X_{[1],n}$ and $X_{[n],n}$, the so-called "extreme value" limiting distributions, which are important in such applications as flood control.

Nor did we treat sums in which the number of terms is itself a random quantity. This area of probability is of great practical importance.

Much research is needed, not so much in proving more limit theorems, but in the development of simple, yet accurate, approximations.

# 8

# Distributions Related to the Normal

This chapter is useful mainly for *reference* purposes in later chapters. It is concerned with distributions that arise as a result of statistical procedures applied to normal random variables.

The first distributions we consider are the Chi square distributions. These distributions arise when we attempt to use independent normal data with mean $\mu$ and variance $\sigma^2$ to learn something about $\sigma^2$.

## 1 Definition: Chi Square Distribution with $n$ Degrees of Freedom

Let $X_1, \ldots, X_n$ be independent normal random variables with $E(X_i) = 0$, $\operatorname{Var} X_i = 1$. The variable $\chi_n^2 = \sum_{i=1}^{n} X_i^2$ is said to have a *Chi square distribution with n degrees of freedom.*

In Exercise 7.13(3)(c), Chapter 6, we found the m.g.f. of $\chi_n^2$ to be given by

**2** $$\varphi_{\chi_n^2}(t) = (1 - 2t)^{-n/2} \qquad \text{for} \qquad t < \tfrac{1}{2}.$$

We can easily verify by direct computation of its m.g.f. that the density

$$f_{\chi_n^2}(z) = \begin{cases} \dfrac{z^{n/2-1}e^{-z/2}}{2^{n/2}\Gamma(n/2)} & \text{for } z \geq 0, \text{ where } \quad \Gamma(\alpha) = \displaystyle\int_0^\infty x^{\alpha-1}e^{-x}\,dx, \\ 0 & \text{for } z < 0, \end{cases}$$

has $\varphi_{\chi^2_n}$ as its m.g.f. Furthermore, it follows from 2 and from Theorem 7.4, Chapter 6, which relates moments and m.g.f's, that

**3** $$E(\chi^2_n) = n, \text{ Var } \chi^2_n = E(\chi^{2^2}_n) - [E(\chi^2_n)]^2 = (n+2)n - n^2 = 2n.$$

The Chi square distribution is usually tabled only for $n = 1, \ldots, 30$, although there are tables available for larger $n$. The tables usually go up only to $n = 30$, because there exists a rather accurate approximation based on the Central Limit Theorem (Theorem 3.1, Chapter 7) for $n \geq 30$. Before deriving this approximation, we remark that a direct application of the Central Limit Theorem to $\chi^2_n = \sum_{i=1}^{n} X_i^2$ apparently does not lead to results as accurate as those we are about to derive.

**4** ***Approximation to*** $P\{\chi^2_n \leq y\}, y > 0$ (from R. A. Fisher)

$$P\{\chi^2_n \leq y\} \cong \int_{-\infty}^{\sqrt{2y}-\sqrt{2n-1}} \frac{1}{\sqrt{2\pi}} e^{-t^2/2}\, dt, \qquad n \geq 30,$$

where $\chi^2_n$ is a random variable having the Chi square distribution with $n$ degrees of freedom.

We shall first show that

$$\lim_{n\to\infty} P\{\sqrt{2\chi^2_n} - \sqrt{2n-1} \leq x\} = \int_{-\infty}^{x} \frac{1}{\sqrt{2\pi}} e^{-t^2/2}\, dt,$$

from which we see that

$$P\{\chi^2_n \leq y\} = P\{2\chi^2_n \leq 2y\} = P\{\sqrt{2\chi^2_n} \leq \sqrt{2y}\}$$

$$= P\{\sqrt{2\chi^2_n} - \sqrt{2n-1} \leq \sqrt{2y} - \sqrt{2n-1}\} \cong \int_{-\infty}^{\sqrt{2y}-\sqrt{2n-1}} \frac{1}{\sqrt{2\pi}} e^{-t^2/2}\, dt.$$

To do this, note that $\chi^2_n = \sum_{i=1}^{n} X_i^2$ where $X_i$ are independent with a standard normal distribution. Hence, we may apply the Central Limit Theorem to see that the standardized variable

$$\frac{\chi^2_n - n}{\sqrt{2n}}$$

has approximately a standard normal distribution function. That is,

$$\lim_{n\to\infty} P\left\{\frac{\chi^2_n - n}{\sqrt{2n}} \leq x\right\} = \int_{-\infty}^{x} \frac{1}{\sqrt{2\pi}} e^{-t^2/2}\, dt.$$

But

$$\lim_{n\to\infty} P\left\{\frac{\chi_n^2 - n}{\sqrt{2n}} \le x\right\} = \lim_{n\to\infty} P\left\{\frac{\chi_n^2 - n}{\sqrt{2n}} \le x + \frac{x^2 - 1}{2\sqrt{2n - 1}}\right\}$$

$$= \lim_{n\to\infty} P\left\{\frac{\chi_n^2 - n}{\sqrt{2n - 1}} \le x + \frac{x^2 - 1}{2\sqrt{2n - 1}}\right\}$$

$$= \lim_{n\to\infty} P\left\{\chi_n^2 \le n + \left[x + \frac{x^2 - 1}{2\sqrt{2n - 1}}\right]\sqrt{2n-1}\right\}$$

$$= \lim_{n\to\infty} P\left\{\chi_n^2 \le n + (\sqrt{2n - 1})x + \frac{x^2 - 1}{2}\right\}$$

$$= \lim_{n\to\infty} P\left\{2\chi_n^2 \le 2n - 1 + 2(\sqrt{2n - 1})x + x^2\right\}$$

$$= \lim_{n\to\infty} P\{\sqrt{2\chi_n^2} \le \sqrt{2n - 1} + x\}$$

$$= \lim_{n\to\infty} P\{\sqrt{2\chi_n^2} - \sqrt{2n - 1} \le x\}.\dagger$$

So,

$$\lim_{n\to\infty} P\{\sqrt{2\chi_n^2} - \sqrt{2n - 1} \le x\} = \int_{-\infty}^{x} \frac{1}{\sqrt{2\pi}} e^{-t^2/2}\, dt,$$

as asserted.

## 5 *Example:* $P\{\chi_{25}^2 \le 41.566\}$

We present this example to give an idea of the approximation just presented in 4. From tables of the Chi square distribution, we see that $P\{\chi_{25}^2 \le 41.566\} = .98$. From Approximation 4, we have

$$P\{\chi_{25}^5 \le 41.566\} \cong \int_{-\infty}^{\sqrt{83.132}-7} \frac{1}{\sqrt{2\pi}} e^{-t^2/2}\, dt$$

$$= \int_{-\infty}^{2.1177} \frac{1}{\sqrt{2\pi}} e^{-t^2/2}\, dt \cong .983.$$

A direct Central Limit Theorem approximation yields

$$P\{\chi_{25}^2 \le 41.566\} = P\left\{\frac{\chi_{25}^2 - 25}{\sqrt{50}} \le \frac{16.566}{\sqrt{50}}\right\}$$

$$P\left\{\frac{\chi_{25}^2 - 25}{\sqrt{50}} \le \frac{16.566}{7.0711}\right\} = P\left\{\frac{\chi_{25}^2 - 25}{\sqrt{50}} \le 2.3427\right\} \cong .9904,$$

which is not nearly as good as the result with Approximation 4. △

† All these arguments make use of both the continuity of the limiting normal distribution and the uniform convergence as shown by the Berry error term (Equation 3.3, Chapter 7).

**6 Definition: Non-Central Chi Square.** If $X_1, \ldots, X_n$ are independent normal random variables with $E(X_i) = \mu_i$ and $\operatorname{Var} X_i = 1$, then

$$Y = \sum_{i=1}^{n} X_i^2$$

is said to have a *non-central Chi square distribution with n degrees of freedom and non-centrality parameter* $\mathcal{N} = \frac{1}{2} \sum_{i=1}^{n} \mu_i^2$.

Although the density of $Y$ does not have a simple formula, it is easy to compute its m.g.f. as follows:

$$\varphi_{\chi_i^2}(t) = \int_{-\infty}^{\infty} e^{z^2 t} \frac{1}{\sqrt{2\pi}} e^{-(z-\mu_i)^2/2} \, dz$$

$$= \frac{1}{(1-2t)^{1/2}} e^{t\mu_i^2/(1-2t)}, \quad t \leq \tfrac{1}{2}$$

using the same completion of the square as in Example 3.3 in Chapter 5. Hence, by applying Theorem 7.11 of Chapter 6, on the m.g.f. of linear combinations of independent random variables, suitably often, we find

$$\varphi_Y(t) = \prod_{i=1}^{n} \left\{ \frac{1}{(1-2t)^{1/2}} e^{t\mu_i^2/(1-2t)} \right\}$$

$$= \frac{1}{(1-2t)^{n/2}} e^{(\mu_1^2 + \cdots + \mu_n^2)t/(1-2t)}$$

that is,

$$\varphi_Y(t) = \frac{1}{(1-2t)^{n/2}} e^{2\mathcal{N}t/(1-2t)}. \tag{7}$$

This shows that this distribution depends on the $\mu_i$ only through the non-centrality parameter, $\mathcal{N} = \frac{1}{2} \sum_{i=1}^{n} \mu_i^2$.

Distributions that arise when we want to learn something about the mean when the variance is unknown are the $t$ distributions. They were first used by W. S. Gosset, writing under the pseudonym of "Student" in 1908.

**8 Definition: Student $t$ Distribution.** Let $Y$ be a random variable with standard normal distribution ($E(Y) = 0$, $\operatorname{Var} Y = 1$), and let $\chi_n^2$ be independent of $Y$ and have a Chi square distribution with $n$ degrees of freedom. The random variable $T_n = Y/\sqrt{\chi_n^2/n}$ is said to have a Student $t$ distribution with $n$ degrees of freedom.

Using the techniques of Section 5, Chapter 5, it can be shown that

$$f_{T_n}(t) = \frac{\Gamma\left(\frac{n+1}{2}\right)}{\sqrt{n\pi}\,\Gamma\left(\frac{n}{2}\right)} \left(1 + \frac{t^2}{n}\right)^{-(n+1)/2} \quad \text{for } t \text{ real}$$

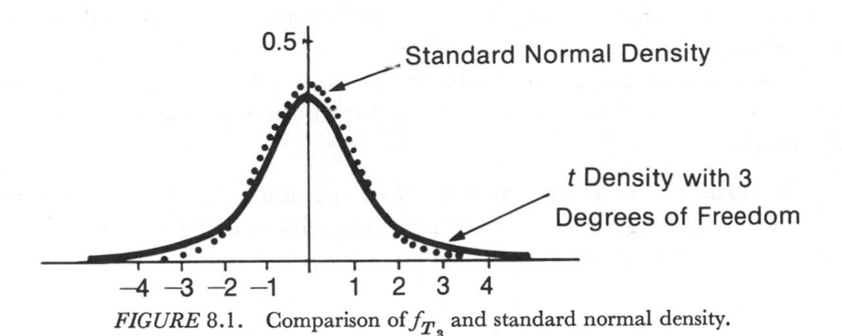


*FIGURE* 8.1. Comparison of $f_{T_3}$ and standard normal density.

where $\Gamma(\alpha) = \int_0^\infty x^{\alpha-1}e^{-x}\,dx$. Now, since $X_i$ are standard normal, $E(X_i^2) = 1$. So

$$\chi_n^2/n = \frac{1}{n}\sum_{i=1}^{n} X_i^2$$

should be tending to 1 as $n$ becomes large. Hence, the distribution of $T_n$ should be close to that of $Y$, the standard normal, for large $n$. This is in fact the case, although because of the greater variability (due to the non-constant denominator $\sqrt{\chi_n^2/n}$ of $T_n$) we find that $f_{T_n}$ is more spread out than the standard normal density. In fact, $T_n$ does not possess any moments beyond the $(n-1)$st, and hence has no moment generating function. In Figure 8.1, we compare $f_{T_3}$ with the standard normal density.

In Chapter 14 we will show that if $Y_1, \ldots, Y_n$ are independent normal random variables with $E(Y_i) = \mu$, $\text{Var } Y_i = \sigma^2$, then

$$\overline{Y}_{(n)} = \frac{1}{n}\sum_{i=1}^{n} Y_i \quad \text{and} \quad S_n^2 = \frac{1}{n-1}\sum_{i=1}^{n}(Y_i - \overline{Y}_{(n)})^2$$

are independent. Furthermore, $(n-1)S_n^2/\sigma^2$ has a Chi square distribution with $n-1$ degrees of freedom. Hence, the random variable

**9**
$$\frac{\sqrt{n}(\overline{Y}_{(n)} - \mu)/\sigma}{\sqrt{S_n^2/\sigma^2}} = \frac{\sqrt{n}(\overline{Y}_{(n)} - \mu)}{S_n}, \quad \text{where} \quad S_n = \sqrt{S_n^2}$$

has the $t$ distribution with $n-1$ degrees of freedom.

**10 Definition: Non-Central *t* Distribution.** Let $Y$ and $\chi_n^2$ be given as in 8, and let $\mu$ be a constant. The random variable

$$T_{n,\mu} = \frac{Y + \mu}{\sqrt{\chi_n^2/n}}$$

is said to have a non-central $t$ distribution with $n$ degrees of freedom and non-centrality parameter $\mu$.

It is easily seen that for large $n$, the random variable $T_{n,\mu}$ should be approximately normal with mean $\mu$ and unit variance.

Regression analysis is concerned with the determination of relationships between measured quantities. A fundamental class of distributions in this study are Fisher's "$F$" distributions.

**11 Definition: *F* Distribution.** Let $\chi_m^2$ and $\chi_n^2$ denote independent random variables having Chi square distributions with $m$ and $n$ degrees of freedom, respectively. The random variable

$$\mathscr{F}_{m,n} = \frac{\chi_m^2/m}{\chi_n^2/n}$$

is said to have the $F$ distribution with $(m, n)$ degrees of freedom.

Using the methods of Section 5, Chapter 5, it can be shown that the density $f$ of $\mathscr{F}_{m,n}$ is given by

$$f(x) = \begin{cases} \dfrac{m}{n} \dfrac{\Gamma\left(\dfrac{m+n}{2}\right)}{\Gamma\left(\dfrac{m}{2}\right)\Gamma\left(\dfrac{n}{2}\right)} \left(\dfrac{m}{n} x\right)^{(m/2)-1} \left(1 + \dfrac{m}{n} x\right)^{-(m+n)/2} \\ 0 \quad \text{for} \quad x \le 0. \end{cases}$$

**12 Definition: Non-Central *F* Distribution.** Let $\chi_{m,\mathscr{N}}^2$ and $\chi_n^2$ denote independent random variables having the non-central Chi square distribution with $m$ degrees of freedom and non-centrality parameter $\mathscr{N}$, and the Chi square distribution with $n$ degrees of freedom, respectively. The ratio

$$\mathscr{F}_{m,n,\mathscr{N}} = \frac{\chi_{m,\mathscr{N}}^2/m}{\chi_n^2/n}$$

is said to have a non-central $F$ distribution with $(m, n)$ degrees of freedom and non-centrality parameter $\mathscr{N}$.

# 9

# Introduction to Statistical Inference

## SECTION 1
## Introduction

This chapter is meant to serve only as a preview of the subject of statistical inference. Therefore, our objectives are rather limited. We want (i) to formulate some of the kinds of problems that are most frequently encountered in statistics and (ii) to introduce a few of the concepts and approaches for handling these problems.

Mainly, we hope to give the reader a taste of statistics by presenting a small sample of statistical situations, problems, and solutions.

Since we will not be developing much theoretical structure in this chapter, the reader should not *yet* expect to be able to attack statistical problems in a very systematic fashion. We will work on developing this ability in some of the later chapters. However, if we succeed in the immediate aims of this chapter, the reader will no longer feel strange or uncomfortable when confronted by a statistical problem.

As we noted before, problems of probability are primarily concerned with computing probabilities $P(A)$ of events $A$ or approximations to such probabilities. These computations make use of related information, such as other assigned probabilities, conditional probabilities, or (as in the Chebyshev inequality) certain moments. In probability problems, it is assumed that we are working with a fixed probability measure about which we have partial or complete information.

In statistical problems, on the other hand, we assume that we are faced with a *class* $\mathscr{P}$ of probability measures, one of which describes (or "governs") the phenomenon that we are going to observe. Since the particular $P$ in $\mathscr{P}$ is generally unknown to us, we shall attempt to use observed data to narrow down the class $\mathscr{P}$ of possible governing measures, that is, to help discover something about the "true" governing $P$. We illustrate these ideas in the following example.

### 1.1 *Example: Probability Contrasted with Statistics*

In *probability* we might ask "*What is the probability of at least* 5300 *heads in* 10,000 *tosses of a fair coin?*"

In *statistics* we might ask "*If, in* 10,000 *tosses of a coin,* 5300 *heads are observed, is it reasonable to believe that the coin is fair?*" △

Note that in the probability question we are dealing with a single known probability measure, namely the binomial distribution with parameters $n = 10{,}000$ and $p = 1/2$. In the statistical question, we may take as our class $\mathscr{P}$ the set of all binomial distributions with $n = 10{,}000$ and $0 \leq p \leq 1$.

We observe in the above statistical problem that the value of the parameter $p = P\{H\}$ is likely to be reflected in the values of the data. If $p$ is large, we expect to see a high proportion of heads. As we shall see, it is this fact that allows the observed number of heads to give us information about the value of the governing $p$.

Viewed in this fashion, statistics may be thought of as the inverse of probability. Since the governing $P$ in $\mathscr{P}$ determines which data to expect, we may use the actual observations to determine which $P$ in $\mathscr{P}$ are reasonable descriptions. In a sense, this is the essence of statistics.

In Example 1.1, we should observe that the statistical problem is much less clearly phrased than the probabilistic one. Since this is so often the case, part of our task in developing statistical theory is to clarify such questions.

There are several other features of statistical problems that contribute to the difficulty of the subject. Since we are allowed to use data to help narrow down the class of governing probability measures, it follows that we may have some choice about which data to observe. That is, we may have a choice as to which experiments to perform. This may involve a choice among different measuring instruments, as well as a choice of the amount of data and of the method by which we determine when to stop observing data.

Unfortunately, adequate criteria for making such decisions are usually not available in most realistic situations. For instance, in choosing which experiments to perform, we may be forced to balance our time and effort against financial costs. Worse yet, in some situations, such as the testing of a new insecticide, the costs may involve long-term effects on the environment that are not predictable from short-term measurements. Furthermore, the importance to be attached to any given effect, such as thermal pollution, is highly controversial. Even in apparently simple situations, we may encounter formidable obstacles. For example, it may be extremely difficult to obtain a representative sample from the population of interest. The public opinion pollsters in the 1948 Presidential election learned this the hard way—they confidently predicted a Dewey victory, but Truman won by over 2,000,000 votes (out of about 45,000,000). The pollsters had consistently picked a sample biased toward Republicans. Overcoming this bias was not at all simple. Also, a representative sample for a public opinion poll may be difficult to obtain if some of the people refuse to reply, or if they reply falsely. It was the assumption that people tell the truth about their auto preferences that was almost catastrophic for Chrysler Corporation in the early 1950's.

Another difficulty is that independence of the observations $X_1, X_2, \ldots$ may be very hard to come by. In a sequence of experiments, it is always tempting to "correct" for past "errors," thus destroying independence and making analysis much more difficult.

These difficulties should discourage the reader from attempting any cookbook applications of the theory to be developed. In fact, in light of this discussion, the reader should be aware that we are solving only a very limited class of problems in the remainder of this text; the importance of our results lies more in the approaches set forth than in the specific formulas. These approaches form the foundation for the attempts at solution of the more complicated problems we have discussed.

In the remainder of this chapter we shall lay the foundations for our classification of statistical problems and methods. We shall also present a few of the most frequently encountered elementary techniques and their associated mathematical models.

## SECTION 2
## Parametric and Non-Parametric Families

In the previous section, we noted how the *family* of binomial probability distributions was used to describe our knowledge of tosses of a coin of unknown bias. Generally, we use some family $\mathscr{P}$ of probability measures to describe a phenomenon of interest. To keep our approach to statistics as simple as possible, we find it convenient to make a rough division of families of distributions into two types.

### Type A: Parametric Families

Roughly speaking, we shall call a family of densities or probability functions *parametric* if there exist a finite number of scalar values, "parameters," (that is, a vector quantity) whose values completely determine a unique member of this family. That is, we are given the form of the distribution with the exception of the value of some unknown scalar or vector parameter.

### *2.1 Example: Binomial Family*

If $n$ is a given positive integer (sample size), then specification of $p$ (probability of success on any given trial) determines precisely one binomial distribution, which we denote by $B(n, p)$. In this case, the parameter $p$ is a scalar. This family might arise if we make a known number $n$ of repetitions of an experiment with unknown probability $p$ of success. △

### *2.2 Example: Binomial Family*

Someone may choose $n$ and $p$ (unknown to you) and furnish you with a sequence of results $S_1, S_2, \ldots, S_k$ each of which has a binomial distribution with parameters

$n$ and $p$. For instance, you may be given the total number of votes in a sequence of elections in a district with a stable population of $n$ voters, where the probability is $p$ that a voter will actually vote in any given election. From the values of $S_1, \ldots, S_k$, you may be asked to determine something about $n$ and $p$. For example, the data looking like 1, 0, 1, 0, 1, 2, 2, 0, 1, 0, 2 might very well suggest $n = 2$, and

$$p = \frac{1}{11} \sum_{j=1}^{11} \frac{S_j}{2} = \frac{10}{22}.$$

In this case, the family under consideration is the binomial family with vector parameter $(n, p)$. △

### 2.3 *Example: The Normal Family*

We recall that the normal density with mean $\mu$ and variance $\sigma^2$ is defined by

$$f(x, \mu, \sigma^2) = \frac{1}{\sqrt{2\pi\sigma^2}} e^{-(x-\mu)^2/(2\sigma^2)}.$$

Specification of $\mu$ and $\sigma^2$ completely determines a normal density.

If a measurement arises as a sum of many independent contributions, all roughly equal in size, then the central limit theorem furnishes a theoretical basis for assuming the measurement to be normally distributed.

In attempting to measure some physical constant with an instrument of known quality, we might know $\sigma^2$ but want to determine $\mu$. On the other hand, in attempting to calibrate a new measuring instrument by making measurements on some object with known attributes (such as a standard cell furnished by the Bureau of Standards), we might know $\mu$ but not $\sigma^2$. In describing the widths of skulls from a newly discovered burial ground, it is possible that both $\mu$ and $\sigma^2$ would be unknown. △

In statistical problems related to parametric families, we are usually interested in determining something about the parameter or about one of its coordinates.

### 2.4 *Example: Acceptance Sampling*

In deciding whether to accept (and pay for) a shipment of goods, we may test $n$ of the units. If our tests are of the simple type that classify a unit as either defective or non-defective, we may want to determine from our test whether or not the binomial parameter $p$ = probability of a good unit satisfies $p \geq .8$. △

### 2.5 *Example: Scientific Measurement*

We may have an instrument for measuring electrical resistance such that when the resistance of a unit is actually $\mu$, the measurement $X$ has a normal distribution with expectation $E(X) = \mu$ and variance $\sigma_X^2 = 24$. Here the expectation is determined solely by the resistance of the unit being measured; the variance is a property of the measuring instrument. That is, assuming the resistance to be a fixed constant, the measurements of this resistance would still vary because of uncontrollable experimental error. Based on one or more independent measurements, we may want to pin down the value of $\mu$ to an accuracy of .1. △

If we feel that we do not have enough information to have faith in a parametric model, we may go to a wider class of distributions. Such classes are our second type.

### Type B: Non-Parametric Families

Roughly speaking, a family is called *non-parametric* if, no matter how many (finite number) scalar parameters are specified, such specification does not generally uniquely determine a member of the family.

### *2.6 Example: Densities with Unique Median*

Recall that a median $\theta$ of a density $f$ is a value such that

$$\int_{-\infty}^{\theta} f(x)\, dx = \frac{1}{2}.$$

Problems often arise in which knowledge concerning the median is needed (for example, whether the median of some population exceeds a given acceptable value). In this case, it is common to treat the family of all densities that have a unique median. Here, the class is so large that no matter how many parameters we may specify, we are not, in general, able to narrow down the family to just one density. To get an idea of how large this class is, we note that it includes the normal family, the family of all uniform densities, the family of all everywhere positive densities, and so on. Hence, this family is classified as non-parametric. △

### *2.7 Example: Family of Unimodal Densities*

Of some interest is the family of all densities $f(x)$ that have a unique positive relative maximum. Such a density is illustrated in Figure 9.1. We classify this family as non-parametric for the same reasons as before. For example, specification of the median does not uniquely specify a single density of this family. Figure 9.2 shows many unimodal densities with the same median.

If all we know about our data is that it is governed by an unknown unimodal density, we may wish to know something about the median. For instance, we might want to determine if median income exceeds \$6000 in a population, and we might only be justified in assuming that income over the population has a unimodal density. △

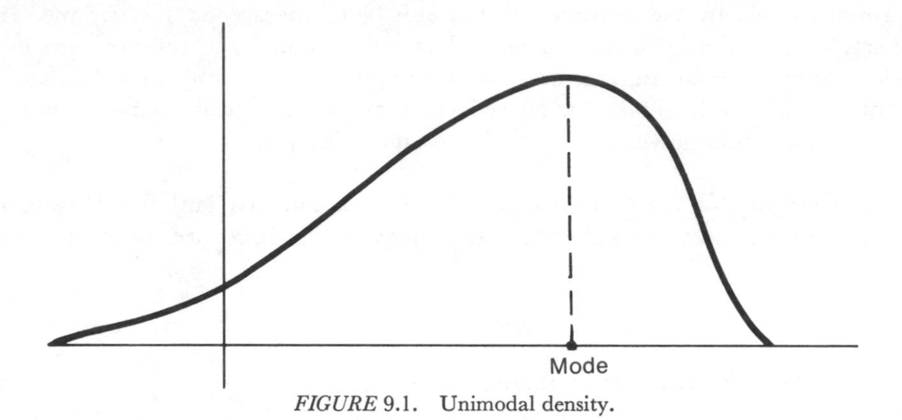


*FIGURE* 9.1. Unimodal density.

The reason for the rough classification just given is that the inference techniques that have been developed are in general different for the two types. Many efficient methods have been developed that require the assumption of a parametric class. However, it frequently happens that we encounter a situation in which we are forced to be more conservative and accept a non-parametric family, that is, we are faced with a situation in which the assumption that our observations are from some density in a non-parametric family is the only reasonable one. Even in these cases, we may be able (at some loss in efficiency over techniques applicable to parametric families) to learn something about some quantity determined by the true distribution, such as the median.

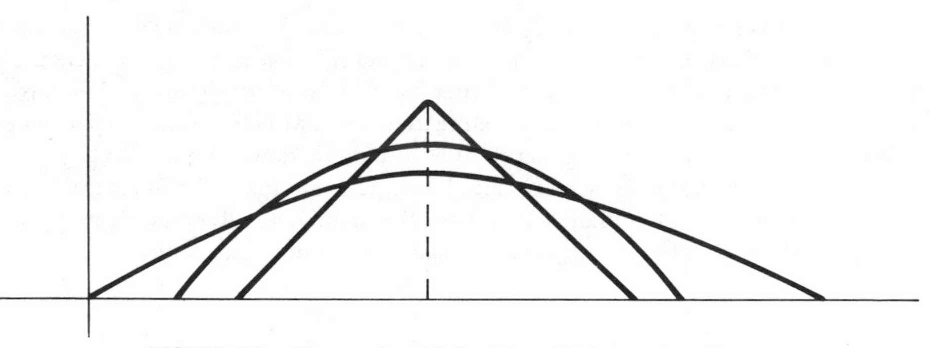

*FIGURE* 9.2. Three unimodal densities with the same median.

## 2.8 *EXERCISES*

In the following decide which families you would classify as parametric and which ones you would classify as non-parametric.

1.° The family of all Poisson distributions. (Usually we want to determine something about $\lambda = E(X)$ when dealing with the Poisson family.)

2.° The family of symmetric densities (that is, densities $f$ satisfying $f(c + x) = f(c - x)$ for some given $c$, and all $x$) that are everywhere positive. (Here we might be interested in finding out something about the value $c$ associated with the governing distribution.)

3. The family of uniform densities on $[0, \theta]$. (Here we frequently want to learn something about the mean, $\theta/2$.)

4.° The family of all continuous distribution functions. (If we have reason to believe that the true distribution $F$ is close to a given distribution $F_0$, we might be interested in

$$d(F, F_0) \equiv \sup_{\text{all } x} |F(x) - F_0(x)|$$

which is essentially the maximum vertical distance between the graphs of $F$ and $F_0$). The quantity $d(F, F_0)$ measures in some sense how far $F$ is from $F_0$; $d(F, F_0) = 0$ only if $F = F_0$.

5. The family of all Weibull distribution functions

$$F(x) = \begin{cases} 1 - e^{-(x-\theta)^m/\mu} & \text{for} \quad x \geq \theta \\ 0 & \text{for} \quad x < \theta \end{cases}$$

with $\theta$ real, $m > 0$ and $\mu > 0$. (The Weibull family is used as a possible description for lifetimes of components. Frequently we want to learn something about the expected lifetime $E(X)$.)

6.° The family of all exponential distribution functions

$$F(x) = \begin{cases} 1 - e^{-x/\theta} & \text{for} \quad x \geq 0 \\ 0 & \text{for} \quad x < 0 \end{cases}$$

where $\theta > 0$. (Here $E(X)$ is usually of interest.)

7. The family of all life distribution functions $F$ that have a positive density $f$ for all $x > 0$ and that have a non-decreasing failure rate, that is, intuitively those life distributions in which wear-out is evidenced. Mathematically $f(x) = 0$ for $x < 0$ and

$$\frac{f(x)}{1 - F(x)}$$

is non-decreasing in $x$ for $x > 0$ (see Definition 2.28, Chapter 4). (We frequently desire to learn something about the expected lifetime of a random quantity with such a distribution.)

---

## 2.9 PROBLEMS

Furnish a class of distributions that you feel is large enough to include at least one that adequately describes the given variable. If possible, you might also furnish a non-parametric class whenever you have supplied a parametric one. The answers here are of course not unique; if possible, they should be based on previous theoretical results. If you can, try to justify your choice. You might find the index helpful here.

1. The number of phone calls initiated between 2:00 and 3:00 P.M. on a weekday.

2.° The number of phone calls initiated between 4:30 and 5:30 P.M. on a Thursday.

3. Roundoff error on one computation on a digital computer.

4.° The roundoff errors $E_1, \ldots, E_n$ in $n$ computations on a digital computer.

5.° The length of time it takes a metal bar to break in a fatigue test.

6. The number of phone calls $X$ initiated between 2:00 and 3:00 P.M. and the number of phone calls $Y$ initiated between 3:00 and 4:30 P.M. weekdays.

7. The lifetimes of lightbulbs that have been pretested for 10 hours.

8.° The lifetimes of lightbulbs fresh off the assembly line.

9. The values of measurements of the speed of light.

10.° Measurements of the percentage yield in an organic chemistry reaction.

11. The time between successive radioactive decays.

Hint: See 2.7 in Chapter 7 and Problem 4.9(3) in Chapter 5.

12.° The heart rate of an adult at rest selected at random from the population of the United States.

13. The heart rate of a man at rest selected at random from the population of the United States.

14. The amount of time between successive heart beats of an individual at rest.

15. The amount of time a given type of cell spends before splitting into two cells.

## SECTION 3
## Point Estimation

In studying either parametric or non-parametric families, it is often reasonable to try to narrow down the class being considered by using the data to learn something about one of the parameters. Whether or not such a process is useful depends on the particular problem being considered. In trying to learn something about a parameter, we might first want some reasonable estimate of its value. We have at our disposal the data, that is, the observed values of certain random variables $X_1, X_2, \ldots, X_n$. Any function $g(X_1, \ldots, X_n)$ that can be computed without knowledge of the value of the unknown parameter is a candidate for an estimate of the parameter. For future reference we formalize this concept in the following definition.

**3.1 Definition: Statistic, Point Estimate.** A *statistic* (also called an *observable statistic*) is an observable function $g(X_1, \ldots, X_n)$ of the data $X_1, \ldots, X_n$. That is to say, we can compute the value of $g(X_1, \ldots, X_n)$ directly from the values of $X_1, \ldots, X_n$, without knowledge of any of the unknown parameters. A statistic used to estimate a parameter is called a *point estimate* of that parameter.

Subsequently, we shall develop criteria that indicate which statistics make good estimators, but just now we shall give some examples of situations in which a reasonable estimate is not difficult to present.

### *3.2 Example: Mean of Normal Random Variables*

If $X_1, X_2, \ldots, X_n$ are independent random variables that are normally distributed with mean $\mu$ and variance $\sigma^2$, we claim that a reasonable estimate of $\mu$ is given by

$$g(X_1, \ldots, X_n) = \bar{X}_{(n)} \equiv \frac{1}{n} \sum_{i=1}^{n} X_i.$$

The reasons are as follows. We note that for $n$ sufficiently large, $\bar{X}_{(n)}$ is "close" to $\mu$, in the sense that

$$P\{|\bar{X}_{(n)} - \mu| \leq \delta\} = \int_{-\sqrt{n}\delta/\sigma}^{\sqrt{n}\delta/\sigma} \frac{1}{\sqrt{2\pi}} e^{-t^2/2}\, dt$$

approaches 1 as $n$ becomes large (see Theorem 7.16, Chapter 6). We note further that $E(\bar{X}_{(n)}) = \mu$ no matter what the values of $\mu$ and $\sigma$ are (see Corollary 4.3, Chapter 6). This gives us somewhat more confidence in $\bar{X}_{(n)}$ as an estimator of $\mu$. Note that for large $\sigma$, we need a large value of $n$ to obtain given "accuracy" in estimation. If $\sigma$ is known, as it may be when $\sigma$ is a characteristic of a calibrated measuring instrument, then $n$ can usually be predetermined so as to obtain certain desired characteristics of the estimator. However, if $\sigma$ is not known, then a preliminary first stage sample may be useful for learning something about $\sigma$. This preliminary sample may be thought of as a "calibration stage." With the knowledge gained in the first stage, we can usually design a second stage of sampling in which precise information about $\mu$ can then be obtained.

Another candidate that seems reasonable as an estimator for $\mu$ is the sample median (the "middle observation") [see Definition 9.6 and Theorem 9.8, both in Chapter 6], since in the normal case the population median and population mean are the same. Later we shall consider various reasons for deciding which estimator we wish to use.

Another candidate, which might appear reasonable at first glance, is $(X_{[1],n} + X_{[n],n})/2$, the average of the two extreme values (largest and smallest values). It can be shown that this candidate is not a satisfactory estimate of $\mu$ in this case. △

### 3.3 *Example: Mean of Uniform Distribution*

Suppose that $X_1, X_2, \ldots, X_n$ are independent random variables all having a common density that is uniform on the interval $[0, \theta]$, $\theta > 0$, but otherwise unknown. There are several reasonable estimates for $\theta$, all of which tend to be close to $\theta$ for large $n$:

(a) $2\bar{X}_{(n)} = \frac{2}{n}\sum_{i=1}^{n} X_i$

(b) $2X_{(1/2),n}$ where $X_{(1/2),n}$ is the sample median.

(c) $X_{[n],n}$ the largest observation. △

### 3.4 *Example: Parameter of Poisson Distribution*

If $X_1, X_2, \ldots, X_n$ are radiation counts, then one reasonable model is to assume that they are independent Poisson random variables with

$$P\{X_i = k\} = e^{-\lambda}\frac{\lambda^k}{k!}$$

where $\lambda$ is a positive constant (see Theorem 2.1 in Chapter 7). Since $\lambda = E(X_i)$, we may use the Chebyshev inequality (Theorem 3.14 in Chapter 6) to show that

$$\bar{X}_{(n)} = \frac{1}{n}\sum_{i=1}^{n} X_i$$

is likely to be close to $\lambda$ for sufficiently large $n$. Hence, $\bar{X}_{(n)}$ is one reasonable estimate for $\lambda$. Since we also have

$$\lambda = \text{Var } X_i = E[(X_i - \lambda)^2] \equiv E(Z_i),$$

another reasonable estimate for $\lambda$ is

$$\frac{1}{n}\sum_{i=1}^{n} Z_i$$

where $Z_i = (X_i - \bar{X}_{(n)})^2$. △

People often use point estimates solely as "educated guesses" for parameters, claiming "anything is better than nothing," or "we have to make a decision, and it's the best we've got." Actually, point estimators are a basis for meaningful and useful solutions to certain problems in testing and interval estimation as we shall see in the two chapters to follow.

---

### *3.5 PROBLEMS*

1. Suppose that $X_1, X_2, \ldots, X_n$ are independent Bernoulli random variables, that is, 0 with probability $1 - p$ and 1 with probability $p$. What function(s) might be reasonable estimators for $p$?

2. Suppose that $X_1, X_2, \ldots, X_n$ are independent normal random variables with mean $\mu$ and variance $\sigma^2$.
   (a)° What seems to be a good estimator of $\sigma^2$ if $\mu = 0$?
   (b) What seems to be a reasonable estimator of $\sigma$ if $\mu = 0$?
   (c) If $\mu$ is known, what might be a good estimator of $\sigma^2$?
   (d)° If $\mu$ is unknown, what might be a good estimator of $\sigma^2$?
Hint: Replace $\mu$ by a reasonable estimate of $\mu$.

3.° Let $X_1, \ldots, X_n$ be independent Bernoulli variables, as in Problem 1. What is a reasonable estimator of $p(1 - p)$?

4. Let $X_1, \ldots, X_n$ be independent with density

$$f(x) = \frac{1}{\pi(1 + [x - \theta]^2)}$$

where $\theta$ is an unknown real parameter. What is a reasonable estimator for $\theta$?
Hint: $P\{X_i \leq \theta\} = 1/2$.

5. Suppose that $X_1, X_2, \ldots$ is a sequence of independent Bernoulli random variables and that you wish to find an estimate of $1/p$ where $0 < p = P\{X_i = 1\}$. You may take

as many observations as you wish (although you must stop sampling eventually). How would you sample and, given your sample, what is your estimate for $1/p$?
Hint: Your estimator must be well defined. But, if you simply take $n$ observations and let $g(X_1, \ldots, X_n) = 1/\bar{X}_{(n)}$ you may have $\bar{X}_{(n)} = 0$.

## SECTION 4
## Interval Estimation, Confidence Intervals

In most scientific and engineering problems, an educated guess of the value of an unknown parameter is not an acceptable answer. Being told that the speed of light is approximately 299,773 kilometers/sec. is usually not sufficient for someone who needs to make use of this parameter in some investigation, such as determining distance by means of radar. To illustrate this situation, we quote a well-known physics text: "In an exhaustive analysis of all work since 1928, E. N. Dorsey of the National Bureau of Standards concludes that the best value to date [of the speed of light in a vacuum] is $c = 299{,}773$ kilometers per second, which *he believes correct* to within 10 kilometers per second." Notice that a whole range of values is given, and that the conclusion is not asserted with certainty, but rather with high credibility.

This approach to presenting scientific results is today almost universally accepted. We go one step further and actually assign a numerical value to the degree of credibility, that is, we shall actually give a lower bound for the probability of correctness of our interval assertion about the unknown parameter. This amounts to giving an interval estimate together with the probability of being correct. To formalize this concept, suppose that the data being observed is governed by some probability measure $P$ that is known only to be a member of some given family $\mathscr{P}$. We are interested in the value of some scalar parameter $\theta$ that would be known if we knew $P$. Recall from Definition 3.1 that a statistic is an observable function of the data. Let $\mathbf{X}$ denote the data (which may be a vector) and let $\alpha$ be given with $0 < \alpha < 1$. Then we have the following definition.

**4.1 Definition: $1 - \alpha$ Level Confidence Intervals and Confidence Limits.** Let $g_L(\mathbf{X})$ and $g_R(\mathbf{X})$ be statistics such that ***for all P in*** $\mathscr{P}$,

$$P\{g_L(\mathbf{X}) \leq \theta \leq g_R(\mathbf{X})\} \geq 1 - \alpha.$$

Then the (random) interval $[g_L(\mathbf{X}), g_R(\mathbf{X})]$ is called a *confidence interval of level*† $1 - \alpha$ for the parameter $\theta$ (relative to the class $\mathscr{P}$). If $P\{\theta \leq g_R(\mathbf{X})\} \geq 1 - \alpha$ for all $P$ in $\mathscr{P}$, then we call $g_R(\mathbf{X})$ an upper $1 - \alpha$ level *confidence limit.* Lower confidence limits are defined similarly.

So if we observe the data $\mathbf{x}$ and form the interval $[g_L(\mathbf{x}), g_R(\mathbf{x})]$, then $[g_L(\mathbf{X}), g_R(\mathbf{X})]$ is a $1 - \alpha$ level confidence interval for $\theta$ provided that the assertion $\theta \in [g_L(\mathbf{X}), g_R(\mathbf{X})]$ has probability at least $1 - \alpha$ of being correct for all $P$ under

† Some statisticians insist that

$$(*) \qquad \inf_{P \in \mathscr{P}} \{g_L(\mathbf{X}) \leq \theta \leq g_R(\mathbf{X})\} = 1 - \alpha$$

in order that the *level* be defined as $1 - \alpha$. We find it more convenient to define the level as in 4.1, and to say that the *exact level* is $1 - \alpha$ if (*) holds. If for all $P$ in $\mathscr{P}$ we have $P\{g_L(\mathbf{X}) \leq \theta \leq g_R(\mathbf{X})\} = 1 - \alpha$, then we will say that the level is *identically* $1 - \alpha$.

consideration (that is, all $P$ in $\mathscr{P}$). Note that $g_L$ and $g_R$ must in no way be determined by $P$, that is, $g_L(\mathbf{X})$ and $g_R(\mathbf{X})$ are *statistics* determined from the data alone. Thus, roughly $100(1 - \alpha)$ per cent (or more) of assertions of the form "$\theta \in [g_L(\mathbf{X}), g_R(\mathbf{X})]$" should be correct in the long run. We associate the probability $1 - \alpha$ with sequences of such assertions rather than with any specific one. The fact that the proportion of correct assertions is so high in the long run is what gives us faith in the correctness of any specific one.

At this stage, we present some examples in order to obtain a better understanding of this concept.

### 4.2 *Example: Confidence Interval for Normal Mean*

Suppose that we have a measuring instrument to measure the speed of sound in miles per hour. As our model, let us assume that when the speed of sound is $\mu$, the instrument registers the value of a normal random variable $X$ where $E(X) = \mu$ (unknown to us) and $\operatorname{Var} X = 1$. Then $X - \mu$ (the instrument error that is unobservable and hence is *not* a statistic) has the completely known density

$$f(t) = \frac{1}{\sqrt{2\pi}} e^{-t^2/2},$$

so that

$$P\{-1.96 \le X - \mu \le 1.96\} = \int_{-1.96}^{1.96} \frac{1}{\sqrt{2\pi}} e^{-t^2/2}\, dt \simeq .95.$$

Figure 9.3 illustrates this.
But we may rewrite the event

$$\{-1.96 \le X - \mu \le 1.96\} = \{X - 1.96 \le \mu \le X + 1.96\}.$$

Therefore, no matter which normal distribution with variance 1 governs $X$, we have

$$P\{X - 1.96 \le \mu \le X + 1.96\} \simeq .95.$$

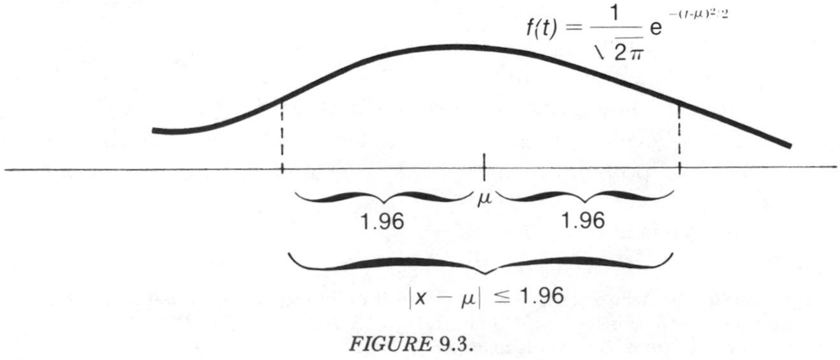

*FIGURE* 9.3.

Thus we see that the random interval $[X - 1.96, X + 1.96]$ is a .95 level confidence interval for $\mu$. That is, the assertion

$$\text{``}\mu \text{ is in } [X - 1.96, X + 1.96]\text{''}$$

has probability not less than .95 of being correct, no matter which $\mu$ is the correct value of the parameter. Here

$$g_L(\mathbf{X}) = X - 1.96, \qquad g_R(\mathbf{X}) = X + 1.96.$$

Note that no matter what the speed $\mu$ of sound is, the distribution of $X$ is such that $X$ always has probability .95 of being within 1.96 units of $\mu$.

Such a result is not at all surprising if measurements are made with an instrument whose error is independent of what is being measured. We shall see later that confidence intervals can be constructed in many cases even when the measurement error depends on the value of the quantity being measured. △

In the example just presented, we first determined an event whose probability is high for all $P$ in $\mathscr{P}$. This event involves the unknown parameter $\mu$. We rewrote this event so that the unknown parameter was the center term of an inequality and the two extreme terms were observable statistics. The extreme terms are the end points of the confidence interval. Essentially, this is the type of procedure we go through whenever we construct a confidence interval.

### 4.3 *Example: Confidence Interval for Binomial p*

In public opinion polls, we often wish to obtain a confidence interval for the proportion $p$ of voters who prefer a given candidate. In the theory of learning, we might desire a confidence interval for the probability $p$ that a rat learns a given maze in some specified time. We usually perform $n$ independent experiments ("sampling with replacement" in the polling case)†, each of which has probability $p$ of success. If we let

$$X_i = \begin{cases} 0 \text{ if } i\text{th experiment fails} \\ 1 \text{ if } i\text{th experiment succeeds,} \end{cases}$$

then

$$\frac{1}{n}\sum_{i=1}^{n} X_i$$

is the proportion of successes in the $n$ experiments. We shall see how we may apply the Chebyshev inequality to

$$\frac{1}{n}\sum_{i=1}^{n} X_i = \bar{X}_{(n)}$$

to obtain a confidence interval for $p$.

---

† In a large population, it is simpler and cheaper to sample with replacement, rather than go to the trouble of guarding against the unlikely event that some person will be polled more than once.

Since

$$E(\bar{X}_{(n)}) = p, \quad \mathrm{Var}(\bar{X}_{(n)}) = \frac{p(1-p)}{n},$$

we see from the Chebyshev inequality that

$$(*) \qquad P\left\{|\bar{X}_{(n)} - p| \le t\sqrt{\frac{p(1-p)}{n}}\right\} \ge 1 - \frac{1}{t^2}.$$

But

$$t\sqrt{\frac{p(1-p)}{n}} \le t\sqrt{\frac{1/4}{n}} = \frac{t}{2\sqrt{n}},$$

since $p(1-p) \le 1/4$ for $0 \le p \le 1$. Hence

$$\left\{|\bar{X}_{(n)} - p| \le t\sqrt{\frac{p(1-p)}{n}}\right\} \subseteq \left\{|\bar{X}_{(n)} - p| \le \frac{t}{2\sqrt{n}}\right\},$$

and therefore

$$(**) \qquad P\left\{|\bar{X}_{(n)} - p| \le t\sqrt{\frac{p(1-p)}{n}}\right\} \le P\left\{|\bar{X}_{(n)} - p| \le \frac{t}{2\sqrt{n}}\right\}.$$

Combining (*) and (**), we obtain

$$P\left\{|\bar{X}_{(n)} - p| \le \frac{t}{2\sqrt{n}}\right\} \ge 1 - \frac{1}{t^2}.$$

We may rewrite

$$\left\{|\bar{X}_{(n)} - p| \le \frac{t}{2\sqrt{n}}\right\} = \left\{-\frac{t}{2\sqrt{n}} \le \bar{X}_{(n)} - p \le \frac{t}{2\sqrt{n}}\right\}$$
$$= \left\{\bar{X}_{(n)} - \frac{t}{2\sqrt{n}} \le p \le \bar{X}_{(n)} + \frac{t}{2\sqrt{n}}\right\}.$$

Therefore, no matter what the value of $p$ is, we have

$$P\left\{\bar{X}_{(n)} - \frac{t}{2\sqrt{n}} \le p \le \bar{X}_{(n)} + \frac{t}{2\sqrt{n}}\right\} \ge 1 - \frac{1}{t^2}.$$

Hence, we see that

$$\left\{\bar{X}_{(n)} - \frac{t}{2\sqrt{n}}, \bar{X}_{(n)} + \frac{t}{2\sqrt{n}}\right\}$$

is a level

$$1 - \frac{1}{t^2}$$

confidence interval for $p$.

If we choose $t = 10$ and perform $n = 10{,}000$ experiments, then our confidence interval is of level .99 (the probability that our interval contains $p$ is $\geq .99$) and of length .1. Alternatively, if we keep $n = 10{,}000$ but choose $t = 5$, then our confidence interval has level .96 (not quite as sure of coverage) and length .05. △

With a substantial extra amount of algebraic manipulation, we can improve this result somewhat. We do so in the following optional part.

### *4.4 Example: Improved Confidence Interval for Binomial p*

From the previous example, we already know that

$$P\left\{|\bar{X} \quad - p| \leq t\sqrt{\frac{p(1-p)}{n}}\right\} \geq 1 - 1/t^2.$$

We manipulate the event whose probability is being bounded as follows:

$$|\bar{X}_{(n)} - p| \leq t\sqrt{\frac{p(1-p)}{n}} \quad \text{if and only if} \quad |\bar{X}_{(n)} - p|^2 \leq \frac{t^2 p(1-p)}{n}.$$

But this latter is equivalent to

$$\bar{X}_{(n)}^2 - 2\bar{X}_{(n)} p + p^2 \leq \frac{t^2}{n} p - \frac{t^2}{n} p^2$$

or

$$\left(1 + \frac{t^2}{n}\right) p^2 - \left(2\bar{X}_{(n)} + \frac{t^2}{n}\right) p + \bar{X}_{(n)}^2 \leq 0.$$

However, this inequality holds if and only if $p$ lies between the two roots of the corresponding quadratic equation

$$\left(1 + \frac{t^2}{n}\right) p^2 - \left(2\bar{X}_{(n)} + \frac{t^2}{n}\right) p + \bar{X}_{(n)}^2 = 0$$

(see Figure 9.4)

$$p_\ell = \frac{2\bar{X}_{(n)} + \frac{t^2}{n} - \sqrt{\left[2\bar{X}_{(n)} + \frac{t^2}{n}\right]^2 - 4\left(1 + \frac{t^2}{n}\right)\bar{X}_{(n)}^2}}{2\left(1 + \frac{t^2}{n}\right)}$$

$$p_u = \frac{2\bar{X}_{(n)} + \frac{t^2}{n} + \sqrt{\left[2\bar{X}_{(n)} + \frac{t^2}{n}\right]^2 - 4\left(1 + \frac{t^2}{n}\right)\bar{X}_{(n)}^2}}{2\left(1 + \frac{t^2}{n}\right)}$$

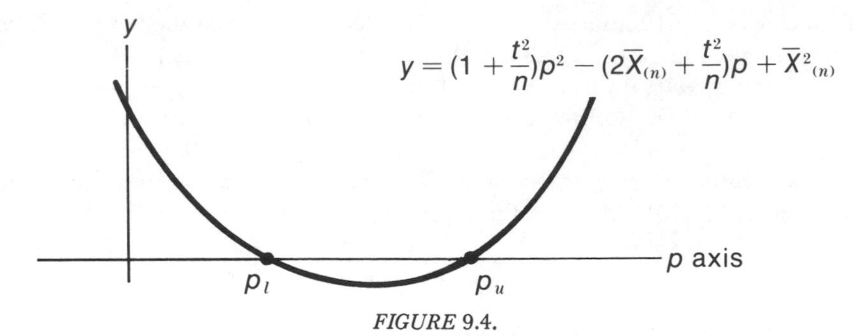


*FIGURE* 9.4.

or

$$p_\ell = \frac{\bar{X}_{(n)}}{\left(1 + \frac{t^2}{n}\right)} + \frac{\frac{t^2}{n} - \sqrt{4\frac{t^2}{n}\bar{X}_{(n)}(1 - \bar{X}_{(n)}) + \frac{t^4}{n^2}}}{2\left(1 + \frac{t^2}{n}\right)}$$

$$p_u = \frac{\bar{X}_{(n)}}{\left(1 + \frac{t^2}{n}\right)} + \frac{\frac{t^2}{n} + \sqrt{4\frac{t^2}{n}\bar{X}_{(n)}(1 - \bar{X}_{(n)}) + \frac{t^4}{n^2}}}{2\left(1 + \frac{t^2}{n}\right)}.$$

So we find that

$$P\{p_\ell \leq p \leq p_u\} \geq 1 - \frac{1}{t^2}.$$

Thus, our confidence interval for $p$ of level $1 - 1/t^2$ is $[p_\ell, p_u]$ where $p_\ell$ and $p_u$ are given above.

Observe that $\bar{X}_{(n)}$ is an estimate of $p$ that we know should be close to $p$.

If $\bar{X}_{(n)}$ is not "too near" 0 or 1 and $t^2/n$ is very small, then

$$p_\ell \cong \bar{X}_{(n)} - t\sqrt{\frac{\bar{X}_{(n)}(1 - \bar{X}_{(n)})}{n}}$$

$$p_u \cong \bar{X}_{(n)} + t\sqrt{\frac{\bar{X}_{(n)}(1 - \bar{X}_{(n)})}{n}}.$$

This is not too surprising in light of the fact that $\sqrt{\bar{X}_{(n)}(1 - \bar{X}_{(n)})/n}$ is an estimate of $\sqrt{p(1 - p)/n} = \sigma_{\bar{X}_{(n)}}$. $\Delta$

---

## 4.5 EXERCISES

1. Find a .99 level confidence interval for the speed $\mu$ of sound of Example 4.2.

2. Suppose that in Example 4.2, the normal mean, we change the conditions to $E(X) = \mu$, $\text{Var } X = 9$. (a)° Find a .95 level confidence interval for $\mu$.

Hint: See Example 2.25, Chapter 4.

(b) Find a .99 level confidence interval for $\mu$.

Hint: See Example 2.25, Chapter 4.

3. (a)° Find a .99 level confidence interval for binomial $p$ of length .02.

(b) Using the Chebyshev inequality, how many observations are required in order that $[\bar{X}_{(n)} - .01, \bar{X}_{(n)} + .01]$ be such an interval?

(c)° Suppose we knew that $p \leq .18$. What changes would you make?

4. Suppose that $X_1, X_2, \ldots, X_n$ are independent normal random variables with $E(X_i) = \mu$ and $\text{Var } X_i = 4$.

(a)° Construct a .95 level confidence interval for $\mu$ based on $\bar{X}_{(25)} = \frac{1}{25}\sum_{i=1}^{25} X_i$.

(b) Construct a .99 level confidence interval for $\mu$ based on $\bar{X}_{(100)} = \frac{1}{100}\sum_{i=1}^{100} X_i$.

(c)° Suppose that we are given the following values assumed by $X_1, \ldots, X_{25}$, assuming $E(X_i) = \mu$ and $\text{Var } X_i = 4$. What is the value assumed by the .95 level confidence interval centered at the value† $\bar{x}_{(25)}$ of $\bar{X}_{(25)}$?

| $x_1 = 4.890$ | $x_6 = 2.232$ | $x_{11} = 1.927$ | $x_{16} = .402$ | $x_{21} = 4.337$ |
|---|---|---|---|---|
| $x_2 = 1.267$ | $x_7 = 1.875$ | $x_{12} = 3.779$ | $x_{17} = 1.859$ | $x_{22} = -1.180$ |
| $x_3 = 7.716$ | $x_8 = 1.889$ | $x_{13} = 2.801$ | $x_{18} = -.856$ | $x_{23} = 2.918$ |
| $x_4 = .963$ | $x_9 = 1.131$ | $x_{14} = 1.830$ | $x_{19} = 1.343$ | $x_{24} = 3.352$ |
| $x_5 = 1.321$ | $x_{10} = 4.1197$ | $x_{15} = 6.772$ | $x_{20} = 2.105$ | $x_{25} = 1.965$ |

(Data from $\mu = 3$.)

(d)° How large should $n$ be so that the confidence interval based on $\bar{X}_{(n)}$ should have length not exceeding .01 and be of level .99?

## 4.6 PROBLEMS

1. Suppose that $X_1, X_2, \ldots, X_n$ are independent normal random variables with $E(X_i) = 0$ and such that $\text{Var } X_i = \sigma^2$ is unknown.

(a)° Construct a confidence interval for $\sigma^2$ based on the fact that, since $X_i/\sigma$ are standard normal (see Example 2.25, Chapter 4) it follows that

$$\frac{1}{\sigma^2}\sum_{i=1}^{n} X_i^2$$

has a Chi square distribution with $n$ degrees of freedom (see Definition 1 in Chapter 8). For simplicity, let the left end point of this interval be 0.

(b) What are the length and the expected length of this interval?

2.° In a given country there are $n$ automobiles with distinct license plates numbered $1, 2, \ldots, n$. An automobile is chosen at random with license plate the value of a random variable $X$. Obtain a .95 level confidence interval for the number $n$ of automobiles, of the form $[0, kX]$, where $k$ is a suitably chosen constant. That is, show how to choose $k$ so that no matter what $n$, $P\{n \leq kX\} \geq .95$.

† Generally we use capital letters to denote random variables, the corresponding lower case letters to denote possible values of these random variables and the type face x, y to stand for actual observed data.

3. Suppose that $X_1, \ldots, X_n$ are independent Poisson random variables with unknown mean $\lambda$. Construct a confidence interval of level

$$1 - \frac{1}{t^2}$$

based on $\bar{X}_{(n)}$ using the Chebyshev inequality.
Hint: $\lambda$ satisfies

$$|\bar{X}_{(n)} - \lambda| \le t \frac{\sqrt{\lambda}}{\sqrt{n}}$$

if and only if

$$\lambda^2 - \left(2\bar{X}_{(n)} + \frac{t^2}{n}\right)\lambda + \bar{X}_{(n)}^2 \le 0.$$

## SECTION 5
## Testing Hypotheses

From a conceptual point of view, the simplest class of statistical problems is that in which only two actions are possible. Such problems arise in acceptance sampling (where the two possible actions are called "accept" and "reject"), in radar detection (where the two possible actions are called "target" and "no target"), and in various other situations. We present a few examples of problems of this nature, called *testing hypotheses* problems.

### 5.1 *Example: Good Table of Normal Random Numbers*

Experiments whose performance is to be simulated on a computer often require values $x_1, x_2, \ldots$ of independent random variables $X_1, X_2, \ldots$ with a specified distribution. One of the most important cases arises when the $X_i$ are to have a standard normal distribution. Let us suppose that we have an electronic random number generator that is supposed to produce numbers $x_1, x_2, \ldots$ that are the values of independent random variables $X_1, X_2, \ldots$ having a normal distribution with $E(X_i) = 0$ and $\text{Var}\, X_i = 1$. Unfortunately, the number generator sometimes tends to drift, and instead of producing the values of independent normal random variables with 0 mean and unit variance, it is known only to produce values of independent normal random variables with mean $E(X_i) = \mu \ge 0$ and unit variance.

Now, suppose that we are given a sample $x_1, x_2, \ldots, x_{36}$ produced by our number generator and want to decide whether or not the sample is acceptable as arising from the standard normal distribution. By what we have already said, we want to determine whether $\mu = 0$ or $\mu > 0$. For our purposes, it may only be desirable to distinguish $\mu \le .1$ from $\mu > .1$. In this case, we might instead try to answer the simpler question, "Which furnishes a better description of our data,

$$\mu = 0, \qquad \text{or} \qquad \mu = .2\text{?"}$$

Thus we pose the problem in the form,

"Is $\mu = 0$, or is $\mu = .2$?"

The justification for phrasing the problem in this particular manner will become apparent as we try to ensure that the test procedure we develop exhibits the desired behavior. Note that although we are acting as if $\mu = 0$ and $\mu = .2$ are the only possibilities, we do this only to determine which furnishes the better description. Let us call the hypothesis that $\mu = 0$ the *null hypothesis* and denote it by $H_0$, and let $H_1$ denote the *alternative hypothesis* that $\mu = .2$.

Now the observable statistic $\bar{X}_{(36)} \equiv \frac{1}{36}\sum_{i=1}^{36} X_i$ is normal with $E(\bar{X}_{(36)}) = \mu$ and $\text{Var } \bar{X}_{(36)} = 1/36$ (see Theorem 7.16, Chapter 6). Therefore, it does not seem unreasonable to use the observed value $\bar{\mathbf{x}}_{(36)}$ of $\bar{X}_{(36)}$ as the basis for our decision as to whether to accept $H_0$ or to reject $H_0$ in favor of $H_1$. In fact, because $\bar{X}_{(36)}$ tends to increase as $\mu$ increases, it is natural to choose some value $c$ and to

**5.2** accept the hypothesis $H_0$ that $\mu = 0$ if $\bar{\mathbf{x}}_{(36)} \equiv \frac{1}{36}\sum_{i=1}^{36} \mathbf{x}_i \leq c$

and to reject $H_0$ in favor of the hypothesis $H_1$ that $\mu = .2$ if $\bar{\mathbf{x}}_{(36)} > c$.

The immediate question that confronts us is how to choose $c$. Our criterion is the following: Choose $c$ so that the probability of rejecting $H_0$ when in fact $H_0$ is true is some acceptably small value $\alpha$, say $\alpha = .05$. That is, taking account of statement 5.2, we choose $c$ so that

**5.3**
$$P\left\{\frac{1}{36}\sum_{i=1}^{36} X_i > c\right\} = \alpha$$

when $X_1, \ldots, X_{36}$ are independent normal with $\mu = E(X_i) = 0$ and $\text{Var } X_i = 1$. Note that we can do this because $H_0$ specifies precisely one distribution, and hence under $H_0$ the distribution of $\frac{1}{36}\sum_{i=1}^{36} X_i$ is completely determined. To carry out the choice of $c$ dictated by Equation 5.3, we notice that when $\mu = 0$, the random variable

$$\frac{1}{\sqrt{36}}\sum_{i=1}^{36} X_i$$

has a standard normal distribution (see Theorem 7.16(c), Chapter 6). Hence

$$P\left\{\frac{1}{36}\sum_{i=1}^{36} X_i > c\right\} = P\left\{\frac{1}{\sqrt{36}}\sum_{i=1}^{36} X_i > 6c\right\} = \int_{6c}^{\infty} \frac{1}{\sqrt{2\pi}} e^{-t^2/2}\, dt.$$

Thus, by Equation 5.3, we want to choose $c$ so that

$$\int_{6c}^{\infty} \frac{1}{\sqrt{2\pi}} e^{-t^2/2}\, dt = \alpha.$$

For $\alpha = .05$, we see from tables of the standard normal distribution that we must have $6c \cong 1.645$ or $c \cong .274$. Note that this value of $c$ appears to be reasonable, since it almost lies in [0, .2] and $\frac{1}{36}\sum_{i=1}^{n} X_i$ is an estimate of $\mu$.

***To summarize***: In our attempt to determine whether data $x_1, \ldots, x_{36}$ are better described as the values of independent unit variance normal random variables with mean $\mu = 0$ or with mean $\mu = .2$, we accept the hypothesis

$$H_0 \text{ that } \mu = 0 \quad \text{if} \quad \frac{1}{36}\sum_{i=1}^{36} x_i \leq .274.$$

We reject $H_0$ in favor of the hypothesis

$$H_1 \text{ that } \mu = .2 \quad \text{if} \quad \frac{1}{36}\sum_{i=1}^{36} x_i > .274.$$

This choice of $c = .274$ guarantees that the probability of rejecting $H_0$ when $H_0$ is in fact true is only .05. Note also that this procedure could have been carried out with any number of observations. That is, given $n$ and $\alpha$, we can choose $c$ so that, by our method, the probability of rejecting $H_0$ when it is in fact true is precisely $\alpha$. △

### 5.4 *Example: Discriminating Power and Sample Size for Normal Random Numbers Test*

In order to decide whether the test developed in Example 5.1 is adequate, we must determine whether it discriminates well enough between $H_0$ and $H_1$. We know already that when $H_0$ is true, $P\{\text{reject } H_0\} = .05$ for $c = .274$. To see whether this test has enough discriminating power, we must determine $P\{\text{reject } H_0\}$ when $H_1$ is true. That is, by statement 5.2, we must compute

$$P\left\{\frac{1}{36}\sum_{i=1}^{36} X_i > .274\right\}$$

when $\mu = E(X_i) = .2$. To do this, we notice that when $X_1, \ldots, X_{36}$ are independent normal with $E(X_i) = \mu$ and unit variance, then

$$\frac{1}{\sqrt{36}}\sum_{i=1}^{36} (X_i - \mu)$$

is a standard normal random variable (see Theorem 7.16, Chapter 6). Hence, when the hypothesis $H_1$ (that $\mu = .2$) is true,

$$\begin{aligned} P\{\text{reject } H_0\} &= P\left\{\frac{1}{36}\sum_{i=1}^{36} X_i > .274\right\} \\ &= P\left\{\frac{1}{36}\sum_{i=1}^{36}(X_i - .2) > .274 - .2\right\} \\ &= P\left\{\frac{1}{\sqrt{36}}\sum_{i=1}^{36}(X_i - .2) > 6(.274 - .2)\right\} \\ &= P\left\{\frac{1}{\sqrt{36}}\sum_{i=1}^{36}(X_i - .2) > .444\right\} \\ &= \int_{.444}^{\infty} \frac{1}{\sqrt{2\pi}} e^{-t^2/2}\,dt \\ &\cong .329 \end{aligned}$$

as determined from the standard normal tables.

Hence, we see that a sample size of 36 does not yield very much discriminating power. In order to overcome this deficiency, we may be willing to increase the sample size. Let us denote by $n$ the sample size needed to achieve the conditions

**5.5**

$$P\{\text{reject } H_0\} = .05 \qquad \text{when the hypothesis } H_0 \text{ (that } \mu = 0\text{) is true}$$

and

$$P\{\text{reject } H_0\} = .9 \qquad \text{when the hypothesis } H_1 \text{ (that } \mu = .2\text{) is true.}$$

Certainly, 5.5 corresponds to much better discriminating power.

We shall still use a test of the type given in statement 5.2, namely, to reject $H_0$ if and only if

**5.6**

$$\frac{1}{\sqrt{n}}\sum_{i=1}^{n} x_i > c,$$

but now we must choose both $n$ and $c$ so as to satisfy Condition 5.5. We see from 5.6 that Condition 5.5 translates into the conditions

**5.7**

$$P\left\{\frac{1}{n}\sum_{i=1}^{n} X_i > c\right\} = .05 \quad \text{when} \quad \mu = 0$$

$$P\left\{\frac{1}{n}\sum_{i=1}^{n} X_i > c\right\} = .9 \quad \text{when} \quad \mu = .2.$$

Since we know, by Theorem 7.16 in Chapter 6, that

**5.8**

$$\frac{1}{\sqrt{n}}\sum_{i=1}^{n}(X_i - \mu)$$

is a standard normal random variable, by computations similar to those in Example 5.1, we want

**5.9**

$$.05 \cong P_{H_0}\left\{\frac{1}{n}\sum_{i=1}^{n} X_i > c\right\} = P_{H_0}\left\{\frac{1}{\sqrt{n}}\sum_{i=1}^{n} X_i > \sqrt{n}\, c\right\} = \int_{\sqrt{n}c}^{\infty} \frac{1}{\sqrt{2\pi}} e^{-t^2/2}\, dt$$

$$.9 \cong P_{H_1}\left\{\frac{1}{n}\sum_{i=1}^{n} X_i > c\right\} = P_{H_1}\left\{\frac{1}{\sqrt{n}}\sum_{i=1}^{n} (X_i - .2) > \sqrt{n}\, c - \sqrt{n}\, .2\right\}$$

$$= \int_{\sqrt{n}(c-.2)}^{\infty} \frac{1}{\sqrt{2\pi}} e^{-t^2/2}\, dt.$$

Here we use the subscripts $H_0$ and $H_1$ to indicate which distribution was used for the given computation. From tables of the standard normal distribution, we see that 5.9 yields $\sqrt{n}\, c \cong 1.645$, $\sqrt{n}(c - .2) \cong -1.28$. Dividing the second equation by the first yields

$$1 - \frac{.2}{c} = -\frac{1.28}{1.645} \cong -.778$$

or

**5.10**
$$c \cong \frac{.2}{1.778} \cong .1125.$$

Substituting into $\sqrt{n}\, c = 1.645$ yields $\sqrt{n} \cong 14.61$ or

**5.11**
$$n = 214.$$

Note that we are able to determine $n$ in a simple manner because of the fact that $H_1$ also is specified by a single distribution. We thus see how the simple formulation of Example 5.1 ($\mu = 0$ versus $\mu = .2$, each specifying a single distribution) permitted easy determination of a test with the desired discriminating power.

*We have shown the following: When $X_1, X_2, \ldots$ are independent normal random variables with unit variance and $E(X_i) = \mu$, the test that rejects the hypothesis $H_0$ that $\mu = 0$ in favor of the hypothesis $H_1$ that $\mu = .2$ when $\frac{1}{214}\sum_{i=1}^{214} x_i > .1125$ satisfies*

$$P_{H_0}\{\text{reject } H_0\} \cong .05 \qquad \text{and} \qquad P_{H_1}\{\text{reject } H_0\} \cong .9.$$

(Here the subscript $H_i$ means that the probability is computed for cases when hypothesis $H_i$ holds.)

We know that $\mu = 0$ and $\mu = .2$ are not the only possible descriptions of our data, and so we should be interested in the behavior of our test procedure for other reasonable descriptions. For reasons of convenience, we shall discuss only the performance of our test when $X_1, X_2, \ldots$ are independent, normal with unit variance and mean value $\mu$. For example, we shall not consider the behavior of the test when $\sigma^2 \neq 1$. The behavior of our test is described by the *power function* $\Pi$ given by

$$\Pi(\mu) = P_{\mu}\{\text{reject } H_0\},$$

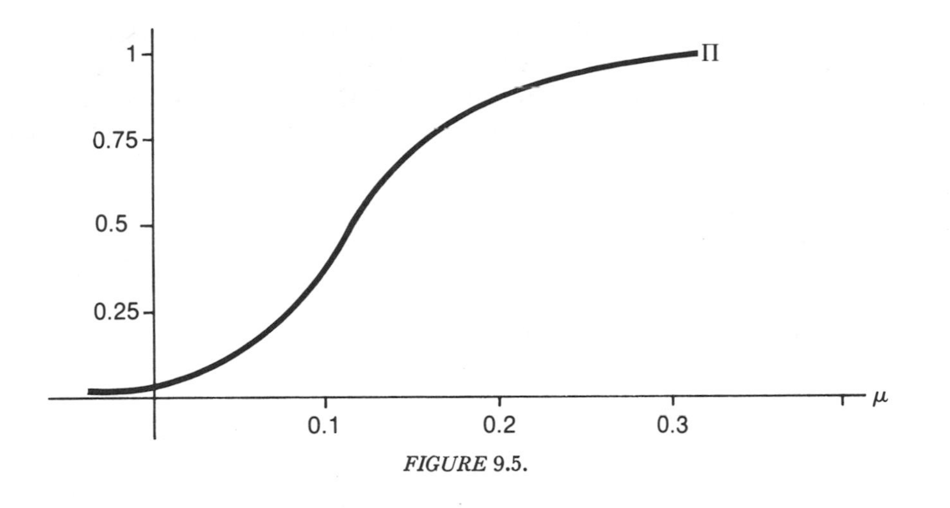


*FIGURE* 9.5.

where the subscript $\mu$ indicates that the computation is done for $E(X_i) = \mu$. Here, using the $c$ from 5.6 and the $n$ from 5.11 in the test given by 5.6, we see that

**5.12**
$$\Pi(\mu) = P_\mu\left\{\frac{1}{214}\sum_{i=1}^{214} X_i > .1125\right\}$$
$$= P_\mu\left\{\frac{1}{\sqrt{214}}\sum_{i=1}^{214}(X_i - \mu) > \sqrt{214}(.1125 - \mu)\right\}$$
$$= \int_{\sqrt{214}(.1125-\mu)}^{\infty} \frac{1}{\sqrt{2\pi}} e^{-t^2/2}\, dt \quad \text{by 5.8.}$$

From 5.12 and the standard normal tables we can compute $\Pi(\mu)$. Roughly speaking, the graph of $\Pi$ has the shape given in Figure 9.5. △

### *5.13 Example: Alternative Approach to Normal Random Number Test*

We already know that in the normal case treated in the previous example, the mean $\mu$ is also the median. This suggests that a test based on the sample median might also be reasonable. In some respects, such a test might be preferred because the sample median is not too sensitive to the occurrence of values with extremely great magnitude. Hence, it would probably perform in a more acceptable manner when the governing distribution looks normal near its median but looks more like a Cauchy distribution far from its median.

Let us see how a test based on the sample median $X_{[108],215}$ of a sample of 215 observations behaves. For reasons analogous to those in Example 5.1 we reject the hypothesis $H_0$ that $\mu = 0$ in favor of $H_1$ that $\mu = .2$ if and only if the given value of the sample median, $\mathrm{x}_{[108],215} > d$. For the purpose of comparison with the test of Example 5.4, we want to choose the constant $d$ so that when $\mu = 0$, we have $P\{\text{reject } H_0\} \cong .05$, that is,

**5.14**
$$P\{X_{[108],215} > d\} \cong .05.$$

In theory, we can determine $d$ from the exact density of $X_{[108],215}$ when $\mu = 0$, but such a determination presents a formidable numerical problem. As a simple approximation, we make use of Theorem 4.1 in Chapter 7, which states that if $X_1, \ldots, X_n$ are independent with a continuous density $f$ having a unique $p$ quantile $\theta_p$, then if $f(\theta_p) > 0$,

$$\lim_{n\to\infty} P\left\{\sqrt{\frac{n}{p(1-p)}}\, f(\theta_p)[X_{(p),n} - \theta_p] \le y\right\} = \int_{-\infty}^{y} \frac{1}{\sqrt{2\pi}} e^{-u^2/2}\, du.$$

Here $X_{(p),n}$ is the sample $p$-quantile. For the case in question,

$$X_{(p),n} = X_{(\frac{1}{2}),215} = X_{[108],215} \quad \text{and} \quad \theta_p = \theta_{1/2} = \mu.$$

Since

$$f(x) = \frac{1}{\sqrt{2\pi}} e^{-(x-\mu)^2/2},$$

we have

$$f(\theta_{1/2}) = \frac{1}{\sqrt{2\pi}}.$$

Hence, the result of Theorem 4.1, Chapter 7, here specializes to

**5.15** $$\lim_{n\to\infty} P\left\{2\sqrt{n}\frac{1}{\sqrt{2\pi}}[X_{[108],215} - \mu] \le y\right\} = \lim_{n\to\infty} P\left\{\sqrt{2n/\pi}\,[X_{[108],215} - \mu] \le y\right\}$$
$$= \int_{-\infty}^{y} \frac{1}{\sqrt{2\pi}} e^{-u^2/2}\, du.$$

To make use of this result, we rewrite the event in Equation 5.14 as follows:

$$\{X_{[108],215} > d\} = \{X_{[108],215} - \mu > d - \mu\} = \{\sqrt{2n/\pi}\,[X_{[108],215} - \mu]$$
$$> \sqrt{2n/\pi}\,(d - \mu)\}.$$

Hence

**5.16** $$P_\mu\{X_{[108],215} > d\} = P\{\sqrt{2n/\pi}\,[X_{[108],215} - \mu] > \sqrt{2n/\pi}\,(d-\mu)\}$$
$$\cong \int_{\sqrt{2n/\pi}(d-\mu)}^{\infty} \frac{1}{\sqrt{2\pi}} e^{-u^2/2}\, du \quad \text{by 5.15.}$$

To satisfy Equation 5.14, we put $\mu = 0$, $n = 215$ and use the standard normal tables to see that we want $\sqrt{430/\pi}\, d \cong 1.645$. Hence, $d \cong .1405$. *We see that if $X_1, \ldots, X_{215}$ are independent random variables normally distributed with unit variance and $E(X_i) = \mu$, then the test that rejects the hypothesis $H_0$ that $\mu = 0$ in favor of $H_1$ that $\mu = .2$ whenever* $X_{[108],215} > .1405$ satisfies the approximate equalities

$$P_{H_0}\{\text{reject } H_0\} \cong .05$$

and (substituting $d = .1405$ into Equation 5.16)

$$P_\mu(\text{reject } H_0\} \cong \int_{\sqrt{430/\pi}(.1405-\mu)}^{\infty} \frac{1}{\sqrt{2\pi}} e^{-u^2/2}\, du.$$

In particular, when $H_1$ is true ($\mu = .2$), this test yields a probability of .757 of rejecting $H_0$.

From Equation 5.12, we see that the test of $H_0$ versus $H_1$ based on the mean $\bar{X}_{(214)}$ had a probability of .898 of rejecting $H_0$ when $H_1$ is true ($\mu = .2$). Thus, the test based on the mean is superior to that based on the median when the data is normally distributed. So when normality holds, we pay a price in discriminating power if we want a test that is better behaved under certain types of departures from normality. △

## 5.17 *Example: Radiation Danger Test*

Measurements $X_1, \ldots, X_n$ of radiation are made. It is assumed that these measurements are independent Poisson random variables with common parameter $\lambda$. Here $\lambda$ is the expected number of decays in a time interval during which a measurement is made and is unknown to us. All such intervals are assumed to have the same duration. The known value $\lambda_1$ is assumed to correspond to a situation that we will label "safe," whereas the known value $\lambda_0 > \lambda_1$ corresponds to a radiation level that we will designate as the minimum dangerous one.† First we may oversimplify the problem, as before, by asking:

**5.18** Is $\lambda = \lambda_0$ *or* is $\lambda = \lambda_1$?

That is, which is a better theoretical description, $\lambda = \lambda_0$ or $\lambda = \lambda_1$? Again, it is simpler to act initially as if $\lambda = \lambda_0$ and $\lambda = \lambda_1$ are the only two possibilities. We shall later investigate the behavior of the procedure resulting from this assumption for various other values of $\lambda$ (and perhaps for distributions other than Poisson). In order to answer the question in 5.18, the following procedure can be shown to be reasonable, because

$$\frac{1}{n} \sum_{i=1}^{n} X_i$$

is a reasonable estimate of $\lambda$.

## 5.19 Test Procedure

Choose a number $\alpha$, such that $0 < \alpha < 1$, where $\alpha$ is the risk you are willing to take in asserting $\lambda = \lambda_1$ (safe level) when in fact $\lambda = \lambda_0$ (unsafe level). To be precise,

† Actually, both levels may be dangerous, and might more properly be labeled "acceptable" and "unacceptable," or even "intolerable" and "too intolerable."

$\alpha$ is the probability of asserting $\lambda = \lambda_1$ when actually $\lambda = \lambda_0$. We choose a number $c$, and if the observed value

$$\frac{1}{n}\sum_{i=1}^{n} x_i > c, \quad \text{we assert } \lambda = \lambda_0 \text{ (unsafe)},$$

whereas if

$$\frac{1}{n}\sum_{i=1}^{n} x_i \leq c, \quad \text{we assert } \lambda = \lambda_1 \text{ (safe)}.$$

Here $c$ is determined by the condition that when $\lambda = \lambda_0$ we want

$$P\left\{\frac{1}{n}\sum_{i=1}^{n} X_i \leq c\right\} = P\left\{\sum_{i=1}^{n} X_i \leq nc\right\} \leq \alpha.$$

Since $\sum_{i=1}^{n} X_i$ has a Poisson distribution with parameter $n\lambda$ (see Example 4.5 in Chapter 5 or Example 7.12 in Chapter 6), this amounts to saying

**5.20**
$$\sum_{k \leq nc} e^{-n\lambda_0}\frac{(n\lambda_0)^k}{k!} \leq \alpha.$$

Note that we may not be able to achieve exact equality in 5.20, so we choose the largest $c$ that gives us the inequality 5.20. The value $P\left\{\sum_{i=1}^{n} X_i \leq nc\right\}$ computed for $\lambda = \lambda_0$ is called the *size* of this test; it is the probability of asserting that we are safe when in fact we are not.

In order to determine how reasonable the test procedure given in 5.19 is, we look at the probability of asserting $\lambda = \lambda_1$ (safe) *as a function of* $\lambda$. Although we feel that there is a true $\lambda$ (or that there is a $\lambda$ yielding a reasonable description), since we do not know this true value, we must ask ourselves what our actions are likely

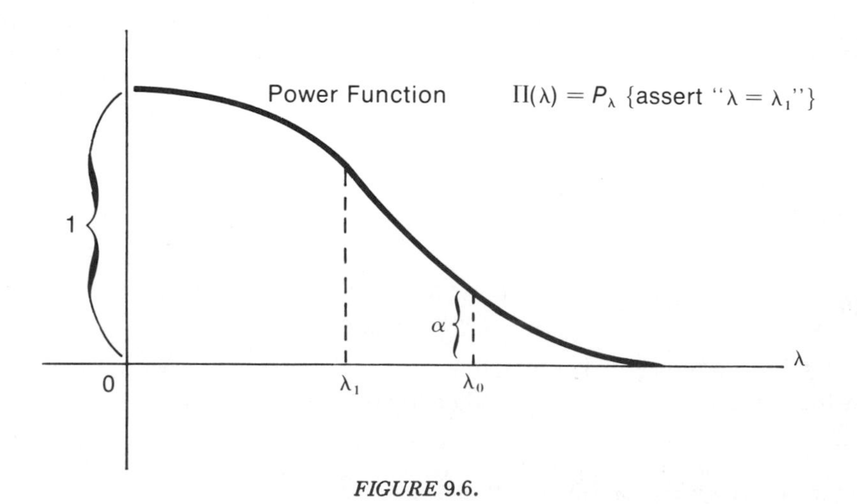


*FIGURE* 9.6.

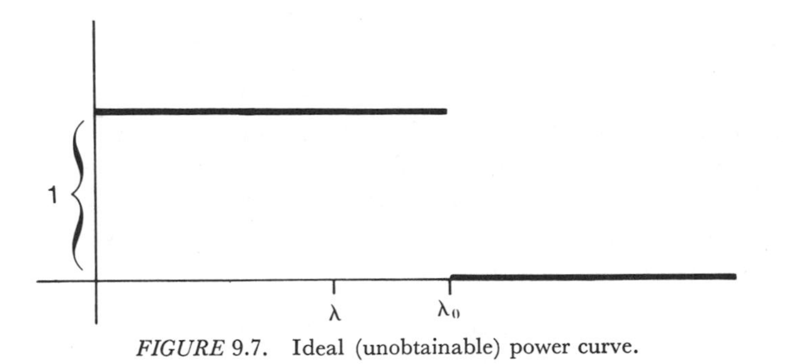


*FIGURE* 9.7. Ideal (unobtainable) power curve.

to be for the various possible $\lambda$'s. That is, we must examine

$$\prod(\lambda) = P_\lambda\{\text{asserting safe radiation}\}$$
$$= P_\lambda\left\{\sum_{i=1}^{n} X_i \leq nc\right\} = \sum_{k\leq nc} e^{-n\lambda}\frac{(n\lambda)^k}{k!}.$$

The subscript $\lambda$ is present to indicate that the probabilities are being computed when $\lambda$, not necessarily either $\lambda_0$ or $\lambda_1$, is the true parameter. This function $\Pi$ is called the *power function* of the test 5.19 and has roughly the shape indicated in Figure 9.6. Note that the graph of $\Pi$ shows that when $\lambda$ is large (that is, when radiation is likely to be high), it is very unlikely that we will assert that we are "safe," that is, that $\lambda = \lambda_1$. Note that the ideal power curve (one giving perfect discrimination) would be Figure 9.7, since ideally when $\lambda \geq \lambda_0$ we want to give the danger signal (unsafe), but for $\lambda < \lambda_0$ we are willing to accept the radiation level. Note further that if we are also able to choose $n$, then we can shape the power function more to our liking. In fact, we can then fix the value of the power function at $\lambda_1$ at any specified value $1 - \beta$ (where $\beta$ is small, $0 < \beta < 1$), $1 - \beta > \alpha$. In this case we we want to satisfy

**5.21**
$$P_{\lambda_1}\left\{\sum_{i=1}^{n} X_i \leq nc\right\} \geq 1 - \beta$$

as well as Equation 5.20. Thus, if $n$ is at our disposal, we try to choose $n$ and $c$ so as to satisfy both 5.20 and 5.21. The latter becomes

**5.22**
$$\sum_{k\leq nc} e^{-n\lambda_1}\frac{(n\lambda_1)^k}{k!} \geq 1 - \beta.$$

You may use a table of the Poisson distribution to help choose $n$ and $c$ so that 5.20 and 5.22 are satisfied for any particular choice of $\lambda_0$ and $\lambda_1$. Again, the main reason for setting up our model in the form $\lambda = \lambda_0$ versus $\lambda = \lambda_1$ is that we can shape the power curve by controlling its values at the two points, $\lambda = \lambda_0$ and $\lambda = \lambda_1$, by means of our choice of the two values $n$ and $c$.

It is not difficult to show, using the Chebyshev inequality, that it is possible to choose $n$ and $c$ so as to ensure satisfaction of both 5.20 and 5.22 (see Exercise 5.25(2)).

In this proof it is sensible to choose $c$ to be midway between $\lambda_0$ and $\lambda_1$. Since $\frac{1}{n}\sum_{i=1}^{n} X_i$ is an estimate of $\lambda$, we might reasonably expect the $c$ obtained for numerical solutions of 5.20 and 5.22 to lie between $\lambda_0$ and $\lambda_1$ where $\alpha$ and $\beta$ are small. If not, we would have reason to be suspicious of our numerical computations.

By showing monotonicity of the power function, we can thus show that it is possible to obtain a power curve with good discriminating ability; one that in some sense is close to the ideal one of Figure 9.7. △

It is not always possible to set up a test procedure with good discriminating power. We illustrate this somewhat less pleasant situation in our next example.

### 5.23 *Example: Plant Yield Comparison*

Measurements are made on pairs of fruit trees in which only one of the pair is given special treatment (perhaps a new type of fertilizer). The difference in yield for the $i$th pair is denoted by $X_i$. Because the overall variability may arise from a variety of contributing sources, we shall assume that $X_1, \ldots, X_n$ are independent normal random variables with $E(X_i) = \mu$ and $\text{Var}\, X_i = \sigma^2$, both unknown. The assumption that $\sigma^2$ is unknown may arise, for example, from our lack of knowledge of the variations in soil conditions. It is desired to determine whether $\mu = 0$ (both treatments have the same effect) or $\mu > 0$ (new treatment is better). We are assuming that there are biological grounds for believing that the new treatment cannot result in a smaller yield than the standard, on the average; that is, that we cannot have $\mu < 0$.

One might be tempted to take the variables $X_1, \ldots, X_n$ and assert that $\mu = 0$ if the observed value $\sum_{i=1}^{n} x_i \leq c$, and that $\mu > 0$ if $\sum_{i=1}^{n} x_i > c$.

As in the previous example, the question of the choice of $c$ arises. If we try to choose $c$ so that $P\{\sum_{i=1}^{n} X_i > c\} \leq \alpha < 1/2$ whenever $\mu = 0$, we shall be unsuccessful. The reason for this difficulty is that the value of $\mu$ alone is not sufficient to determine the distribution of $\sum_{i=1}^{n} X_i$. As we know from Theorem 7.16 in Chapter 6, the random variable $\sum_{i=1}^{n} X_i$ is normal, with mean $n\mu$ and variance $n\sigma^2$, and the unknown $\sigma^2$ influences the behavior of the observations quite strongly. In fact, when $\mu = 0$,

$$P\left\{\sum_{i=1}^{n} X_i \geq c\right\} = P\left\{\frac{1}{\sqrt{n}} \sum_{i=1}^{n} \frac{X_i}{\sigma} > \frac{c}{\sigma\sqrt{n}}\right\} = \int_{c/(\sigma\sqrt{n})}^{\infty} \frac{1}{\sqrt{2\pi}} e^{-t^2/2}\, dt$$

and for any fixed $c$, $n$ if $\sigma$ is large, this probability is near 1/2. We therefore call $\sigma^2$ a *nuisance parameter*. Later, we shall investigate several ways to handle the nuisance parameter problem. We briefly mention here two possible ways that might be used for the particular case being considered. In the first method, we notice that

$$\sum_{i=1}^{n} \frac{X_i}{\sigma}$$

has a distribution that does not depend on $\sigma$ when $\mu = 0$. This suggests using

$$\sum_{i=1}^{n} \frac{X_i}{S_n},$$

where

$$S_n = \sqrt{\frac{1}{n-1} \sum_{i=1}^{n} (X_i - \bar{X}_{(n)})^2}$$

is an estimate of $\sigma$, in place of

$$\sum_{i=1}^{n} \frac{X_i}{\sigma}.$$

From Example 8 in Chapter 8, it is known that when $\mu = 0$, the statistic

$$\sum_{i=1}^{n} \frac{X_i}{S_n\sqrt{n}}$$

has the $t$ distribution with $n - 1$ degrees of freedom. Consequently, we may choose $c$ such that if we
assert $\mu = 0$ whenever the observed value

$$\sum_{i=1}^{n} \frac{x_i}{s_n\sqrt{n}} \leq c$$

and

$$\text{assert } \mu > 0 \text{ whenever } \sum_{i=1}^{n} \frac{x_i}{s_n\sqrt{n}} \geq c,$$

then

$$P\{\text{asserting } \mu > 0\} = P\left\{\sum_{i=1}^{n} \frac{X_i}{S_n\sqrt{n}} > c\right\} = \alpha$$

in case $\mu = 0$. It can be shown that the performance of this test depends only on $\mu/\sigma$. This is undesirable if we want to discriminate well between $\mu = 0$ and $\mu = 2$, since large $\sigma$ will make such discrimination effectively unattainable by this scheme.

In the second method, we allow two stages of sampling. In the first we get an estimate for $\sigma^2$ and in the second we choose a large sample size whenever the first stage estimate for $\sigma^2$ is large. We shall investigate this problem further in the next chapter. △

---

## 5.24 *EXERCISES*

1.° For the test of a good table of standard normal random numbers developed in Example 5.1, what is the probability of rejecting $H_0$ when $E(X_i) = \mu$, $\text{Var } X_i = \sigma^2$?

2. Answer Exercise 1 for the test developed in Example 5.13.

3. For a test of $\mu = 0$ versus $\mu = .2$, as in Example 5.13, based on the sample median, how many observations would be needed to satisfy $P_{H_0}\{\text{reject } H_0\} \cong .05$, $P_{H_1}\{\text{reject } H_0\} \cong .9$ when $\sigma^2 = 1$? (Recall that the test of Example 5.4 satisfied these conditions with $n = 214$.)

4. Suppose that we observe the values of independent normal random variables $X_1, \ldots, X_{25}$, representing the excess hardness (in grains) of water from a particular cheaper source than the normal one. Assuming that the mean value $E(X_i) = \mu$ and the variance $\text{Var } X_i = 9$:

(a) Construct a reasonable test of the hypothesis $\mu = 0$ versus the alternative $\mu \geq 2$, based on $\bar{X}_{(25)} = \frac{1}{25}\sum_{i=1}^{25} X_i$, such that the probability of asserting $\mu \geq 2$ when in fact $\mu = 0$ is exactly .05. Plot the probability of asserting $\mu \geq 2$ as a function of $\mu$.

(b) Given the following sets of data, what is your action in applying the test of part (a)?

(i)

| | | | | |
|---|---|---|---|---|
| $x_1 = -.864$ | $x_6 = 4.737$ | $x_{11} = -1.473$ | $x_{16} = -.438$ | $x_{21} = 8.127$ |
| $x_2 = .561$ | $x_7 = 3.270$ | $x_{12} = .657$ | $x_{17} = -5.094$ | $x_{22} = -.171$ |
| $x_3 = 2.355$ | $x_8 = 1.344$ | $x_{13} = -.507$ | $x_{18} = 3.123$ | $x_{23} = -.9$ |
| $x_4 = .582$ | $x_9 = -1.371$ | $x_{14} = 3.288$ | $x_{19} = -3.087$ | $x_{24} = -1.782$ |
| $x_5 = -.774$ | $x_{10} = 2.880$ | $x_{15} = 3.617$ | $x_{20} = 1.437$ | $x_{25} = -3.041$ |

(Data from $\mu = 0$, $\sigma = 3$.)

(ii)

| | | | | |
|---|---|---|---|---|
| $x_1 = 5.304$ | $x_6 = -1.004$ | $x_{11} = 4.890$ | $x_{16} = 4.272$ | $x_{21} = 3.411$ |
| $x_2 = 4.125$ | $x_7 = 2.139$ | $x_{12} = 4.125$ | $x_{17} = 4.779$ | $x_{22} = -4.578$ |
| $x_3 = 1.467$ | $x_8 = 3.483$ | $x_{13} = -1.260$ | $x_{18} = 5.586$ | $x_{23} = 1.938$ |
| $x_4 = 3.876$ | $x_9 = -1.038$ | $x_{14} = 2.547$ | $x_{19} = 3.705$ | $x_{24} = 1.584$ |
| $x_5 = 6.078$ | $x_{10} = 6.750$ | $x_{15} = 2.073$ | $x_{20} = .441$ | $x_{25} = 1.335$ |

(Data from $\mu = 3$, $\sigma = 3$.)

5.° (a) Same as Exercise 4(a), but based on $X_{[13]}$, the sample median, with "exactly .05" changed to "approximately .05." (See Theorem 4.1, Chapter 7.)

(b) Same as Exercise 4(b), applied to part (a) of this exercise.

6. Suppose we try to determine whether a measuring instrument is sufficiently precise for some purpose. We observe $X_1, X_2, \ldots, X_n$, which are measurements on an object called a standard with a precisely known stable numerical characteristic $\mu$. We assume that the instrument has been properly calibrated so that $Y_i = X_i - \mu$ are normal with $E(Y_i) = 0$, but $\text{Var } Y_i = \sigma^2$ is unknown. Based on $Y_1, \ldots, Y_n$, we desire to test the hypothesis $\sigma^2 \leq 49$ (instrument sufficiently precise) versus the alternative $\sigma^2 \geq 64$ (instrument not precise enough). Give a reasonable test procedure for this problem.

Hint:

$$\frac{1}{n}\sum Y_i^2$$

is a reasonable estimator for $\sigma^2$ and also $\sum_{i=1}^{n} (Y_i/\sigma)^2$ has a $\chi_n^2$ distribution (Chi square distribution with $n$ degrees of freedom).

(a) Choose $n$ so that the following two conditions are satisfied:

(i) when $\sigma^2 = 49$, the probability of asserting $\sigma^2 \geq 64$ does not exceed .01.

(ii) when $\sigma^2 = 64$, the probability of asserting $\sigma^2 \leq 49$ does not exceed .05.

(b) Plot the probability of asserting $\sigma^2 \geq 64$ as a function of $\sigma^2$.

(c)° You now relax requirements (i), (ii) to:

(i′) when $\sigma^2 = 36$, the probability of asserting $\sigma^2 \geq 81$ does not exceed .25.

(ii′) when $\sigma^2 = 81$, the probability of asserting $\sigma^2 \leq 36$ does not exceed .25.

Find $n$ and the desired test.

Hint: Here you will have to "hunt" in tables of Chi square for the smallest reasonable $n$ and check your answer to see if it makes sense.

(d)° Given the following data as values of $Y_1, \ldots, Y_n$, what would your actions be for the test of part c?

(i)

| | | |
|---|---|---|
| $y_1 = 5.645$ | $y_4 = -6.975$ | $y_7 = 1.270$ |
| $y_2 = -3.085$ | $y_5 = -3.225$ | |
| $y_3 = 4.800$ | $y_6 = -4.910$ | (data from $\sigma^2 = 25$.) |

(ii)

| | | |
|---|---|---|
| $y_1 = .945$ | $y_4 = -1.233$ | $y_7 = 2.079$ |
| $y_2 = 2.565$ | $y_5 = -.126$ | |
| $y_3 = -7.668$ | $y_6 = -8.748$ | (data from $\sigma^2 = 81$.) |

---

## 5.25 *PROBLEMS*

1. To test its effectiveness, Salk polio vaccine was given to about 1,000,000 people. The incidence of polio prior to the development of the vaccine was 19 per 100,000. A test of the hypothesis that the Salk vaccine is effective was the following: Let $Y$ be the number of people among the 1,000,000 who were inoculated and who contract polio. The test is to accept the hypothesis that the vaccine is effective provided $y < 140$ where $y$ is the observed value of $Y$.

(a) Using the Central Limit Theorem, compute an approximation to $P\{Y < 140\}$:

(i) if the "true" incidence using the vaccine is 19/100,000 (that is, vaccine not effective).

(ii) if the "true" incidence using the vaccine is 10/100,000.

(b)° Do part (a) using the Uspensky correction term (Theorem 3.16, Chapter 7), giving upper and lower bounds for the desired probabilities.

2.° Show, using the Chebyshev inequality, that it is possible to choose $n$ and $a$ so as to satisfy both 5.20 and 5.22.

# 10

# Estimation

## SECTION 1
## Introduction

Much of scientific work is repeated experimentation in an attempt to learn something about an unknown parameter. The following are some examples of parameters we may wish to estimate:

(a) the probability of a defective from some manufacturing process.

(b) the probabilities of various symptoms in patients afflicted with a given disease.

(c) the average recovery time after a given type of surgery.

(d) the average remaining survival time of persons who have undergone a given treatment for a given disease.

(e) the median income of the population.

(f) the correlation coefficient between a pair of attributes (smoking and lung cancer).

(g) the value of an important physical constant; for example, the half-life of plutonium (the time it takes for 50 per cent of a batch of plutonium atoms to split).

In order to obtain such information, we usually repeat some relevant experiment as many times as the situation requires or permits. For instance, in trying to estimate the median income, we might select at random $n$ people from the population and use the median income in this sample as our estimate of the median income of the population.

The theoretical framework in which we treat estimation consists, therefore, of independent random variables or random vectors all governed by some distribution $P$ known only to come from some given class $\mathscr{P}$. We seek to estimate some parameter $\theta$, scalar or vector, that is determined by $P$. This chapter is devoted to a systematic development of the theory of estimation.

## SECTION 2
## Consistency

In Chapter 9, we defined a point estimator as an *observable statistic* that serves as an educated guess for the parameter of interest. In certain specific cases, we exhibited estimators that appeared to be reasonable. We will now investigate in some detail properties of point estimators that might be desirable. The property that we shall consider in this section is called *consistency*. Think of any quantities, such as those listed in the previous section, that you might want to estimate. It is surely desirable that with high probability the estimator should be close to the parameter being estimated. At the very least, this should be the case if we are permitted to repeat the experiment as often as we wish. This property is embodied in the following definition.

**2.1 Definition: Consistent Sequence of Estimators.** Let $X_1, X_2, \ldots$ be independent random variables having common distribution $P$ known only to come from some given class $\mathscr{P}$, and let $\theta$ be some parameter determined by $P$. The sequence of point estimators $g_1(X_1)$, $g_2(X_1, X_2)$, ... will be called *consistent for* $\theta$ (relative to $\mathscr{P}$) if for each $\varepsilon > 0$ and for each fixed $P \in \mathscr{P}$,

$$\lim_{n\to\infty} P\{|g_n(X_1, \ldots, X_n) - \theta| \leq \varepsilon\} = 1.$$

We remark that a sequence of random variables $Z_1, Z_2, \ldots$ is said to *converge in probability* to a random variable $Z$ if for each $\varepsilon > 0$,

$$\lim_{n\to\infty} P\{|Z_n - Z| \leq \varepsilon\} = 1.$$

Thus, consistency of $g_1(X_1), g_2(X_1, X_2), \ldots$ for $\theta$ is simply convergence of this sequence in probability to the constant random variable $\theta$.

### *2.2 Example: Consistent Estimate of Probability*

Suppose we repeatedly run an experiment with probability $p$ of success. Let $X_i$ denote the number of successes on the $i$th trial. Then

$$\frac{1}{n}\sum_{i=1}^{n} X_i$$

is the proportion of successes among the first $n$ trials. It is therefore not surprising that the sequence whose $n$th element is

$$g_n(X_1, \ldots, X_n) = \frac{1}{n}\sum_{i=1}^{n} X_i$$

should be consistent for $p$.

To prove the asserted consistency, note that the $X_i$ are independent with $E(X_i) = p$, Var $X_i = p(1 - p)$ (see Example 2.5 and 3.9, in Chapter 6). It follows from the weak law of large numbers, Theorem 6.5 in Chapter 6, that

$$\lim_{n\to\infty} P\left\{\left|\frac{1}{n}\sum_{i=1}^{n} X_i - p\right| \le \varepsilon\right\} = 1,$$

establishing consistency. △

The preceding example is a special case of the following result.

### 2.3 *Example: Consistent Estimation of Expectation*

If $X_1, X_2, \ldots$ are independent with common expectation $E(X_i) = \mu$ and common variance Var $X_i = \sigma^2$ then, from the weak law of large numbers, we know that

$$\lim_{n\to\infty} P\left\{\left|\frac{1}{n}\sum_{i=1}^{n} X_i - \mu\right| \le \varepsilon\right\} = 1.$$

Hence, the sequence given by

$$g_n(X_1, \ldots, X_n) = \frac{1}{n}\sum_{i=1}^{n} X_i$$

is consistent for $\mu$. △

Once we have found a consistent sequence of estimators, we may feel that the problem of accurate estimation of $\theta$ is solved, for if we just "choose $n$ large enough, then $g_n(X_1, \ldots, X_n)$ is likely to be close to $\theta$." Unfortunately, this is not the case, because the sample size $n$ required to achieve high accuracy may vary with $P$. In fact, arbitrarily large sample sizes may be needed. If we knew the true $P$ governing the data, of course we could determine the required sample size. However, if we knew $P$, then the problem would be academic, for then we could determine $\theta$ exactly. So the difficulty comes from the fact that we do not know the $P$ in $\mathscr{P}$ that governs the data, and some $P \in \mathscr{P}$ may require quite large sample sizes for accurate estimation of $\theta$ from the sample.† We illustrate this unfortunate situation in the following two examples.

### 2.4 *Example: Normal Mean*

Let $\mu$ be the speed of sound, which we try to estimate by repeated measurements. Assume that our measurements $X_1, X_2, \ldots, X_n$ are independent normal random

† The mathematically oriented student will recognize that this difficulty arises from a lack of uniform convergence. This same problem also arises in numerical solution of algebraic and differential equations. In these subjects, we try to approximate an unknown function $f$ by some computed approximation $f_n$. Unfortunately, for any given $n$, $f_n(x)$ may be a poor approximation to $f(x)$ for some $x$. This can occur even though we may know that for all $x$, $\lim_{n\to\infty} f_n(x) = f(x)$.

variables with $E(X_i) = \mu$ and Var $X_i = \sigma^2$, where $\sigma^2$ is unknown. We know that

$$X_1, \tfrac{1}{2}(X_1 + X_2), \ldots, \frac{1}{n}\sum_{i=1}^{n} X_i, \ldots$$

form a consistent sequence for $\mu$, and in fact

$$P\left\{\left|\frac{1}{n}\sum_{i=1}^{n} X_i - \mu\right| \le \varepsilon\right\} = \int_{-\sqrt{n}\varepsilon/\sigma}^{\sqrt{n}\varepsilon/\sigma} \frac{1}{\sqrt{2\pi}} e^{-t^2/2}\, dt.$$

Note, however, that no matter how $n$ is chosen in advance, for each given $\varepsilon > 0$ there are values of $\sigma$ such that

$$\frac{\sqrt{n}\varepsilon}{\sigma}$$

is close to 0, and hence

$$\int_{-\sqrt{n}\varepsilon/\sigma}^{\sqrt{n}\varepsilon/\sigma} \frac{1}{\sqrt{2\pi}} e^{-t^2/2}\, dt$$

is close to 0. That is, no matter how large we choose $n$, or how small we choose $\varepsilon$, there are situations in which it is *highly unlikely* that

$$\frac{1}{n}\sum_{i=1}^{n} X_i$$

will be within $\varepsilon$ of $\mu$.

Suppose $\sigma$ is unknown and we use

$$\frac{1}{n}\sum_{i=1}^{n} X_i$$

to estimate the speed $\mu$ of sound. Under the given model, we cannot plan in advance a fixed number $n$ of measurements for accurate determination of $\mu$. This model corresponds to a measuring instrument of unknown "quality." The situation may be radically altered if we know something about the quality of our measuring device. For instance, suppose that we know an upper bound for $\sigma$, that is, a value $\sigma_u$ such that $\sigma \le \sigma_u$. Then for any given $\varepsilon$, we see that

$$\frac{\sqrt{n}\varepsilon}{\sigma} \ge \frac{\sqrt{n}\varepsilon}{\sigma_u}.$$

Thus, if $0 < \alpha < 1/2$, we may choose $n$ to satisfy

$$\frac{\sqrt{n}\varepsilon}{\sigma_u} \ge \varphi_\alpha,$$

where

$$\int_{-\varphi_\alpha}^{\varphi_\alpha} \frac{1}{\sqrt{2\pi}} e^{-t^2/2}\, dt = 1 - \alpha.$$

Then

$$P\left\{\left|\frac{1}{n}\sum_{i=1}^{n} X_i - \mu\right| \le \varepsilon\right\} = \int_{-\sqrt{n}\varepsilon/\sigma}^{\sqrt{n}\varepsilon/\sigma} \frac{1}{\sqrt{2\pi}} e^{-t^2/2}\,dt$$
$$\ge \int_{-\varphi_a}^{\varphi_a} \frac{1}{\sqrt{2\pi}} e^{-t^2/2}\,dt = 1 - \alpha.$$

We have shown that if we are dealing with a normal model and we know an upper bound on $\sigma$, then we can choose in advance the required number $n$ of measurements, and

$$\frac{1}{n}\sum_{i=1}^{n} X_i$$

will be close to $\mu$ with at least the desired probability. △

At this point, we might hope that some other estimator of $\mu$ would not suffer this dependence on the uncontrollable parameter $\sigma^2$. Since the median and the mean are the same for normal distributions, we might be tempted to try the sample median as an estimator for the mean. However, we shall show in the next chapter that this will not help with a sample size $n$ fixed in advance. In fact, for given $\varepsilon > 0$ and $0 < \alpha < 1$, an estimator $g(X_1, \ldots, X_n)$ of $\mu$ that satisfies

$$P\{|g(X_1, \ldots, X_n) - \mu| \le \varepsilon\} \ge 1 - \alpha$$

for all $\mu$ and $\sigma^2 > 0$ *cannot be found.*

### 2.5 *Example: Mean of Uniform Random Variables*

Suppose that test drillings yield independent measurements $X_1, \ldots, X_n$ on the amount $\theta$ of oil available from some oil field. As a model let us assume that $X_i$ is uniform on $[0, 2\theta]$; that is,

$$f_{X_i}(x) = \begin{cases} 1/(2\theta) & \text{for } 0 \le x \le 2\theta \\ 0 & \text{otherwise.} \end{cases}$$

Note that if $[a_1, b_1], \ldots, [a_n, b_n]$ are subintervals of $[0, 2\theta]$, then, by independence,

$$P\{a_1 \le X_1 \le b_1, \ldots, a_n \le X_n \le b_n\} = (b_1 - a_1)(b_2 - a_2)\cdots(b_n - a_n)\frac{1}{(2\theta)^n}.$$

It is desired to find a precise estimate of the amount $\theta$ of available oil. We propose as a consistent sequence for $\theta$ the one given by

$$g_n(X_1, \ldots, X_n) = \frac{\max(X_1, \ldots, X_n)}{2}.$$

The consistency of this sequence follows from the fact that, for $0 < \varepsilon < \theta$,

$$P\left\{\theta - \varepsilon \leq \frac{\max(X_1, \ldots, X_n)}{2} \leq \theta\right\}$$

$$= P\{2\theta - 2\varepsilon \leq \max(X_1, \ldots, X_n) \leq 2\theta\}$$

$$= 1 - P\{\max(X_1, \ldots, X_n) \leq 2\theta - 2\varepsilon\}$$

$$= 1 - P\{X_1 \leq 2\theta - 2\varepsilon, \ldots, X_n \leq 2\theta - 2\varepsilon\}$$

$$= 1 - \left(\frac{2\theta - 2\varepsilon}{2\theta}\right)^n = 1 - \left(1 - \frac{\varepsilon}{\theta}\right)^n.$$

Hence

$$\lim_{n\to\infty} P\left\{\left|\frac{\max(X_1, \ldots, X_n\}}{2} - \theta\right| \leq \varepsilon\right\}$$

$$= \lim_{n\to\infty} P\left\{\theta - \varepsilon \leq \frac{\max(X_1, \ldots, X_n)}{2} \leq \theta\right\}$$

$$= \lim_{n\to\infty}\left[1 - \left(1 - \frac{\varepsilon}{\theta}\right)^n\right] = 1 - \lim_{n\to\infty}\left(1 - \frac{\varepsilon}{\theta}\right)^n = 1,$$

which establishes consistency.

Unfortunately, no matter what $n$ and $\varepsilon$ are given, there are values of $\theta$ sufficiently large that

$$P\left\{\left|\frac{\max(X_1, \ldots, X_n)}{2} - \theta\right| \leq \varepsilon\right\} = 1 - \left(1 - \frac{\varepsilon}{\theta}\right)^n$$

is close to 0. Thus, exactly as in the previous example, when using the given estimator, we cannot ensure accurate estimation of $\theta$ based on a pre-assigned number $n$ of observations. However, in this case, *we can ensure a small relative error* by a pre-assigned (before observing the data) choice of $n$. If $\hat{\theta}$ is an estimator of $\theta$, then the *relative error* of estimation is defined to be

$$\left|\frac{\hat{\theta} - \theta}{\theta}\right|.$$

In many cases of interest, it is the relative error rather than the absolute error that is important. This might well be the case in this problem, where a larger yield would reasonably allow a greater error. Here, the event that the relative error does not exceed $k$ is

$$\left|\frac{\hat{\theta} - \theta}{\theta}\right| = \left|\frac{\max(X_1, \ldots, X_n)}{2\theta} - 1\right| \leq k.$$

But

$$P\left\{\left|\frac{\max(X_1, \ldots, X_n)}{2\theta} - 1\right| \leq k\right\} = P\left\{\left|\frac{\max(X_1, \ldots, X_n)}{2} - \theta\right| \leq k\theta\right\}$$

$$= 1 - \left(1 - \frac{k\theta}{\theta}\right)^n = 1 - (1 - k)^n.$$

Thus, we see that we can choose $n$ so that it is highly probable that the relative error lies below any specified level $k$. △

## 2.6 EXERCISES

The following observation is of some help in solving some of the exercises in this set: If for large $n$ the variable $Z_n$ is likely to be near $\mu$, then $G(Z_n)$ is likely to be near $G(\mu)$, provided that $G$ is continuous. More precisely, suppose that $G$ is continuous at $\mu$. If for each $\delta > 0$, we have

$$\lim_{n \to \infty} P\{|Z_n - \mu| \leq \delta\} = 1,$$

then for each $\varepsilon > 0$ we have

$$\lim_{n \to \infty} P\{|G(Z_n) - G(\mu)| \leq \varepsilon\} = 1.$$

This follows from the fact that continuity of $G$ at $\mu$ provides a $\delta$ for each $\varepsilon$ with the property that when $|Z_n - \mu| \leq \delta$, then $|G(Z_n) - G(\mu)| \leq \varepsilon$.

1.° Let $X_1, X_2, \ldots, X_n$ be independent random variables representing the results of repeated tosses of a die. Let

$$Y_i = \begin{cases} 1 & \text{if } X_i = 1 \\ 0 & \text{otherwise.} \end{cases}$$

Which of the following sequences are consistent for the probability of a 1 in tossing the die? Justify.

(a) The sequence whose $n$th term is $\frac{1}{n} \sum_{i=1}^{n} X_i$.

(b) The sequence whose $n$th term is $Y_n$.

(c) The sequence whose $n$th term is $\frac{1}{n} \sum_{i=1}^{n} Y_i$.

(d) The sequence whose $n$th term is $\frac{1}{n} \sum_{i=1}^{n} Y_i^2$.

2.° Referring to the previous question, suppose that we wanted to estimate the probability of two successive 1's. Which of the following are consistent sequences for this probability? Justify.

The sequence whose $n$th term is:

(a) $\left(\frac{1}{n}\sum_{i=1}^{n} X_i\right)^2$.

(b) $\frac{1}{n}\sum_{i=1}^{n}(X_i^2)$.

(c) $\left(\frac{1}{n}\sum_{i=1}^{n} Y_i\right)^2$.

(d) $\frac{1}{n}\sum_{i=1}^{n}(Y_i^2)$.

(e) $\frac{1}{n}\sum_{i=1}^{n} Y_{2i-1}Y_{2i}$.

(Use the fact that if $W$ and $Z$ are independent, then $E(WZ) = E(W)E(Z)$; see Lemma 4.8 in Chapter 6.)

3. Suppose $X_1, \ldots, X_n$ are independent error measurements. We assume that $X_i$ is normal with mean 0 and variance $\sigma^2$. Which of the following are consistent sequences of estimators for $\sigma^2$?

The sequence whose $n$th term is:

(a) $\frac{1}{n}\sum_{i=1}^{n}(X_i^2)$.

(b) $\left(\frac{1}{n}\sum_{i=1}^{n} X_i\right)^2$.

(c) $\frac{1}{n}\sum_{i=1}^{n}(X_i - \bar{X}_{(n)})^2$ where $\bar{X}_{(n)} = \frac{1}{j}\sum_{j=1}^{n} X_j$.

Of those which are consistent, which one do you prefer?

4.° Suppose $X_1, X_2, \ldots$ represent measurements by a new instrument on some standard. As a model, assume $X_1, X_2, \ldots$ to be independent normal random variables with *known* expectation $E(X_i) = \mu$ and unknown variance $\text{Var } X_i = \sigma^2$. Find a consistent sequence of estimators for:

(a) $\sigma^2$.

(b) $\sigma$.

Hints: Remember that $\text{Var } X_i = E[(X_i - \mu)^2]$, and see 4 of Chapter 8.

5. Suppose that radiation measurements $X_1, X_2, \ldots$ each taken over a time interval of 5 seconds duration are independent Poisson random variables with mean $\lambda$. Find a consistent sequence of estimators for:

(a) $\lambda$.

(b) $P\{X_i = 0\}$.

6.° Show that in attempting to estimate $\theta$ in Example 2.5, (the oil-drilling example), if we know an upper bound $\theta_u$ for $\theta$, then we can choose a pre-assigned number $n$ of observations to ensure high accuracy estimation of $\theta$.

7. Let $T_1, T_2, \ldots$ be lifetimes of transistors, assumed independent with density

$$f_{T_i}(t) = \begin{cases} \dfrac{1}{\theta} e^{-t/\theta} & \text{for } t \geq 0 \\ 0 & \text{otherwise.} \end{cases}$$

Find two distinct consistent sequences of estimators for the *median* lifetime.

---

## 2.7 *PROBLEMS*

1. Answer Exercise 2.6(4) when $\mu$ is unknown without resorting to any other specific distribution computations than those already done for the answer to 2.6(4). Hints: Replace $\mu$ by a good estimator and use the fact that $P(AB)$ is large if both $P(A)$ and $P(B)$ are large.

2. Let $X_1, X_2, \ldots$ be independent normal random variables with $E(X_i) = \mu$, Var $X_i = \mu^2$ where $\mu$ is unknown. Let

$$g_n(X_i, \ldots, X_n) = \frac{1}{n} \sum_{i=1}^{n} X_i.$$

Can a pre-assigned $n$ be chosen so as to achieve any given relative error in estimation of $\mu$ with any desired high probability? (See Example 2.5 for a definition of relative error.)

3. Suppose that $g_n(X_1, \ldots, X_n)$ is a consistent sequence of estimators for $\theta$ such that a pre-assigned $n$ can be chosen to yield high accuracy estimation of $\theta$. Is it necessarily true that a pre-assigned $n$ can be chosen to yield high accuracy estimation of $\theta^2$? What is the answer to this question if it is known in advance that $|\theta| \leq \theta_u$ where $\theta_u$ is a known number? Exhibit the sequence whose existence is asserted.

4. Show that if $g_n(X_1, \ldots, X_n)$ is consistent for $\theta$, and if $G$ is a given function such that for each $y_0$,

$$\lim_{y \to y_0} G(y) = G(y_0),$$

then there is a consistent sequence of estimators for $G(\theta)$, and give such a sequence.

## SECTION 3
## Numerical Criteria for Judging Estimators

In order to choose between various estimators, it is quite useful to have some numerical measure of the worth of an estimate. It is not difficult to construct a reasonable numerical measure for this purpose under certain circumstances. In particular, suppose that for each pair $(\theta, y)$ we know some numerical measure $C(\theta, y)$ of the cost to be expected when the number $y$ is used as an estimate of the true parameter value $\theta$. Sometimes a reasonable "cost function" $C$ will be easy to determine, but in other cases it may be meaningless. We illustrate the first situation in the following example.

### 3.1 *Example: Anticipated Cost of a Guarantee*

Suppose that $\eta$ is the expected lifetime of a certain type of picture tube. We assume that tube lifetimes follow an exponential distribution with mean $\eta$ weeks (see Example 2.6, Chapter 4). The manufacturer has produced 100,000 such tubes. He will price each tube at $y$ dollars, with a "double your money pro-rated refund" of $\$2(y - x)$ if the tube lasts only $x$ weeks (for $x < y$). From these assumptions, it is really not difficult to determine a reasonable measure $C(\theta, y)$ of the extra cost to be expected when:

(a) the price for each picture tube is $y$ dollars.

(b) $\theta$ denotes the optimal price that the manufacturer could choose if he possessed only the knowledge that expected tube lifetime is $\eta$.

The details of determination of $C(\theta, y)$ are somewhat tedious. We will present them as an optional section.

When the manufacturer prices his tubes at $y$ dollars and a tube lasts $x$ weeks, his return on this sale is

**3.2**
$$r_y(x) = \begin{cases} y & \text{if } x > y \\ y - 2(y - x) & \text{if } x \leq y. \end{cases}$$

Note that the actual return to the manufacturer on any particular sale is the value of a random variable. To obtain a single overall non-random measure of anticipated cost, it is reasonable to take expectations. From 3.2 and our assumptions about the lifetime $X$, we see that the expected return from the sale of a single picture tube is

$$\begin{aligned} \mathscr{E}_\eta(y) &\doteq \int_0^\infty r(x) \frac{1}{\eta} e^{-x/\eta}\, dx \\ &= \int_0^y [y - 2(y - x)] \frac{1}{\eta} e^{-x/\eta}\, dx + \int_y^\infty y \frac{1}{\eta} e^{-x/\eta}\, dx \\ &= -y + 2\eta(1 - e^{-y/\eta}). \end{aligned}$$

The price $y = \theta$ that maximizes $\mathscr{E}_\eta(y)$ is found by setting $\mathscr{E}'_\eta(y) = 0$. We find

**3.3**
$$\theta = \eta \log_e 2.$$

That is, $\theta = \eta \log_e 2$ is the optimal choice of price knowing only $E(X) = \eta$. The expected extra cost per sale arising from the use of $y$ rather than $\theta$ as the price is

**3.4**
$$\begin{aligned} \mathscr{E}_\eta(\theta) - \mathscr{E}_\eta(y) &= -\theta + 2\eta(1 - e^{-\theta/\eta}) - [-y + 2\eta(1 - e^{-y/\eta})] \\ &= y - \theta + \frac{2\theta}{\log_e 2}\left(e^{-\frac{y \log 2}{\theta}} - \frac{1}{2}\right) \text{ by 3.3.} \end{aligned}$$

We know from the weak law of large numbers that the observed average extra cost over $n$ sales,

$$\frac{1}{n}\sum_{i=1}^{n}[r_y(X_i) - r_\theta(X_i)],$$

tends to Expectation 3.4. Since the manufacturer will presumably sell 100,000 tubes, it seems reasonable to take as the numerical measure $C(\theta, y)$ of his ignorance of $\theta$ when he prices tubes at $y$ dollars, the quantity

$$C(\theta, y) = 100{,}000\left[y - \theta + \frac{2\theta}{\log_e 2}\left(e^{-\frac{y\log_e 2}{\theta}} - \frac{1}{2}\right)\right].$$

We graph $C(\theta, y)$ as a function of $y$ for $\theta = 1$ in Figure 10.1. From this, it appears that overpricing by a given amount is likely to prove more costly than underpricing by the same amount.

In order to set his price, the manufacturer needs to estimate the optimal price $\theta = \eta \log_e 2$. Since $\eta$ is the expected lifetime of a picture tube of the given type, the manufacturer will very likely use the lifetimes $X_1, \ldots, X_n$ of $n$ tubes of this type to obtain an estimate $g(X_1, \ldots, X_n)$ for $\theta$. His price will then be the observed value of $g(X_1, \ldots, X_n)$. It is now rather easy to use the cost function $C$ to judge the merit of $g(X_1, \ldots, X_n)$.

A good approximation to the extra cost of using $g(X_1, \ldots, X_n)$ as his price instead of $\theta$ is

$$C(\theta, g(X_1, \ldots, X_n)) + tn$$

where $\eta$ is a tube's expected lifetime and $t$ is the cost of testing a single unit. However, this is a random variable, since it depends on the data $X_1, \ldots, X_n$. In order to obtain a numerical measure of the worth of $g(X_1, \ldots, X_n)$ when expected lifetime is $\eta$, we use the expectation

$$R_P(g) = E_P[C(\theta, g(X_1, \ldots, X_n))] + tn.$$

Here $P$ is the governing distribution of $X_1, \ldots, X_n$ and $\theta = \eta \log_e 2$ is the value associated with $P$. Note that the risk $R_P(g)$ for any fixed $g$ depends on which $P \in \mathscr{P}$

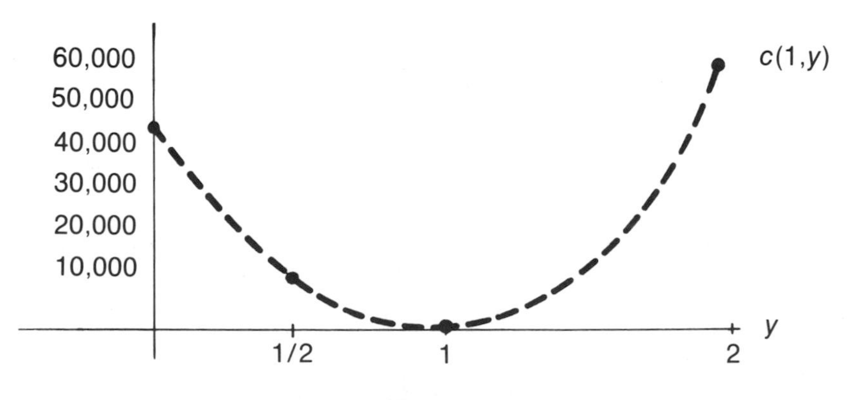


*FIGURE* 10–1.

governs the data. Although this does not tell us what $g$ should be, if the risk $R_P(g)$ is sufficiently small for all $P$ in $\mathscr{P}$, then $g(X_1, \ldots, X_n)$ would certainly seem satisfactory. Similarly, if for all $P$ in $\mathscr{P}$ we have $R_P(g^*) < R_P(g)$, then $g^*(X_1, \ldots, X_n)$ would seem preferable to $g(X_1, \ldots, X_n)$ for our purposes. $\wedge$

The essence of the above example is that if we can find a numerical measure $C(\theta, y)$ of the cost of using the numerical value $y$ to estimate the parameter $\theta$, then it is reasonable to judge the merit of the estimator $g(X_1, \ldots, X_n)$ of $\theta$ by means of the quantity

$$R_P(g) = E_P[C(\theta, g(X_1, \ldots, X_n)] + tn.$$

Here $\theta$ is the parameter value determined by $P$. The quantity $R_P(g)$, called "the risk using $g$ at $P$," is a measure of the cost of using $g(X_1, \ldots, X_n)$ to estimate $\theta$ when $P$ is the governing distribution of the data. We note that the cost $C(\theta, y)$ may very well be such that the penalty for overestimation differs greatly from that for comparable underestimation.

---

### 3.5 *EXERCISES*

1.° Find the risk $R_P(g)$ in Example 3.1 (cost of a guarantee) when $n = 1, g(X) = X \log_e 2$, $t = 50$, and $\eta = 40$.

2. Find the risk $R_P(g)$ in Example 3.1 when $n = 1, g(X) = X \log_e 2$, and $t = 50$.

3.° Find the risk $R_P(g)$ in Example 3.1 for general $n$ when

$$g(X_1, \ldots, X_n) = \left(\frac{1}{n} \sum_{i=1}^{n} X_i\right) \log_e 2$$

and $t = 50$.

4.° If you know that $40 \leq \eta \leq 60$:
(a) would you prefer $n = 1$ or $n = 10$ for the conditions of Exercise 3?
(b) what value of $n$ would you choose, and why?

In economic situations, we can sometimes derive a convincing cost function $C(\theta, y)$, but in a scientific problem this is not likely to be possible. For instance, if we were attempting to estimate the speed of light, we doubt that anyone could justifiably assign any sort of cost to an error of 5 miles per hour. In these cases, we still may be able to choose a risk function $R_P(g)$ that may lead to meaningful results.

We introduce here the most commonly used such risk function, the mean square error, and study some of its properties.

**3.6 Definition: Mean Square Error.** If $g(X_1, \ldots, X_n)$ is an estimator for the parameter $\theta$, then its *mean square error* $m_P(g)$ is defined by

$$m_P(g) = E_P[(g(X_1, \ldots, X_n) - \theta)^2].$$

Here the subscript $P$ is used to indicate the dependence of this expectation on the particular $P \in \mathscr{P}$ used in computing the expectation.

It should be noted that the mean square error criterion gives much more emphasis to large deviations of $g(X_1, \ldots, X_n)$ from $\theta$ than to small ones. Furthermore, it treats positive and negative deviations identically. It may be observed that $m_P(g)$ can arise from a cost function $C(\theta, y) = (\theta - y)^2$. Such a cost function may have no operational meaning in a scientific situation. However, for scientific purposes, the following result, proved in a manner identical to the proof of the Chebyshev inequality, is useful.

### 3.7 *Theorem: Operational Meaning of Mean Square Error*

Provided that $m_P(g)$ is finite, we have

$$P\{|g(X_1, \ldots, X_n) - \theta| \geq t\sqrt{m_P(g)}\} \leq \frac{1}{t^2}.$$

The proof is left as an exercise.

Although there are other measures of quality with a similar operational meaning, the mean square error is one of the most tractable. Note that if $E_P[g(X_1, \ldots, X_n)] = \theta$ for all $P$, then

$$m_P(g) = \text{Var}[g(X_1, \ldots, X_n)],$$

and Theorem 3.7 reduces to the Chebyshev inequality.

### 3.8 *Example: Mean Square Error of Waiting Time Estimate*

Suppose that waiting times $T_1, T_2, \ldots, T_n$ between successive radioactive disintegrations in a nuclear reaction are independent. (This describes a bomb very poorly, but it describes an atomic power plant reasonably well.) Assume also that each $T_i$ has an exponential density with unknown mean $\lambda > 0$. That is,

$$P\{T_i \leq t\} = \begin{cases} 1 - e^{-t/\lambda} & \text{for} \quad t \geq 0 \\ 0 & \text{otherwise.} \end{cases}$$

Let

$$g(T_1, \ldots, T_n) = \frac{1}{n} \sum_{i=1}^{n} T_i.$$

Then we see that the mean square error of the estimator $g(X_1, \ldots, X_n)$ of the parameter $\lambda$ is

$$m_P(g) = E_P\left[\left(\frac{1}{n} \sum_{i=1}^{n} T_i - \lambda\right)^2\right] = \frac{1}{n} \text{Var}_P\, T_i$$

$$= \frac{1}{n}\left[\int_0^\infty \frac{t^2}{\lambda} e^{-t/\lambda}\, dt - \lambda^2\right] = \frac{1}{n}[2\lambda^2 - \lambda^2] = \frac{\lambda^2}{n}.$$

If we know that, say, $\lambda \leq 4$ hours and $n = 14{,}400$, then we would have

$$m_P(g) \leq \frac{16}{14{,}400} = \frac{1}{900}.$$

Now, using Theorem 3.7, we could conclude that the chance that

$$\frac{1}{14{,}400} \sum_{i=1}^{14{,}400} T_i$$

will deviate from $\lambda$ by more than 5/30 does not exceed 1/25, because

$$P\left\{\left|\frac{1}{14{,}400}\sum_{i=1}^{14{,}400} T_i - \lambda\right| \geq \frac{5}{30}\right\} = P\left\{\left|\frac{1}{14{,}400}\sum_{i=1}^{14{,}400} T_i - \lambda\right| \geq \underbrace{5}_{t}\underbrace{\sqrt{\frac{1}{900}}}_{\uparrow\ \sqrt{m_P(g)}}\right\} \leq \underbrace{\frac{1}{25}}_{\uparrow\ \frac{1}{t^2}}$$

Thus, under the condition that the average waiting time $\lambda \leq 4$, we can estimate $\lambda$ to within 5/30 hr. (10 minutes) with confidence $>.96$, using 14,400 observations. If we know an upper bound for $\lambda$, we can choose a pre-assigned $n$ large enough to keep the mean square error small for *all* values of $\lambda$ being considered. This would permit us to make useful probability statements by means of Theorem 3.7. △

*3.9 Remark on Modified Mean Square Error.* For many problems, a modified version of the mean square error criterion might be preferable. For example, if relative error is more meaningful than absolute error, we might use

$$\mathscr{M}_P(g) = E_P\left[\left(\frac{g(X_1, \ldots, X_n) - \theta}{\theta}\right)^2\right] = \frac{m_P(g)}{\theta^2}.$$

## *3.10 EXERCISES*

1. Suppose we are told that the mean square error $m_P(g) \leq 9$.
   (a)° Find a lower bound for

$$P\{|g(X_1, \ldots, X_n) - \theta| \leq 72\}.$$

   (b) Find a lower bound for

$$P\{|g(X_1, \ldots, X_n) - \theta| \leq 20\}.$$

2. Suppose you are given that the mean square error $m_P(g(X_1, \ldots, X_n)) = \theta/n$, where $0 \leq \theta \leq 1$. How large an $n$ is needed to guarantee that:
   (a)° $P\{|g(X_1, \ldots, X_n) - \theta| \leq .01\} \geq .95$.
   (b) $P\{|g(X_1, \ldots, X_n) - \theta| \leq .05\} \geq .99$.

3.° Let $X_1, X_2, \ldots, X_n$ be independent random variables where $X_i$ is the number of successes on the $i$th trial of some experiment. We assume that $P\{X_i = 1\} = p$ and $P\{X_i = 0\} = 1 - p$, where $p$ is unknown. Let

$$g(X_1, \ldots, X_n) = \frac{1}{n} \sum_{i=1}^{n} X_i$$

be the proportion of successes in $n$ trials. Find the mean square error $m_P(g)$.

4. Let $X_1, \ldots, X_n$ stand for the results obtained on successive independent tosses of a (possibly loaded) die. Find the mean square error $m_P(g)$ for

$$g(X_1, \ldots, X_n) = \frac{1}{n} \sum_{i=1}^{n} X_i \qquad \text{where} \qquad \theta = E(X_i).$$

5. Let $X_1, \ldots, X_n$ be independent normal random variables representing repeated measurements of some given physical constant.

We assume that $E(X_i) = \mu$, $\operatorname{Var} X_i = \sigma^2$.

(a) If

$$g(X_1, \ldots, X_n) = \frac{1}{n} \sum_{i=1}^{n} X_i$$

is used as an estimator of $\mu$, what is the mean square error $m_P(g)$?

(b) If $\mu$ is known and we use

$$g(X_1, \ldots, X_n) = \frac{1}{n} \sum_{i=1}^{n} (X_i - \mu)^2$$

as an estimator of $\sigma^2$, what is the mean square error $m_P(g)$?

6.° Suppose that repeated weighings are made on an empty scale. We assume that $X_1, \ldots, X_n$ are independent and normal with $E(X_i) = \mu$, $\operatorname{Var} X_i = \sigma^2$, both unknown. We want to estimate $\sigma^2$, and we let

$$g(X_1, \ldots, X_n) = \frac{1}{n} \sum_{j=1}^{n} \left(X_i - \frac{1}{n} \sum_{j=1}^{n} X_j\right)^2.$$

What is the mean square error $m_P(g)$?

Hint: You may assume, as given, that

$$\sum_{i=1}^{n} \left(X_i - \frac{1}{n} \sum_{j=1}^{n} X_j\right)^2 \Big/ \sigma^2$$

has a Chi square distribution with $n - 1$ degrees of freedom under the given conditions (see Chapter 8). This will be proved in Chapter 14.

7.° (a) What is the mean square error $m_P(g)$ for the estimator

$$Y = \frac{\max(X_1, \ldots, X_n)}{2}$$

of Example 2.5?

Hint: First find $F_Y$, and from it find $f_Y$.

(b) Using the mean square error criterion, which seems preferable as an estimator of the mean of these variables,

$$Y \text{ or } \bar{X}_{(n)} = \frac{1}{n} \sum_{i=1}^{n} X_i?$$

### *3.11 PROBLEMS*

1. Prove Theorem 3.7 on the operational meaning of mean square error.

2.° Referring to Exercise 3.10(6), if we look at all estimators $q(X_1, \ldots, X_n)$ of the form

$$q(X_1, \ldots, X_n) = c \sum_{i=1}^{n} \left(X_i - \frac{1}{n} \sum_{j=1}^{n} X_j\right)^2,$$

which value of $c$ minimizes the mean square error $m_P(q)$?

The mean square error criterion standing alone furnishes only a means of judging the quality of proposed estimators. It does not by itself yield a method of construction of estimates, because in general there is no estimator of $\theta$ with smallest mean square error for all $\theta$. We can see this by noting that the estimator given by $g(X_1, \ldots, X_n) = c$ (constant) satisfies $m_P(g) = 0$ whenever $\theta = c$. Thus, this estimator has minimum mean square error when $\theta = c$. Except in trivial cases, $g(X_1, \ldots, X_n) = c$ is better than any other estimate of $\theta$ whenever $\theta = c$. However, such an estimator is not reasonable, since generally its mean square error will be too large for other values of $\theta$. If we want to use smallness of mean square error as a criterion for construction of estimators, we must exclude possibilities like the one mentioned. One way of doing this is to impose additional restrictions on the estimators. We might, for example, require for some $k$ chosen by us that $m_P(g) \leq k$ for all $P$ in $\mathscr{P}$. Another requirement that does not seem too unreasonable, and that has frequently led to useful results, is given in the following definition.

**3.12 Definition: Unbiased Estimator.** Let $X_1, \ldots, X_n$ be random variables whose joint distribution is $P$, known only to come from some class $\mathscr{P}$. Let $\theta$ be a parameter determined by the distribution $P$. We say that the observable statistic $g(X_1, \ldots, X_n)$ is an *unbiased estimator of* $\theta$ (relative to $\mathscr{P}$) if for each $P \in \mathscr{P}$,

$$E_P[g(X_1, \ldots, X_n)] = \theta.$$

Here the subscript $P$ indicates that computations are made under the assumption that $P$ is the governing distribution. The value $\theta$ appearing on the right is the one determined by $P$ on the left.

### *3.13 Example: Uniform Random Variables*

Let $X_1, \ldots, X_n$ be independent random variables with a uniform distribution on $[\theta - 1/2, \theta + 1/2]$. This might be a model for an instrument that is adjusted to within 1/2 a unit of accuracy. Then each $X_i$ is an unbiased estimate of $\theta$. This follows from the fact that $X_i$ is an observable statistic and from the computation:

$$E(X_i) = \int_{\theta-1/2}^{\theta+1/2} x\,dx = \frac{x^2}{2}\Big|_{\theta-1/2}^{\theta+1/2}$$

$$= \frac{(\theta + \frac{1}{2})^2 - (\theta - \frac{1}{2})^2}{2}$$

$$= \frac{(\theta + \frac{1}{2} + \theta - \frac{1}{2})(\theta + \frac{1}{2} - [\theta - \frac{1}{2}])}{2}$$

$$= \theta.$$

Furthermore,

$$\frac{1}{n}\sum_{i=1}^{n} X_i$$

is an unbiased estimate of $\theta$ because it is an observable statistic and

$$E\left[\frac{1}{n}\sum_{i=1}^{n} X_i\right] = \frac{1}{n}E\left[\sum_{i=1}^{n} X_i\right]$$

$$= \frac{1}{n}\sum_{i=1}^{n} E(X_i)$$

$$= \frac{1}{n}\sum_{i=1}^{n} \theta = \theta. \qquad \triangle$$

### *3.14 Example: More on Uniform Random Variables*

Consider the variables $X_1, \ldots, X_n$ of the previous example. We claim that

$$W = \frac{\max\limits_{1\le i\le n} X_i + \min\limits_{1\le i\le n} X_i}{2}$$

is an unbiased estimate of $\theta$. We show this as follows: Let $Y = \max\limits_{1\le i\le n} X_i$ and $Z = \min\limits_{1\le i\le n} X_i$. Then

$$F_Y(y) = P\{Y \le y\}$$

$$= P\{\max_{1\le i\le n} X_i \le y\}$$

$$= P\{Y_1 \le y, Y_2 \le y, \ldots, Y_n \le y\}$$

$$= (y - [\theta - \tfrac{1}{2}])^n$$

for

$$\theta - \tfrac{1}{2} \le y \le \theta + \tfrac{1}{2}.$$

Hence,

**3.15** $$f_Y(y) = n(y - [\theta - \tfrac{1}{2}])^{n-1}, \qquad \text{for} \qquad \theta - \tfrac{1}{2} \le y \le \theta + \tfrac{1}{2}.$$

Similarly,

$$\begin{aligned} F_Z(z) &= P\{Z \le z\} \\ &= 1 - P\{Z > z\} \\ &= 1 - P\{\min_{1 \le i \le n} X_i > z\} \\ &= 1 - P\{X_1 > z, X_2 > z, \ldots, X_n > z\} \\ &= 1 - (\theta + \tfrac{1}{2} - z)^n \qquad \text{for} \qquad \theta - \tfrac{1}{2} \le z \le \theta + \tfrac{1}{2}. \end{aligned}$$

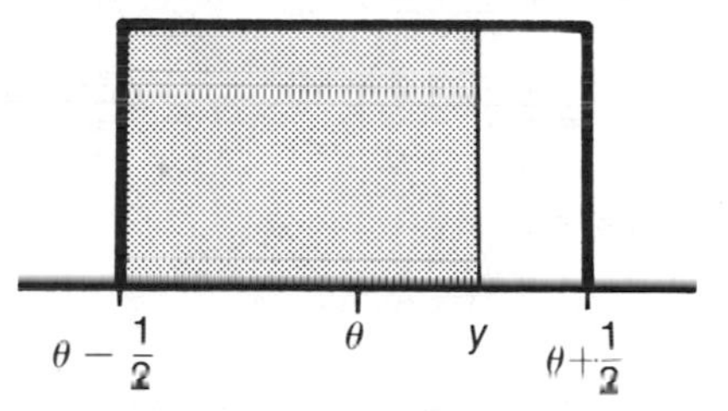


Hence,

**3.16** $$f_Z(z) = n(\theta + \tfrac{1}{2} - z)^{n-1}, \qquad \text{for} \qquad \theta - \tfrac{1}{2} \le z \le \theta + \tfrac{1}{2}.$$

Using Equations 3.15 and 3.16, we find

$$\begin{aligned} E(W) &= E\left(\frac{Y+Z}{2}\right) = \frac{1}{2}E(Y) + \frac{1}{2}E(Z) \\ &= \frac{1}{2}\int_{\theta-1/2}^{\theta+1/2} yn(y - [\theta + \tfrac{1}{2}])^{n-1}\,dy + \frac{1}{2}\int_{\theta-1/2}^{\theta+1/2} zn(\theta + \tfrac{1}{2} - z)^{n-1}\,dz. \end{aligned}$$

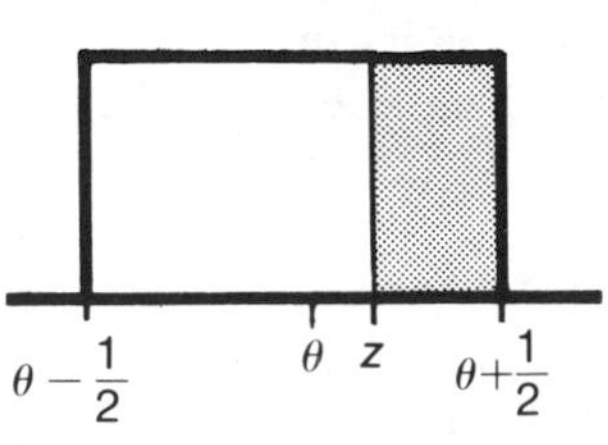

In the first integral we make the substitution $u = y - \theta$, and in the second integral $u = z - \theta$. This yields

$$E\left(\frac{Y+Z}{2}\right) = \frac{1}{2}\int_{-1/2}^{1/2}(u+\theta)n(u+\tfrac{1}{2})^{n-1}\,du + \frac{1}{2}\int_{-1/2}^{1/2}(u+\theta)n(\tfrac{1}{2}-u)^{n-1}\,du$$

$$= \frac{n}{2}\int_{-1/2}^{1/2}(u+\theta)[(\tfrac{1}{2}+u)^{n-1} + (\tfrac{1}{2}-u)^{n-1}]\,du$$

$$= \frac{n}{2}\int_{-1/2}^{1/2}u[(\tfrac{1}{2}+u)^{n-1} + (\tfrac{1}{2}-u)^{n-1}]\,du$$

$$+ \frac{n}{2}\theta\int_{-1/2}^{1/2}[(\tfrac{1}{2}+u)^{n-1} + (\tfrac{1}{2}-u)^{n-1}]\,du.$$

But the first integrand, $g(u) = u[(1/2+u)^{n-1} + (1/2-u)^{n-1}]$, is an odd function—that is, $g(u) = -g(-u)$. Hence $\int_{-1/2}^{1/2} g(u)\,du = 0$. Thus

$$E(W) = E\left(\frac{Y+Z}{2}\right) = \frac{n}{2}\theta\int_{-1/2}^{1/2}[(\tfrac{1}{2}+u)^{n-1} + (\tfrac{1}{2}-u)^{n-1}]\,du$$

$$= \frac{n}{2}\theta\left[\frac{(\tfrac{1}{2}+u)^n}{n} - \frac{(\tfrac{1}{2}-u)^n}{n}\right]\Bigg|_{-1/2}^{u=1/2}$$

$$= \frac{n}{2}\theta\left[\frac{1}{n} - \left(-\frac{1}{n}\right)\right]$$

$$= \theta.$$

This illustrates the fact that in general there may be many unbiased estimators. Of the estimators we considered here, we would naturally prefer to use the one with the smallest mean square error. In the case of unbiased estimators, that would be the one with minimum variance. Here this turns out to be the last one,

$$\frac{Y+Z}{2}. \qquad \triangle$$

We remark that there are circumstances in which unbiasedness is unachievable, or in which the "best" unbiased estimate is poorer in some sense than some other estimate. Our purpose in mentioning this possibility is to present a balanced picture and to help the student avoid an unthinking acceptance of the criteria that we have proposed. Whatever estimate is proposed should be judged according to some reasonable objective criterion, such as its mean square error or its modified mean square error. We now consider a few more examples relating to unbiasedness.

### 3.17 *Example:* $E(X_i) = \theta$

Consider any situation in which we have as the parameter $\theta$ the common expectation of all of the random variables $X_1, \ldots, X_n$. For example, we may make a sequence of measurements on some physical constant using a variety of measuring devices. We might then assume $X_i$ to be normal with mean $E(X_i) = \theta$. Or, the $X_i$ may be measurements of radiation, so that we may assume that each $X_i$ has a Poisson distribution with mean value $\theta$.

Let $a_1, \ldots, a_n$ be any set of real numbers with $\sum_{j=1}^{n} a_j = 1$. Then, by Corollary 4.3 in Chapter 6, the linear combination

$$g(X_1, \ldots, X_n) \equiv \sum_{j=1}^{n} a_j X_j$$

is also an unbiased estimator of $\theta$. Such a linear combination is called a *weighted average.* △

### 3.18 *Example: Best Linear Unbiased Estimators*

Suppose that in repeating an experiment the same instrument is used by $n$ different people. Suppose also that each person averages his results and preserves only this average and the amount of data, $k_i$, on which his average was based. It seems reasonable to assume that the results are independent random variables $X_1, \ldots, X_n$ with $E(X_i) = \mu$ and $\text{Var } X_i = k_i\sigma^2$. Here $\mu$ is unknown, $\sigma^2$ may be known or unknown, and $k_i$ are known.

*Linear estimates* of $\mu$, that is, those of the form $\sum_{i=1}^{n} a_i X_i$, which are

(a) unbiased estimates of $\mu$ and

(b) of minimum variance (among all linear unbiased estimates)

are frequently desired. Such estimates are called *best linear unbiased estimates*. Since

$$E\left(\sum_{i=1}^{n} a_i X_i\right) = \mu \sum_{i=1}^{n} a_i,$$

unbiasedness requires that $\sum_{i=1}^{n} a_i = 1$. Now

$$\text{Var}\left(\sum_{i=1}^{n} a_i X_i\right) = \sigma^2 \sum_{i=1}^{n} a_i^2 k_i,$$

so that minimizing the variance is equivalent to minimizing $\sum_{i=1}^{n} a_i^2 k_i$ subject to the condition $\sum_{i=1}^{n} a_i = 1$. Substituting $a_1 = 1 - \sum_{i=2}^{n} a_i$ in the expression $\sum_{i=1}^{n} a_i^2$ transforms the problem to that of finding $a_2, \ldots, a_n$ that minimize

$$\left(1 - \sum_{i=2}^{n} a_i\right)^2 k_1 + \sum_{i=2}^{n} a_i^2 k_i \equiv G(a_2, \ldots, a_n).$$

From the method that is outlined in the Appendix, we can see that any $a_i$ minimizing $G(a_2, \ldots, a_n)$ must satisfy the $n - 1$ equations

$$\frac{\partial G}{\partial a_i} = 0 \qquad \text{for} \quad i = 2, \ldots, n.$$

In this case, the equations are

$$-2\left(1 - \sum_{i=2}^{n} a_i\right)k_1 + 2a_ik_i = 0.$$

That is, since $1 - \sum_{i=2}^{n} a_i = a_1$, we have $a_ik_i = a_1k_1$ for $i = 2, \ldots, n$. Since

$$1 = \sum_{j=1}^{n} a_j = a_1 + \sum_{j=2}^{n} a_j = a_1 + a_1k_1 \sum_{j=2}^{n} \frac{1}{k_j},$$

we have

$$a_1 = \frac{1}{1 + \sum_{j=2}^{n} \frac{k_1}{k_j}} \qquad \text{and} \qquad a_i = a_1k_1/k_i = \frac{k_1}{k_i} \frac{1}{1 + \sum_{j=2}^{n} \frac{k_1}{k_j}}.$$

That is,

(*)
$$a_i = \frac{k_1}{k_i} \frac{1}{1 + \sum_{j=2}^{n} \frac{k_1}{k_j}} \qquad \text{for all} \quad i = 1, \ldots, n.$$

Thus, the best linear unbiased estimate of $\mu$ when $E(X_i) = \mu$ and $\text{Var } X_i = k_i\sigma^2$ is $\sum_{i=1}^{n} a_iX_i$, where $a_i$ is given by (*). We see that the weights $a_i$ of $X_i$ are inversely proportional to Var $X_i$. That is, less importance is attached to those $X_i$ with larger variances. △

---

## *3.19 EXERCISES*

1.° Suppose that $X_1, \ldots, X_n$ are independent random variables, uniformly distributed on $[0, \theta]$. Which of the following are unbiased estimates of $\theta$?

(a) $2X_i$.

(b) $\frac{2}{n} \sum_{i=1}^{n} X_i$.

(c) $X_i^2$.

(d) $Y = \max_{1 \le i \le n} X_i$.

Hint: Find $F_Y(y)$ first.

(e) $Z = \max_{1 \le i \le n} X_i + \min_{1 \le i \le n} X_i$.

2. (a) Based on the computations of part (d) of the previous Exercise, find the constant $c_n$ such that $c_n \max_{1 \le i \le n} X_i$ is an unbiased estimate of $\theta$.

(b) Which estimator of $\theta$ has the smaller mean square error—the one found in part (a), or

$$\frac{2}{n}\sum_{i=1}^{n} X_i?$$

3. Let $X$ be normal with mean 0 and variance $\sigma^2$. Find an unbiased estimator for $\sigma^2$.
Hint: Compute the first several moments of $X$. It is wise to use the moment generating function $\varphi_X(t) = e^{t^2\sigma^2/2}$.

4. Let $X_1, \ldots, X_n$ be independent normal random variables with $E(X_i) = \mu$ and Var $X_i = 1/i$. Find the best linear unbiased estimator of $\mu$.

5. Let $X$ be uniform on $[0, \theta]$, that is, $f_X(x) = \begin{cases} 1/\theta \text{ for } 0 \leq x \leq \theta \\ 0 \quad \text{otherwise.} \end{cases}$

(a) Find an unbiased estimate of $\theta^2$. (Remember that your estimate must be observable, that is, it cannot depend on $\theta$.)
Hint: Look at $X^2$.

(b)° Find an unbiased estimate of $\theta^n$ where $n$ is any non-negative real number.

6. Let $X$ be Poisson with mean $\lambda$.

(a)° Find an unbiased estimate of $\lambda^2$.
Hint: Use the moment generating function to find $E(X^2)$, and recall that $E(X) = \lambda$.

(b) Find an unbiased estimate of $\lambda^3$.

(c) Show that it is possible to find unbiased estimates of $\lambda^n$ for each counting number $n$.

---

## 3.20 PROBLEMS

1.° Let $X$ be a Bernoulli random variable, that is, $P\{X = 1\} = p = 1 - P\{X = 0\}$ Is there a function $g(X)$ that is an unbiased estimate of $p^2$? Explain your reasoning.
Hint: If $g(X)$ is unbiased, you must have $g(0)[1 - p] + g(1)p = p^2$ for all $p$ in $[0, 1]$. Is this possible?

2. Let $X$ be a binomial random variable with parameters $n$ and $p$.

(a) Find an unbiased estimate of $p^2$ where $n \geq 2$.
Hint: Compute $E(X^2)$.

(b) For given $n$, show that it is impossible to obtain unbiased estimates for $p^k$, all positive integers $k$. In particular, if $k > n$, there is no unbiased estimate of $p^k$.

3. Find a sequence $g_1(X_1)$, $g_2(X_1, X_2)$, ... that is consistent for $\theta$ but for which $E_P[g_n(X_1, \ldots, X_n)] \cong \theta + 1$ for all large $n$ and all $P$ in $\mathscr{P}$.
Hint: Let $X_1, X_2, \ldots$ be independent normal random variables with mean $\mu$ and variance 1 and suppose

$$\int_{-a_n}^{a_n} \frac{1}{\sqrt{2\pi}} e^{-t^2/2}\, dt = \frac{1}{n^2}.$$

Look at

$$g_n(X_1, \ldots, X_n) = \begin{cases} \dfrac{1}{n}\displaystyle\sum_{i=1}^{n-2} X_i + n^2 & \text{if } \dfrac{X_n - X_{n-1}}{\sqrt{2}} \in [-a_n, a_n] \\ \dfrac{1}{n}\displaystyle\sum_{i=1}^{n-2} X_i & \text{otherwise for } n = 3, 4, \ldots. \end{cases}$$

This shows that consistency does not imply unbiasedness.

4. Find a sequence $g_1(X_1)$, $g_2(X_1, X_2)$, . . . , such that

$$E_P[g_n(X_1, \ldots, X_n)] = \theta \text{ for all } n, \text{ and all } P \text{ in } \mathscr{P},$$

that is not consistent for $\theta$. This shows that unbiasedness does not imply consistency.

## SECTION 4
## Sufficiency and Data Reduction

After an experiment has been performed, it is the job of the experimentor to interpret the data he has amassed—and this can involve an enormous number of accumulated results. For instance, data on taste tests for beers are categorized by the age, sex, geographic location, occupation, economic level, and consumption of each individual tested. Each individual is asked a series of questions, and all the answers are recorded also. As another instance, mortality data in census tracts divide individuals into age, sex, location, and various disease categories. One look at a census tract is enough to stagger even the most fearless. On a more personal level, the record of how you have scored on each question of all the tests you have ever taken would be somewhat overwhelming.

In all of the situations just discussed, there is a wealth of information available in the accumulated data. However, any attempt to assimilate information or to draw useful conclusions from a direct examination of such raw data is likely to be unrewarding. Furthermore, the cost of storage, whether it is measured by the number of file drawers needed or by the amount of computer memory and processing time required, is likely to be high. *We therefore would like to "reduce" the amount of data without losing the information in it.* However, if we do not have some well-defined statistical model for our data, we will have no criterion for judging whether or not information is destroyed by data reduction. For this reason, all of our discussion will be in the framework of models such as were introduced in the previous chapter. We shall see how a systematic approach within this type of theoretical framework will lead us to practical, trustworthy results.

For simplicity, we confine ourselves *initially* to the case of a parametric family $\mathscr{P}$ where the parameter $\theta$ that determines $P$ is a scalar. Furthermore, we will start out by dealing only with discrete random variables. Let $P_\theta$ denote the distribution of the sample specified by the parameter $\theta$. Recall that in this context the distribution $P_\theta$ of the sample $X_1, \ldots, X_n$ *usually* varies with $\theta$; this variation is likely to be reflected in the observed values of $X_1, \ldots, X_n$. Hence, we can usually use the observed values of $X_1, \ldots, X_n$ to tell us something about $\theta$. If $P_\theta$ doesn't vary with $\theta$, then certainly the sample $X_1, \ldots, X_n$ will be of no use in learning anything about $\theta$.

Suppose that someone suggests using the observable statistic $g(X_1, \ldots, X_n)$ to extract all of the sample's information about $\theta$. Suppose also that the *conditional* distribution of the sample *given* the value of $g(X_1, \ldots, X_n)$ does not vary with $\theta$. In light of the discussion just presented, we claim that the sample contains no *further* information about $\theta$. More precisely, if, for each $A$ and $k$,

**4.1** $$P_\theta\{(X_1, \ldots, X_n) \in A \mid g(X_1, \ldots, X_n) = k\}$$

does not vary with $\theta$, we feel that $g(X_1, \ldots, X_n)$ possesses all the information about $\theta$ contained in the sample. A statistic with this property is called *sufficient* for $\theta$.

Rather than give an example immediately, we will first perform some manipulation to get a more workable definition.

Because

$$P_\theta\{(X_1, \ldots, X_n) \in A \mid g(X_1, \ldots, X_n) = k\} = \sum_{\substack{\text{all } (x_1, \ldots, x_n) \\ \text{such that} \\ (x_1, \ldots, x_n) \in A}} P_\theta\{X_1 = x_1, \ldots, X_n = x_n \mid g(X_1, \ldots, X_n) = k\},$$

we see that Equation 4.1 will surely hold if, for each $(x_1, \ldots, x_n)$ and $k$,

**4.2** $$P_\theta\{X_1 = x_1, \ldots, X_n = x_n \mid g(X_1, \ldots, X_n) = k\}$$

does not vary with $\theta$. Furthermore, 4.2 must hold if Equation 4.1 holds for each $A$ and $k$, since 4.2 is a special case of 4.1, namely, that where $A$ consists of the single point $(x_1, \ldots, x_n)$.

Thus, we might take 4.2 as our definition. However, we notice further that if $k \neq g(x_1, \ldots, x_n)$, then $P_\theta\{X_1 = x_1, \ldots, X_n = x_n \mid g(X_1, \ldots, X_n) = k\} = 0$, which certainly does not vary with $\theta$. Hence, we give the following equivalent definition.

**4.3 Definition: Sufficient Statistic for Discrete Random Variables.** Let $\mathscr{P}$ be a parametric family of distributions governing the discrete random variables $X_1, \ldots, X_n$. Let $\theta$ be the parameter and $P_\theta$ the element of $\mathscr{P}$ determined by $\theta$. We will say that the observable statistic $g(X_1, \ldots, X_n)$ is *sufficient* for $\theta$ if for all $\theta$ and all choices for the possible values $x_i$ of the random variables $X_i$, $i = 1, \ldots, n$, the value

$$P_\theta\{X_1 = x_1, \ldots, X_n = x_n \mid g(X_1, \ldots, X_n) = g(x_1, \ldots, x_n)\}$$

does not vary with $\theta$.†

All the knowledge we can gain about $\theta$ by looking at the raw data can just as well be obtained by observing only the sufficient statistic.

We note that there was nothing at all in the discussion leading to Definition 4.3 that required either that $\theta$ be a scalar or that $g(X_1, \ldots, X_n)$ be only a scalar. Hence, *in Definition 4.3 we will allow both $\theta$ and $g(X_1, \ldots, X_n)$ to be vectors.* This generality will become more understandable in cases such as the normal distribution, where $\theta = (\mu, \sigma^2)$ and it takes two scalar statistics

$$g(X_1, \ldots, X_n) = \left(\frac{1}{n}\sum X_i, \frac{1}{n}\sum X_i^2\right)$$

to summarize the data's information about $\theta$.

---

† Actually, we should say that $g(X_1, \ldots, X_n)$ is sufficient for $\mathscr{P}$, since the conditional distribution of the sample is then independent of which element of $\mathscr{P}$ governs the observations. Nevertheless, we shall adhere to the more conventional terminology, "sufficient for $\theta$."

### 4.4 *Example: Independent Bernoulli Random Variables*

Let $X_1, \ldots, X_n$ be independent Bernoulli random variables, that is to say, $P\{X_i = 1\} = p = 1 - P\{X_i = 0\}$. Since $X_i$ is the number of successes on the $i$th trial, we might suspect that $\sum_{i=1}^{n} X_i$, the total number of successes, is sufficient for $p$. Let us therefore compute for $x_i = 0, 1$:

$$P_p\left\{X_1 = x_1, \ldots, X_n = x_n \,\middle|\, \sum_{i=1}^{n} X_i = \sum_{i=1}^{n} x_i\right\}$$

$$= \frac{P_p\left\{X_1 = x_1, \ldots, X_n = x_n, \sum_{i=1}^{n} X_i = \sum_{i=1}^{n} x_i\right\}}{P_p\left\{\sum_{i=1}^{n} X_i = \sum_{i=1}^{n} x_i\right\}}$$

$$= \frac{P_p\{X_1 = x_1, \ldots, X_n = x_n\}}{P_p\left\{\sum_{i=1}^{n} X_i = \sum_{i=1}^{n} x_i\right\}}$$

$$= \frac{[p^{x_1}(1-p)^{1-x_1}][p^{x_2}(1-p)^{1-x_2}]\cdots[p^{x_n}(1-p)^{1-x_n}]}{\binom{n}{\sum_{i=1}^{n} x_i} p^{\sum_{i=1}^{n} x_i}(1-p)^{n-\sum_{i=1}^{n} x_i}} = \frac{1}{\binom{n}{\sum_{i=1}^{n} x_i}}.$$

Since this expression does not depend on $p$ and $\sum_{i=1}^{n} X_i$ is an observable statistic, we conclude that in this case $\sum_{i=1}^{n} X_i$ is sufficient for $p$. That is, *in the case of $n$ independent trials, each of which has probability $p$ of success, all the information about $p$ that the sample has to offer is "summarized" by the observed number of successes $\sum X_i$.*

Note here the enormous extent of the data reduction—from having to examine $n$ pieces of data to needing to look at only one such piece. Actually, for storage on a computer, the original data can be stored as an $n$-digit binary, and the sufficient statistic $\sum_{i=1}^{n} X_i$ can be stored as a $1 + \log_2 n$-digit number (since $0 \le \sum_{i=1}^{n} X_i \le n$), which gives an indication of how much data reduction is really accomplished in this case; for $n = 128$ we have $1 + \log_2 n = 8$. △

### 4.5 *Example: A Statistic That Is not Sufficient*

Suppose that we run three independent experiments, each of which has probability $p$ of success. If $X_i$ is the number of successes on trial $i$, then we claim that the statistic

$$X_1 + 2X_2 + 3X_3$$

is not sufficient for $p$. This follows from the computation

$$P_p\{X_1 = 1, X_2 = 1, X_3 = 0 \mid X_1 + 2X_2 + 3X_3 = 3\}$$
$$= \frac{P_p\{X_1 = 1, X_2 = 1, X_3 = 0, X_1 + 2X_2 + 3X_3 = 3\}}{P_p\{X_1 + 2X_2 + 3X_3 = 3\}}$$
$$= \frac{P_p\{X_1 = 1, X_2 = 1, X_3 = 0\}}{P_p(\{X_1 = 1, X_2 = 1, X_3 = 0\} \cup \{X_1 = 0, X_2 = 0, X_3 = 1\})}$$
$$= \frac{p^2(1-p)}{p^2(1-p) + (1-p)^2 p} = \frac{p}{p + [1-p]} = p.$$

Thus

$$P_p\{X_1 = 1, X_2 = 1, X_3 = 0 \mid X_1 + 2X_2 + 3X_3 = 3\} = p,$$

which certainly varies with $p$.

Notice the interesting result that the sufficient statistic $X_1 + X_2 + X_3$ assumes only four possible values, whereas the non-sufficient statistic $X_1 + 2X_2 + 3X_3$ assumes the values 0, 1, 2, 3, 4, 5, and 6. So $X_1 + 2X_2 + 3X_3$ reduces the data less, but loses information; $X_1 + X_2 + X_3$ loses no information about $p$, even though it reduces the data more. △

### 4.6 *Example: n Tosses of a Die*

Suppose that you toss a possibly loaded die independently $n$ times in order to learn something about the probabilities $p_j$ that the die comes up $j$, $j = 1, \ldots, 6$. Intuitively, it would seem that you could reduce the data from a full record of the result of every single toss to just $(N_1, \ldots, N_5)$, where $N_j$ is the total number of tosses resulting in the value $j$. More formally, we feel that $N = (N_1, \ldots, N_5)$ is sufficient for $\theta = (p_1, \ldots, p_6)$. The reader is invited to prove this, using Definition 4.3, but it is a tedious computation. It will be much easier to use the Neyman factorization theorem, which we will establish shortly. △

Because the definition of sufficiency is often hard to use directly, we furnish only a few exercises to help you become familiar with it.

---

## 4.7 *EXERCISES*

1. Let $X_1, \ldots, X_n$ be independent Poisson random variables, each with expectation $\theta$. Show that $\sum_{i=1}^{n} X_i$ is sufficient for $\theta$. In essence, this shows that any statistical decision made concerning $\theta$ can just as well be based on $\sum_{i=1}^{n} X_i$. We shall become more precise about this shortly.

2.° Let $X_1$ be Bernoulli with $P\{X_1 = 1\} = p$, and let $X_2$ independent of $X_1$ be Bernoulli with $P\{X_2 = 1\} = 2p$, where $0 \le p \le 1/2$. Show that $X_1 + X_2$ is not sufficient for $p$.
Hint: Look at $P_p\{X_1 = 1, X_2 = 0 \mid X_1 + X_2 = 1\}$. Note that here the position of the 1, not just the proportion of 1's, is important in finding out something about $p$.

3. In Example 4.5, show that $X_1 + 2X_2 + X_3$ is not sufficient.
Hint: Look at $\{X_1 = 0, X_2 = 1, X_3 = 0\}$.

4.° In Exercise 2, is $X_1 + 7X_2$ sufficient?

5.° In the case of two independent identically distributed Bernoulli random variables, $P\{X_i = 1\} = p = 1 - P\{X_i = 0\}$, where $X_1 + X_2$ and $X_1 + 7X_2$ are both sufficient for $p$, which is preferable, in the sense that it condenses the data more?

*Remarks.* There may be several sufficient statistics for a given parameter. We prefer those that in some sense "reduce the data the most," or assume the fewest values. With respect to Exercise 4.7(4), the statistic $X_1 + 7X_2$ assumes more values than $X_1 + X_2$. Hence, $X_1 + X_2$ is preferred. We shall not pursue this subject further, but the reader should always be looking for the sufficient statistics that condense the data the most, called minimal sufficient statistics.

We note that although the concept of sufficiency tells us how to reduce data for certain statistical models, it does not give any special clues as to how to use the reduced data for estimation or testing.

All the criteria for judging the merit of any given statistical procedure are based on the distribution of the sample. Now we shall show that even if we are allowed to observe only a sufficient statistic (as opposed to being allowed to observe the original sample), we can still generate a sample with the same distribution as the original sample.

All we need in the discrete case besides the observed value of the sufficient statistic is the ability to generate a sample with any pre-assigned discrete distribution. Knowledge of $\theta$ is not required. In Problem 4.9, we will show how to generate a sample with any pre-assigned discrete distribution, from a random variable with the uniform distribution on [0, 1]. (We can do this in the non-discrete case as well, but we lack the machinery to prove it.)

Now, any criteria for judging statistical procedures must assign the original sample and the one with the same distribution generated from the sufficient statistic identical overall performance ratings. Hence, we see that ***with respect to any criterion of judgement, we can do as well basing our procedure on a sufficient statistic as we could if we also had available the raw data.*** We now prove the fundamental theorem that justifies the restriction to sufficient statistics.

## *4.8 Theorem: Simulation of Original Sample*

Let $Y = g(X_1, \ldots, X_n)$ be sufficient for $\theta$ in the sense of Definition 4.3, and let $U$ be a uniform random variable generated independently of $Y$. Then there is a function $G$ of two variables, taking on values in $n$ dimensions, such that $(Z_1, \ldots, Z_n) = G(Y, U)$ and $(X_1, \ldots, X_n)$ have the same distribution for each $P$ in $\mathscr{P}$.

*Proof:* By the sufficiency of $g(X_1, \ldots, X_n)$, we know from Definition 4.3 that for each given sequence $(x_1, \ldots, x_n)$ and each given $y$, the probability

$$P\{X_1 = x_1, \ldots, X_n = x_n \mid g(X_1, \ldots, X_n) = y\} \equiv p(x_1, \ldots, x_n \mid y)$$

is the same for all $P$ in $\mathscr{P}$.

Note that for each fixed possible value $y$ of $Y = g(X_1, \ldots, X_n)$, the values $p(x_1, \ldots, x_n \mid y)$ determine a discrete probability function. Furthermore, note that $p(x_1, \ldots, x_n \mid y)$ is computable, since we may use any $P$ in $\mathscr{P}$ to compute it.

Now, for each possible value $y$ of $Y = g(X_1, \ldots, X_n)$, we are able to construct from $U$ a sample $G(y, U)$ of size $n$, governed by the discrete probabilities $p(x_1, \ldots, x_n \mid y)$ (see Problem 4.9). Define the random vector $(Z_1, \ldots, Z_n)$ to be $G(Y, U)$. That is, when $Y$ assumes the value $y$, then $(Z_1, \ldots, Z_n)$ is governed by $p(x_1, \ldots, x_n \mid y)$. More formally,

$$P\{Z_1 = x_1, \ldots, Z_n = x_n \mid Y = y\} = p(x_1, \ldots, x_n \mid y).$$

But

$$\begin{aligned}
P\{Z_1 = x_1, \ldots, Z_n = x_n\} &= \sum_{\substack{\text{all values } y \text{ of} \\ Y = g(X_1, \ldots, X_n)}} \underbrace{P\{Z_1 = x_1, \ldots, Z_n = x_n \mid Y = y\}}_{\text{this is the same for all } P \text{ in } \mathscr{P}} \underbrace{P\{Y = y\}}_{\text{this varies as } P \text{ ranges over } \mathscr{P}} \\
&= \sum_{\substack{\text{all values } y \text{ of} \\ Y = g(X_1, \ldots, X_n)}} p(x_1, \ldots, x_n \mid y) P\{Y = y\} \\
&= \sum_{\substack{\text{all values } y \text{ of} \\ Y = g(X_1, \ldots, X_n)}} P\{X_1 = x_1, \ldots, X_n = x_n \mid g(X_1, \ldots, X_n) = y\} P\{Y = y\} \\
&= \sum_{\substack{\text{all values} \\ y \text{ of } Y}} P\{X_1 = x_1, \ldots, X_n = x_n \mid Y = y\} P\{Y = y\} \\
&= P\{X_1 = x_1, \ldots, X_n = x_n\}.
\end{aligned}$$

That is, for all $P$ in $\mathscr{P}$,

$$P\{Z_1 = x_1, \ldots, Z_n = x_n\} = P\{X_1 = x_1, \ldots, X_n = x_n\},$$

as asserted.

Note that $P\{Z_1 = x_1, \ldots, Z_n = x_n\}$ varies as $P$ ranges over $\mathscr{P}$ only on account of the variation of $P\{Y = y\}$ with $P$.

---

### 4.9 PROBLEM

Let $P$ be a discrete probability distribution in $n$ dimensions. That is, there is a countable set $V \subseteq \mathscr{R}_n$ whose elements are denoted by $(w_1, \ldots, w_n)$ and a function whose value at $(w_1, \ldots, w_n)$ is $p(w_1, \ldots, w_n) \geq 0$ such that

$$\sum_{(w_1, \ldots, w_n) \in V} p(w_1, \ldots, w_n) = 1.$$

Show how to generate a random vector $W$ with this distribution from a random variable with a uniform distribution on [0, 1].
Hint: Arrange the probabilities $p(w_1, \ldots, w_n)$ in non-increasing order of magnitude $p(w_{11}, \ldots, w_{n1}) \geq p(w_{12}, \ldots, w_{n2}) \geq \ldots$ and look at Figure 10.2.

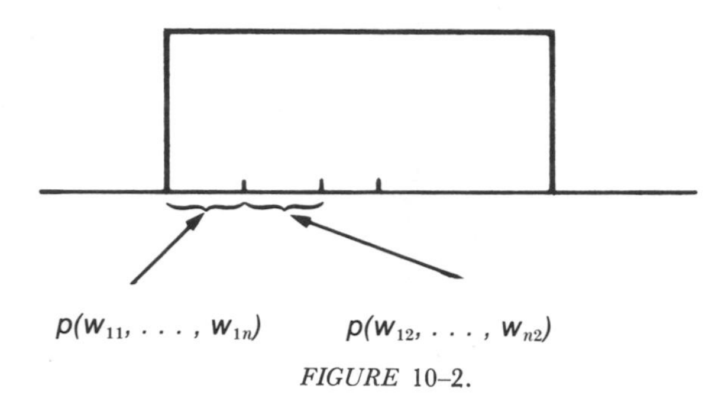


*FIGURE* 10–2.

We notice that nothing in our discussion so far has presented us with a constructive method for determining useful sufficient statistics. However, let us notice that in the case of the independent Bernoulli random variables of Example 4.4 with sufficient statistic $g(X_1, \ldots, X_n) = X_1 + X_2 + \cdots + X_n$, the value $g(x_1, \ldots, x_n)$ appears recognizably in the discrete probability function of the $X_1, \ldots, X_n$, namely,

$$\begin{aligned} P\{X_1 = x_1, \ldots, X_n = x_n) &= p^{x_1+x_2+\cdots+x_n}(1-p)^{n-(x_1+x_2+\cdots+x_n)} \\ &= p^{g(x_1,\ldots,x_n)}(1-p)^{n-g(x_1,\ldots,x_n)}. \end{aligned}$$

We shall see that it is true, in general, that when a sufficient statistic exists, it is often easy to recognize it by direct examination of the probability function.

### *4.10 Theorem: Neyman Factorization Criterion for Sufficient Statistics*

Suppose that $X_1, \ldots, X_n$ are discrete random variables governed by one of the members of a parametric family $\mathscr{P}$ of distributions whose elements are determined by the parameter $\theta$. Then $g(X_1, \ldots, X_n)$ is sufficient for $\theta$ if and only if we can write $p_\theta(x_1, \ldots, x_n) \equiv P_\theta\{X_1 = x_1, \ldots, X_n = x_n\}$ in the form

**4.11**
$$h(x_1, \ldots, x_n) q_\theta(g(x_1, \ldots, x_n)),$$

that is, the probability function factors into a product such that one factor is independent of $\theta$ and the other depends on $(x_1, \ldots, x_n)$ only through $g(x_1, \ldots, x_n)$. Here $x_i$ range over all the possible values of $X_i$, $i = 1, \ldots, n$, and the function $h$ does not depend on $\theta$. The statistic $g(X_1, \ldots, X_n)$ may be a vector and $\theta$ may also be a vector.

*Proof:* To simplify the notation in the proof, let

$$\mathbf{X} \text{ stand for } X_1, \ldots, X_n$$

$$\mathbf{x} \text{ stand for } x_1, \ldots, x_n$$

and

$$\{\mathbf{X} = \mathbf{x}\} \text{ stand for } \{X_1 = x_1, \ldots, X_n = x_n\}.$$

For the first half of our proof, let us examine the case in which it is known that $g(\mathbf{X})$ is sufficient for $\theta$. Then from Definition 4.3,

$$P\{\mathbf{X} = \mathbf{x} \mid g(\mathbf{X}) = g(\mathbf{x})\} = h(\mathbf{x})$$

independent of $\theta$. That is,

**4.12**
$$\frac{P_\theta\{\mathbf{X} = \mathbf{x}\}}{P_\theta\{g(\mathbf{X}) = g(\mathbf{x})\}} = h(\mathbf{x}).$$

But the denominator is a function of $\theta$ and $\mathbf{x}$, which depends on $\mathbf{x}$ only through $g(\mathbf{x})$, since we need only know the value $g(\mathbf{x})$ (and not $\mathbf{x}$ itself) to compute

$$P_\theta\{g(\mathbf{X}) = g(\mathbf{x})\} = \sum_{\substack{\text{all } \mathbf{x}^* \text{ such} \\ \text{that } g(\mathbf{x}^*) = g(\mathbf{x})}} P_\theta(\mathbf{X} = \mathbf{x}^*).$$

Hence, we may write

$$P_\theta\{g(\mathbf{X}) = g(\mathbf{x})\} = q_\theta(g(\mathbf{x})).$$

Substituting this into 4.12 yields the desired result, 4.11, when $g(\mathbf{X})$ is sufficient for $\theta$.

For the remaining half of the theorem, suppose that we can write

$$P_\theta\{\mathbf{X} = \mathbf{x}\} = h(\mathbf{x})q_\theta(g(\mathbf{x})).$$

Let $A = \{\mathbf{x}' \in \mathcal{S}: g(\mathbf{x}') = g(\mathbf{x})\}$. Then

$$\begin{aligned} P_\theta\{g(\mathbf{X}) = g(\mathbf{x})\} &= \sum_{\mathbf{x}^* \in A} P_\theta\{\mathbf{X} = \mathbf{x}^*\} \\ &= \sum_{\mathbf{x}^* \in A} h(\mathbf{x}^*) q_\theta(g(\mathbf{x}^*)) \\ &= q_\theta(g(\mathbf{x})) \sum_{\mathbf{x}^* \in A} h(\mathbf{x}^*). \end{aligned}$$

Hence,

$$\begin{aligned} P_\theta\{\mathbf{X} = \mathbf{x} \mid g(\mathbf{X}) = g(\mathbf{x})\} &= \frac{P_\theta\{\mathbf{X} = \mathbf{x}\}}{P_\theta\{g(\mathbf{X}) = g(\mathbf{x})\}} \\ &= \frac{h(\mathbf{x}) q_\theta(g(\mathbf{x}))}{q_\theta(g(\mathbf{x})) \sum_{\mathbf{x}^* \in A} h(\mathbf{x}^*)} \\ &= \frac{h(\mathbf{x})}{\sum_{\mathbf{x}^* \in A} h(\mathbf{x}^*)}, \end{aligned}$$

which is independent of $\theta$. Thus $P_\theta\{\mathbf{X} = \mathbf{x} \mid g(\mathbf{X}) = g(\mathbf{x})\}$ is independent of $\theta$ whenever $P_\theta\{\mathbf{X} = \mathbf{x}\} = h(\mathbf{x})q_\theta(g(\mathbf{x}))$. Hence, from the definition, $g(\mathbf{X})$ is sufficient whenever $P_\theta\{\mathbf{X} = \mathbf{x}\} = h(\mathbf{x})q_\theta(g(\mathbf{x}))$, concluding the proof.

### 4.13 *Example: Poisson Random Variables*

If $X_1, \ldots, X_n$ are independent Poisson random variables with common expectation $\theta$, then

$$P_\theta\{X_1 = x_1, \ldots, X_n = x_n\} = \frac{e^{-\theta}\theta^{x_1}}{x_1!} \cdots \frac{e^{-\theta}\theta^{x_n}}{x_n!} = \frac{1}{x!\, x_2! \cdots x_n!}\,[e^{-n\theta}\theta^{x_1+x_2+\cdots+x_n}].$$

Notice that where $\theta$ occurs, the $x_i$'s occur only through $x_1 + x_2 + \cdots + x_n$. Hence we may choose

$$h(x_1, \ldots, x_n) = \frac{1}{x_1! \cdots x_n!},$$

$$g(x_1, \ldots, x_n) = x_1 + x_2 + \cdots + x_n,$$

and

$$q_\theta(g(x_1, \ldots, x_n)) = e^{-n\theta}\theta^{g(x_1,\ldots,x_n)},$$

and write

$$P_\theta\{X_1 = x_1, \ldots, X_n = x_n\} = h(x_1, \ldots, x_n)q_\theta(g(x_1, \ldots, x_n)).$$

Then, by Equation 4.11 of the Neyman factorization criterion for sufficient statistics, we see that $g(X_1, \ldots, X_n) = X_1 + X_2 + \cdots + X_n$ is sufficient for $\theta$ in the Poisson case. That is, we may base any inference about $\lambda$ on the observed total radiation with no loss of information. △

---

### 4.14 *EXERCISES Using Theorem 4.10, the Neyman Factorization Criterion for Sufficient Statistics*

1.° Find a sufficient statistic for $p$ when $X_1, \ldots, X_n$ are independent, each with the binomial distribution

$$P\{X_i = k\} = \binom{j}{k} p^j(1 - p)^{j-k}.$$

2. Suppose that $X_1, \ldots, X_n$ are independent radiation measurements with $X_i$ having a Poisson distribution with mean value $m_i\lambda$ where $m_i$ is known. On which scalar statistic is it reasonable to base any inferences about $\lambda$?

3. An experiment with probability $p$ of success is repeated until a success is obtained. Let $N$ represent the number of trials required. If this experiment, whose outcome is $N$, is repeated $n$ times, where $N_i$ denotes the outcome of the $i$th repetition, what is a sufficient statistic for $p$?

4.° Suppose that a possibly loaded die is tossed independently $n$ times. Let $X_i$ be the result of the $i$th toss, and let $p_K$ be the probability that the die comes up $K$ on any given toss. Show that the statistic

$$g(X_1, \ldots, X_n) = (N_1, N_2, N_3, N_4, N_5 N_6)$$

where $N_K$ is the number of times the result $K$ turns up is sufficient for the vector parameter $\theta = (p_1, p_2, p_3, p_4, p_5, p_6)$.

Hint: Let $y_{ij}(x_i) = \begin{cases} 1 & \text{if } x_i = j \quad j = 1, \ldots, 6, \quad i = 1, 2, \ldots, n \\ 0 & \text{otherwise} \end{cases}$

and write

$$P\{X_i = x_i\} = p_1^{y_{i1}(x_i)} p_2^{y_{i2}(x_i)} \ldots p_6^{y_{i6}(x_i)}.$$

Notice that $\sum_{i=1}^{n} y_{ij}(x_i)$ is the number of elements from $(x_1, \ldots, x_n)$ equal to $j$.

We now turn our attention to sufficient statistics in which our observations are governed by a density. Since densities may be thought of as arising from limits of discrete probabilities, our definition of sufficiency should be something like: $g(X_1, \ldots, X_n)$ is sufficient for $\theta$ if

$$\lim_{h \to 0} P_\theta\{(X_1, \ldots, X_n) \in A \mid a \leq g(X_1, \ldots, X_n) \leq a + h\}$$

is independent of $\theta$ for each event $A$ and each $a$. Since we do not have available the precise mathematical machinery to institute a careful investigation of this topic, we shall be content to work by analogy with the discrete case. We shall state the corresponding version of Theorem 4.10 for the density case.

### 4.15 *Theorem: Neyman Factorization Criterion for Densities*

If $X_1, \ldots, X_n$ are random variables governed by one of a parametric family of densities with values $f(x_1, \ldots, x_n; \theta)$, then $g(X_1, \ldots, X_n)$ is sufficient for $\theta$ if we can write $f(x_1, \ldots, x_n; \theta)$ in the form

$$h(x_1, \ldots, x_n) q_\theta(g(x_1, \ldots, x_n)) \text{ as in Expression 4.11.}$$

Here, just as in Expression 4.11, both $\theta$ and $g(X_1, \ldots, X_n)$ may be vectors.

### 4.16 *Example: Waiting Times*

If $T_1, T_2, \ldots, T_n$ are independent waiting times, each with density

$$f_{T_1}(t;\lambda) = \begin{cases} (1/\lambda)e^{-t/\lambda} & \text{for} \quad t \geq 0 \\ 0 & \text{otherwise,} \end{cases}$$

where $\lambda > 0$, then

$$f_{T_1,\ldots,T_n}(t_1, \ldots, t_n;\lambda) = \frac{1}{\lambda^n} \exp\left[-\frac{1}{\lambda}\sum_{i=1}^{n} t_i\right].$$

From the Neyman factorization criterion (Theorem 4.15), we see that the statistic $g(T_1, \ldots, T_n) = \sum_{i=1}^{n} T_i$ is sufficient for $\lambda$ by choosing

$$h(t_1, \ldots, t_n) = 1, \quad g(t_1, \ldots, t_n) = \sum_{i=1}^{n} t_i,$$

and

$$q_\lambda(g(t_1, \ldots, t_n)) = \frac{1}{\lambda^n} e^{-(1/\lambda)g(t_1,\ldots,t_n)}.$$

Thus, we may base any inference concerning the average waiting time on the total observed waiting time. △

### 4.17 *Example: Normal Random Variables, Unknown* $\mu$, $\sigma^2$

Let $X_1, X_2, \ldots, X_n$ be independent and normal with $E(X_i) = \mu$, Var $X_i = \sigma^2$ both unknown. Recall that this is often used as a model for repeated measurements by an instrument of unknown quality, $\sigma^2$, on an unknown constant $\mu$. We take $\theta = (\mu, \sigma^2)$, a vector. Then

$$\begin{aligned} f_{X_1,\ldots,X_n}(x_1, \ldots, x_n;\theta) &= \frac{1}{(2\pi\sigma^2)^{n/2}} \exp\left[-\frac{1}{2\sigma^2}\sum_{i=1}^{n}(x_i - \mu)^2\right] \\ &= \frac{1}{(2\pi\sigma^2)^{n/2}} \exp\left[-\frac{1}{2\sigma^2}\left[\sum_{i=1}^{n} x_i^2 - 2\mu\sum_{i=1}^{n} x_i + n\mu^2\right]\right]. \end{aligned}$$

Let $g_1(x_1, \ldots, x_n) = \sum_{i=1}^{n} x_i^2$ and $g_2(x_1, \ldots, x_n) = \sum_{i=1}^{n} x_i$, and let $g$ be the vector $(g_1, g_2)$. Then we claim that $g(X_1, \ldots, X_n) = (g_1(X_1, \ldots, X_n), g_2(X_1, \ldots, X_n))$ is sufficient for the vector parameter $\theta$ by Example 4.13. Here $h(x_1, \ldots, x_n) = 1$ and

$$\begin{aligned} &q_\theta(g_1(x_1, \ldots, x_n), g_2(x_1, \ldots, x_n)) \\ &\quad = \frac{1}{(2\pi\sigma^2)^{n/2}} \exp\left[-\frac{n\mu^2}{2\sigma^2}\right] \exp\left\{-\frac{1}{2\sigma^2}[g_1(x_1, \ldots, x_n) - 2\mu g_2(x_1, \ldots, x_n)]\right\}. \end{aligned}$$

The reader should be aware that in spite of the fact that

$$\frac{1}{n} g_2(X_1, \ldots, X_n) = \frac{1}{n}\sum_{i=1}^{n} X_i$$

is an estimator of $\mu$, ***it is not correct*** to conclude that in this case $\sum_{i=1}^{n} X_i$ is sufficient for $\mu$; we know only that

$$\left(\sum_{i=1}^{n} X_i^2, \sum_{i=1}^{n} X_i\right)$$

is sufficient for $(\mu, \sigma^2)$. This is not unreasonable in view of the fact that when both $\mu$ and $\sigma^2$ are unknown, the sample variance

$$\frac{1}{n-1}\sum_{i=1}^{n}\left(X_i - \frac{1}{n}\sum_{j=1}^{n} X_j\right)^2 = \frac{1}{n-1}\sum_{i=1}^{n} X_i^2 - \frac{1}{n(n-1)}\left(\sum_{i=1}^{n} X_i\right)^2$$

gives us information about $\mu$ (how close we may expect

$$\frac{1}{n}\sum_{i=1}^{n} X_i$$

to be to $\mu$; see Exercise 2.6(4)). △

We close this section by remarking that the concept of sufficiency is linked to the particular class of distributions being considered. If we have made a poor choice for the class of possible governing distributions (that is, if no distribution in the class furnishes a good description of the situation), then the possible savings afforded by a sufficient statistic may be illusory. In such a case, use of a sufficient statistic may actually destroy information. An example of this kind of destruction would arise if the "true" density were either normal with variance 1 or Cauchy,

$$f_X(x) = \frac{1}{\pi(1 + [x - \mu]^2)}.$$

If we assume normality, then we would condense to the sufficient statistic

$$\frac{1}{n}\sum_{i=1}^{n} X_i$$

(see Exercise 4.18(1)). It can be shown that

$$\frac{1}{n}\sum_{i=1}^{n} X_i$$

is a very poor estimate of $\mu$ in the Cauchy case, since its density is still

$$\frac{1}{\pi(1 + [x - \mu]^2)}$$

(see Problem 9.9(3), Chapter 6). That is,

$$\frac{1}{n}\sum_{i=1}^{n} X_i$$

gives the same information about $\mu$ as does $X_1$ alone.

## *4.18 EXERCISES USING 4.15*

1. If $X_1, \ldots, X_n$ are measurements assumed to be independent normal with known variance $\sigma^2 = 1$, find a scalar statistic that is just as useful as the original sample for statistical purposes, when $E(X_i) = \mu$ unknown.

2.° Suppose that $X_1, \ldots, X_n$ are independent normal random variables with variance 1 and $E(X_i) = i\mu$, where $\mu$ is unknown. Find a sufficient statistic for $\mu$.

3. As a model for a system whose quality is improving, we assume that $T_i$ is a random variable representing the time between the $i$th and the $i + 1$st failure, where $T_1, T_2, \ldots, T_n$ are independent with

$$f_{T_i}(t;\theta) = \begin{cases} \dfrac{1}{i\theta} e^{-t/(i\theta)} & \text{for } t \geq 0 \\ 0 & \text{otherwise.} \end{cases}$$

Here $\theta > 0$. What scalar statistic would you use to learn something about $\theta$?

4. For $X_1, \ldots, X_n$ independent normal random variables with unknown variance $\sigma^2$ and $E(X_i) = i\mu$, $\mu$ unknown, find sufficient statistics for $(\mu, \sigma^2)$.

5.° Suppose that $X_1, \ldots, X_n$ are independent random variables, each with density

$$f_{X_i}(x;\theta) = c(x)k(\theta) \exp \sum_{j=1}^{m} Q_j(\theta)S_j(x)$$

(where $c$, $k$, $Q_j$, and $S_j$, $j = 1, \ldots, m$, are known functions). Show that

$$\left(\sum_{i=1}^{n} S_1(X_i), \ldots, \sum_{i=1}^{n} S_m(X_i)\right)$$

are sufficient for $\theta$.

6. Suppose that $U_1, \ldots, U_n$ are independent uniformly distributed random variables on $[0, \theta]$, $\theta > 0$. Which of the following statistics would you prefer to observe in order to make a decision about $E(U_i)$?

(a) $\dfrac{1}{n}\sum_{i=1}^{n} U_i$.

(b) $\max\limits_{i \leq 1 \leq n} U_i$.

Hint: $f_{U_1,\ldots,U_n}(u_1, \ldots, u_n;\theta) = \begin{cases} 1/\theta^n & \text{if } 0 \leq u_i \leq \theta \text{ for all } i = 1, \ldots, n \\ 0 & \text{otherwise.} \end{cases}$

But $0 \leq u_i \leq \theta$ for all $i = 1, \ldots, n$ if and only if $\max u_i \leq \theta$, provided $u_i$ are all non-negative.

7.° Suppose that $U_1, \ldots, U_n$ are independent and uniformly distributed on $[a, b]$ where $a$ and $b$ are unknown. Find a sufficient statistic for $\theta = (a, b)$.

8.° A Weibull density, given by

$$f(x; c, \alpha) = \begin{cases} \alpha c x^{\alpha-1} e^{-cx^{\alpha}} & \text{for } x > 0, \\ 0 & \text{otherwise,} \end{cases}$$

is widely used to describe the variation in measurements of fatigue in metals. Here $\alpha > 0$ and $c > 0$ are the specifying parameters of this family.

If $\alpha$ is known, find a sufficient statistic for $c$ based on independent measurements $X_1, \ldots, X_n$, each with the same Weibull density.

9. Let $T_1, \ldots, T_n$ be independent, with a Gamma density

$$f_{T_i}(t; \beta) = \begin{cases} \dfrac{1}{(k-1)!\,\beta^k} t^{k-1} e^{-t/\beta} & \text{if } t > 0, \\ 0 & \text{otherwise.} \end{cases}$$

(The Gamma density is used to describe the length of time between the $j$th and the $j + k$th failure in memoryless models (Example 2.6, Chapter 4), that is, one in which an exponential density is assumed to govern the lifetime of objects such as electronics parts.) Find a sufficient statistic for $\beta$.

## SECTION 5
## Constructive Methods of Obtaining Point Estimates

Up to now we have had to use guesswork to find useful estimates. Fortunately, there are systematic approaches for finding sensible estimators. In this section we shall present the two most widely used "practical" methods of obtaining point estimates for parameters.

### The Method of Moments

In a rather large number of commonly used statistical models, the parameter that we desire to estimate is a known function of a given finite number of moments. This should not be surprising, since we have seen that frequently the moments completely determine the distribution. For example, the first moment, $E(X) = np$, completely determines the binomial distribution when $n$ is known. The first and second moments, $E(X)$ and $E(X^2)$, completely determine any univariate normal distribution. For simplicity, we will consider $\theta$ to be a scalar. *Let $m_j$ denote the jth moment about the origin of the distribution P governing each of the independent random variables $X_j$.* Then we are considering a model in which

**5.1** $$\theta = G(m_1, \ldots, m_k),$$

where $\theta$ is the parameter of interest and $G$ is some known function.

**5.2 Definition: Method of Moments.** *The method of moments consists of estimating $\theta$ by the statistic*

$$\hat{\theta}_n = G\left(\frac{1}{n}\sum_{i=1}^{n} X_i, \frac{1}{n}\sum_{i=1}^{n} X_i^2, \ldots, \frac{1}{n}\sum_{i=1}^{n} X_i^k\right).$$

The method extends naturally to estimation of joint moments. For example, when $(X_1, Y_1), \ldots, (X_n, Y_n)$ are independent random vectors, we use

$$\frac{1}{n}\sum_{i=1}^{n} X_i Y_i$$

to estimate $E(XY)$. Suppose we know that for all possible meaningful values $x_1, \ldots, x_k$,

**5.3**
$$\lim_{(x_1', \ldots, x_k') \to (x_1, \ldots, x_k)} G(x_1', \ldots, x_k') = G(x_1, \ldots, x_k).$$

That is, $G$ is a continuous function of $k$ variables. Then it follows from the weak law of large numbers that the sequence given by

$$\frac{1}{n}\sum_{i=1}^{n} X_i^j$$

is consistent for $m_j$; we see that $\hat{\theta}_n$ yields a consistent sequence for $\theta$. Intuitively, this follows from the fact that, with high probability, when $n$ is large enough,

$$\frac{1}{n}\sum_{i=1}^{n} X_i^j$$

is close to $m_j$ for each $j$ separately.

Since the intersection of a fixed finite number of high probability events is a high probability event (Theorem 3.5(e) in Chapter 3) it follows that for all $j = 1, \ldots, k$, each

$$\frac{1}{n}\sum_{i=1}^{n} X_i^j$$

is close to $m_j$ when $n$ is sufficiently large. Therefore, we see from 5.3 that for each given $\theta$ and large enough $n$ (depending on $\theta$), the statistic

$$G\left(\frac{1}{n}\sum_{i=1}^{n} X_i, \ldots, \frac{1}{n}\sum_{i=1}^{n} X_i^k\right)$$

should be close to

$$\theta = G(m_1, \ldots, m_k).$$

***5.4 Example: Standard Deviation***

If $X_1, \ldots, X_n$ are independent random variables with the same distribution, with standard deviation $\sigma$ and expectation $m_1$, then

$$\sigma = \sqrt{m_2 - m_1^2}.$$

Consequently, we see that

$$\hat{\sigma}_n = \sqrt{\frac{1}{n}\sum_{i=1}^{n} X_i^2 - \left(\frac{1}{n}\sum_{i=1}^{n} X_i\right)^2}$$

obtained from the method of moments yields a consistent sequence for $\sigma$. △

### 5.5 *Example: Coefficient of Variation*

If $X$ is a random variable with $E(X) = m_1$, $\text{Var}\, X = \sigma^2 = m_2 - m_1^2$, the coefficient of variation $v$ of $X$ is defined as the ratio

$$v = \frac{\sigma}{m_1} = \frac{\sqrt{m_2 - m_1^2}}{m_1}.$$

Suppose that $X$ is thought of as a measure of $m_1$. If $m_1 > 0$ and $m_1$ is much greater than $\sigma$, then $v$ tells (in some sense) the relative error of $X$ as a measure of $m_1$. If $X_1, \ldots, X_n$ are independent identically distributed measurements with coefficient of variation $v$, then the method of moments yields

$$\hat{v}_n = \frac{1}{n}\sqrt{\sum_{i=1}^{n} X_i^2 - \left(\frac{1}{n}\sum_{i=1}^{n} X_i\right)^2} \Bigg/ \left(\frac{1}{n}\sum_{i=1}^{n} X_i\right)$$

as a consistent sequence of estimators for $v$. △

---

### 5.6 EXERCISES

1. Use the method of moments to estimate the parameter

$$\theta = \frac{2}{a + b}$$

where $X_1, \ldots, X_n$ are independent with a uniform density on $[a, b]$. Here $a$ and $b$ are unknown.

2.° Use the method of moments to estimate the parameter $\theta = b - a$ when $X_1, \ldots, X_n$ are independent and uniformly distributed on $[a, b]$. Here $a$ and $b$ are unknown. Note that $b - a$ is the range of the density and measures its extent.

3.° Let $X_1, \ldots, X_n$ be independent random variables with a Gamma density, that is,

$$f_{X_i}(x; \beta) = \begin{cases} \dfrac{1}{(k-1)!\,\beta^k} x^{k-1} e^{-x/\beta} & \text{for } x > 0 \\ 0 & \text{otherwise} \end{cases}$$

where $k$ is known.

Using the method of moments, find an estimate for $\beta$.
Hint: We may think of $X_i$ as the length of time between the $j$th and $j + k$th failure, where the time between successive failures is exponential with mean $\beta$. See problem 4.9(3) of Chapter 5. Identify $k$ here with $n$ there, and $X_i$ here with $Y_n$ there. Hence $E(X_i) = k\beta$.

4. Let $X_1, \ldots, X_n$ be independent random variables with Gamma density

$$f_{X_i}(x; \alpha, \beta) = \begin{cases} \dfrac{1}{\Gamma(\alpha)\beta^\alpha} x^{\alpha-1} e^{-x/\beta} & \text{for } x > 0 \\ 0 & \text{otherwise.} \end{cases}$$

See 2 of Chapter 8 for the definition of $\Gamma(\alpha)$ if its definition is not obvious from the fact that $f_{X_i}$ is a probability density. Using the method of moments, find estimates of $\alpha$ and $\beta$.
Hint: Show that $E(X_i) = \alpha\beta$ and $E(X_i^2) = (\alpha + 1)\alpha\beta^2$ by using the facts that (a) $\Gamma(\alpha + 1) = \alpha\Gamma(\alpha)$ (which can be shown using integration by parts on the integral defining $\Gamma(\alpha)$) and (b) $f_{X_i}$ is a probability density.

5.° Let $(X_1, Y_1), (X_2, Y_2), \ldots, (X_n, Y_n)$ be independent random vectors with multivariate normal density

$$f_{X_i,Y_i}(x, y) = \frac{1}{2\pi\sqrt{1-\rho^2}} e^{-\frac{1}{2}[(x-\mu_x)^2 - 2\rho(x-\mu_x)(y-\mu_y) + (y-\mu_y^2)]}.$$

From the method of moments find an estimate of $\rho$:
(a) when $\mu_x$ and $\mu_y$ are known.
(b) when $\mu_x$ and $\mu_y$ are unknown.
Hint: Replace $\mu_x$ by

$$\frac{1}{n}\sum_{i=1}^{n} X_i$$

and $\mu_y$ by

$$\frac{1}{n}\sum_{i=1}^{n} Y_i.$$

(c) Do these estimates yield a consistent sequence for $\rho$?
Hint: Recall from 5.1 and 5.3 of Chapter 6 that

$$\rho = E\left[\left(\frac{X-\mu_x}{\sigma_x}\right)\left(\frac{Y-\mu_y}{\sigma_y}\right)\right].$$

Here $\sigma_x = \sigma_y = 1$.

## The Method of Maximum Likelihood

This is the most popular method of point estimation. In essence, the method of maximum likelihood selects that distribution $P \in \mathscr{P}$ that "best explains" the observed data as its estimate of the true governing distribution. We assume that our random variables $X_1, \ldots, X_n$ have joint distribution $P$ from some parametric family $\mathscr{P}$ with parameter $\theta$.

**5.7 Definition: Maximum Likelihood Estimate.** Let $x_i = X_i(s)$ be the actual observed value of the random variable $X_i$. In the discrete case when $P\{X_1 = x_1, X_2 = x_2, \ldots, X_n = x_n\} = p_\theta(x_1, \ldots, x_n)$, *a maximum likelihood estimate* $\hat{\theta}$ is a statistic whose value $\tilde{\theta}$ is any $\theta$ maximizing $p_\theta(x_1, \ldots, x_n)$.†

In the case in which $X_1, \ldots, X_n$ have a joint density $f$, the value of a maximum likelihood estimate of $\theta$ is a value of $\theta$ maximizing $f(x_1, \ldots, x_n; \theta)$. Note that although we have not been explicit about it, the parameter $\theta$ lies in some natural set—such as the collection of all real numbers, or a given interval—that will usually be understood from the context. In most of the cases usually encountered, the maximum likelihood estimate is unique.

As we shall see, the computational problem of determining maximum likelihood estimates is greatly facilitated by use of logarithms and calculus techniques. We shall see here how many of the estimators, which were constructed more or less ad hoc earlier, arise from a systematic approach to estimation.

### 5.8 *Example: Normal Mean*

Suppose that an aptitude test consisting of $n$ parts is administered to an applicant for some job. Let $\theta$ be some measure of the individual's aptitude that we are trying to determine with the aid of the test. As a model, we assume that

$$X_1, \ldots, X_n \text{ are independent normal with } E(X_i) = \theta, \text{ Var } X_i = 10.$$

Then

$$f_{X_1,\ldots,X_n}(x_1, \ldots, x_n; \theta) = \frac{1}{(20\pi)^{n/2}} \exp\left[-\frac{1}{20} \sum_{i=1}^{n} (x_i - \theta)^2\right].$$

To determine $\tilde{\theta}_n$, it is convenient to maximize $\log f = L$ rather than to try to maximize $f$ directly. Now,

$$L(\theta) \equiv \log f_{X_1,\ldots,X_n}(x_1, \ldots, x_n; \theta) = \log \frac{1}{(20\pi)^{n/2}} - \frac{1}{20} \sum_{i=1}^{n} (x_i - \theta)^2.$$

Therefore $L'(\theta) = \frac{1}{10} \sum_{i=1}^{n} (x_i - \theta)$. Recalling elementary calculus, we see that $\tilde{\theta}_n$ must satisfy the equation

$$L'(\tilde{\theta}_n) = 0$$

or

$$\tilde{\theta}_n = \frac{1}{n} \sum_{i=1}^{n} x_i.$$

We verify that

$$\tilde{\theta}_n = \frac{1}{n} \sum_{i=1}^{n} x_i$$

† Unfortunately it is necessary to make a notational distinction among the statistic $\hat{\theta}$, its observed value $\tilde{\theta}$, and the parameter $\theta$. This clumsy notation is standard.

does in fact yield the desired maximum, since $L(\theta)$ goes to $-\infty$ when $|\theta|$ becomes large. Since $L$ is differentiable and $L'(\theta) = 0$ only for $\theta = \tilde{\theta}_n$, this value $(\theta = \tilde{\theta}_n)$ must yield a maximum. △

**5.7′ Extension of Definition of Maximum Likelihood Estimate.** If $G$ is a given function and $\mu = G(\theta)$, then we define a maximum likelihood estimator $\hat{\mu}$ of $\mu$ to be $\hat{\mu} = G(\hat{\theta})$, where $\hat{\theta}$ is a maximum likelihood estimate of $\theta$. We must take some care to show that if $\mu$ also determines the distribution, then this definition still yields a maximum likelihood estimator in the originally defined sense. We will ask the student to do this as a problem.

This result greatly simplifies matters. For example, if we have the maximum likelihood estimator of $\theta = (m, \sigma^2)$ and want a maximum likelihood estimate of $\mu = (m - \sigma^2, m + \sigma^2)$, we needn't go through the whole maximization process; we can just use $\hat{\mu} = (\hat{m} - \hat{\sigma}^2, \hat{m} + \hat{\sigma}^2)$. Note that this simplification is not possible for unbiased estimators, since if $\hat{\theta}$ is an unbiased estimate of $\theta$ it does not necessarily follow that $G(\hat{\theta})$ is an unbiased estimate of $G(\theta)$.

### 5.8′ *Example 5.8 Extended*

If large $\theta$ happens to correspond to a poor aptitude, we might want to find the maximum likelihood estimate of $\mu = 1/\theta$. From 5.7′, we see that the maximum likelihood estimate of $\mu = 1/\theta$ is

$$\hat{\mu}_n = \frac{1}{\hat{\theta}_n} = \frac{n}{\sum_{i=1}^{n} X_i}.$$

Let $f^*_{X_1,\ldots,X_n}(x_1, \ldots, x_n; \mu)$ denote the density of $X_1, \ldots, X_n$ at $x_1, \ldots, x_n$ when $\mu$ is considered the specifying parameter. Then we can obtain $\tilde{\mu}_n$ directly by looking at

$$H(\mu) = \log f^*_{X_1,\ldots,X_n}(x_1, \ldots, x_n; \mu) = \log \frac{1}{(20\pi)^{n/2}} - \frac{1}{20} \sum_{i=1}^{n} \left(x_i - \frac{1}{\mu}\right)^2$$

$$H'(\mu) = -\frac{1}{10\mu^2} \sum_{i=1}^{n} \left(x_i - \frac{1}{\mu}\right).$$

We obtain the result $\tilde{\mu}_n = 1/\tilde{\theta}_n$ as the solution $\tilde{\mu}_n$ of $H'(\mu) = 0$. △

### 5.9 *Example: Normal Mean and Variance*

Suppose that a new process for producing titanium rods is being tested. Measurements $X_1, \ldots, X_n$ are made on the yield strength of $n$ different rods. As a model, we assume that the variables $X_1, \ldots, X_n$ are independent and normal, with mean

$\mu$ and variance $\sigma^2$. We want to determine the maximum likelihood estimates of $\mu$ and $\sigma^2$. Here

$$f_{X_1,\ldots,X_n}(x_1, \ldots, x_n; \mu, \sigma^2) = \frac{1}{(2\pi\sigma^2)^{n/2}} \exp\left[-\frac{1}{2\sigma^2}\sum_{i=1}^{n}(x_i - \mu)^2\right].$$

As before, we will find it easier to choose $\mu$ and $\sigma^2$ to maximize $\log f$. Now

$$L \equiv \log f_{X_1,\ldots,X_n}(x_1, \ldots, x_n; \mu, \sigma^2) = -\frac{n}{2}\log 2\pi - \frac{n}{2}\log \sigma^2 - \frac{1}{2\sigma^2}\sum_{i=1}^{n}(x_i - \mu)^2.$$

To find the desired $\tilde{\mu}$, $\tilde{\sigma}^2$ we solve the equations

$$\frac{\partial L}{\partial \mu} = 0$$

$$\frac{\partial L}{\partial \sigma^2} = 0.$$

These become, respectively,

$$\frac{1}{\tilde{\sigma}^2}\sum_{i=1}^{n}(x_i - \tilde{\mu}) = 0$$

$$-\frac{n}{2\tilde{\sigma}^2} + \frac{1}{2\tilde{\sigma}^4}\sum_{i=1}^{n}(x_i - \tilde{\mu})^2 = 0,$$

and

$$\tilde{\mu} = \frac{1}{n}\sum_{i=1}^{n} x_i = \bar{x}_{(n)}$$

$$\hat{\sigma}^2 = \frac{1}{n}\sum_{i=1}^{n}(x_i - \tilde{\mu})^2 = \frac{1}{n}\sum_{i=1}^{n} x_i^2 - [\bar{x}_{(n)}]^2.$$

As before, it can be shown that these values minimize $L$.

Note that $\bar{X}_{(n)}$ is normal with mean $\mu$ and variance $\sigma^2/n$. Hence, using the fact that $\operatorname{Var}\bar{X}_{(n)} = E(\bar{X}_{(n)}^2) - [E(\bar{X}_{(n)})]^2$ (Lemma 3.8, Chapter 6) we find that

$$E(\bar{X}_{(n)}^2) = \frac{\sigma^2}{n} + \mu^2.$$

Thus

$$E(\hat{\sigma}^2) = \frac{1}{n}\sum_{i=1}^{n}(\sigma^2 + \mu^2) - \left(\frac{\sigma^2}{n} + \mu^2\right) = \frac{n-1}{n}\sigma^2.$$

This shows that the maximum likelihood estimate $\hat{\sigma}^2$ of $\sigma^2$ is not unbiased.

From the formulae for $\tilde{\mu}$ and $\tilde{\sigma}^2$, we find that the maximum likelihood estimates of the mean tensile strength and standard deviation of this tensile strength have the values

$$\tilde{\mu} = \bar{x}_{(n)}$$

and

$$\tilde{\sigma} = \sqrt{\frac{1}{n}\sum_{i=1}^{n}(x_i - \bar{x}_{(n)})^2}. \qquad \triangle$$

### 5.10 *Example: Service Time*

In order to schedule their service calls in a reasonable fashion, a repair company measures the length of time of $n$ independent service calls. As a model, we assume that the $i$th service call duration, $X_i$, is uniformly distributed on $[a, b]$, where $a$ and $b$ are unknown. For simplicity, we put no restriction on $a$ and $b$ in our model, even though we know $a > 0$ and $b \le 8$. This will not cause any problems, and it does make our computation easier. Here

$$f_{X_1,\ldots,X_n}(x_1, \ldots, x_n; a, b) = \begin{cases} \dfrac{1}{(b-a)^n} & \text{if } a \le \min\limits_{1\le i\le n} x_i \le \max\limits_{1\le i\le n} x_i \le b \\ 0 & \text{otherwise.} \end{cases}$$

We see directly that the values of $a$ and $b$ maximizing this density are

$$\tilde{a} = \min_{1\le i\le n} x_i \quad \text{and} \quad \tilde{b} = \max_{1\le i\le n} x_i.$$

This result should not be surprising to you. △

### 5.11 *Example: Binomial Random Variable*

Suppose $X$ is a binomial random variable with unknown parameter $\theta = (n, p)$. The situation of observing such an $X$ arises when you are told the number of successes, but not the number of trials, as is the case in some television commercials, or when you observe the number of dead white cells at an infection, or when you observe the amount of pollutant that escapes from a smoke "scrubber." Let the observed value of $X$ be denoted by $x$. Then the value of the maximum likelihood estimate of $\theta$ is easily seen to be $\tilde{n} = x, \tilde{p} = 1$, since if we let $p_{n,p}(x) = P\{X = x\}$ when the parameter $\theta = (n, p)$, then $p_{n,p}(x) = 1$. Certainly these are the values of $n$ and $p$ that maximize the likelihood of what you observed. However, you would not put much faith in such estimates, since $\tilde{p} = 1$ always. △

---

### 5.12 *EXERCISES*

1.° Let $X_1, \ldots, X_n$ represent independent measurements, assumed to be normal, with $E(X_i) = \mu$ and Var $X_i = \sigma^2$, both unknown.
Find the maximum likelihood estimate of the coefficient of variation $\sigma/\mu$. (This coefficient is a measure of the relative error in determining $\mu$, and is meaningful mainly if $\mu > 0$ and $\sigma$ is small compared to $\mu$.)

2. Let $X_1, \ldots, X_n$ be random variables, $X_i$ representing the number of fatal penicillin reactions in some year $i$. If these random variables are independent Poisson with $E(X_i) = \lambda$, find the maximum likelihood estimate of $\lambda$.

3.° Suppose that $X_1, \ldots, X_n$ represent the number of traffic fatalities in $n$ different cities. Suppose further that $E(X_i) = k_i\lambda$ where $k_i$ is the (known) population of the $i$th city, and that the $X_i$ are independent with a Poisson distribution. Find the maximum likelihood estimate of $\lambda$. ($\lambda$ is the mean number of fatalities per person.)

4. The interarrival times between buses, $T_1, \ldots, T_n$, are independent with density

$$f_{T_i}(t) = \begin{cases} \frac{1}{\lambda} e^{-t/\lambda} & \text{for } t > 0 \\ 0 & \text{otherwise.} \end{cases}$$

Find the maximum likelihood estimate of $\lambda$, the mean interarrival time.

5. Let $X_1, \ldots, X_n$ be uniformly distributed on $[0, \theta]$ and independent. Find the maximum likelihood estimate of $\theta$.

6.° Find the maximum likelihood estimate of $\theta$ when the random variables $X_1, \ldots, X_n$ are independent and uniformly distributed on $[\theta - 1/2, \theta + 1/2]$.

7. Suppose that the probability of a defective reel of magnetic tape is some unknown value $p$. If $n$ reels of tape are examined, what is the maximum likelihood estimate of:

(a) $p$?

(b) $p^3$ (the probability of three successive defectives)?

(c) $\frac{1}{p}$ (the expected number of reels until finding a defective)?

What assumptions did you make to come to your conclusions? Discuss their validity.

8. Let $X_1, \ldots, X_n$ be independent with $P\{X_i = K\} = (1 - p)^{K-1}p$ for $K = 1, 2, \ldots$. Find the maximum likelihood estimate of $p$. Is your result reasonable?

---

## 5.13 PROBLEMS

1. Let $g$ be a given function and suppose that if $\mu = G(\theta)$, the value of $\mu$ uniquely determines the value of $\theta$ (scalar or vector). Write the density of $X_1, \ldots, X_n$ at $x_1, \ldots, x_n$ as $f_{X_1,\ldots,X_n}(x_1, \ldots, x_n; \theta)$ when written in terms of $\theta$ and in the form $f^*_{X_1,\ldots,X_n}(x_1, \ldots, x_n; \mu)$ when written in terms of $\mu$. Let $\tilde{\theta}_n$ be the value of $\theta$ maximizing $f_{X_1,\ldots,X_n}(x_1, \ldots, x_n; \theta)$ and let $\tilde{\mu}_n$ be a value of $\mu$ maximizing the expression $f^*_{X_1,\ldots,X_n}(x_1, \ldots, x_n; \mu)$. Show that $\tilde{\mu}_n = G(\tilde{\theta}_n)$.
Hint: $f_{X_1,\ldots,X_n}(x_1, \ldots, x_n; \theta) = f^*_{X_1,\ldots,X_n}(x_1, \ldots, x_n; G(\theta))$.

2. Prior to some suspected pollution, a lake is known to have contained a stable population of about 100,000 fish. We are to estimate the present fish population to see if any change has occurred. In order to do this, 1000 fish are tagged and then returned to the lake. A week later, 600 fish are caught. Let $N$ be the number of fish in the lake, and let $X$ be the number of the 600 fish that are tagged. Find the maximum likelihood estimate of $N$.

Hint:

$$P\{X = x\} = \frac{\binom{1000}{x}\binom{N-1000}{600-x}}{\binom{N}{600}} = P_x(N).$$

Examine $P_x(N+1)/P_x(N)$.

3. Show that, given a sufficient statistic $g(X_1, \ldots, X_n)$, there is a maximum likelihood estimate for $\theta$ that is a function of $g(X_1, \ldots, X_n)$ when the m.l.e. exists.

We make some final comments on the method of maximum likelihood. It can be shown that, under reasonable hypotheses, this method yields a consistent sequence of estimators. Furthermore, in most of the cases we are likely to meet, it yields better estimates than does the method of moments. However, as mentioned before, in any given case it is best to judge any estimate by some specific reasonable criterion, such as its mean square error.

## SECTION 6
## Confidence Limits and Confidence Interval Estimation

From the practical point of view, a scientist usually requires interval estimates of a parameter. To mention just a few examples, in reliability we often want an upper bound for the probability of system failure. That is, we want an interval $[0, p_u]$ that we are reasonably confident includes the unknown probability $p$ of system failure. The engineer working on radar wants an accurate interval estimate of the speed of radio waves; the metallurgist needs an interval estimate of the hardness of various alloys.

In this section we will develop the subject of confidence bounds and confidence interval estimation systematically. We base our construction on a given scalar statistic $u(X_1, \ldots, X_n)$, using a fixed number, $n$, of observations. We treat two sub-cases, the regular case usually associated with continuous distributions and the semi-regular case, associated normally with discrete distributions.

### 6.1 The Regular Case

Let $\theta$ be the parameter of interest and suppose that $\hat{\theta} = u(X_1, \ldots, X_n)$ is a scalar statistic (often an estimator of $\theta$) whose distribution is determined completely by $\theta$. In order to construct confidence limits, we need to define upper and lower points of a distribution.

**6.2 Definition: Upper and Lower $1 - \alpha$ Points.** Let the distribution of $\hat{\theta}$ be denoted by $P_\theta$ when $\theta$ is the specifying parameter. The smallest number $x$ such that $P_\theta\{\hat{\theta} \le x\} \ge 1 - \alpha$ is called the *upper* $1 - \alpha$ *point of* $P_\theta$, and will be denoted

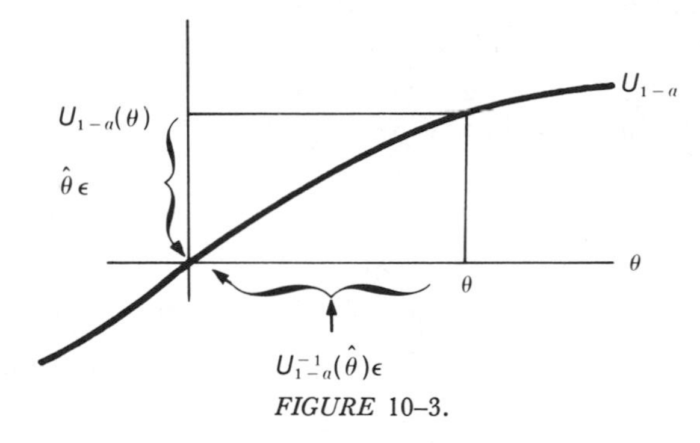

*FIGURE* 10–3.

by $U_{1-\alpha}(\theta)$. The largest value $x$ for which $P_\theta\{\hat{\theta} \geq x\} \geq 1 - \alpha$ will be called the *lower* $1 - \alpha$ *point of* $P_\theta$, and will be denoted by $L_{1-\alpha}(\theta)$.

Note that if the distribution function $F_\theta$ is continuous, then $F_\theta(U_{1-\alpha}(\theta)) = 1 - \alpha$ and $F_\theta(L_{1-\alpha}(\theta)) = \alpha$.

In the simpler regular case that we are examining first, we are only considering circumstances under which the following conditions are satisfied.

**6.3 Definition: Regularity Conditions.** For given $\alpha$ in $(0, 1)$ we shall say that the upper $1 - \alpha$ point $U_{1-\alpha}$ is *regular* if $U_{1-\alpha}$ is strictly monotone increasing as a function of $\theta$ and continuous at each $\theta$. (The same definition is given for regularity of the lower $1 - \alpha$ point $L_{1-\alpha}$.)

Note that it is natural for $U_{1-\alpha}(\theta)$ and $L_{1-\alpha}(\theta)$ to increase with $\theta$ if $\hat{\theta}$ is an estimator of $\theta$.

Figure 10.3 illustrates a regular upper $1 - \alpha$ point.

From the definition of $U_{1-\alpha}$, we know that

**6.4** $$P_\theta\{\hat{\theta} \leq U_{1-\alpha}(\theta)\} \geq 1 - \alpha.$$

But if $U_{1-\alpha}$ is regular, it follows that the inverse function $U_{1-\alpha}^{-1}$ exists. In fact, applying $U_{1-\alpha}^{-1}$ to both sides of the inequality $\hat{\theta} \leq U_{1-\alpha}(\theta)$ yields

**6.5** $$\hat{\theta} \leq U_{1-\alpha}(\theta) \quad \text{if and only if} \quad U_{1-\alpha}^{-1}(\hat{\theta}) \leq \theta.$$

(This is illustrated in Figure 10.3.) Thus, by 6.4,

$$P_\theta\{U_{1-\alpha}^{-1}(\hat{\theta}) \leq \theta\} \geq 1 - \alpha.$$

That is, when $U_{1-\alpha}$ is regular, then $U_{1-\alpha}^{-1}(\hat{\theta})$ is a lower $1 - \alpha$ level confidence limit for $\theta$ (Definition 4.1 in Chapter 9). A similar result holds when $L_{1-\alpha}$ is regular.

We summarize our findings in the following theorem.

### 6.6 *Theorem: Confidence Limits*

If $U_{1-\alpha}$ and $L_{1-\alpha}$ are regular, as given by Definition 6.3, then
(i) $U_{1-\alpha}^{-1}(\hat{\theta})$ is a $1 - \alpha$ level lower confidence limit for $\theta$ and
(ii) $L_{1-\alpha}^{-1}(\hat{\theta})$ is a $1 - \alpha$ level upper confidence limit for $\theta$.

We next see how to derive confidence intervals from confidence limits. Suppose we are in the regular case, Definition 6.3. Then

$$P_\theta\{\theta \le L_{1-\beta}^{-1}(\hat{\theta})\} \ge 1-\beta \quad \text{and} \quad P_\theta\{\theta \ge U_{1-\gamma}^{-1}(\hat{\theta})\} \ge 1-\gamma.$$

Hence,

$$\begin{aligned} P_\theta\{U_{1-\gamma}^{-1}(\hat{\theta}) \le \theta \le L_{1-\beta}^{-1}(\hat{\theta})\} &= 1 - P_\theta[\{U_{1-\gamma}^{-1}(\hat{\theta}) > \theta\} \cup \{L_{1-\beta}^{-1}(\hat{\theta}) < \theta\}] \\ &= 1 - P_\theta\{U_{1-\gamma}^{-1}(\hat{\theta}) > \theta\} - P_\theta\{L_{1-\beta}^{-1}(\hat{\theta}) < \theta\} \\ &\qquad \text{(due to the disjointness)} \\ &= P_\theta\{U_{1-\gamma}^{-1}(\hat{\theta}) \le \theta\} + P_\theta\{L_{1-\beta}^{-1}(\hat{\theta}) \ge \theta\} - 1 \\ &\ge 1-\gamma+1-\beta-1 = 1-\gamma-\beta. \end{aligned}$$

That is,

$$P_\theta\{U_{1-\gamma}^{-1}(\hat{\theta}) \le \theta \le L_{1-\beta}^{-1}(\hat{\theta})\} \ge 1-\beta-\gamma.$$

Recalling Definition 4.1 in Chapter 9, we see that this is simply another way of writing the level $1-\beta-\gamma$ confidence interval $[U_{1-\gamma}^{-1}(\hat{\theta}), L_{1-\beta}^{-1}(\hat{\theta})]$ for $\theta$. Thus we have proved the following theorem.

### *6.7 Theorem: Confidence Intervals*

In the regular case (Definition 6.3), the interval

**6.8**
$$[U_{1-\gamma}^{-1}(\hat{\theta}), L_{1-\beta}^{-1}(\hat{\theta})]$$

is a $1-\beta-\gamma$ level confidence interval for $\theta$.

Recall that the ***interpretation*** of Theorem 6.7 is that the ***long-run proportion of correct assertions of the form*** $U_{1-\gamma}^{-1}(\tilde{\theta}) \le \theta \le L_{1-\beta}^{-1}(\tilde{\theta})$ (where $\tilde{\theta}$ is the observed value of $\hat{\theta}$) is at least $1-\beta-\gamma$. That is, our long-run "batting average" of correct assertions about $\theta$ will be at least $1-\beta-\gamma$. ***It does not assert*** that with a given probability $\theta$ has taken on a value in $[U_{1-\gamma}^{-1}(\tilde{\theta}), L_{1-\beta}^{-1}(\tilde{\theta})]$, since $\theta$ is not considered to be the value of a random variable. (Such a situation will be considered in the next section.) Rather, Theorem 6.7 is simply an infinity of probability statements, one for each $\theta$, which are of great practical use. For $1-\beta-\gamma = 1-\alpha$, we see that there are many possible choices for $\beta$ and $\gamma$. The upper $1-\alpha$ confidence limit corresponds to $\gamma = 0$, $\beta = \alpha$, whereas the lower $1-\alpha$ confidence limit corresponds to $\gamma = \alpha$ and $\beta = 0$. ***Reasonably short*** $1-\alpha$ ***level confidence intervals are usually obtained by choosing*** $\gamma = \beta = \alpha/2$. We give a few of the simplest examples that can be handled with this approach.

### *6.9 Example: Waiting Time*

Suppose that you are allowed one observation on the length of time you wait at a store's checkout counter, and want a .95 level upper confidence limit for the mean waiting time. We take an exponential model, namely

$$f_X(x; \lambda) = \begin{cases} \dfrac{1}{\lambda} e^{-x/\lambda} & \text{for} \quad x \geq 0 \\ 0 & \text{otherwise.} \end{cases}$$

Here $\lambda$ is the mean waiting time. To obtain the desired confidence limit, we first find $L_{.95}(\lambda)$. In this case, because of the continuity of $F_X(\ ; \lambda)$, we must satisfy

$$P_\lambda\{L_{.95}(\lambda) \leq X\} = .95.$$

But

**6.10**
$$P_\lambda\{L_{.95}(\lambda) \leq X\} = \int_{L_{.95}(\lambda)}^{\infty} \frac{1}{\lambda} e^{-x/\lambda}\, dx = e^{-(L_{.95}(\lambda))/\lambda}.$$

In order to carry out our aim, we must set $e^{-\frac{L_{.95}(\lambda)}{\lambda}} = .95$ or

$$L_{.95}(\lambda) = -\lambda \log_e .95 \cong .0513\lambda.$$

Note, therefore, that we are in the regular case (Definition 6.3). Rather than appeal to Theorem 6.6, we simply rewrite 6.10 as $P_\lambda\{-\lambda \log_e .95 \leq X\} = .95$ or

$$P_\lambda\left\{\lambda \leq \frac{X}{-\log_e .95}\right\} = .95$$

and recognize

$$\frac{-X}{\log_e .95} \simeq 19.5X$$

as the desired .95 level upper confidence limit for the mean waiting time $\lambda$. We can use the value assumed by this confidence limit to compute probable upper limits for the amount of time we have to wait in line, with a 95 per cent assurance that we are being conservative. (Note that the level here is identically .95; see the footnote to Definition 4.1 in Chapter 9.) △

### *6.11 Example: Normal Mean*

Measurements $X_1, \ldots, X_n$ are made on the breaking strength of metal beams to be used in airplane wings. Because of the amount of variation in the tempering process, we assume these measurements to be independent and normally distributed with known variance $\sigma^2$ and unknown mean $\theta$. We have good reason to be interested in a $1 - \gamma$ level lower confidence limit for $\theta$.

From our assumptions, we know that

$$\hat{\theta} = T = \frac{1}{n} \sum_{i=1}^{n} X_i$$

is in many senses the best estimate of $\theta$ (see Definition 5.7). Then $T$ is normal with $E(T) = \theta$, Var $T = \sigma^2/n$. From the continuity of the distribution function of $T$, we know that we want to choose $U_{1-\gamma}$ such that

$$P_\theta\{T \leq U_{1-\gamma}(\theta)\} = 1 - \gamma.$$

That is, we want to choose $U_{1-\gamma}(\theta)$ so that

$$\int_{-\infty}^{U_{1-\gamma}(\theta)} \frac{1}{\sqrt{2\pi(\sigma^2/n)}} \exp\left[-\frac{(t-\theta)^2}{2\sigma^2/n}\right] dt = 1 - \gamma.$$

If the above integral were tabled, we might simply use the table; since only the standard normal distribution function is available, we make the natural change of variable

$$w = \frac{t-\theta}{\sigma/\sqrt{n}}, \qquad dw = \frac{dt}{\sigma/\sqrt{n}}$$

and the corresponding change of upper limit of integration to

$$\frac{U_{1-\gamma}(\theta) - \theta}{\sigma/\sqrt{n}}$$

(since this is the value of $w$, then $t = U_{1-\gamma}(\theta)$). Our objective then becomes to choose $U_{1-\gamma}(\theta)$ so that

$$\int_{-\infty}^{(U_{1-\gamma}(\theta)-\theta)/(\sigma/\sqrt{n})} \frac{1}{\sqrt{2\pi}} e^{-w^2/2}\, dw = 1 - \gamma.$$

If $\varphi_\gamma$ is the value for which

**6.12**
$$\int_{-\infty}^{\varphi_\gamma} \frac{1}{\sqrt{2\pi}} e^{-w^2/2}\, dw = 1 - \gamma,$$

then we want

$$\frac{U_{1-}(\theta) - \theta}{\sigma/\sqrt{n}} = \varphi_\gamma$$

or

$$U_{1-\gamma}(\theta) = \frac{\sigma}{\sqrt{n}}\varphi_\gamma + \theta.$$

Then the inequality $\hat{\theta} \leq U_{1-\gamma}(\theta)$ in this case is

$$T \leq \frac{\sigma}{\sqrt{n}}\varphi_\gamma + \theta.$$

Hence

$$P_\theta\{\hat{\theta} \leq U_{1-\gamma}(\theta)\} \equiv P_\theta\left\{T \leq \frac{\sigma}{\sqrt{n}}\varphi_\gamma + \theta\right\} = 1 - \gamma.$$

We may write this very simply in the desired form $P_\theta\{U_{1-\gamma}^{-1}(\hat{\theta}) \leq \theta\} = 1 - \gamma$, by subtracting

$$\frac{\sigma}{\sqrt{n}}\varphi_\gamma$$

from both sides of the inequality sign. This gives

$$P_\theta\left\{T - \frac{\sigma}{\sqrt{n}}\varphi_\gamma \leq \theta\right\} = 1 - \gamma.$$

Thus

$$\frac{1}{n}\sum_{i=1}^{n} X_i - \frac{\sigma}{\sqrt{n}}\varphi_\gamma = T - \frac{\sigma}{\sqrt{n}}\varphi_\gamma$$

is a 1 $\gamma$ ***level lower confidence bound for the mean $\theta$ of independent random variables $X_1, \ldots, X_n$, which are normal with known variance*** $\sigma^2$, where $\varphi_\gamma$ is given by 6.12. Note that the level is identically $1 - \gamma$, as in 6.10. △

Sometimes it is not possible for us to find a statistic for which the distribution is completely determined by the parameter of interest. However, we may be able to find a random variable that may not be observable, but that involves only the parameter of interest, and for which the distribution is completely determined by this parameter. That is, the random variable and its distribution do not involve any other unknown parameters. In this case, it may be possible to modify the procedure just given so as to obtain a confidence interval. We illustrate this in the following example.

### *6.13 Example: Normal Mean with Unknown Variance*

It is of some interest to obtain a confidence interval based on the available data for the expected lifetime of people in our own population. If $X_1, X_2, \ldots, X_n$ are random variables whose values are lifetimes of $n$ individuals selected in a representative way, then it might be reasonable to assume them independent and normal with $E(X_i) = \mu$, Var $X_i = \sigma^2$, where both $\mu$ and $\sigma^2$ are unknown. In Chapter 14, we will show that the random variable

$$T = \frac{\sqrt{n}(\bar{X}_{(n)} - \mu)}{\sqrt{S_n^2}}$$

with

**6.14** $$\bar{X}_{(n)} = \frac{1}{n}\sum_{i=1}^{n} X_i \quad \text{and} \quad S_n^2 = \frac{1}{n-1}\sum_{i=1}^{n}(X_i - \bar{X}_{(n)})^2,$$

has a $t$ distribution (see 8 in Chapter 8) with $n - 1$ degrees of freedom. Note that $T$ is not an observable statistic. If $f_{n-1}$ denotes the density of this distribution and

$(t_{l,\alpha}, t_{u,\alpha})$ are chosen so that

**6.15** $$\int_{t_{l,\alpha}}^{t_{u,\alpha}} f_{n-1}(x)\, dx = 1 - \alpha,$$

then

$$P_{(\mu,\sigma)}\left\{t_{l,\alpha} \le \frac{\sqrt{n}(\bar{X}_{(n)} - \mu)}{\sqrt{S_n^2}} \le t_{u,\alpha}\right\} = 1 - \alpha.$$

(Here $t_{l,\alpha}$ and $t_{u,\alpha}$ do not even involve $\mu$, but this is only fortuitous.) In this case, we may proceed to rewrite the event

$$\left\{t_{l,\alpha} \le \frac{\sqrt{n}(\bar{X}_{(n)} - \mu)}{\sqrt{S_n^2}} \le t_{u,\alpha}\right\}$$

by noting that

$$t_{l,\alpha} \le \frac{\sqrt{n}(\bar{X}_{(n)} - \mu)}{\sqrt{S_n^2}}$$

is equivalent to

$$\frac{\sqrt{S_n^2}}{\sqrt{n}} t_{l,\alpha} \le \bar{X}_{(n)} - \mu$$

or

$$\mu \le \bar{X}_{(n)} - \frac{\sqrt{S_n^2}}{\sqrt{n}} t_{l,\alpha};$$

while

$$\frac{\sqrt{n}(\bar{X}_{(n)} - \mu)}{\sqrt{S_n^2}} \le t_{u,\alpha}$$

is equivalent to

$$\bar{X}_{(n)} - \frac{\sqrt{S_n^2}}{\sqrt{n}} t_{u,\alpha} \le \mu.$$

Thus

$$\left\{t_{l,\alpha} \le \frac{\sqrt{n}(\bar{X}_{(n)} - \mu)}{\sqrt{S_n^2}} \le t_{u,\alpha}\right\} = \left\{\bar{X}_{(n)} - \frac{\sqrt{S_n^2}}{\sqrt{n}} t_{u,\alpha} \le \mu \le \bar{X}_{(n)} - \frac{\sqrt{S_n^2}}{\sqrt{n}} t_{l,\alpha}\right\}.$$

Hence ***the*** $1 - \alpha$ ***level confidence intervals for the mean*** $\mu$ ***of independent normal random variables with unknown variance*** $\sigma^2$ ***are of the form***

$$\left[\bar{X}_{(n)} - \frac{\sqrt{S_n^2}}{\sqrt{n}} t_{u,\alpha}, \bar{X}_{(n)} - \frac{\sqrt{S_n^2}}{\sqrt{n}} t_{l,\alpha}\right]$$

where $\bar{X}_{(n)}$, $S_n^2$ are given by 6.14 and $t_{l,\alpha}$, $t_{u,\alpha}$ are any pair of numbers satisfying 6.15. (Note that $t_{l,\alpha}$ is usually negative.) For short intervals, we usually choose $t_{l,\alpha} = -t_{u,\alpha}$.

△

## *6.16 EXERCISES*

1.° Suppose that the numbers listed here represent the lifetimes of 16 individuals selected in a representative way from some human population (actually, the authors selected the data from an appropriate table of random numbers, but the fiction sounds more interesting). Supposing the data to be normally distributed with unknown mean $\mu$ and unknown variance $\sigma^2$, find the values of:

(a) a reasonably short confidence interval of level .9 for the mean lifetime.
(b) a level .9 upper confidence limit for the mean lifetime.
(c) a level .9 lower confidence limit for the mean lifetime.
(d) Repeat (a), (b), (c) when it is known that $\sigma = 10$.

| | | | |
|---|---|---|---|
| 60.73 | 78.65 | 32.44 | 62.04 |
| 61.21 | 68.11 | 75.09 | 75.24 |
| 57.03 | 51.94 | 70.77 | 55.42 |
| 52.69 | 65.45 | 52.00 | 50.57 |

2. Do the same Exercise as 1 with the data:

| | | | |
|---|---|---|---|
| 60.04 | 47.25 | 42.07 | 50.14 |
| 46.37 | 51.20 | 58.42 | 51.69 |
| 51.87 | 46.55 | 65.00 | 56.82 |
| 55.68 | 70.45 | 67.33 | 71.64 |

3.° Find a reasonably short $1 - \alpha$ level confidence interval for the variance $\sigma^2$ of normal random variables $X_1, \ldots, X_n$:

(a) When $E(X_i) = \mu$ is known.

Hint: $[\sum_{i=1}^{n}(X_i - \mu)^2]/\sigma^2$ has a Chi square distribution with $n$ degrees of freedom. If we let $\chi_n^2$ denote a random variable having the Chi square distribution with $n$ degrees of freedom, and $Q_n^2 = \sum_{i=1}^{n}(X_i - \mu)^2$, then it helps to write $Q_n^2 = \sigma^2\chi_n^2$.

(b) When $E(X_i) = \mu$ is unknown.

Hint: $[\sum_{i=1}^{n}(X_i - \bar{X}_{(n)})^2]/\sigma^2$ has a Chi square distribution with $n - 1$ degrees of freedom.

(c) What is the expected length for the interval in (a)?

(d) Can you give an upper bound for the number of observations needed to keep the expected length below 6.1 if it is known that $\sigma^2 \leq 4$, at a confidence level of .9?

Hint: Look at a table of the Chi square distribution.

(e) Can you compute a simple theoretical upper bound for the smallest number $n$ of observations that would be needed for a confidence interval for $\sigma^2$ of length not exceeding $l$ and level $1 - \alpha$, knowing $\sigma^2 \leq \sigma_0^2$ where $\sigma_0^2$ is given, $\mu = 0$?

Hint: $Y_n = \sum_{i=1}^{n} X_i^2$ has expectation $n\sigma^2$ and variance $2n\sigma^4$ (since $E(X_i^4) = 3\sigma^4$, and thus $\text{Var } X_i^2 = 2\sigma^4$, computable from the moment generating function or directly). Use Chebyshev's inequality.

4. Based on the data and assumptions of Exercise 1, construct the values of:

(a) a reasonably short confidence interval for $\sigma$ of level .9.

Hint: See the results of the previous question.

(b) a level .9 upper confidence limit for $\sigma$.
(c) the same as (a) and (b) for $\mu = 60$.

5.° Repeat Exercise 4 for the data of Exercise 2.

6. Suppose that $X_1, \ldots, X_n$ are independent random variables uniform on $[0, \theta]$, representing the lengths of busy periods of some service operation. Then

$$f_{X_i}(x) = \begin{cases} 1/\theta \text{ for } 0 \leq x \leq \theta \\ 0 \text{ otherwise.} \end{cases}$$

(a) Find a confidence interval for the mean busy period length $\theta$, of level $1 - \alpha$, whose length goes to 0 as $n$ gets large. Use $\max(X_1, \ldots, X_n)$ to obtain the interval.

(b) Determine how large $n$ should be in order that at level .95 the length of the interval is less than .01 if $\theta \leq 5$.

(c) Using the statistic

$$\frac{1}{n}\sum_{i=1}^{n} X_i$$

instead of $\max(X_1, \ldots, X_n)$, repeat the computations of part (b), with the aid of the Central Limit Theorem.

7. Let $X_1, \ldots, X_n$ be independent random variables uniformly distributed on $[\theta_1, \theta_2]$, where $\theta_i$ are unknown. Here $X_i$ may represent the length of time of the $i$th service call.

(a) Show that if $Y_n = \min(X_1, \ldots, X_n)$, $Z_n = \max(X_1, \ldots, X_n)$, then $(Y_n, Z_n)$ is sufficient for $\theta = (\theta_1, \theta_2)$.

(b) Determine a pair of confidence intervals $I_1$, $I_2$ for $\theta_1$ and $\theta_2$ such that the probability $P\{\theta_1 \in I_1, \theta_2 \in I_2\} \geq 1 - \alpha$.

(c) Determine how large $n$ has to be such that, when $\theta_2 - \theta_1 \leq l$, the length of $I_1$ is $\leq l_1$ and that of $I_2$ is $\leq l_2$, where $l$, $l_1$ and $l_2$ are assumed given.

8. Let $X_1, \ldots, X_n$ be independent, each with density

$$f_{X_i}(x) = \frac{1}{\pi[1 + (x - \theta)^2]}.$$

Using the limiting distribution of the sample median (Equation 4.2, Chapter 7) find an approximate level $1 - \alpha$ confidence interval for $\theta$.

9.° Using the data of Exercise 1, assuming $\sigma = 10$, find an approximate level .9 reasonably short confidence interval for the median (= mean here) using the limiting distribution of the sample median. Compare your result with 1(a).

10. Same as Exercise 9 for the data of Exercise 2.

---

## *6.17 PROBLEMS*

1.° Let $n$ people be chosen at random from the population and let $X_i$ be the average number of cigarettes per day smoked by the $i$th individual and $Y_i$ his length of life. We would like a confidence interval for the correlation coefficient

$$\rho = \frac{E([X - E(X)][Y - E(Y)])}{\sqrt{\operatorname{Var} X \operatorname{Var} Y}}$$

based on the fact that the random variable

$$Z = \tfrac{1}{2} \log \frac{1 + r_n}{1 - r_n} \quad \text{where } r_n = \frac{\sum_{i=1}^{n} (X_i - \bar{X}_{(n)})(Y_i - \bar{Y}_{(n)})}{\sqrt{\left[\sum_{i=1}^{n} (X_i - \bar{X}_{(n)})^2\right]\left[\sum_{i=1}^{n} (Y_i - \bar{Y}_{(n)})^2\right]}}$$

$$\bar{X}_{(n)} = \frac{1}{n} \sum_{i=1}^{n} X_i, \qquad \bar{Y}_{(n)} = \frac{1}{n} \sum_{i=1}^{n} Y_i,$$

known as "*Fisher's Z*," is approximately normal, with

$$E(Z) \cong \tfrac{1}{2} \log \frac{1 + \rho}{1 - \rho} + \frac{\rho}{2(n-1)}, \quad \text{Var } Z \cong \frac{1}{n-3}.$$

Such a confidence interval is of great interest in determining the effects of smoking on length of life. How many observations would be needed to determine $\rho$ to within .01 with confidence .95?

2. A city has $n$ cars with distinct license plates numbered $1, 2, \ldots, n$. We choose a random sample of $k$ cars. Based on this sample, find an upper $1 - \alpha$ level confidence limit for $n$.

3.° Suppose $X_1, X_2, \ldots$ are independent normal with unknown mean $\mu$ and unknown variance $\sigma^2$. Based on two stages of sampling, obtain a confidence interval of preassigned length $l$ and confidence $1 - \alpha$ for $\mu$.
Hint: Let $\beta + \gamma = \alpha$. First obtain a $1 - \beta$ upper confidence limit $\hat{\sigma}_u^2$ for $\sigma^2$. Then use the sample size corresponding to $\sigma^2 = \sigma_u^2$ to obtain an $l$ length $1 - \gamma$ level confidence interval for $\mu$. Justify your results. (Use $P(A \cap B) \geq 1 - P(A^c) - P(B^c)$.)

4. Using the same method as in the previous problem, obtain a $1 - \alpha$ level $l$ length confidence interval for $\sigma^2$ based on independent normal random variables with unknown mean $\mu$ and variance $\sigma^2$.

## 6.18 The Semi-Regular Case

Recall that we obtained confidence intervals for the binomial parameter $p$ by Chebyshev's inequality. However, these confidence intervals were much longer than they needed to be. Also, we obtained approximate confidence intervals by means of the Central Limit Theorem. However, when the sample size is small, the correctness of these intervals was very much suspect. It is for this reason that we treat in some detail the subject of confidence intervals when the regularity conditions of Definition 6.3 are not satisfied. Since this generally happens in the discrete case, we will assume that the statistic $\hat{\theta}$ on which we base our construction can take on only non-negative integer values. We do need to assume that $U_{1-\alpha}(\theta)$ and $L_{1-\alpha}(\theta)$ in some sense "try to follow" $\theta$. However, even in the tractable discrete cases, we can only require that $U_{1-\alpha}(\theta)$ and $L_{1-\alpha}(\theta)$ do not decrease as $\theta$ increases.

**6.19 Definition: Semi-Regularity.** We will say that $U_{1-\alpha}$ is *semi-regular* if $U_{1-\alpha}(\theta)$ does not decrease as $\theta$ increases. (The same definition is given for $L_{1-\alpha}$ to be semi-regular.)

Now, we would like to manipulate the event $\hat{\theta} \leq U_{1-\alpha}(\theta)$, just as we did in 6.5, to obtain a lower $1 - \alpha$ level confidence limit for $\theta$. However, because of the lack of strict monotonicity of $U_{1-\alpha}$, we must resort to a more sophisticated manipulation, which we will explain in the section below. Then we will state the main theorem.

It may help if you think of $\hat{\theta}$ as $\sum X_i$, where $\sum X_i$ is the number of successes in $n$ Bernoulli trials, and think of $\theta$ as $p$, the probability of success on one trial.

Let us now define a function $\Theta_L$ by letting $\Theta_L(y)$ be any value (of $\theta$) for which

**6.20** $$U_{1-\alpha}[\Theta_L(y)] < y \text{ for all } y \text{ in the range of } \hat{\theta}.$$

(This can be found from tables of the distributions of $\hat{\theta}$.)

### *6.21 Lemma*

When $U_{1-\alpha}$ is semi-regular, the condition

(*) $$y \leq U_{1-\alpha}(\theta)$$

implies the condition

(**) $$\theta > \Theta_L(y).$$

Before proving this lemma, let us note that the implication, (*) implies (**), replaces the equivalence

$$y \leq U_{1-\alpha}(\theta) \quad \text{if and only if} \quad \theta \geq U_{1-\alpha}^{-1}(y)$$

used essentially in 6.5. The implication that (*) implies (**) shows that

$$\{\hat{\theta} \leq U_{1-\alpha}(\theta)\} \subseteq \{\theta > \Theta_L(\hat{\theta})\}.$$

Hence

$$P_\theta\{\theta > \Theta_L(\hat{\theta})\} \geq P_\theta\{\hat{\theta} \leq U_{1-\alpha}(\theta)\} \geq 1 - \alpha,$$

so that $\Theta_L(\hat{\theta})$ is a lower $1 - \alpha$ level confidence limit for $\theta$.

*Proof of* 6.21. We demonstrate this lemma both geometrically and analytically. The geometry is similar to that of the regular case (see Figure 10.1). However, only semi-regularity is used. First, *geometrically*, for any $\theta$ you would choose, in order that (*) be satisfied, we see from Figure 10.4 that $y$ must lie in

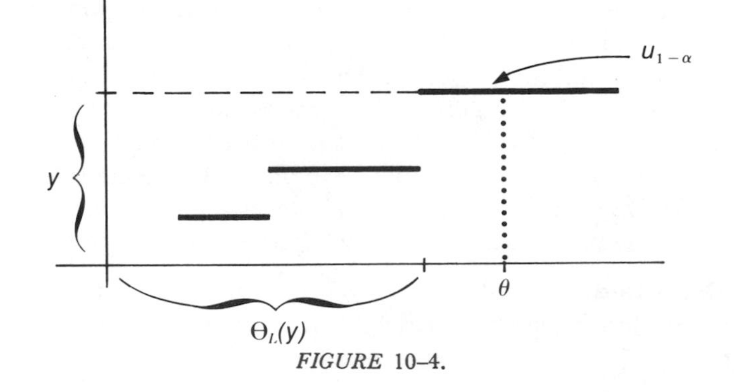


*FIGURE* 10–4.

the indicated interval along the vertical axis. Using the semi-regularity 6.21 yields the result that $\Theta_L(y)$ must lie in the interval indicated on the horizontal axis in Figure 10.4, from which (**) follows.

*Analytically,* we note that combining 6.20 and (*) yields

$$U_{1-\alpha}[\Theta_L(y)] < U_{1-\alpha}(\theta).$$

Semi-regularity implies that $U_{1-\alpha}$ is non-decreasing, which now yields (**), proving what we asserted.

We next find a simple way to characterize $\Theta_L(\tilde{\theta})$ (which will be the value of our lower $1 - \alpha$ level confidence limit for $\theta$ when $\tilde{\theta}$ is the observed value of $\hat{\theta}$). We claim that if $\theta'$ satisfies

**6.22** $$P_{\theta'}\{\hat{\theta} \leq \tilde{\theta} - 1\} \geq 1 - \alpha,$$

then $\theta'$ will serve as $\Theta_L(\tilde{\theta})$. We prove this as follows: From 6.20, we see that we need only show that if 6.22 holds, then

**6.23** $$U_{1-\alpha}(\theta') < \tilde{\theta}.$$

But

**6.24** $$P_{\theta'}\{\hat{\theta} \leq U_{1-\alpha}(\theta')\} \geq 1 - \alpha$$

by the definition of $U_{1-\alpha}$ (6.2). However, by 6.2 we know that $U_{1-\alpha}(\theta')$ is the *smallest* value $x$ such that $P_{\theta'}\{\hat{\theta} \leq x\} \geq 1 - \alpha$. Comparing 6.24 and 6.22 yields

$$U_{1-\alpha}(\theta') \leq \tilde{\theta} - 1 < \tilde{\theta}.$$

This gives the desired 6.23.

Now we turn to upper confidence limits. A similar development when $L_{1-\alpha}$ is semi-regular shows that if $\Theta_U(\tilde{\theta})$ is any $\theta'$ such that

$$P_{\theta'}\{\hat{\theta} \geq \tilde{\theta} + 1\} \geq 1 - \alpha, \quad \text{that is,} \quad P_{\theta'}\{\hat{\theta} \leq \tilde{\theta}\} < \alpha$$

then $\Theta_U(\hat{\theta})$ is a $1 - \alpha$ level upper confidence limit for $\theta$.

We have thus proved the following theorem.

### 6.25 *Theorem: Confidence Limits, Semi-Regular Case*

Let $\hat{\theta}$ be an integer valued random variable whose distribution is specified by the parameter $\theta$. Suppose $U_{1-\alpha}$ (see 6.2) is semi-regular (6.19). Let $\tilde{\theta}$ be the observed value of $\hat{\theta}$. Denote by $\Theta_L(\tilde{\theta})$ any value of $\theta$ (preferably as large as possible) for which the event $\{\hat{\theta} \leq \tilde{\theta} - 1\}$ has probability at least $1 - \alpha$, that is,

$$P_{\Theta_L(\tilde{\theta})}\{\hat{\theta} \leq \tilde{\theta} - 1\} \geq 1 - \alpha.$$

Then $\Theta_L(\tilde{\theta})$ is the value of a $1 - \alpha$ level lower confidence limit for $\theta$. Similarly, if $L_{1-\alpha}$ is semi-regular and $\Theta_U(\tilde{\theta})$ is any value of $\theta$ such that

$$P_{\Theta_U(\tilde{\theta})}\{\hat{\theta} \leq \tilde{\theta}\} \leq \alpha,$$

then $\Theta_U(\tilde{\theta})$ is the value of a $1 - \alpha$ level upper confidence limit for $\theta$. (We try to choose $\Theta_U(\tilde{\theta})$ as small as possible.)

We can also prove the following theorem as we did Theorem 6.7.

### 6.26 *Theorem: Construction of Confidence Intervals from Confidence Limits*

If $\Theta_{U,\beta}(\hat{\theta})$ is a $1 - \beta$ level upper confidence limit for $\theta$ and $\Theta_{L,\gamma}(\hat{\theta})$ is a $1 - \gamma$ level lower confidence limit for $\theta$, then

$$[\Theta_{L,\gamma}(\hat{\theta}), \Theta_{U,\beta}(\hat{\theta})]$$

is a $1 - \beta - \gamma$ level confidence interval for $\theta$.

We will now illustrate Theorem 6.25.

### 6.27 *Example: Lower Confidence Limit for Bernoulli Trials*

Suppose that we repeat an experiment independently 10 times and let $\tilde{\theta}$ be the observed number of successes. Based on $\tilde{\theta}$, we want to determine a lower .8 level confidence limit for $\theta$, the probability of success. In Table 10.1, we reproduce an

**TABLE 10.1 Family of Binomial Distribution Functions for $n = 10$**

| $\theta = p$ \ $k$ | 0 | 1 | 2 | 3 | 4 | 5 | 6 | 7 | 8 | 9 | 10 |
|---|---|---|---|---|---|---|---|---|---|---|---|
| 1/8 | .26 | .64 | .88 | .97 | 1.0 | 1.0 | 1.0 | 1.0 | 1.0 | 1.0 | 1.0 |
| 2/8 | .06 | .24 | .53 | .78 | .92 | .98 | 1.0 | 1.0 | 1.0 | 1.0 | 1.0 |
| 3/8 | .009 | .06 | .21 | .45 | .69 | .87 | .96 | .99 | 1.0 | 1.0 | 1.0 |
| 4/8 | .001 | .01 | .05 | .17 | .38 | .62 | .83 | .95 | .99 | 1.0 | 1.0 |
| 5/8 | 0 | 0 | .008 | .04 | .13 | .31 | .55 | .79 | .94 | .99 | 1.0 |
| 6/8 | 0 | 0 | 0 | .003 | .02 | .08 | .22 | .47 | .76 | .94 | 1.0 |
| 7/8 | 0 | 0 | 0 | 0 | 0 | .004 | .03 | .12 | .36 | .74 | 1.0 |

abbreviated table of the family of binomial distribution functions for $n = 10$.

Suppose for simplicity $\tilde{\theta} = 4$. We then look in the table for a value of $\theta$ such that $P_\theta\{\hat{\theta} \leq 3\} \geq .8$. The largest such value is seen to be $\theta = 1/8$. (Actually it is nearer to 2/8, but the crudeness of our table forces us to use 1/8.) Thus $\Theta_L(\tilde{\theta}) = 1/8$ is, in this case, the value of a .8 level lower confidence limit for $\theta$. △

---

*6.28 EXERCISES*

1.° Using Table 10.1, construct an upper .8 level confidence limit for $p$ when the binomial $\tilde{\theta} = 3$ with $n = 10$.

2. (a) Using Table 10.1, construct a lower .9 level confidence limit for $p$ when $\tilde{\theta} = 3$ with $n = 10$.

(b) Combine the results of Exercises 1 and 2(a) to obtain a .7 level confidence interval for $\theta$ when $\tilde{\theta} = 3$ with $n = 10$.

3.° Suppose that $X_1, X_2, X_3$ are independent Poisson random variables representing radiation counts with mean value $\lambda$.

(a) Using tables of the Poisson family of distribution functions, construct a .85 level upper confidence limit for $\lambda$ when the observed values $x_i$ of $X_i$ are $x_1 = 1$, $x_2 = 2$, $x_3 = 3$.

(b) Construct a .85 level lower confidence limit for $\lambda$.

(c) Combine these to form a .7 level confidence interval for $\lambda$.

4.° Use the Central Limit Theorem to obtain approximate confidence intervals for binomial $p$.

5. Use the Central Limit Theorem to obtain approximate confidence intervals for the Poisson parameter $\lambda$.

A particularly interesting practical application of the theory of confidence intervals for estimation of a binomial parameter $p$ occurs in the attempt to determine the probability distribution of disease symptoms. Suppose, for example, that some disease can give rise to any of five symptoms (any subset of which can occur). It is seen that there are thus $2^5 = 32$ possible symptom sets, call them $y_1, \ldots, y_{32}$. For diagnostic purposes it may be of interest to determine the probability distribution $p_1, \ldots, p_{32}$. It is assumed that a later test, such as a culture test or an autopsy, can determine with near certainty whether or not the given disease was present, so that we can restrict ourselves to the population suffering from the disease. Thus, suppose we have a random sample of $n$ individuals from the population of those with the disease of interest. To estimate $p_j$, we may simply define the random variable $Z_j$ to be the number of individuals in our sample with symptom set $y_j$. Then $Z_j$ is binomial with parameters $n$ and $p_j$. From this we may obtain a $1 - \alpha_j$ level confidence interval for $p_j$. We may assert that all $p_j$ are included in their respective intervals and we may conclude that the probability that this assertion is correct is at least $1 - \sum_{j=1}^{32} \alpha_j$. (Recall that $P(AB) \geq 1 - P(A^c) - P(B^c)$.) If we take $\alpha_j = 1/320$ and use a normal approximation to the binomial, it might be interesting to determine the number $n$ needed to obtain confidence intervals all of length $\leq .01$. Later we shall present a more efficient method to solve this problem.

## SECTION 7
## Bayesian Intervals

In the previous section, we acted as if we had no information at all about the parameter $\theta$ that we desire to estimate (other than the information that $\theta$ specified the true distribution of the data). In some situations it may be reasonable to describe the governing $\theta$ as the value of a random variable $\Theta$ whose distribution is known. There is much controversy as to how one determines which distribution to use for describing $\Theta$. Some individuals feel that it should be determined by how much a person is willing to bet in favor of various alternatives. Our feeling is that it should be based on objective information as much as possible. In developing the empirical Bayes method, Robbins has shown how to use a certain past experience in a precisely justified manner, but unfortunately we have neither the space nor the machinery to do justice to the empirical Bayes method. We will therefore present a few examples in which it seems appropriate to describe $\theta$ as the value of a random variable with a known distribution.

Suppose that over many years of experience with shipments of 1000 transistors, the percentage of good units seemed to vary almost uniformly between 90 and 100 per cent. Then we might describe the probability $\theta$ of a good transistor in any one shipment as the value of a random variable with a uniform density on $[.9, 1]$. Or, suppose that every day we attempt to reset a measuring instrument to some given setting $\theta_0$. From Central Limit Theorem arguments, it may be reasonable to postulate that the true setting is $\theta$, where $\theta$ is the value of a normal random variable with mean $\theta_0$ and variance $\sigma_0^2$. By making many attempts to attain setting $\theta_0$, and many measurements on a known standard for each such attempt, we could obtain an accurate estimate of $\sigma_0^2$ (see Problems 6.17(3) and 6.17(4)). The actual measurements each day may reasonably be described as normal with mean $\theta$ and variance $\sigma^2$ for many such instruments.

Here we will treat the case of random variables $X_1, \ldots, X_n$ with a density whose value at $x_1, \ldots, x_n$ is $f(x_1, \ldots, x_n; \theta)$, the discrete case being similar. Then the density value $f(x_1, \ldots, x_n; \theta)$ can be considered as a conditional density $f_{X_1,\ldots,X_n|\Theta}(x_1, \ldots, x_n \mid \theta)$. In such cases, if $\Theta$ has density $f_\Theta$ (known as the *prior* density of $\Theta$), then the joint density $f_{X_1,\ldots,X_n,\Theta}$ of $X_1, \ldots, X_n$ and $\Theta$ is given by the formula

$$f_{X_1,\ldots,X_n,\Theta}(x_1, \ldots, x_n, \theta) = f_{X_1,\ldots,X_n|\Theta}(x_1, \ldots, x_n \mid \theta) f_\Theta(\theta)$$

(see Definition 7.1, Chapter 5). The density of $\Theta$, given that $X_1 = x_1, \ldots, X_n = x_n$, is given by

**7.1**
$$f_{\Theta|X_1,\ldots,X_n}(\theta \mid x_1, \ldots, x_n) = \frac{f_{X_1,\ldots,X_n,\Theta}(x_1, \ldots, x_n, \theta)}{f_{X_1,\ldots,X_n}(x_1, \ldots, x_n)}$$
$$= \frac{f_{X_1,\ldots,X_n,\Theta}(x_1, \ldots, x_n, \theta)}{\int_{-\infty}^{\infty} f_{X_1,\ldots,X_n,\Theta}(x_1, \ldots, x_n, \theta')\, d\theta'}$$
$$= \frac{f_{X_1,\ldots,X_n|\Theta}(x_1, \ldots, x_n \mid \theta) f_\Theta(\theta)}{\int_{-\infty}^{\infty} f_{X_1,\ldots,X_n|\Theta}(x_1, \ldots, x_n \mid \theta') f_\Theta(\theta')\, d\theta'}.$$

We see that once we know the values $x_1, \ldots, x_n$ assumed by $X_1, \ldots, X_n$, we can compute the conditional density $f_{\Theta|X_1,\ldots,X_n}(\theta \mid x_1, \ldots, x_n)$ of $\Theta$, for each $\theta$. (This computation essentially re-proves Bayes Theorem, Theorem 4.12 in Chapter 3, explaining the name of this section.) From this density we can surely construct an

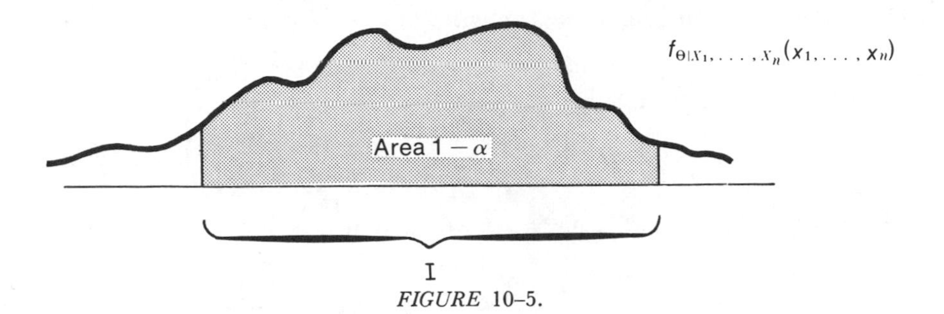


*FIGURE* 10–5.

interval $I$ (see Figure 10.5) that has probability at least $1 - \alpha$ of including the (unobservable) value of $\theta$.

**7.2 Definition: $1 - \alpha$ Level Bayes Interval.** An interval $I$, constructed in the manner just described, is called a $1 - \alpha$ *level Bayes interval for* $\Theta$. If the interval consists of all $\Theta$ values to the left of some value, this value is called a $1 - \alpha$ level Bayes upper limit for $\Theta$ (and similarly for a $1 - \alpha$ level Bayes lower limit for $\Theta$).

### *7.3 Example: Speed Trap*

Suppose that the local gendarmes have set up a radar surveillance unit outside the town. However, let's also suppose that they want an airtight case against any of the unfortunates who happen to be caught. So, they make some tests on an automobile known to be going 35 miles per hour. They postulate that conditionally given $\Theta = \theta$, the errors $X_1, \ldots, X_n$ are independent normal with variance 1 and mean value $\theta$. (If $\theta > 0$, then generally the measured speed will exceed the true speed.) The errors $X_i$ arise from several sources, such as variability in the reflections from the different parts of the autos, dust, and other small variations. The mean $\theta$ might be thought of as arising from errors in setting the controls, and the general change in ambient conditions that occurs from day to day. It is thus not unreasonable to assume that $\theta$ is the value of a random variable $\Theta$, which is normal. Here we will assume $\Theta$ to have mean 0 and variance 1.

Based on $X_1, \ldots, X_n$, the police want to construct an interval that contains $\theta$ with probability $1 - \alpha$. To do this, they first use Equation 7.1 to find $f_{\Theta|X_1,\ldots,X_n}$. Now

$$f_{\Theta|X_1,\ldots,X_n}(\theta \mid x_1, \ldots, x_n) = \frac{f_{X_1,\ldots,X_n|\Theta}(x_1, \ldots, x_n \mid \theta) f_\Theta(\theta)}{\int_{-\infty}^{\infty} f_{X_1,\ldots,X_n|\Theta}(x_1, \ldots, x_n \mid \theta') f_\Theta(\theta')\, d\theta'}$$

$$= \frac{\dfrac{1}{(2\pi)^{n/2}} e^{-(1/2)\sum_{i=1}^{n}(x_i-\theta)^2} \dfrac{1}{(2\pi)^{1/2}} e^{-\theta^2/2}}{\int_{-\infty}^{\infty} \dfrac{1}{(2\pi)^{n/2}} e^{-(1/2)\sum_{i=1}^{n}(x_i-\theta')^2} \dfrac{1}{(2\pi)^{1/2}} e^{-\theta'^2/2}\, d\theta'}.$$

Using completion of the square, we can write $f_{\Theta|X_1,\ldots,X_n}(\theta \mid x_1, \ldots, x_n)$ in the form

$$G(x_1, \ldots, x_n)\exp\left\{-\frac{1}{2\left(\frac{1}{n+1}\right)}\left[\theta - \frac{1}{n+1}\sum_{i=1}^{n} x_i\right]^2\right\}.$$

Thus, given $X_1 = x_1, \ldots, X_n = x_n$, we see that $\Theta$ is normal,

$$\text{with mean} \quad \frac{1}{n+1}\sum_{i=1}^{n} x_i \quad \text{and variance} \quad \frac{1}{n+1}.$$

Hence, an interval in which $\Theta$ will assume its value with probability .95 is

$$\left[\frac{1}{n+1}\sum_{i=1}^{n} x_i - \frac{1.96}{\sqrt{n+1}}, \frac{1}{n+1}\sum_{i=1}^{n} x_i + \frac{1.96}{\sqrt{n+1}}\right].$$

Note that the .95 level *confidence* interval for $\theta$ is a random interval whose value is

$$\left[\frac{1}{n}\sum_{i=1}^{n} x_i - \frac{1.96}{\sqrt{n}}, \frac{1}{n}\sum_{i=1}^{n} x_i + \frac{1.96}{\sqrt{n}}\right].$$

We see that the Bayes interval is a bit shorter than the corresponding confidence interval, as expected, since we assumed more information. Also, its center is closer to the origin, since the initial information assigns more probability to $\Theta$ values close to 0. If we had assumed more prior information about $\Theta$, say a more concentrated distribution, our Bayes interval would have been much smaller. △

In order to have the flexibility we need in choosing a prior density, that is, in choosing a model that "fits" whatever prior information is available for our next example, we introduce the family of *Beta distributions*.

Recall that the Gamma function $\Gamma$ is defined by

$$\Gamma(\lambda) = \int_0^\infty u^{\lambda-1}e^{-u}\,du \equiv \lim_{A\to\infty}\int_0^A u^{\lambda-1}e^{-u}\,du \quad \text{for} \quad \lambda > 0.$$

Using integration by parts, we also know that for $n = 1, 2, 3, \ldots, \Gamma(n) = (n-1)!$.

**7.4 Definition: Beta Density.** The *Beta density with parameters* $\alpha$ *and* $\beta$ is given by the formula

$$f(y, \alpha, \beta) = \begin{cases} \dfrac{\Gamma(\alpha+\beta)}{\Gamma(\alpha)\Gamma(\beta)} y^{\alpha-1}(1-y)^{\beta-1} & \text{for} \quad 0 < y < 1 \\ 0 & \text{otherwise,} \end{cases}$$

where $\alpha > 0$ and $\beta > 0$. It can be shown that if $Y$ has such a Beta density, then

**7.5** $$E(Y) = \frac{\alpha}{\alpha+\beta} \quad \text{and} \quad \operatorname{Var} Y = \frac{\alpha\beta}{(\alpha+\beta)^2(\alpha+\beta+1)}.$$

Some tedious algebra shows that $E(Y)$ and Var $Y$ determine $\alpha$ and $\beta$; in fact, if $E(Y) = \mu$ and Var $Y = \sigma^2$, then

**7.6** $$\alpha = \mu[(1 - \mu)\mu/\sigma^2 - 1], \qquad \beta = (1 - \mu)[(1 - \mu)\mu/\sigma^2 - 1].$$

By suitably choosing $\mu$ and $\sigma^2$, we can control the "center" and "concentration about the center" of the prior distribution in [0, 1] that we use as a mathematical model of the information presented to us, using 7.5 to determine $\alpha$ and $\beta$ from $\mu$ and $\sigma^2$.

The Beta distribution is a bit unpleasant to handle exactly for large $\alpha$ or $\beta$. However, suppose that $Y_m$ has a Beta distribution with parameters $\alpha_m = mp$ and $\beta_m = mq$. Let

$$\mu_m = \frac{\alpha_m}{\alpha_m + \beta_m} = \frac{p}{p + q} = E(Y_m)$$

$$\sigma_m^2 = \frac{\alpha_m \beta_m}{(\alpha_m + \beta_m)^2(\alpha_m + \beta_m + 1)} = \frac{pq}{(p + q)^2 m\left(p + q + \frac{1}{m}\right)} = \text{Var } Y_m.$$

Then it can be shown (see Cramer 6) that

**7.7** $$\lim_{m \to \infty} P\left\{\frac{Y_m - \mu_m}{\sigma_m} \leq y\right\} = \int_{-\infty}^{y} \frac{1}{\sqrt{2\pi}} e^{-t^2/2}\, dt.$$

We therefore usually approximate the distribution of a Beta random variable with large $\alpha$ and $\beta$, where $\alpha/\beta$ and $\beta/\alpha$ are of the same order of magnitude, by the corresponding normal distribution, that is, by the distribution with mean and variance given by 7.5. (For legitimate use, some upper bound should, of course, be obtained for the error in this approximation.)

### 7.8 *Example: Estimating q for Transistors*

Suppose that every week a shipment of 10,000 transistors from the weekly production run at some factory is shipped to us. Furthermore, suppose that over the years the proportion of good units in our lots has been roughly uniform over the interval [.9, .95]. A new shipment arrives and we want a Bayes interval of level .99 for the week's probability $q$ that any given unit is a good one. Since the uniform distribution on [.9, .95] has a mean of .925 and a variance of 1/96,000, we will use a Beta distribution with this same mean and variance to describe the distribution of $Q$. (Note that $q$ is the value of $Q$.) To find the parameters of this Beta distribution, we use 7.6, which yields

**7.9** $$\alpha_{\text{prior}} \cong 6150, \qquad \beta_{\text{prior}} \cong 499.$$

Now suppose we test 1000 transistors from this lot. Let the number of good ones be denoted by $X$. Then the distribution of $X$, given $Q = q$, is binomial with parameters $n = 1000$ and $q$. By computations analogous to those leading to 7.1, we find

$$f_{Q|X}(q \mid x) = c_x q^{\alpha+x-1}(1 - q)^{\beta+n-x-1}.$$

Since $f_{Q|X}$ is a Beta density, we know that

$$c_x = \frac{\Gamma(\alpha + \beta + n)}{\Gamma(\alpha + x)\Gamma(\beta + n - x)}.$$

Thus, if in our sample of 1000 we observe $x$ good units, we see from Definition 7.4 that the conditional distribution of $Q$ is Beta with parameters $\alpha_{\text{post}} = \alpha_{\text{prior}} + x = 6150 + x$ and $\beta_{\text{post}} = \beta_{\text{prior}} + 1000 - x = 1499 - x$. Let us determine what our .99 level Bayes interval would be if our sample of 1000 had only 100 good units. Then $\alpha_{\text{post}} = 6250$ and $\beta_{\text{post}} = 1399$. Hence, by 7.5,

$$\mu_{\text{post}} = \frac{6250}{7649} \cong .819, \quad \sigma^2_{\text{post}} = \frac{1399 \times 6250}{7649^2(7650)} \cong \frac{1}{19{,}520} \cong \left(\frac{1}{140}\right)^2.$$

Thus, by Equation 7.7, the posterior distribution of $Q$ given $X = 100$ is approximately normal with mean .819 and standard deviation 1/140. From tables of the standard normal distribution, we find that the Bayes .99 level interval has the approximate value

$$\left[.819 - \frac{2.58}{140}, \quad .819 + \frac{2.58}{140}\right] = [.8005, .8375].$$

Of course, this indicates that something may have gone drastically wrong with this week's production. We might not believe in the model if we obtained such a result. △

We make some final remarks concerning Bayes intervals. First, as we saw in Example 7.8, it is wise not to stretch this model too far—it has an elastic limit. This arises from the fact that our prior distribution is at best a convenient approximation. Experience and good judgement are needed for proper use. Secondly, it should be noticed that the problem treated here is really one of probability, not statistics, because there is only one overall probability measure, namely, $f_{Q,X}$. However, it looks statistical because we are unable to observe the value of $Q$.

Finally, questions of a convenient and useful class of prior distributions are taken up in M. H. De Groot's *Optimal Statistical Decisions* (McGraw-Hill, 1970).

## 7.10 EXERCISES

1. In the speed trap example, 7.3, suppose that only one observation, 39 miles per hour, is made. Find a Bayes .95 level short interval for $\theta$.

2.° In Example 7.3 (speed trap), if only one observation, 36 miles per hour, has been made:

(a) find a Bayes .975 level upper limit $\theta_u$ for $\Theta$.

(b) Using the result just found and assuming a measurement on a passing car of 41 miles per hour, find a .95 level lower limit for the actual speed.
Hints: Let $Y$ be the measured speed and $s$ the true speed. Then $Y - s = X$, which

is normal, with mean $\Theta$ and variance 1. Hence, if

$$\int_{-\infty}^{\varphi_\beta} \frac{1}{\sqrt{2\pi}} e^{-t^2/2}\, dt = 1 - \beta,$$

then

$$P\{Y - s - \Theta \leq \varphi_\beta\} = 1 - \beta, \text{ that is, } P\{s \geq Y - \Theta - \varphi_\beta\} = 1 - \beta.$$

Thus, if $\Theta \leq \theta_u$, then

$$P\{s \geq Y - \theta_u - \varphi_\beta\} \geq 1 - \beta.$$

If $P\{\Theta \leq \Theta_u\} = 1 - \delta$, then

$$P\{s \geq Y - \Theta_u - \varphi_\beta\} \geq 1 - \beta - \delta$$

using $P(AB) \geq 1 - P(A^c) - P(B^c)$.

3. In Example 7.3, find $f_{\Theta|X_1,\ldots,X_n}(\theta \mid x_1, \ldots, x_n)$ if $f_\Theta$ is a normal density with mean 0 and variance $\sigma_0^2$.

4. In Example 7.8, (estimating $q$ for transistors), find the Bayes .99 level "short" intervals for $q$ when:
(a) $x = 990$.
(b)° $x = 900$.
(c) $x = 700$.

# 11

# Testing Hypotheses

## SECTION 1
## Introduction

In contrast to the problem of estimation, in which we are trying to find the value of a parameter, there are many situations in which we are forced to choose between two possible actions. For example, after the Salk vaccine was developed, extensive tests had to be made to see whether it would really prevent polio. The two possible actions were either recommending or discouraging the use of the vaccine. As another example, when air frames are built, the wing spars have to undergo some rugged strength tests. Those that break cannot be accepted for use. The two actions taken here are to accept a spar for further testing or to consign it to the scrap heap. A sample of the unbroken spars might then be tested to failure on a fatigue machine in order to decide whether to accept the remaining ones (action 1) or reject them (action 2). When taste tests are made on two brands of coffee, the two actions are the two possible preferences. If we test a pair of dice for "fairness," our two possible actions are to accept the dice as fair or to reject them as being loaded. In radar collision warning systems, the two possible actions are to sound an alarm and take evasive action or to continue on course—they might be labelled "alarm" and "no alarm."

We remark here that the value of the statistical theory we are about to develop lies in the aid it gives us in making decisions when the results are not obvious. This was the case when the Salk polio vaccine was being tested. The low incidence of polio in the population made it much more difficult to judge the effectiveness of any proposed vaccine. Using statistical methods, the sample size was kept down to the minimum needed, since even a small percentage increase in the sample size would have cost a great deal. (Less care is needed for a more prevalent disease.) Similarly, target-detecting radar is designed according to statistical principles in order to get the most information from limited data for making decisions. In World War II, such care was not necessary; the slow speed of aircraft at that time permitted enough visual observation of the radar screen to allow a correct decision in adequate time.

Although many testing situations can be handled from a confidence interval viewpoint (see the Remarks at the end of this chapter), the whole area is significant enough to warrant its own theory. The simplest problem arises when we observe data assumed to have arisen from one of two completely specified distributions—that is, when we are only willing to describe our measurements in one of two specified ways. Our problem here is to determine which of the two distributions is preferred as a model for the observed data. More generally, it is often assumed that the data is governed by a distribution from one of two specified disjoint classes of distributions. In this case, our object is to use the data to try to determine which class contains the governing distribution. In other words, we want to decide which class contains the most appropriate description of our data.

In these problems, the statement that the governing distribution belongs to a given one of the two classes of distributions is referred to as a hypothesis. To distinguish between these hypotheses, we refer to one of these statements as the *null hypothesis* and to the other as the *alternative* or *alternative hypothesis*, although the particular choice is often not meaningful. When a hypothesis specifies a single distribution, it is called *simple*, otherwise it is called *composite*.

In the next two sections, we will discuss two simple hypotheses. Subsequently, we will consider more complicated testing situations. A rather considerable preliminary buildup of concepts is needed before we can clearly phrase and solve some of the problems of hypothesis tests.

## SECTION 2
## Simple Hypothesis versus Simple Alternative

In this section we assume that our observations are governed by one of two known distributions. We want to use the actual observed data to decide which distribution furnishes the better description. For example, suppose that we determine from a preliminary diagnosis that an individual has one of two given diseases, $d_0$ or $d_1$. Further clinical tests are made, and the outcome of these tests are the values of random variables $X_1, \ldots, X_n$. Assume that we have concluded that the distribution of $X_1, \ldots, X_n$ is $P_i$ when the individual has disease $d_i$, $i = 0, 1$. We would like to derive a good procedure for distinguishing between these diseases. In terms of our model, we are trying to determine which of the two distributions, $P_0$ or $P_1$, actually is the distribution of $X_1, \ldots, X_n$, based on the observed values $\mathsf{x}_1, \ldots, \mathsf{x}_n$ of these random variables. We may let our sample space $\mathcal{S}$ consist of all sequences $(x_1, \ldots, x_n)$ of possible values of $X_1, \ldots, X_n$. In order to see how best to use the data, we must first consider the possible consequences of any procedure for discriminating between these two hypotheses. Within the framework of our present model, there are two possible wrong decisions that can be made.

**2.1 Definition: Errors of Type I and Type II.** Asserting that the distribution $P_1$ governs $X_1, \ldots, X_n$ when in fact $X_1, \ldots, X_n$ are distributed according to $P_0$ is called an *error of type I*. On the other hand, asserting that $P_0$ governs the random variables $X_1, \ldots, X_n$ when $P_1$ is the true distribution is called an *error of type II*.

*Ideally, we would like both types of error to have probability* 0. *Except in trivial cases, we shall have to be content with keeping these two error probabilities at an acceptably low level.*†

In order to accomplish this goal, we observe the data and, based on what we see, try to draw a sensible conclusion as to which distribution better explains our observations. That is, for each possible set of data $x_1, \ldots, x_n$ we must make a decision whether $P_0$ or $P_1$ furnishes the better explanation for this data. This leads us to

**2.2 Definition: Test of Hypothesis.** A test of the hypothesis $H_0$ (that $P_0$ governs the data) against the alternative $H_1$ (that $P_1$ governs the data) is a partition of the sample space $\mathcal{S}$ into the union of two disjoint sets, $S_0$ and $S_1$. If the data $(x_1, \ldots, x_n) \in S_0$, then we accept $H_0$; if $(x_1, \ldots, x_n) \in S_1$ then we accept $H_1$. The set $S_0$ is called the *acceptance region for* $H_0$; $S_1$ is called the *rejection region for* $H_0$. The hypothesis $H_0$ is often called the null hypothesis.

In the older statistical literature, and indeed even among many statisticians today, the null hypothesis $H_0$ occupied a favored position. It was the hypothesis that the scientist was trying to prove, and as such should only be rejected in the face of overwhelming evidence against it. Our view here is more flexible. If it is reasonable to be especially conservative and favor one of the two hypotheses, we shall do so explicitly in setting up the test. We shall see how to do this when we discuss the choice of sample size.

***2.3 Example: Fair Dice Test***

Suppose that we have just purchased a pair of allegedly fair dice, but have reason to suspect that in fact each die is biased toward high values, with

**2.4**
$$p_1 = p_2 = p_3 = 2/15,$$
$$p_4 = p_5 = p_6 = 1/5.$$

We therefore want to test the hypothesis $H_0$—that the dice are fair, $p_i = 1/6$—against the hypothesis $H_1$—that the dice turn up as in 2.4.

Let $X$ be the result of the toss of a single die. Under $H_0$, we have $E(X) = 3.5$, whereas under $H_1$ we see that $E(X) = 3.8$. We therefore make 300 tosses of each die, for a total of 600 tosses, letting $X_i$ be the result of the $i$th toss, $i = 1, \ldots, 600$. Note that $\frac{1}{600}\sum_{i=1}^{600} X_i$ is likely to be close to $E(X)$. Therefore, it is sensible to accept $H_0$ if $\frac{1}{600}\sum_{i=1}^{600} x_i \leq 3.6$ and to accept $H_1$ if $\frac{1}{600}\sum_{i=1}^{600} x_i > 3.6$, where $x_i$ is the observed value of $X_i$. As our sample space, $\mathcal{S}$, it is natural to choose the set of all sequences $(x_1, \ldots, x_{600})$ with $x_i \in \{1, 2, 3, 4, 5, 6\}$. Note that in terms of $\mathcal{S}$ the random variables

† In this regard we mention a device for detecting firearms on airplane passengers, which was claimed to be almost 100 per cent effective. Not so much publicized was the fact that this clever device had an 87 per cent false alarm rate—that is, the alarm rang for 87 per cent of people who were unarmed. In our terminology, we could say that although the probability of type I error was low, the probability of type II error was unacceptably high.

$X_i$ are the coordinate variables defined by $X_i(x_1, \ldots, x_{600}) = x_i$. Furthermore,

$$S_0 = \left\{(x_1, \ldots, x_{600}) \in \mathcal{S}: \frac{1}{600}\sum_{i=1}^{600} x_i < 3.6\right\} = \left\{\frac{1}{600}\sum_{i=1}^{600} X_i \leq 3.6\right\};$$

$$S_1 = \left\{(x_1, \ldots, x_{600}) \in \mathcal{S}: \frac{1}{600}\sum_{i=1}^{600} x_i > 3.6\right\} = \left\{\frac{1}{600}\sum_{i=1}^{600} X_i > 3.6\right\}.$$

Let $P_i$ denote the probability distribution of $X_1, \ldots, X_{600}$ under $H_i$. Then the probability of type I error is simply the probability of saying the die is biased when it is fair. This translates into

$$P_0\{S_1\} = \underset{\substack{\text{distribution}\\\text{when die is}\\\text{fair}}}{P_0}\underbrace{\left\{\frac{1}{600}\sum_{i=1}^{600} X_i > 3.6\right\}}_{\substack{\text{saying die}\\\text{is biased}}}.$$

Under $H_0$, we calculate that

$$E(X_i) = 3.5 \qquad \text{and} \qquad \operatorname{Var} X_i = 105/36.$$

Hence by the Central Limit Theorem (Theorem 3.1, Chapter 7),

$$\begin{aligned}
P_0\left\{\frac{1}{600}\sum_{i=1}^{600} X_i > 3.6\right\} &= P_0\left\{\frac{1}{600}\sum_{i=1}^{600}(X_i - 3.5) > .1\right\} \\
&= P_0\left\{\frac{1}{\sqrt{600}}\sum_{i=1}^{600}\frac{X_i - 3.5}{\sqrt{105/36}} > \frac{.1\sqrt{600}}{\sqrt{105/36}}\right\} \\
&\cong P_0\left\{\frac{1}{\sqrt{600}}\sum_{i=1}^{600}\frac{X_i - 3.5}{\sqrt{105/36}} > 1.432\right\} \\
&\cong \int_{1.432}^{\infty}\frac{1}{\sqrt{2\pi}}e^{-t^2/2}\,dt \\
&\cong .0761.
\end{aligned}$$

Hence, the probability of type I error is about .0761. Under $H_1$, $E(X_i) = 3.8$ and $\operatorname{Var} X_i = 636/225$. The probability of type II error is (by the Central Limit Theorem)

$$\begin{aligned}
P_1\left\{\frac{1}{600}\sum_{i=1}^{600} X_i \leq 3.6\right\} &= P_1\left\{\frac{1}{600}\sum_{i=1}^{600}(X_i - 3.8) \leq -.2\right\} \\
&= P_1\left\{\frac{1}{\sqrt{600}}\sum_{i=1}^{600}\frac{X_i - 3.8}{\sqrt{636/225}} \leq \frac{-.2\sqrt{600}}{\sqrt{636/225}}\right\} \\
&\cong P_1\left\{\frac{1}{\sqrt{600}}\sum_{i=1}^{600}\frac{X_i - 3.8}{\sqrt{636/225}} \leq -2.92\right\} \\
&\cong \int_{-\infty}^{-2.92}\frac{1}{\sqrt{2\pi}}e^{-t^2/2}\,dt \\
&\cong .0018.
\end{aligned}$$

So, our intuitive test procedure appears to work fairly well. But *does it work as well as it could—that is, is there a way to handle the data to obtain even smaller error probabilities?* As we shall see in Example 2.27, there is a way to improve our results with no increase in the amount of data. After we develop Lemma 2.9, it will be obvious how to accomplish this improvement. △

### 2.5 *Example: Meter Test*

Suppose that $X_1, \ldots, X_n$ are random variables whose values are the measurement errors made by some meter. The claim may have been made that the meter is biased in the positive direction. As a model, we assume that $X_1, \ldots, X_n$ are independent normal random variables with variance 1 and $E(X_i)$ all equal either to 0 or to 1. The normality assumption might arise from knowledge that the error is a sum of many small independent errors, to which the Central Limit Theorem might apply. Such a model is, of course, a simplification, as is just about every model. It is likely to be adequate if it is not important to distinguish biases smaller than 1. We might accept $H_0$: $E(X_i) = 0$ if

$$\frac{1}{n}\sum_{i=1}^{n} \mathsf{x}_i \leq \tfrac{1}{2},$$

and accept $H_1$: $E(X_i) = 1$ if

$$\frac{1}{n}\sum_{i=1}^{n} \mathsf{x}_i > \tfrac{1}{2},$$

where $\mathsf{x}_i$ is the observed value of $X_i$. If we choose as the sample space $\mathcal{S}$ the set of all sequences $(x_1, \ldots, x_n)$ of real numbers, then, as in the previous example, $X_i(x_1, \ldots, x_n) = x_i$. In that case,

$$S_0 = \left\{(x_1, \ldots, x_n) \in \mathcal{S}\colon \frac{1}{n}\sum_{i=1}^{n} x_i \leq 1/2\right\} = \left\{\frac{1}{n}\sum_{i=1}^{n} X_i \leq 1/2\right\}$$

and

$$S_1 = \left\{(x_1, \ldots, x_n) \in \mathcal{S}\colon \frac{1}{n}\sum_{i=1}^{n} x_i > 1/2\right\} = \left\{\frac{1}{n}\sum_{i=1}^{n} X_i > 1/2\right\}.$$

Note that to carry out this test, we observe the outcome $(\mathsf{x}_1, \ldots, \mathsf{x}_n)$ and decide whether

$$(\mathsf{x}_1, \ldots, \mathsf{x}_n) \in S_0 \qquad \left(\text{if } \frac{1}{n}\sum_{i=1}^{n} \mathsf{x}_i \leq 1/2\right),$$

in which case we accept the hypothesis $H_0$, that the meter is unbiased, *or*

$$(\mathsf{x}_1, \ldots, \mathsf{x}_n) \in S_1 \qquad \left(\text{if } \frac{1}{n}\sum_{i=1}^{n} \mathsf{x}_i > 1/2\right),$$

in which case we accept the hypothesis $H_1$, that the meter is biased.

The probabilities of the two types of error furnish the criteria to determine whether or not this test is satisfactory. Type I error here consists of classifying an unbiased meter (governed by $P_0$) as biased. Hence, the probability of type I error is

$$\underbrace{P_0}_{E(X_i)=0} \underbrace{\{S_1\}}_{\text{sample points that classify the meter as biased}}.$$

Similarly, the probability of type II error is

$$\underbrace{P_1}_{E(X_i)=1} \underbrace{\{S_0\}}_{\text{sample points that classify the meter as unbiased}}.$$

But here $P_0\{S_1\}$ is the probability that

$$\frac{1}{n}\sum_{i=1}^{n} X_i > \tfrac{1}{2}$$

when $X_i$ are independent standard normal random variables. $P_1\{S_0\}$ is the probability that

$$\frac{1}{n}\sum_{i=1}^{n} X_i \leq \tfrac{1}{2}$$

when $X_i$ are independent normal random variables with $E(X_i) = 1$ and unit variance. Probabilities such as these have been computed many times earlier in the text. △

***Let us briefly summarize our development thus far:*** We suppose that our observed data $x_1, \ldots, x_n$ are the values of random variables $X_1, \ldots, X_n$. We assume these random variables to be distributed according to one of two known distributions, $P_0$ or $P_1$. The hypothesis $H_i$ specifies that $P_i$ is the governing distribution, $i = 0, 1$. We let the sample space consist of sequences $(x_1, \ldots, x_n)$ of possible values of $(X_1, \ldots, X_n)$. To choose between these hypotheses, we divide the sample space into two exclusive sets, $S_0$ and $S_1$. If the observed data $(x_1, \ldots, x_n)$ fall in $S_0$, we accept $H_0$. If $(x_1, \ldots, x_n)$ fall in $S_1$, we accept $H_1$.

To judge the merit of the test, we examine the error probabilities: (1) *probability of type I error* $= P_0(S_1) = P_0\{(X_1, \ldots, X_n) \in S_1\}$, the probability of choosing $H_1$ when $H_0$ holds; and (2) *probability of type II error* $= P_1(S_0) = P_1\{(X_1, \ldots, X_n) \in S_0\}$, the probability of choosing $H_0$ when $H_1$ holds.

---

## *2.6 EXERCISES*

1. Let $X_1, X_2$ represent independent measurements on the idling engine temperature of a car. A new type of thermostat has been installed and the mean temperature it allows has been adjusted to be 130 degrees. The important characteristic by which the thermostat is judged is the variance $\sigma^2$ it allows. To determine if the new unit is adequate, we want to test the hypothesis

$$H_0, \quad \text{that} \quad \sigma = 20$$

against the hypothesis

$$H_1 \quad \text{that} \quad \sigma = 30.$$

We will do this by rejecting $H_0$ if

$$\sqrt{(x_1 - 130)^2 + (x_2 - 130)^2} > 30.$$

(a)° Pick a suitable sample space $\mathcal{S}$ for the experiment.
(b)° Describe $S_0$ in terms of $\mathcal{S}$ and graph $S_0$.
(c)° Find the probability of type I error.
(d) Find the probability of type II error.

2. A pair of independent phase angle measurements $\Theta_1$, $\Theta_2$ are known to be either
   (i) uniform on [0, 1]

or

   (ii) uniform on [0, 1.4].

In order to determine their source, we want to find whether (i) or (ii) holds. Let $H_0$ be the hypothesis that they come from source (i) and $H_1$ the hypothesis that they come from source (ii). In order to test $H_0$ against $H_1$, we reject $H_0$ if and only if

$$\max(\theta_1, \theta_2) > .9, \text{ where } \theta_i \text{ is the observed value of } \Theta_i.$$

(a) Pick a suitable sample space $\mathcal{S}$ for the experiment.
(b)° Describe $S_0$ in terms of $\mathcal{S}$ and graph $S_0$
(c)° Find the probability of type I error
(d) Find the probability of type II error.

Now that we have some idea of what is meant by a test of hypotheses, and have seen some criteria for judging the merit of a test, we can focus on the last preliminaries to our main development.

In many testing situations, there are certain natural constraints that greatly simplify construction of a test with the desired properties. The set of constraints that we first consider arises if we are given the sample size $n$ and are forced to accept an upper bound $\alpha$ for the probability of type I error. There are many examples of such restrictions. They might be economic, such as budgetary limitations on the money available for a public opinion poll, or practical—in the case of medical diagnosis, there may only be $n$ clinical tests that are of any proven use. In attempting to set up an aircraft collision warning system, only a specified amount of data can be observed before a decision whether or not to take evasive action must be made. Or, it might be that if more than a certain proportion, $\alpha$, of people who actually have disease $d_0$ are mistakenly classified (and hence treated) as having disease $d_1$, a dangerous epidemic is likely to develop. A salesman is sometimes required to guarantee that the probability of acceptance of a bad lot of the units he sells be suitably small. To ensure this, the customer may require a test on a sample from the lot with the property that the probability of accepting an unsuitable lot (type I error) be less than or equal to a given $\alpha$. Note that this does not guarantee good lots—it just ensures small probability of acceptance of bad lots. So if all lots are bad, most will be rejected.†

---

† We shall see later (when we discuss Bayes procedures) that we can sometimes give good assurance of high quality of accepted lots if we have prior information about the proportion of "good" lots.

We are now in a position to handle our first problem, namely ***the construction of best tests when the sample size and an upper bound on the probability of type I error are specified.***

In order to simplify the problem, we shall assume that the random variables $X_1, \ldots, X_n$ to be observed are governed by the known density $f_0$ or by the known density $f_1$. Let $x_1, \ldots, x_n$ denote the observed values of $X_1, \ldots, X_n$. Certainly, if $f_1(x_1, \ldots, x_n)$ is much greater than $f_0(x_1, \ldots, x_n)$, we should say that $f_1$ "explains" the data better than does $f_0$. This intuitive reasoning leads us to the basic definition that we are about to introduce.

Let $P_i$ refer to probabilities computed under the assumption that $f_i$ governs the data.

**2.7 Definition: Likelihood Ratio Test of Size α of a Simple Hypothesis.** Suppose there is a value $c_\alpha$ such that

$$P_0\{f_1(X_1, \ldots, X_n) > c_\alpha f_0(X_1, \ldots, X_n)\} = \alpha, \qquad 0 < \alpha < 1.$$

Then the *likelihood ratio test (L.R. test) of size* α, *of the simple hypothesis*

$$H_0\text{: } f_0 \text{ is the true density}$$

*against the hypothesis*

$$H_1\text{: } f_1 \text{ is the true density,}$$

is the procedure:
Accept $H_1$ if and only if $f_1(x_1, \ldots, x_n) > c_\alpha f_0(x_1, \ldots, x_n)$, where $x_1, \ldots, x_n$ are the observed values of $X_1, \ldots, X_n$.†

All the definitions just given extend to the discrete case if we replace $f_i(x_1, \ldots, x_n)$ with the value $p_i(x_1, \ldots, x_n)$ of the discrete probability function.

Note that in terms of Definition 2.2 regarding a test of hypothesis, we have, as the L.R. test of size α,

**2.8**
$$S_1 = \{(x_1, \ldots, x_n) \in \mathcal{S}\text{: } f_1(x_1, \ldots, x_n) > c_\alpha f_0(x_1, \ldots, x_n)\};$$
$$S_0 = \{(x_1, \ldots, x_n) \in \mathcal{S}\text{: } f_1(x_1, \ldots, x_n) \leq c_\alpha f_0(x_1, \ldots, x_n)\}.$$

We now show that ***in a very meaningful sense, the L.R. tests are the best possible ones.***

### *2.9 Theorem: Fundamental Lemma of Neyman and Pearson*

1. The likelihood ratio test of size α minimizes the probability of type II error among all tests based on $X_1, \ldots, X_n$ whose probability of type I error does not exceed α.

2. If there is a likelihood ratio test of size α, it is essentially the only test with the property specified in 1. That is, any other test whose probability of type I error is $\leq\alpha$

† This terminology arises by defining $L(x_1, \ldots, x_n) = \dfrac{f_1(x_1, \ldots, x_n)}{f_0(x_1, \ldots, x_n)}$ as the *likelihood ratio* provided that this expression is meaningful. Then the L.R. test of size α is equivalent to accepting $H_1$ if and only if the likelihood ratio $L(x_1, \ldots, x_n) > c_\alpha$.

and whose probability of type II error does not exceed that of the L.R. test of size $\alpha$ makes the same decision as the L.R. test with probability 1 under both $f_0$ and $f_1$.

These two assertions hold for the discrete case also.

For those who do not care to go into the details of the more precise proof to follow, we offer the following arguments, which we hope are convincing.

Imagine building up the rejection region $S_0^c$ for $H_0$ by successively adding in small $n$-dimensional intervals.† At any given stage, suppose that we look at any two small intervals $[\mathbf{x}; \mathbf{x} + d\mathbf{x}]$ and $[\mathbf{x}^*; \mathbf{x}^* + d\mathbf{x}^*]$ chosen so that $f_0(\mathbf{x})\, d\mathbf{x} = f_0(\mathbf{x}^*)\, d\mathbf{x}^*$ (where the sum of this small probability and the probability of type I error already built up does not exceed $\alpha$). As far as the type I error is concerned, it does not matter which of these two intervals is assigned to the rejection region $S_0^c$ for $H_0$, since they both contribute the same amount to the probability of type I error. Thus, we should choose to add to $S_0^c$ the interval that adds the most probability of making a correct decision when $f = f_1$, that is, to $P_1(S_0^c)$. Thus, we choose $[\mathbf{x}; \mathbf{x} + d\mathbf{x}]$ over $[\mathbf{x}^*; \mathbf{x}^* + d\mathbf{x}^*]$ if

$$f_1(\mathbf{x})\, d\mathbf{x} > f_1(\mathbf{x}^*)\, d\mathbf{x}^*.$$

Since

$$f_0(\mathbf{x})\, d\mathbf{x} = f_0(\mathbf{x}^*)\, d\mathbf{x}^*,$$

this means that we prefer $[\mathbf{x}, \mathbf{x} + d\mathbf{x}]$ if

$$\frac{f_1(\mathbf{x})}{f_0(\mathbf{x})} > \frac{f_1(\mathbf{x}^*)}{f_0(\mathbf{x}^*)},$$

that is, we prefer to add $S_0^c$, the small region that is almost a point and has the greater likelihood ratio. We now present the precise proof. Since the proof is quite technical, we recommend that most readers skip it and proceed to Example 2.19.

*Proof:* We shall prove Theorem 2.9(1) for densities (the discrete case being similar). We haven't got the tools available to prove (2) for the density case (although this can be done with the aid of certain measure theoretic results). Thus, we will be content to prove (2) for the discrete case only.

The test accepts $H_1$ if the observed outcome $(x_1, \ldots, x_n)$ satisfies $f_1(x_1, \ldots, x_n) > c_\alpha f_0(x_1, \ldots, x_n)$. Type I error means accepting $H_1$ when we should accept $H_0$; that is, accepting $H_1$ when $f_0$ governs the data. Hence, the probability of type I error is $P_0\{f_1(X_1, \ldots, X_n) > c_\alpha f_0(X_1, \ldots, X_n)\}$. (The subscript "0" is what makes this an *error* probability, since $P_1\{f_1(X_1, \ldots, X_n) > c_\alpha f_0(X_1, \ldots, X_n)\}$ is the probability of a correct action.) By Definition 2.7, the likelihood ratio test has probability of type I error equal to $\alpha$.

Now let us look at any competing test, $T'$, whose probability of type I error does not exceed $\alpha$. Suppose that we denote the acceptance and rejection regions for $T'$ by $S_0'$ and $S_1'$, respectively. Since the L.R. test has probability of

† Here we are using the abbreviations $\mathbf{x} = (x_1, \cdots, x_n)$, $d\mathbf{x} = dx_1 \ldots dx_n$, $[\mathbf{x}; \mathbf{x} + d\mathbf{x}] = \{(y_1, \ldots, y_n) \in R^n\colon \text{ for all } i,\ x_i \leq y_i \leq x_i + dx_i\}$. This set is called an $n$-dimensional interval. In two dimensions it is a rectangle with sides parallel to the coordinate axes. Note that each interval has two possible probability assignments, one from $f_0$ and the other from $f_1$.

type I error equal to $\alpha$, we have

$$\int_{S_1} \cdots \int f_0(x_1, \ldots, x_n)\, dx_1 \cdots dx_n = \alpha.$$

Similarly, since $T'$ has probability of type I error not exceeding $\alpha$, we have

$$\int_{S_1'} \cdots \int f_0(x_1, \ldots, x_n)\, dx_1 \cdots dx_n \leq \alpha.$$

For the sake of more readable notation, we let $f_i(x_1, \ldots, x_n)$ be denoted by $f_i(\mathbf{x})$ and $dx_1 \cdots dx_n$ by $d\mathbf{x}$. Finally, we use only a single $\int$ sign even for multiple integrals. Then the previous pair of relations becomes

**2.10**
$$\int_{S_1} f_0(\mathbf{x})\, d\mathbf{x} = \alpha$$

and

**2.11**
$$\int_{S_1'} f_0(\mathbf{x})\, d\mathbf{x} \leq \alpha.$$

In the following computations, we make use of the fact that an integral over a disjoint union is the sum of the integrals over the sets forming this union. We write

$$S_1 = (S_1 \cap S_1') \cup (S_1 \cap S_0')$$
$$S_1' = (S_1 \cap S_1') \cup (S_0 \cap S_1').$$

Subtracting $\int_{S_1 \cap S_1'} f_0(\mathbf{x})\, d\mathbf{x}$ from both 2.10 and 2.11 it follows that

**2.12**
$$\int_{S_0 \cap S_1'} f_0(\mathbf{x})\, d\mathbf{x} \leq \int_{S_0' \cap S_1} f_0(\mathbf{x})\, dx.$$

Now, the probability of type II error for the L.R. test is

$$\begin{aligned}
\int_{S_0} f_1(\mathbf{x})\, d\mathbf{x} &= \int_{S_0 \cap S_0'} f_1(\mathbf{x})\, d\mathbf{x} + \int_{S_0 \cap S_1'} f_1(\mathbf{x})\, d\mathbf{x} \\
&\leq \int_{S_0 \cap S_0'} f_1(\mathbf{x})\, d\mathbf{x} + \int_{S_0 \cap S_1'} c_\alpha f_0(\mathbf{x})\, d\mathbf{x} \qquad \text{(by 2.8)} \\
&\leq \int_{S_0 \cap S_0'} f_1(\mathbf{x})\, d\mathbf{x} + \int_{S_0' \cap S_1} c_\alpha f_0(\mathbf{x})\, d\mathbf{x} \qquad \text{(by 2.12)} \\
&\leq \int_{S_0 \cap S_0'} f_1(\mathbf{x})\, d\mathbf{x} + \int_{S_0' \cap S_1} f_1(\mathbf{x})\, d\mathbf{x} \qquad \text{(by 2.8)} \\
&= \int_{S_0'} f_1(\mathbf{x})\, d\mathbf{x}.
\end{aligned}$$

Since the last expression is the probability of type II error for the test $T'$, we have shown that

**2.13** $$P\{\text{type II error for L.R. test of size } \alpha\} \leq P\{\text{type II error for test } T'\}$$

where $T'$ is any test with probability of type I error not exceeding $\alpha$. The proof for the discrete case is almost the same. Simply replace any expression of the form

$$\int_A f_i(\mathbf{x})\, d\mathbf{x}$$

with

$$\sum_{\mathbf{x} \in A} p_i(\mathbf{x}).$$

To prove Theorem 2.9(2) for the discrete case, we must show that if $T'$ is "as good as" the L.R. test, then

**2.14** $$0 = P_0\{S_0 \cap S_1'\} = P_0\{S_0' \cap S_1\} = P_1\{S_0 \cap S_1'\} = P_1\{S_0' \cap S_1\}.$$

That is, we must prove that under both hypotheses, the probability of $T'$ taking action different from the L.R. test is 0. Asserting that $T'$ is at least as good as the L.R. test of size $\alpha$ is to say that

**2.15** $$P_{T'}\,\{\text{type I error}\} \leq \alpha,$$

and that equality holds in Statement 2.13. Examining the proof of 2.13 shows that equality holds there if and only if all of the following hold:

**2.16** $$\sum_{\mathbf{x} \in S_0 \cap S_1'} p_1(\mathbf{x}) = c_\alpha \sum_{\mathbf{x} \in S_0 \cap S_1'} p_0(\mathbf{x})$$

**2.17** $$c_\alpha \sum_{\mathbf{x} \in S_0 \cap S_1'} p_0(\mathbf{x}) = c_\alpha \sum_{\mathbf{x} \in S_0' \cap S_1} p_0(\mathbf{x})$$

**2.18** $$\sum_{\mathbf{x} \in S_0' \cap S_1} c_\alpha p_0(\mathbf{x}) = \sum_{\mathbf{x} \in S_0' \cap S_1} p_1(\mathbf{x}).$$

Now we note that we must have $c_\alpha \geq 0$, since $c_\alpha < 0$ would cause violation of 2.7 (because $0 < \alpha < 1$). We consider first the case $c_\alpha = 0$. In this case, $S_0 = \{\mathbf{x} \in \mathcal{S}\colon p_1(\mathbf{x}) = 0\}$, and hence $P_1\{S_0 \cap S_1'\} = 0$. By 2.18, we also have $P_1\{S_0' \cap S_1\} = 0$. Now, by 2.12, we know that

$$P_0\{S_0 \cap S_1'\} \leq P_0\{S_0' \cap S_1\},$$

so that to complete the case $c_\alpha = 0$, we need only show that $P_0\{S_0' \cap S_1\} = 0$. Suppose $P_0\{S_0' \cap S_1\} \neq 0$. From this, it follows that $S_1'$ must *fail* to include at least one point of $S_1$. That is, $S_1'$ fails to include at least one point $\mathbf{x}$ such that

$p_1(\mathbf{x}) > 0$.† Hence, $P_1\{S_1'\} < 1$. But since $S_1 = \{\mathbf{x} \in S: p_1(\mathbf{x}) > 0\}$, we have $P_1\{S_1\} = 1$. Thus, we see that if $P_0\{S_0' \cap S_1\} \neq 0$, then $P_1\{S_1'\} < P_1\{S_1\}$. This inequality implies that it is false that $T'$ is at least as good as the L.R. test. Thus, we have shown that if $T'$ is at least as good as the L.R. test, then 2.14 holds for the $c_\alpha = 0$. Finally, we handle the case $c_\alpha > 0$. Now, since $S_1 = \{\mathbf{x} \in \mathcal{S}: p_1(\mathbf{x}) > c_\alpha p_0(\mathbf{x})\}$, we see that 2.18 can only hold if there aren't any points in $S_0' \cap S_1$. But this implies that $P_1\{S_0' \cap S_1\} = P_0\{S_0' \cap S_1\} = 0$. Using this last equality, 2.17, and the fact that $c_\alpha > 0$ yields $P_0\{S_0 \cap S_1'\} = 0$. Now, it follows from 2.16 that $P_1\{S_0 \cap S_1'\} = 0$. Thus, we have shown that if $T'$ is at least as good as the L.R. test, then 2.14 holds for the case $c_\alpha > 0$, proving part (2) of Theorem 2.9.

### 2.19 *Example: Anthropology*

On an archaeological dig, a collection of skeletons is discovered. This collection is known to have been from one of two given races. On each skeleton a series of measurements are made, and then combined into one "summary" measurement. Let $X_i$ denote the summary measurement on the $i$th skeleton and suppose that if the skeletons are from race $j$, where $j = 0$ or 1, all $i$, then the $X_i$ are independent normal, with $E(X_i) = \mu_j$, Var $X_i = \sigma^2$. The normality assumption is being used here on the grounds that everybody believes it, so the journal to which the results will be sent is more likely to accept them. Suppose we have 100 skeletons and want the probability of classifying them as being from race 1 when they are in fact from race 0 equal to .05. If $\mu_0 = 10$, $\mu_1 = 11.2$, and $\sigma^2 = 49$, the likelihood ratio

$$\frac{f_1(x_1, \ldots, x_n)}{f_0(x_1, \ldots, x\ )} = \frac{\dfrac{1}{(2\pi\sigma^2)^{50}} \exp\left[-\dfrac{1}{2\cdot 49}\sum_{i=1}^{100}(x_i - 11.2)^2\right]}{\dfrac{1}{(2\pi\sigma^2)^{50}} \exp\left[-\dfrac{1}{2\cdot 49}\sum_{i=1}^{100}(x_i - 10)^2\right]}$$

$$= \exp\left[-\frac{1}{2\cdot 49}\sum_{i=1}^{100}[(x_i - 11.2)^2 - (x_i - 10)^2]\right]$$

$$= \exp\left[-\frac{1}{2\cdot 49}\sum_{i=1}^{100}[-2.4x_i + 25.44]\right]$$

$$= \exp\left[-\frac{2544}{98}\right]\exp\left[\frac{1.2}{49}\sum_{i=1}^{100}x_i\right].$$

† For the density case, we may assume $\mathcal{S} = \{\mathbf{x}: f_0(\mathbf{x}) > 0 \text{ or } f_1(\mathbf{x}) > 0\}$. The only measure theoretic result that is needed to let us prove Theorem 2.9(2) for densities is the following:
If $f_1(\mathbf{x}) > 0$ for all $\mathbf{x} \in A$ and

$$\int_{B\cap A} f_0(\mathbf{x})\, d\mathbf{x} > 0 \text{ (that is, } P_0\{B \cap A\} > 0),$$

then

$$\int_{B\cap A} f_1(\mathbf{x})\, d\mathbf{x} > 0 \text{ (that is, } P_1\{B \cap A\} > 0).$$

*Thus, the likelihood ratio test is to reject $H_0$ if*

$$e^{-(2544/98)}\exp\left[\frac{1.2}{49}\sum_{i=1}^{100} x_i\right] > c_{.05}$$

where $c_{.05}$ is chosen so that

$$P_0\left\{e^{-(2544/98)}\exp\left(\frac{1.2}{49}\sum_{i=1}^{100} X_i\right) > c_{.05}\right\} = .05.$$

Rather than compute $c_{.05}$, we rewrite the event "reject $H_0$" $= S_1 =$

$$\left\{e^{-(2544/98)}\exp\left(\frac{1\cdot 2}{49}\sum_{i=1}^{100} X_i\right) > c_{.05}\right\}$$

as

$$\left\{\frac{1\cdot 2}{49}\sum_{i=1}^{100} X_i > \log[c_{.05}e^{(2544/98)}]\right\}.$$

We write this in the form

$$\left\{\frac{1}{10}\sum_{i=1}^{100}\frac{X_i - 10}{7} > \frac{49}{7\cdot 12}\log[c_{.05}e^{(2544/98)}] - 100/7\right\}.$$

Call

$$\frac{49}{7\cdot 12}\log[c_{.05}e^{(2544/98)}] - \frac{100}{7} = d_{.05}.$$

*Then the event "reject $H_0$" is*

$$\left\{\frac{1}{10}\sum_{i=1}^{100}\frac{X_i - 10}{7} > d_{.05}\right\}.$$

Rather than finding $c_{.05}$ so that the probability of type I error is .05, it is simpler to choose $d_{.05}$ to accomplish this, since

$$\frac{1}{10}\sum_{i=1}^{100}\frac{X_i - 10}{7}$$

has a standard normal distribution when $H_0$ is true. Thus, our test is to reject $H_0$ if

$$\frac{1}{10}\sum_{i=1}^{100}\frac{x_i - 10}{7} \geq 1.645;$$

the value 1.645 is obtained from a table of the standard normal distribution.

Under these circumstances the probability of type II error is obtained as follows: Type II error consists of accepting $H_0$ when we should accept $H_1$, that is, of accepting

$H_0$ when $f_1$ is the true density. Type II error is the event

$$\frac{1}{10}\sum_{i=1}^{100}\frac{X_i - 10}{7} \leq 1.645$$

where $X_1, \ldots, X_n$ have density $f_1$. We rewrite this event as

$$\left\{\frac{1}{10}\sum_{i=1}^{100}\frac{X_i - 11.2}{7} \leq 1.645 - \frac{1}{10}\sum_{i=1}^{100}\frac{1.2}{7}\right\},$$

since

$$\frac{1}{10}\sum_{i=1}^{100}\frac{X_i - 11.2}{7}$$

has a standard normal distribution under $H_1$. Because

$$1.645 - \frac{1}{10}\sum_{i=1}^{100}\frac{1.2}{7} = 1.645 - {}^{120}\!/_{70} = -.069,$$

we see that in this case the probability of type II error is .472. This may be unacceptably high, so that it might be preferable to allow a higher probability of type I error or to find a few more skeletons. △

A few comments on this example are in order. First, since $\mu$ admits a sufficient statistic when $\sigma^2$ is known, we could have used the simpler likelihood ratio of the sufficient statistic to derive this test. In fact, as we shall show in Exercise 5.10(9), the L.R. is a function of every sufficient statistic. Thus, by Theorem 2.9(2) we know that the exact same size $\alpha$ test is obtained no matter whether we use the likelihood ratio principle on the sufficient statistic or on the raw data.

Second, we should realize that because of the way in which we have formulated this testing problem, our conclusions do not admit any assertions concerning the probability that the true density is $f_0$. Such assertions could be made under a different model—one in which in addition to the data, we had prior information concerning the probabilities of $f_0$ and $f_1$. We will discuss this situation shortly, when we treat Bayes procedures. Finally, the reader should notice that in this example (as in many others), the main difficulty is in rewriting events in a way that makes the computation of the probabilities a simple matter. Since several distributions are involved, care must be taken to decide when to use $f_0$ and when to use $f_1$ to calculate the appropriate probabilities.

### *2.20 Example: Allergy Test*

A drug in wide use unfortunately causes an allergic reaction in certain individuals. In attempting to keep the number of allergic reactions down, a series of allergy tests are made on each individual's blood in contact with the drug. For simplicity, the following model is used. If an individual is not allergic, the probability of a positive reaction on any one of the tests is $p_1 = 4/16$. If he is allergic, the probability of a positive reaction is $p_0 = 7/16$. The tests are independent and the series consists of 25 tests. If it is necessary to keep the probability of administering the drug to an allergic person down to .1, what should the procedure be?

$$\text{Let } X_i = \begin{cases} 1 & \text{if the reaction to the } i\text{th test is positive} \\ 0 & \text{otherwise.} \end{cases}$$

Then

$$P_i\{X_1 = x_1, \ldots, X_{25} = x_{25}\} = p_i^{\sum_{j=1}^{25} x_j}(1 - p_i)^{25 - \sum_{j=1}^{25} x_j}.$$

The likelihood ratio is

$$\frac{p_1^{\sum_{j=1}^{25} x_j}(1 - p_1)^{25 - \sum_{j=1}^{25} x_j}}{p_0^{\sum_{j=1}^{25} x_j}(1 - p_0)^{25 - \sum_{j=1}^{25} x_j}} = (4/7)^{\sum_{j=1}^{25} x_j}(12/9)^{25}(9/12)^{\sum_{j=1}^{25} x_j}$$

$$= (36/84)^{\sum_{j=1}^{25} x_j}(12/9)^{25}.$$

As in the previous example, we do not try to choose $c_\alpha$ directly to satisfy

$$P_0\left\{(36/84)^{\sum_{j=1} X_j}(12/9)^{25} > c_\alpha\right\} \le \alpha = .1.$$

Instead we rewrite the event in question as

$$\left\{\sum_{j=1}^{25} X_j < \log[(9/12)^{25}c_\alpha]/\log(36/84)\right\},$$

the inequality being reversed since $\log(36/84) < 0$. It is natural to let $\log[(9/12)^{25}c_\alpha]/\log(36/84) = d_\alpha$ and try to choose $d_\alpha$ so that $P_0\left\{\sum_{j=1}^{25} X_j < d_\alpha\right\} \le .1$. We may find $d_\alpha$ by going directly to tables of the binomial distribution with $n = 25$, $p = p_0 = 7/16$. For $\alpha = .1$, we obtain $d_\alpha = 8$. In this case, we are not able to achieve an exact probability .1 of type I error because of the discreteness of the binomial distribution. The actual probability of type I error is .0810, and we have the best test whose probability of type I error does not exceed .0810. There are ways to achieve a best test whose probability of type I error is exactly .1, but we shall not treat them here. The probability of type II error is

$$P_1\left\{\sum_{j=1}^{25} X_j \ge d_\alpha\right\} = 1 - P_1\left\{\sum_{j=1}^{25} X_j < 8\right\} = .2735.$$

Let us see what answer we obtain if we use the Central Limit Theorem as an approximation to help choose $d_\alpha$.

$$P_0\left\{\sum_{j=1}^{25} X_j < d_\alpha\right\} = P_0\left\{\frac{1}{5}\sum_{j=1}^{25} \frac{X_j - (7/16)}{\sqrt{7/16 \cdot 9/16}} < \frac{16}{5\sqrt{63}}\,d_\alpha - \frac{35}{\sqrt{63}}\right\}$$

$$= P_0\left\{\frac{1}{5}\sum_{j=1}^{25} \frac{X_j - (7/16)}{\sqrt{7/16 \cdot 9/16}} < .403\,d_\alpha - 4.4\right\}.$$

Now, from the Central Limit Theorem, when $H_0$ is true, the random variable

$$\frac{1}{5}\sum_{j=1}^{25}\frac{X_j-(7/16)}{\sqrt{7/16\cdot 9/16}}$$

has approximately a standard normal distribution. To make the type I error probability approximately .1, we see that we should choose

$$.403\, d_\alpha - 4.4 \cong -1.282 \qquad \text{or} \qquad d_\alpha \cong 7.7,$$

which gives us the same answer previously obtained because the data is discrete. For the probability of type II error, we find as an approximation

$$\begin{aligned}
P_1\left\{\sum_{j=1}^{25} X_j \ge d_\alpha\right\} &= 1 - P_1\left\{\sum_{j=1}^{25} X_j < 7.7\right\}\\
&= 1 - P_1\left\{\frac{1}{5}\sum_{j=1}^{25}\frac{X_j-(4/16)}{\sqrt{4/16\cdot 12/16}} < \frac{4\cdot 7.7}{5\sqrt{3}} - \frac{5}{\sqrt{3}}\right\}\\
&= 1 - P_1\left\{\frac{1}{5}\sum_{j=1}^{25}\frac{X_j-(4/16)}{\sqrt{4/16\cdot 12/16}} < .57\right\}\\
&\cong 1 - \int_{-\infty}^{.57}\frac{1}{\sqrt{2\pi}}e^{-t^2/2}\,dt \quad \text{by the Central Limit Theorem}\\
&\cong 1 - .715 = .285.
\end{aligned}$$

This is not too far from the exact answer of .2735. △

## 2.21 EXERCISES

1.° A supposedly fair die is tossed 60 times. It is suspected, however, that

$$(*) \qquad \begin{aligned} P\{2\} &= P\{4\} = P\{6\} = 2/9\\ P\{1\} &= P\{3\} = P\{5\} = 1/9. \end{aligned}$$

That is, each of the even outcomes is twice as likely as each of the odd outcomes. Find the best size $\cong .05$ test of the hypothesis $H_0$, that the die is fair, versus the hypothesis $H_1$, that (*) holds.

2. A pair of supposedly fair dice is tossed 60 times. The sum is all that is recorded. It is suspected that each die follows (*) of Exercise 1. Based only on the recorded 60 sums, find the best size $\alpha$ test of the hypothesis $H_0$, that the dice (independently) each satisty (*).

3. Suppose that we want to determine whether or not a given archaeological object is of organic origin or not. The following model is assumed. In an organic object, the radiation level of given independent measurements is Poisson with mean $\lambda_0 = 10$. In a non-organic object, the measurements are Poisson with mean $\lambda_1 = 5$. Thirty measurements are made. It is desired that the probability of type I error not

exceed .05. Determine the proper test and the probability of type II error. You may require use of the Central Limit Theorem to determine $d_\alpha$, and the probability of type II error.

4.° In trying to determine whether a given instrument is of sufficiently high quality, repeated measurements $X_1, \ldots, X_n$ are made on an object of known characteristics. It is assumed that the measurements are independent and normal, with $E(X_i) = 6$ and variance $\sigma^2$. For simplicity, only two possibilities are assumed—$H_0: \sigma^2 = \sigma_0^2 = 9$ or $H_1: \sigma^2 = \sigma_1^2 = 16$.

(a) If $n = 25$, determine the best size $\alpha = .05$ test of $H_0$ vs. $H_1$ and find the probability of type II error.

(b) Apply the results of part (a) to the following data ($\mu = 6$, $\sigma^2 = 9$):

| | | | | |
|---|---|---|---|---|
| 12.318 | 7.137 | 8.244 | 4.833 | 4.509 |
| 8.277 | 7.455 | 1.221 | 5.619 | 12.036 |
| 9.846 | 4.224 | 3.009 | 9.750 | 1.158 |
| 6.912 | 12.213 | 5.268 | 3.489 | 6.711 |
| 6.681 | 6.636 | 5.520 | 7.560 | .009 |

5.° Suppose that, based on theoretical considerations, we assume radar measurements $X_1, \ldots, X_{100}$ to be independent normal random variables, with $E(X_i) = \mu$, $\text{Var } X_i = \sigma^2$, and that when a target is present, $\mu = 2$ and $\sigma^2 = 4$. When no target is present, we assume that $\mu = 0$, $\sigma^2 = 1$. Using reasonable Central Limit Theorem approximations, find the best size $\alpha$ test of the hypothesis $H_0$ that there is no target. Determine the specific test when $\alpha = .05$ as well as the probability of type II error. Which error corresponds to false alarm and which to missing the target?

6. Suppose that $X_1, \ldots, X_{100}$ are independent uniform random variables on $[\theta_1, \theta_2]$. Test $H_0: \theta_1 = 1$, $\theta_2 = 2$ versus $H_1: \theta_1 = 1.7$, $\theta_2 = 2.3$, with probability 0 of type I error. What is the probability of type II error?

7. Referring to Example 2.19, the anthropology example, find the test that:

(a) minimizes the maximum of the two probabilities of error.

(b) minimizes the sum of the two error probabilities.

(c) minimizes $2P\{\text{type I error}\} + 3P\{\text{type II error}\}$.

Thus far we have treated the problem of determining the best test of $H_0$ vs. $H_1$ based on $X_1, \ldots, X_n$, when we are given $n$ and the probability $\alpha$ of type I error. It is sometimes more natural to have prescribed an upper bound $\alpha$ for the probability of type I error and an upper bound $\beta$ for the probability of type II error. We then try to find a test that requires the smallest sample size $n^*$ of those that achieve these pre-assigned error probabilities.

### 2.22 *Corollary: Pre-Assigned Error Probabilities*

Suppose that for each $n$, we can achieve a probability of type I error of exactly $\alpha$ with a likelihood ratio test $L_n$ based on $n$ observations. Let $n_0$ be the smallest $n$ such that $L_n$ has probability $\leq\beta$ of type II error. Then $n^* = n_0$.

*Proof:* We shall show that if $T_m$ is *any* test (not necessarily a likelihood ratio test) satisfying the $\alpha$, $\beta$ constraints and based on a sample size of $m$, then $n_0 \leq m$. This follows from the fact that by construction $L_m$ satisfies the $\alpha$ constraint, and by the Neyman Pearson Lemma (2.9) $L_m$ satisfies the $\beta$ constraint. But by construction,

$n_0$ is the smallest $n$ such that $L_n$ satisfies the $\alpha$, $\beta$ constraints. Thus $n_0 \leq m$. Consequently, the test with the smallest $n$ must be a likelihood ratio test and $n^* = n_0$.

We may, if we are lucky, be able to find a simple formula for the required size. In other cases, we may have to resort to some numerical approximation scheme, such as bisection or some version of Newton's method. It can be shown (see Problem 2.40) that if $f_0$ and $f_1$ are really distinct, then for any $\alpha$ and $\beta$ strictly between 0 and 1 we can find a test whose probabilities of error do not exceed $\alpha$ and $\beta$, respectively. We will now illustrate these ideas with a few examples.

### 2.23 *Example: Choosing Sample Size*

A lot of newly manufactured magnetic recording tape is to be tested to see whether the process that bonds the oxide layer to the mylar backing is successful. A signal is recorded in a special way on the tape, and measurements $X_1, X_2, \ldots, X_n$ are made on the amount of signal that actually is imprinted on the tape at intervals of one foot. It is assumed that this process is satisfactory when the $X_i$ can be described as independent normal random variables with $E(X_i) = 5$, $\mathrm{Var}\, X_i = 1$. When the process is not satisfactory, we assume that they can be described as being independent normal with $E(X_i) = 4$, $\mathrm{Var}\, X_i = 1$.

We want to test $H_0$: $E(X_i) = 4$ versus $H_1$: $E(X_i) = 5$, with probability of type I error .01 and the probability of type II error .05. Here we have

$$\frac{f_1(x_1, \ldots, x_n)}{f_0(x_1, \ldots, x_n)} = \exp\left(-\frac{1}{2}\left[\sum_{i=1}^{n} (x_i - 5)^2 - \sum_{i=1}^{n} (x_i - 4)^2\right]\right)$$

$$= \exp\left(-\frac{9n}{2}\right) \cdot \exp\left(\sum_{i=1}^{n} x_i\right).$$

*Therefore, the test we want is of the form*

$$\text{reject } H_0 \text{ if } \sum_{i=1}^{n} x_i \text{ is too large,}$$

where $x_i$ is the observed value of $X_i$. To simplify, we normalize under the null hypothesis so that the desired test is of the form

reject $H_0$ if and only if

$$\frac{1}{\sqrt{n}}\sum_{i=1}^{n} (x_i - 4) > d_\alpha, \qquad \mathbf{2.24}$$

where $d_\alpha$ is the upper $1 - \alpha$ point of the standard normal distribution. If $\alpha = .01$, then $d_\alpha = 2.326$.

Note that since the left side of 2.24 is the value of a standard normal random variable *under* $H_0$, the probability of type I error is $\alpha = .01$ *for every n*. Now we would like the probability of type II error to equal $\beta = .05$. Hence, we want

$$P_1\left\{\frac{1}{\sqrt{n}}\sum_{i=1}^{n} (X_i - 4) \leq 2.326\right\} = .05.$$

Now

$$\frac{1}{\sqrt{n}}\sum_{i=1}^{n}(X_i - 5)$$

is a standard normal random variable under $H_1$, and

$$\frac{1}{\sqrt{n}}\sum_{i=1}^{n}(X_i - 4) = \frac{1}{\sqrt{n}}\sum_{i=1}^{n}(X_i - 5 + 1) = \frac{1}{\sqrt{n}}\sum_{i=1}^{n}(X_i - 5) + \sqrt{n}.$$

Therefore, we want

$$P_1\left\{\frac{1}{\sqrt{n}}\sum_{i=1}^{n}(X_i - 5) \le 2.326 - \sqrt{n}\right\} = .05.$$

That is, we choose $n$ so that $2.326 - \sqrt{n}$ does not exceed the upper .05 point of the standard normal distribution. Hence, $2.326 - \sqrt{n} \le -1.645$, that is, $\sqrt{n} \ge 3.971$, $n = 16$ observations are needed. The test is to reject $H_0$ if and only if

$$\frac{1}{4}\sum_{i=1}^{16}(x_i - 4) > 2.326. \qquad \triangle$$

### 2.25 *Example: Too Large a Sample Size Required*

A new process for making transistors is being tested. It is known that the old process yielded a probability of .7 that any given unit would function properly. Let $p$ be the probability that a unit manufactured under the new process functions properly. Based on technical reasons, it is assumed that $p \ge .7$. It is also known that even if $p \ge .8$, the extra profit that can be realized from conversion to the new process is at most $10,000. Suppose it costs $50 to test each unit. Before a potential purchaser is willing to commit himself, he insists on a test having error probabilities $\le .01$. The question is whether or not it pays to make the test of $H_0$, that $p = .7$, versus $H_1$, that $p = .8$, with $\alpha =$ probability of type I error $\le .01$ and $\beta =$ probability of type II error $\le .01$. To find the answer, we first note that it doesn't pay to test more than 200 units, since testing 200 units would eat up all of the possible extra profit derived from converting to the new process. We shall therefore determine the best test of $H_0$, that $p = .7$, versus $H_1$, that $p = .8$, having both probabilities of error not exceeding .01. If the required sample size exceeds 200, then we know that it is not worth making the test. We assume that $X_1, X_2, \ldots, X_n$ are random variables that are independent and

$$X_i = \begin{cases} 1 & \text{if the } i\text{th transistor is good} \\ 0 & \text{otherwise} \end{cases}$$

with $P\{X_i = 1\} = p$, $P\{X_i = 0\} = 1 - p$. The probability function when $p = .7$ is

$$p_{.7}(x_1, \ldots, x_n) = .7^{x_1+\cdots+x_n}.3^{n-(x_1+\cdots+x_n)}.$$

The probability function when $p = .8$ is

$$p_{.8}(x_1, \ldots, x_n) = .8^{x_1+\cdots+x_n}.2^{n-(x_1+\cdots+x_n)}.$$

Hence, the likelihood ratio is

$$\frac{p_{.8}(x_1, \ldots, x_n)}{p_{.7}(x_1, \ldots, x_n)} = (8/7)^{x_1+\cdots+x_n}(2/3)^{n-(x_1+\cdots+x_n)}.$$

Thus

$$\frac{p_{.8}(x_1, \ldots, x_n)}{p_{.7}(x_1, \ldots, x_n)} > d_{.01}$$

when

$$(8/7 \times 3/2)^{x_1+\cdots+x_n}(2/3)^n > d_{.01}.$$

Therefore, the desired test takes the form of accepting the hypothesis $H_1$ that $p = .8$ when $x_1 + \cdots + x_n > c_{.01}$. Here $c_{.01}$ and $n$ must be chosen to achieve the desired error probabilities. To get approximations for $n$ and $c_{.01}$, we will make use of the Central Limit Theorem. Under $H_0$, we want

$$\alpha = .01 = P_{H_0}\{X_1 + \cdots + X_n > c_{.01}\} = P\left\{\frac{X_1 + \cdots + X_n - .7n}{\sqrt{n(.7)(.3)}} > \frac{c_{.01} - .7n}{\sqrt{n(.7)(.3)}}\right\}$$

$$\cong \int_{\frac{c_{.01}-.7n}{\sqrt{.21n}}}^{\infty} \frac{1}{\sqrt{2\pi}} e^{-u^2/2}\, du.$$

Under $H_1$, we want

$$\beta = .01 = P_{H_1}\{X_1 + \cdots + X_n < c_{.01}\}$$

$$= P\left\{\frac{X_1 + \cdots + X_n - .8n}{\sqrt{n(.8)(.2)}} \le \frac{c_{.01} - .8n}{\sqrt{n(.8)(.2)}}\right\}$$

$$\cong \int_{-\infty}^{(c_{.01}-.8n)/\sqrt{.16n}} \frac{1}{\sqrt{2\pi}} e^{-u^2/2}\, du.$$

Since

$$\int_{2.33}^{\infty} \frac{1}{\sqrt{2\pi}} e^{-u^2/2}\, dn = .01 = \int_{-\infty}^{-2.33} \frac{1}{\sqrt{2\pi}} e^{-u^2/2}\, dn,$$

we see that we want $c_{.01}$ and $n$ to satisfy

**2.26**

$$\frac{c_{.01} - .7n}{\sqrt{.21n}} = 2.33$$

$$\frac{c_{.01} - .8n}{\sqrt{.16n}} = -2.33.$$

Dividing the first equality by the second yields

$$\frac{c_{.01} - .7n}{c_{.01} - .8n}\sqrt{16/21} = -1.$$

Hence, solving for $c_{.01}$ gives

$$c_{.01} = \frac{n(.8 + .7\sqrt{16/21})}{(\sqrt{16/21} + 1)}.$$

Substituting this in 2.26 and solving for $\sqrt{n}$ gives

$$\sqrt{n} = \sqrt{.21} \times 2.33 \frac{1}{\left[\frac{(.8 + .7\sqrt{16/21})}{\sqrt{16/21} + 1}\right] - .7}$$

$$= \frac{(\sqrt{.21} \times 2.33)(\sqrt{16/21} + 1)}{.1}$$

$$\cong \frac{1.048 \times 1.873}{.1}$$

$$\cong 19.6,$$

or

$$n \cong 400.$$

We see that the required $n$ appears to be much too large. Note that our conclusion is based on a Central Limit Theorem approximation. To be absolutely certain that the desired test cannot be achieved using $n \leq 200$ would take either direct numerical calculation on the binomial distributions involved or an analytic investigation. Such an investigation might be based on the Uspensky result (Theorem 3.16, Chapter 7) by putting $n = 200$ and showing it impossible to achieve the desired error probabilities even if $\varepsilon$ "aids" us. △

### 2.27 *Example: Fair Dice Test Revisited*

In Example 2.3, we developed a test of whether a pair of dice were fair ($H_0$) versus a probability distribution for each die of $p_1 = p_2 = p_3 = 2/15$, $p_4 = p_5 = p_6 = 1/5$ ($H_1$). The test was to accept $H_0$ if $\frac{1}{600}\sum_{i=1}^{600} x_i \leq 3.6$, where $x_i$ is the observed result of the $i$th toss (300 tosses of die 1 followed by 300 tosses of die 2). We now ask for the test that will achieve the same result with the smallest number of tosses.

To find this test, we first examine the likelihood ratio for a test based on $n$ tosses ($n/2$ tosses of each die). Still letting $x_i$ denote the observed result of the $i$th toss, we let

$$y_{ij} = \begin{cases} 1 & \text{if } x_i = j, \quad j = 1, 2, 3, 4, 5, 6 \\ 0 & \text{otherwise.} \end{cases}$$

The joint probability function of $X_1, \ldots, X_n$ at $x_1, \ldots, x_n$ is

$$\left(\prod_{j=1}^{6} p_j\right)^{\sum_{i=1}^{n} y_i}.$$

(Here $\prod_{i=1}^{6} a_j$ is the product $a_1 a_2 \cdots a_j$.) Hence, the likelihood ratio is

$$L(x_1, \ldots, x_n) = \left(\frac{2/15}{1/6}\right)^{\sum_i y_{i1}} \left(\frac{2/15}{1/6}\right)^{\sum_i y_{i2}} \left(\frac{2/15}{1/6}\right)^{\sum_i y_{i3}} \left(\frac{1/5}{1/6}\right)^{\sum_i y_{i4}}$$

$$\times \left(\frac{1/5}{1/6}\right)^{\sum_i y_{i5}} \left(\frac{1/5}{1/6}\right)^{\sum_i y_{i6}}$$

$$= (4/5)^{\sum_{i=1}^{n} (y_{i1}+y_{i2}+y_{i3})} (6/5)^{\sum_{i=1}^{n} (y_{i4}+y_{i5}+y_{i6})}$$

Thus, $L(x_1, \ldots, x_n) > c_\alpha$ if and only if

$$\log L(x_1, \ldots, x_n) = (\log 4/5) \sum_{i=1}^{n} (y_{i1} + y_{i2} + y_{i3})$$

$$+ (\log 6/5) \sum_{i=1}^{n} (y_{i4} + y_{i5} + y_{i6}) > \log c_\alpha \equiv d_\alpha.$$

The best test of $H_0$ versus $H_1$ must have the form

reject $H_0$ if and only if

$$(\log 4/5) \sum_{i=1}^{n} (y_{i1} + y_{i2} + y_{i3}) + (\log 6/5) \sum_{i=1}^{n} (y_{i4} + y_{i5} + y_{i6}) > d_\alpha.$$

Note that

$\sum_{i=1}^{n} (y_{i1} + y_{i2} + y_{i3})$ is the total observed number of 1's, 2's, and 3's in $n$ tosses,

whereas

$\sum_{i=1}^{n} (y_{i4} + y_{i5} + y_{i6})$ is the total observed number of 4's, 5's, and 6's in $n$ tosses.

This suggests that we let

$v_i$ be the total observed number of 1's, 2's, and 3's on toss $i$,

$z_i$ be the total observed number of 4's, 5's, and 6's on toss $i$,

and

$$u_i = (\log 4/5) v_i + (\log 6/5) z_i.$$

In terms of $u_i$, every L.R. test has the form

reject $H_0$ if and only if

**2.28** $$\sum_{i=1}^{n} u_i > d_\alpha.$$

To determine the performance of this test, we compute the approximate distribution of $\sum_{i=1}^{n} U_i$ under $H_0$ and under $H_1$. For this, we need to determine $E(U_i)$ and Var $U_i$ under these hypotheses. Under $H_0$, we see that

**2.29**
$$E(U_i) = (\log 4/5) \cdot 1/2 + (\log 6/5) \cdot 1/2 = 1/2 \log 24/25 \cong -.0204,$$
$$E(U_i^2) = (\log 4/5)^2 E(V_i^2) + (\log 6/5)^2 E(Z_i^2) + 2(\log 4/5)(\log 6/5) E(V_i Z_i).$$

But

$$V_i = \begin{cases} 1 & \text{with probability } 1/2 \\ 0 & \text{with probability } 1/2 \end{cases}$$

and $Z_i = 1 - V_i$. Hence, under $H_0$,

$$E(U_i^2) = \tfrac{1}{2}(\log 4/5)^2 + \tfrac{1}{2}(\log 6/5)^2.$$

Thus, using the relation Var $U = E(U^2) - [E(U)]^2$ (Lemma 3.8, Chapter 6), we see that under $H_0$,

**2.30** $$\text{Var } U_i = \tfrac{1}{4}[\log 4/5 - \log 6/5]^2 = \tfrac{1}{4}(\log 2/3)^2 = (\tfrac{1}{2}\log 3/2)^2 \cong [.203]^2.$$

Similar computations show that under $H_1$,

**2.31** $$E(U_i) = \tfrac{2}{5}\log 4/5 + \tfrac{3}{5}\log 6/5 \cong .0203$$

and

**2.32** $$\text{Var } U_i = \tfrac{6}{25}[\log 4/5 - \log 6/5]^2 = \left[\frac{\sqrt{6}}{5}\log 3/2\right]^2 \cong [.199]^2.$$

Recall now that the test of Example 2.3 achieved approximately a probability of type I error equal to 0.761 and a probability of type II error equal to .0018, based on 600 tosses. Let us see how many tosses are needed to approximately achieve these error probabilities, using the best L.R. test.

We want to determine $n$ and $d_\alpha$ so that, under $H_0$,

**2.33** $$P\left\{\sum_{i=1}^{n} U_i > d_\alpha\right\} \cong .0761,$$

and under $H_1$,

**2.34** $$P\left\{\sum_{i=1}^{n} U_i \le d_\alpha\right\} \cong .0018.$$

Now, both under $H_0$ and under $H_1$, the $U_i$ are independent random variables with the same distribution and finite variance. Hence, we may apply the Central Limit

Theorem. Under $H_0$, using 2.29 and 2.30, we find

**2.35**
$$P\left\{\sum_{i=1}^{n} U_i > d_\alpha\right\} = P\left\{\frac{1}{\sqrt{n}}\sum_{i=1}^{n}\frac{U_i + .0204}{.203} > \frac{d_\alpha + .0204n}{.203\sqrt{n}}\right\}$$
$$\cong \int_{(d_\alpha+.0204n)/.203\sqrt{n}}^{\infty} \frac{1}{\sqrt{2\pi}} e^{-u^2/2}\, du.$$

Similarly, using the Central Limit Theorem, 2.31, and 2.32, we find

**2.36**
$$P\left\{\sum_{i=1}^{n} U_i \le d_\alpha\right\} \cong \int_{-\infty}^{(d_\alpha-.0203n)/.199\sqrt{n}} \frac{1}{\sqrt{2\pi}} e^{-u^2/2}\, du.$$

By 2.35, we see from the standard normal tables that to satisfy 2.33, we must have

**2.37**
$$\frac{d_\alpha + .0204n}{.203\sqrt{n}} \cong 1.432.$$

To satisfy 2.34, we see from 2.36 and standard normal tables that we need

**2.38**
$$\frac{d_\alpha - .0203n}{.199\sqrt{n}} \cong 2.92.$$

Solving 2.37 and 2.38 yields

$$n \cong 460, \qquad d_\alpha \cong 99.94.$$

Thus, whereas the test of Example 2.3 used $n = 600$ tosses of the dice, the best such test having the same error probabilities needed only $n \cong 460$ using 2.28. △

---

## 2.39 *EXERCISES*

1. Referring to Exercise 2.21(3), on the origin of an archeological object, what is the best test of $H_0{:}\lambda = 10$ versus $H_1{:}\lambda = 5$, where we require $\alpha = .05$ and $\beta = P\{\text{type II error}\} \le .012$.

2.° Determine the best test of $H_0$, that the instrument of Exercise 2.21(4) is of high enough quality ($\sigma^2 = 9$) versus the alternative that it is not ($\sigma^2 = 16$), where $\alpha = .01$ and $\beta = P\{\text{type II error}\} \le .1$. In particular, if only 50 observations are available, is it possible to find such a test?

3. In the radar situation of Exercise 2.21(5), suppose that technical limitations permit us only 25 observations. Is it possible for us to test $H_0$, that there is no target, versus $H_1$, that there is, with $\alpha = .05$ and $\beta = P\{\text{type II error}\} \le .05$? (Note that if there is no such test, then an engineering improvement of the radar characteristics may still be possible to achieve the desired results.)

4. Find the best test of $H_0$ versus $H_1$ of Exercise 2.21(6) with $\alpha = .1$ and $\beta = P\{\text{type II error}\} \le .05$.

---

*2.40 PROBLEM*

Suppose that $f_0$ and $f_1$ are univariate densities which are distinct, that is, for some $A$

$$P_{f_0}(A) \neq P_{f_1}(A).$$

Show that if $X_1, X_2, \ldots$ are independent random variables, all with density $f_i$, then for any given $\alpha > 0$, $\beta > 0$ there is a sample size $n$ and a test based on $X_1, \ldots, X_n$ of $H_0$: that $f_0$ is the density of the $X_i$, versus $H_1$, that $f_1$ is the density, with

$$P\{\text{Type I error}\} \leq \alpha$$

and

$$P\{\text{Type II error}\} \leq \beta.$$

Hint: Find a simpler test based on the Chebyshev inequality.

## SECTION 3
## Bayes Procedures

In this section, we will briefly investigate what our test procedures should be in testing a simple hypothesis $H_0$ against a simple alternative $H_1$, when we have some additional prior information about the probabilities of $H_0$ and $H_1$ being true. The hypothesis $H_i$ will be that the density $f$ of $\mathbf{X} = (X_1, \ldots, X_n)$ is $f_i$, $i = 0$ or $i = 1$, where $f_0$ and $f_1$ are known densities. Results when $f$ and $f_i$ are replaced by discrete probability functions $p$ and $p_i$ are formally identical. The additional information that we are trying to utilize is that for some known $q$,

$$P\{f = f_0\} = q \qquad \text{and} \qquad P\{f = f_1\} = 1 - q.$$

Thus, we are now considering $f$ to be a random quantity with a known distribution. Our additional prior knowledge concerning $f$ may be gathered empirically, as in the situation when we are trying to distinguish between two diseases, or it might be based on theoretical grounds, as when we know that some radiation process *should* emit twice as many electrons as it does protons. Suppose that we lose an amount $a$ dollars when we falsely reject $H_0$ and $b$ dollars when we falsely reject $H_1$; in any other circumstances we lose nothing. In this situation it is reasonable to choose that test procedure that minimizes the expected loss. But since you lose $a$ dollars when you reject $H_0$ and $H_0$ is true (that is, when $f = f_0$) and $b$ dollars when you reject $H_1$ and $H_1$ is true (that is, $f = f_1$), the expected loss is

**3.1** $$P\{\text{rejecting } H_0 \text{ and } H_0 \text{ is true}\} \cdot a + P\{\text{rejecting } H_1 \text{ and } H_1 \text{ is true}\} \cdot b$$
$$= P\{\text{Rej } H_0, f = f_0\}a + P\{\text{Rej } H_1, f = f_1\}b$$
$$= P\{\mathbf{X} \in S_1, f = f_0\}a + P\{\mathbf{X} \in S_0, f = f_1\}b$$
$$= P\{\mathbf{X} \in S_1 \mid f = f_0\}P\{f = f_0\}a + P\{X \in S_0 \mid f = f_1\}P\{f = f_1\}b$$
$$= \alpha q a + \beta(1 - q)b.$$

We want to choose $S_0$ and $S_1 = S_0^c$ to minimize this expression, where $\alpha$ and $\beta$ are the probabilities of type I and type II errors, respectively, associated with the acceptance region $S_0$. Now $q$, $a$, and $b$ are fixed, and for given probability $\alpha$ of type I error, we know that a size $\alpha$ likelihood ratio test minimizes $\beta$, the probability of type II error. Thus, provided a size $\alpha$ L.R. test exists for each $\alpha$, we see that there is some L.R. test that minimizes our expected loss. We could use a numerical approximation scheme to find the "best" such test. This was essentially what we required in Exercise 2.21(7); we will not repeat the computations here. An alternative, which is much simpler, will now be presented for the discrete case; the analogous results hold for the continuous case. We want to minimize

$$P\{\mathbf{X} \in S_1, p = p_0\}a + P\{\mathbf{X} \in S_0, p = p_1\}b,$$

where $p$ is the discrete probability function of $\mathbf{X}$, by properly choosing $S_0$ and $S_1 = S_0^c$.

But

$$\begin{aligned} P\{\mathbf{X} \in S_1, p = p_0\}a &+ P\{\mathbf{X} \in S_0, p = p_1\}b \\ &= \sum_{\text{all } \mathbf{x} \in S} P\{\mathbf{X} \in S_1, p = p_0, \mathbf{X} = \mathbf{x}\}a + \sum_{\text{all } \mathbf{x} \in S} P\{\mathbf{X} \in S_0, p = p_1, \mathbf{X} = \mathbf{x}\}b \\ &= \sum_{\mathbf{x} \in S_1} P\{p = p_0, \mathbf{X} = \mathbf{x}\}a + \sum_{\mathbf{x} \in S_0} P\{p = p_1, \mathbf{X} = \mathbf{x}\}b \\ &= \sum_{\mathbf{x} \in S_1} P\{\mathbf{X} = \mathbf{x} \mid p = p_0\}P\{p = p_0\}a + \sum_{\mathbf{x} \in S_0} P\{\mathbf{X} = \mathbf{x} \mid p = p_1\}P\{p = p_1\}b \\ &= \sum_{\mathbf{x} \in S_1} p_0(\mathbf{x})qa + \sum_{\mathbf{x} \in S_0} p_1(\mathbf{x})(1 - q)b. \end{aligned}$$

In order to choose $S_0$ and $S_1 = S_0^c$ to *minimize* the preceding, it now follows that we should assign $\mathbf{x}$ to $S_0$ if and only if

$$p_1(\mathbf{x})(1 - q)b < p_0(\mathbf{x})qa.$$

This procedure is called a *Bayes procedure* corresponding to the prior probabilities $q = P\{p = p_0\}$, $1 - q = P\{p = p_1\}$ and weights $a$ and $b$. It can be shown that in the density case we can minimize the overall expected loss

$$P\{\mathbf{X} \in S_1, f = f_0\}a + P\{\mathbf{X} \in S_0, f = f_1\}b$$

by assigning $\mathbf{x}$ to $S_0$ if and only if

$$f_1(\mathbf{x})(1 - q)b < f_0(\mathbf{x})qa.$$

We summarize the results just derived in the following theorem.

### 3.2 *Theorem: Bayes Procedures*

Suppose that the random vector $\mathbf{X} = (X_1, \ldots, X_n)$ has discrete probability function $p$ with $P\{p = p_0\} = q$, $P\{p = p_1\} = 1 - q$, where $p_0$, $p_1$ and $q$ are known. Suppose that in testing $H_0$, that $p = p_0$, versus $H_1$, that $p = p_1$, we lose $a$ dollars in rejecting $H_0$ when $H_0$ is true, and we lose $b$ dollars in rejecting $H_1$ when $H_1$ is true (we lose nothing otherwise). Let $S_0$ be the acceptance region for $H_0$. That is, $S_0$ is the subset of $\mathcal{S}$ containing those outcomes leading to acceptance of $H_0$. In order to minimize the overall expected loss, $P\{\text{rej } H_0, H_0 \text{ is true}\} \cdot a + P\{\text{rej } H_1, H_1 \text{ is true}\} \cdot b$, we assign $\mathbf{x} = (x_1, \ldots, x_n)$ to $S_0$ if and only if

$$p_1(\mathbf{x})(1 - q)b < p_0(\mathbf{x})qa,$$

or equivalently, if

$$\frac{p_1(\mathbf{x})}{p_0(\mathbf{x})} < \frac{qa}{(1 - q)b}.$$

The analogous result holds when $p$, $p_0$, and $p_1$ are replaced by densities $f$, $f_0$, and $f_1$.

Note that in the cases being considered in this section, it is meaningful to speak of $P\{p = p_0 \mid \mathbf{X} = \mathbf{x}\}$. In fact

$$\begin{aligned} P\{p = p_0 \mid \mathbf{X} = \mathbf{x}\} &= \frac{P\{p = p_0, \mathbf{X} = \mathbf{x}\}}{P\{\mathbf{X} = \mathbf{x}\}} \\ &= \frac{P\{\mathbf{X} = \mathbf{x} \mid p = p_0\}P\{p = p_0\}}{P\{\mathbf{X} = \mathbf{x} \mid p = p_0\}P\{p = p_0\} + P\{\mathbf{X} = \mathbf{x} \mid p = p_1\}P\{p = p_1\}} \\ &= \frac{p_0(\mathbf{x})q}{p_0(\mathbf{x})q + p_1(\mathbf{x})(1 - q)}. \end{aligned}$$

Thus,

**3.3**
$$P\{p = p_0 \mid \mathbf{X} = \mathbf{x}\} = \frac{p_0(\mathbf{x})q}{p_0(\mathbf{x})q + p_1(\mathbf{x})(1 - q)},$$

and similarly, in the density case,

$$P\{f = f_0 \mid \mathbf{X} = \mathbf{x}\} = \frac{f_0(\mathbf{x})q}{f_0(\mathbf{x})q + f_1(\mathbf{x})(1 - q)}.$$

### *3.4 Example: Acceptance Sampling*

A manufacturer has a process that can be in one of two states:

$$s_{\text{in}}: \text{ in control},$$

or

$$s_{\text{out}}: \text{ out of control}.$$

Suppose that when the true state is $s = s_{\text{in}}$, the process produces units independently that have probability .9 of being good. If $s = s_{\text{out}}$, we assume that the process produces units independently whose probability of being good is .75. It is assumed that for each lot of goods either $s = s_{\text{in}}$ or $s = s_{\text{out}}$. Suppose that the guarantees made on units and the profits to be made lead to the following situation. When a customer tests 20 units from each lot, the cost to the customer of rejecting a manufactured lot when $s = s_{\text{in}}$ is $b = \$30$, whereas the cost to the customer of accepting a lot with $s = s_{\text{out}}$ is $a = \$50$. Suppose also that the probability that the process is in control, $P(s = s_{in}) \equiv 1 - q$, is known to be .7. We ask for the acceptance procedure based on testing 20 units from a lot that minimizes the customer's expected loss. We make use of the Bayes Procedure, Theorem 3.2, to conclude that in order to minimize his overall expected loss, the customer should *reject* the lot if and only if

$$p_{in}(\mathbf{x})(.7)(30) < p_{out}(\mathbf{x})(.3)(50).$$

Here $\mathbf{x}$ is the value assumed by $\mathbf{X} = (X_1, \ldots, X_{20})$, and $X_i = 1$ if the $i$th tested unit is good, and equals 0 otherwise. But for $\mathbf{x} = (x_1, \ldots, x_{20})$,

$$p_{in}(\mathbf{x}) = .9^{\sum_{i=1}^{20} x_i} (.1)^{20 - \sum_{i=1}^{20} x_i}$$

$$p_{out}(\mathbf{x}) = .75^{\sum_{i=1}^{20} x_i} (.25)^{20 - \sum_{i=1}^{20} x_i}.$$

Hence, the customer should reject the lot if and only if

$$21(.9)^{\Sigma x_i}(.1)^{20 - \Sigma x_i} < 15(.75)^{\Sigma x_i}(.25)^{20 - \Sigma x_i}.$$

Equivalently, he rejects the lot (that is, assigns $\mathbf{x} = (x_1, \ldots, x_{20})$ to $S_0$, asserting that the process is out of control) if and only if

$$\sum_{i=1}^{20} x_i < \frac{\log {}^{15}\!/_{21} + 20 \log({}^{25}\!/_{10})}{\log 3} = 16.25$$

where $\sum_{i=1}^{20} x_i$ is the total number of good units of the 20 sampled. The customer rejects the lot if it does not possess at least 17 good units. △

We make a final observation on the subject of quality assurance. Early in Section 2, we saw that in the absence of prior information we could not assure high quality by acceptance testing, because there is always a possibility that only inferior lots are being submitted. In such a case, the few that get through are still inferior. However, in the presence of prior information of the sort just discussed, we can assure quality by keeping the probabilities of the two types of error small enough. To see this, we suppose that the data concerning a lot have either discrete probability function $p_{out}$, corresponding to a bad lot, or discrete probability function $p_{in}$, corresponding to a good lot. Suppose we know that

$$P\{p = p_{in}\} = 1 - q,$$

where $0 < q < 1$ is known and $p$ is the true probability function. Let $S_0$ be the acceptance region for the hypothesis $H_0$, that $p = p_{out}$, in which case we *reject* the lot. Put

$$\alpha = P\{\text{reject } H_0\colon p = p_{out} \mid p = p_{out}\} = P\{\mathbf{X} \notin S_0 \mid p = p_{out}\}$$

$$\beta = P\{\text{reject } H_1\colon p = p_{in} \mid p = p_{in}\} = P\{\mathbf{X} \in S_0 \mid p = p_{in}\}.$$

Then

$$\begin{aligned}
P\{p = p_{\text{in}} \mid \text{lot is accepted}\} \\
&= P\{p = p_{\text{in}} \mid \mathbf{X} \notin S_0\} \\
&= \frac{P\{p = p_{\text{in}}, \mathbf{X} \notin S_0\}}{P\{\mathbf{X} \notin S_0\}} \\
&= \frac{P\{\mathbf{X} \notin S_0 \mid p = p_{\text{in}}\}P\{p = p_{\text{in}}\}}{P\{\mathbf{X} \notin S_0 \mid p = p_{\text{in}}\}P\{p = p_{\text{in}}\} + P\{\mathbf{X} \notin S_0 \mid p = p_{\text{out}}\}P\{p = p_{\text{out}}\}} \\
&= \frac{(1-\beta)(1-q)}{(1-\beta)(1-q) + \alpha q} = \frac{1}{1 + \dfrac{\alpha}{1-\beta}\dfrac{q}{1-q}}.
\end{aligned}$$

That is,

**3.5**
$$P\{p = p_{\text{in}} \mid \text{lot is accepted}\} = \frac{1}{1 + \dfrac{\alpha}{1-\beta}\dfrac{q}{1-q}}.$$

But $P\{p = p_{\text{in}} \mid \text{lot is accepted}\}$ is just the probability of good quality among the accepted lots. For any given $q$ with $0 < q < 1$, we can surely make

$$\frac{1}{1 + \dfrac{\alpha}{1-\beta}\dfrac{q}{1-q}}$$

as close to 1 as we desire by choosing $\alpha$ and $\beta$ close enough to 0. Thus we can assure quality by obtaining small enough error probabilities, when we have the prior information about $P\{p = p_{\text{in}}\}$. △

---

### *3.6 EXERCISES*

1.° In (the acceptance sampling) Example 3.4, compute approximately:
(a) $P\{\text{lot is good} \mid \text{lot is accepted}\}$.
Hint: See 3.5 and note that

$$\alpha = P\{\mathbf{X} \notin S_0 \mid p = p_{\text{out}}\} = P\left\{\sum_{i=1}^{20} X_i > 16\right\} \quad \text{when} \quad p = p_{\text{out}},$$

$$\beta = P\{\mathbf{X} \in S_0 \mid p = p_{\text{out}}\} = P\left\{\sum_{i=1}^{20} X_i \le 16\right\} \quad \text{when} \quad p = p_{\text{in}}.$$

(b) What is

$$P\{\text{lot is good} \mid 19 \text{ out of the 20 tested units are good}\}.$$

Hint: See 3.3, where "lot is good" means $p = p_{in}$ of Example 3.4.
(c) What is the overall expected loss in Example 3.4? (See 3.1.)

2. Suppose that we have the conditions of Example 3.4, but we are allowed to test 40 units. Answer the previous question in this case. (Note that you must first derive the optimal test for a sample of size 40.)

3.° Suppose that you are given a measuring instrument to test for accuracy, and that you are allowed 50 independent measurements $X_1, \ldots, X_{50}$, in which $X_1, \ldots, X_{50}$ are all independent normal, with $E(X_i) = 0$, Var $X_i = 9$ or with $E(X_i) = 0$, Var $X_i = 16$. You want to test whether or not the instrument is sufficiently accurate (Var $X_i = 9$). Accepting a poor instrument costs you \$100, whereas rejecting a good instrument costs \$250. On the average, four out of five instruments are poor.
(a) What test procedure do you use?
(b) Given that the instrument is accepted, what is the probability that it is good?
(c) Given that $\sum_{i=1}^{50} X_i^2 = 445$, what is the probability that the instrument is good?
(d) On the average, what does each test cost you?

---

## 3.7 *PROBLEMS*

1. Show that the Bayes procedures depend essentially on $P\{p = p_{out} \mid X = x\}$, the so-called posterior probability of $p_{out}$.

2. Show what the Bayes procedure should be if:
rejecting $H_0$ when $H_0$ is true costs $a$ dollars.
rejecting $H_0$ when $H_0$ is false costs $c$ dollars.
accepting $H_0$ when $H_1$ is false costs $d$ dollars.
accepting $H_0$ when $H_1$ is true costs $b$ dollars.
(Here $d < a$ and $c < b$ are all that is required. None of these numbers need be positive. Can you explain why?)

# SECTION 4
# Composite Hypotheses

Most commonly encountered testing situations involve composite hypotheses. For example, we may want to know whether one insecticide is stronger than another. This may lead to a test of $E(X_i) = \mu_0$ versus $E(X_i) > \mu_0$, where $\mu_0$ is the known potency of the old insecticide and $E(X_i)$ the potency of the newer product. At least the alternative hypothesis, $E(X_i) > \mu_0$, is certainly composite. In an election poll, we may want to test $H_0$: $p < 1/2$ versus $H_1$: $p \geq 1/2$, both composite hypotheses. If $\sigma^2$ is the measure of quality of a given machine, and $\sigma_a^2$ is the largest value of $\sigma^2$ corresponding to an acceptable machine, then we might want to test $H_0$: $\sigma \leq \sigma_a^2$ versus $H_1$: $\sigma > \sigma_a^2$.

In order to handle the composite testing situations, let us introduce the following notation.

### 4.1 Notation

Let $\mathscr{P}_0$ and $\mathscr{P}_1$ be a pair of disjoint parametric families, where the parameter $\theta$ specifies a unique element of $\mathscr{P}_0$ or $\mathscr{P}_1$. Further suppose that the set $\Theta_0$ of $\theta$ values corresponds to $\mathscr{P}_0$, whereas $\Theta_1$ corresponds to $\mathscr{P}_1$. Let $P$ be the distribution of the random variables $X_1, \ldots, X_n$ to be observed. We desire to test the hypothesis $H_0$: $\theta \in \Theta_0$ (that $P \in \mathscr{P}_0$) versus $H_1$: $\theta \in \Theta_1$ (that $P \in \mathscr{P}_1$). An extension to the idea of the size of a test of a simple hypothesis is given in the following definition.

**4.2 Definition: Level $\alpha$ Test of $H_0$.** A test of the hypothesis $H_0$: $\theta \in \Theta_0$ is said to be of *level* $\alpha$ if, for each $\theta \in \Theta_0$,

$$P_\theta\{\text{reject } H_0\} \leq \alpha,$$

and for each $\delta < \alpha$, there exists at least one $\theta \in \Theta_0$ such that

$$P_\theta\{\text{reject } H_0\} > \delta.$$

That is, $\alpha = \sup_{\theta \in \Theta_0} P_\theta\{\text{rej } H_0\}$. Here $P_\theta$ is the probability distribution determined by $\theta$. By now, we already know that we must be concerned not only with how well a test performs under the hypothesis $H_0$, but also with its behavior under other possibilities. For this reason, we introduce the following definition.

**4.3 Definition: Power Function.** In a test of hypotheses, the power function $\Pi$ is defined for all values of the specifying parameter $\theta$ by

$$\Pi(\theta) = P_\theta\{\text{reject } H_0\} = P_\theta\{S_1\}.$$

It is clearly desirable that $\Pi(\theta)$ be small for $\theta \in \Theta_0$ and large for $\theta \in \Theta_1$. We will be using the power function both as a theoretical and as a practical tool for judging the merit of test procedures.

The first substantial topic that we shall consider here conerns the situation when there is a clearly "best" level $\alpha$ test of $H_0$ versus $H_1$. Intuitively, this only occurs when *all* best tests of hypotheses of the form

$$\theta = \theta_0$$

against hypotheses of the form

$$\theta = \theta_1$$

lead to the same type of action (here $\theta_0$ and $\theta_1$ stand for arbitrary elements of $\Theta_0$ and $\Theta_1$, respectively). For example, they may all lead to rejecting $\theta = \theta_0$ whenever $\sum_{i=1}^{n} x_i$ is too large.

More formally, we give the following definition.

**4.4 Definition: Uniformly Most Powerful Level $\alpha$ Test.** A test $T$ of the hypothesis $H_0$: $\theta \in \Theta_0$ is said to be *uniformly most powerful* of level $\alpha$ against the hypothesis $H_1$: $\theta \in \Theta_1$ if:

(a) it is a level $\alpha$ test of $H_0$

(b) for each $\theta \in \Theta_1$, it minimizes the probability of accepting $H_0$ (type II error) among all level $\alpha$ tests of $H_0$.

We shall use the abbreviation U.M.P. for *uniformly most powerful.*

The Neyman-Pearson Lemma furnished us with U.M.P. tests in the case of a simple hypothesis versus a simple alternative. If we consider the case of a simple hypothesis versus a composite alternative, when a U.M.P. level $\alpha$ test exists we can use the Neyman-Pearson Lemma to find it.

## 4.5 *Theorem: U.M.P. Test of Simple Versus Composite*

Suppose the likelihood ratio test of size $\alpha$ of the simple hypothesis $\theta = \theta_0$ against the alternative $\theta = \theta_1$ is the same for each $\theta_1 \in \Theta_1$. That is, the acceptance region $S_0$ for $\theta = \theta_0$ does not depend on $\theta_1$. Then this test, with $S_0$ as the acceptance region, is a U.M.P. size $\alpha$ test of $H_0$: $\theta = \theta_0$ against $H_1$: $\theta \in \Theta_1$. Furthermore, it is essentially the *unique* such U.M.P. size $\alpha$ test of $H_0$ versus $H_1$, in the following sense. Let $T'$ be any other size $\alpha$ test of $H_0$ versus $H_1$, with $P_{\theta_1}\{\text{acc } H_0 \text{ using } T'\} \leq P_{\theta_1}\{\text{acc } H_0 \text{ using L.R. test of size } \alpha\}$ for some $\theta_1 \in \Theta_1$. Then $T'$ and the L.R. test take the same actions, with probability 1 for all $\theta$ being considered. If the size $\alpha$ likelihood ratio test of $\theta = \theta_0$ versus $\theta = \theta_1$ is not independent of which $\theta_1 \in \Theta_1$ is chosen, then there is no U.M.P. size $\alpha$ test of $H_0$ against $H_1$.

*Proof:* We note that when $H_0$ is a simple hypothesis, a level $\alpha$ test is the same as a size $\alpha$ test. This explains our reference to a U.M.P. *size* $\alpha$ test.

The proof is almost a direct consequence of the Neyman-Pearson Lemma. Certainly, if the size $\alpha$ likelihood ratio test of $\theta = \theta_0$ versus $\theta = \theta_1$ is the same for *all* $\theta_1 \subset \Theta_1$, this test is a U.M.P. test of $H_0$: $\theta = \theta_0$ versus $H_1$: $\theta \in \Theta_1$. The uniqueness follows from Theorem 2.9(2). We now show that there is no U.M.P. size $\alpha$ test of $H_0$ versus $H_1$ when the size $\alpha$ likelihood ratio tests of $\theta = \theta_0$ versus $\theta = \theta_1$ vary with $\theta_1$.

Suppose that there are two values, $\theta_1^*$ and $\theta_1^{**}$, in $\Theta_1$ with the following property: the size $\alpha$ likelihood ratio test, $T^*$, of $\theta = \theta_0$ versus $\theta = \theta_1^*$ is different from the size $\alpha$ likelihood ratio test, $T^{**}$, of $\theta = \theta_0$ versus $\theta = \theta_1^{**}$. Then a U.M.P. size test of $H_0$ versus $H_1$ would have to be at least as good as each of these tests. But, by the uniqueness property in Theorem 2.9(2), this would mean that such a U.M.P. test would have to be the same as $T^*$ and as $T^{**}$, and this is an impossibility under the given assumptions. Hence, if the L.R. tests of $\theta = \theta_0$ versus $\theta = \theta_1$ vary with $\theta \in \Theta_1$, we see that there is no U.M.P. size $\alpha$ test of $\theta = \theta_0$ versus $\theta \in \Theta_1$, concluding the proof.

## 4.6 *Example: U.M.P. Test of* $\boldsymbol{\theta} = \boldsymbol{\theta}_0$ *versus* $\boldsymbol{\theta} > \boldsymbol{\theta}_0$

Suppose that an "improved" version of an old vaccine has just been developed, and that we want to determine whether the new vaccine is really better than the old one. The new vaccine is administered to $n$ people and the antibody level is measured in each person. Let $X_j$ be a random variable whose observed value $x_j$ is the antibody level of the $j$th person to whom the vaccine is given. Suppose that $X_1, \ldots, X_n$ are independent normal random variables with variance 4. Furthermore, suppose that the expected antibody level for the old vaccine is $\theta_0$. Let us assume that we know the new vaccine to be at least as good as the old one. Then we want to test $H_0$: $E(X_i) \equiv \theta = \theta_0$ versus $H_1$: $E(X_i) \equiv \theta > \theta_0$.

We first check to determine whether all size $\alpha$ L.R. tests of $\theta = \theta_0$ versus $\theta = \theta_1 > \theta_0$ are the same. From computations like those in Example 2.19 on anthropology, we conclude that all these tests are of the form

reject $H_0$ if and only if

**4.7**
$$\sum_{i=1}^{n} x_i > d_\alpha,$$

where $d_\alpha$ is chosen so that

**4.8**
$$P_{\theta_0}\left\{\sum_{i=1}^{n} X_i > d_\alpha\right\} = \alpha.$$

Hence, in this case we can see from Theorem 4.5 that there is a U.M.P. size $\alpha$ test of $H_0$: $\theta = \theta_0$ versus $H_1$: $\theta > \theta_0$. It is given by 4.7, subject to 4.8.

In order to judge the quality of this test, and in particular to judge how large a sample size is needed for adequate discrimination, we look at the power function

$$\Pi(\theta) = P_\theta\left\{\sum_{i=1}^{n} X_i > d_\alpha\right\} = P_\theta\left\{\frac{1}{\sqrt{n}} \sum \frac{X_i - \theta}{2} > \frac{d_\alpha}{2\sqrt{n}} - \frac{\sqrt{n}\,\theta}{2}\right\}.$$

We note that when $E(X_i) = \theta$, then

$$\frac{1}{\sqrt{n}} \sum_{i=1}^{n} \frac{X_i - \theta}{2}$$

is a standard normal random variable. Since we want $\Pi(\theta_0) = \alpha$, we see that

$$\frac{d_\alpha}{2\sqrt{n}} - \frac{\sqrt{n}\,\theta_0}{2} = \varphi_\alpha,$$

where $\varphi_\alpha$ is the upper $1 - \alpha$ point of the standard normal distribution. Hence, $d_\alpha = 2\sqrt{n}\,\varphi_\alpha + n\theta_0$ and

$$\Pi(\theta) = 1 - \Phi\left(\varphi_\alpha + \frac{\sqrt{n}\,\theta_0}{2} - \frac{\sqrt{n}\,\theta}{2}\right)$$

where $\Phi$ is the standard normal distribution function. We plot this for $\alpha = .05$,

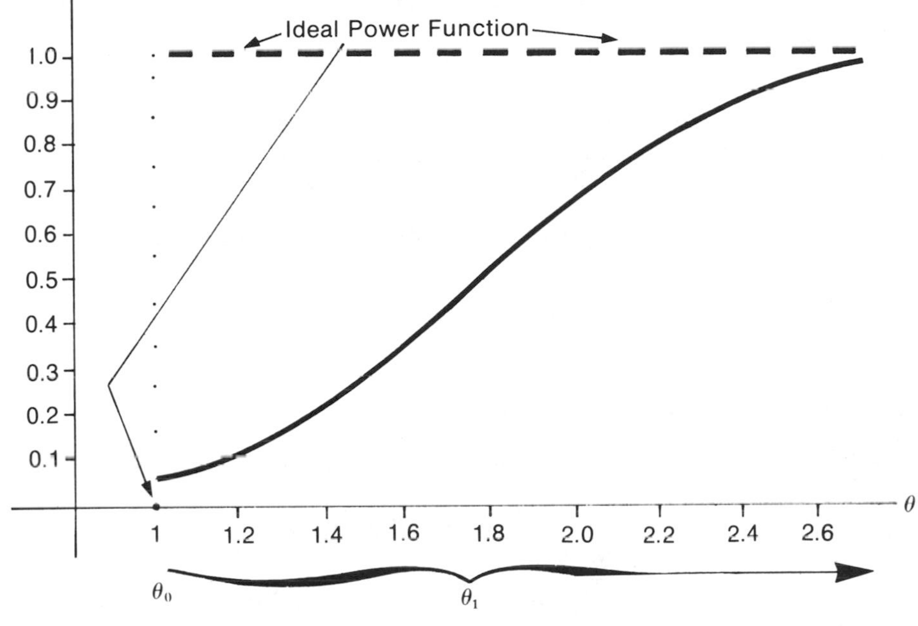


*FIGURE* 11–1. Power function for test of Example 4.6, $\alpha = .05$, $\theta_0 = 1$, $n = 16$.

$\theta_0 = 1$, $n = 16$, in Figure 11.1. Note that for $\theta > \theta_0$ near $\theta_0$ the power is not too much like the desired ideal, which is drawn in dashed lines. However, notice that if we look at values of $\theta$ fairly far away from $\theta_0$, then the test discriminates rather well. For $\theta$ near 2.6, the power of the test is almost 1. △

We next present an example in which no U.M.P. test exists.

### 4.9 *Example: No U.M.P. Test of* $\sigma^2 = \sigma_0^2$ *versus* $\sigma^2 \neq \sigma_0^2$

In trying to decide whether a given class is more or less typical, and hence whether it is reasonable to use it as a test group in an experimental program, a standardized test† is administered. The individual scores $x_1, \ldots, x_n$ are assumed to be the values of independent normal random variables $X_1, \ldots, X_n$ with known mean $E(X_i) = 75$ and Var $X_i = \sigma^2$. It is assumed that the class is "typical" if $\sigma^2 = 25$. Hence, it is desired to construct a size $\alpha$ test of the hypothesis $H_0$ that $\sigma^2 = 25$ versus the hypothesis $H_1$ that $\sigma^2 \neq 25$. We shall show that there is no U.M.P. size $\alpha$ test of $H_0$ versus $H_1$. To see this, let $\sigma_1^2 < 25$ and $\sigma_2^2 > 25$. Let us

† The really difficult question of whether a standardized test actually measures what is desired, is not treated here. Without a thorough verification of this requirement, the authors would be very skeptical of any conclusions drawn from the statistics alone.

look at the size $\alpha$ L.R. tests of $\sigma^2 = 25$ versus $\sigma^2 = \sigma_i^2$ $(i = 1, 2)$. The likelihood ratios are

$$\frac{f_{\sigma_i^2}(x, \ldots, x_n)}{f_{25}(x_1, \ldots, x_n)} = \frac{\dfrac{1}{(2\pi\sigma_i^2)^{n/2}} \exp\left(-\dfrac{1}{2\sigma_i^2} \sum_{i=1}^{n} (x_i - 75)^2\right)}{\dfrac{1}{(2\pi \cdot 25)^{n/2}} \exp\left(-\dfrac{1}{2 \cdot 25} \sum_{i=1}^{n} (x_i - 75)^2\right)}$$

$$= \left(\frac{25}{\sigma_i^2}\right)^{n/2} \exp\left(-1/2\left[\frac{1}{\sigma_i^2} - 1/25\right] \sum_{i=1}^{n} (x_i - 75)^2\right).$$

We see that

$$-\frac{1}{2}\left[\frac{1}{\sigma_1^2} - 1/25\right] < 0,$$

and hence the size $\alpha$ L.R. test of $\sigma^2 = 25$ versus $\sigma^2 = \sigma_1^2 < 25$ is of the type that rejects $\sigma^2 = 25$ if $\sum_{i=1}^{n} (x_i - 75)^2 < d_{1\alpha}$. Since

$$-\frac{1}{2}\left[\frac{1}{\sigma_2^2} - 1/25\right] > 0,$$

we see that the size $\alpha$ L.R. test of $\sigma^2 = 25$ versus $\sigma^2 = \sigma_2^2$ is of the type that rejects $\sigma^2 = 25$ if $\sum_{i=1}^{n} (x_i - 75)^2 > d_{2\alpha}$.

Hence, it follows from Theorem 4.5 that there is no U.M.P. size $\alpha$ test of $H_0$: $\sigma^2 = 25$ versus $H_1$: $\sigma^2 \neq 25$. This is essentially because there are two distinct "best" ways to handle the problem, one for $\sigma^2 < 25$ and another for $\sigma^2 > 25$. △

---

## *4.10 EXERCISES*

1.° Suppose measurements $X_1, X_2, \ldots, X_{25}$ on an "improved" instrument are assumed to be normal and independent with mean value $\mu = 3.27$ and variance $\sigma^2$. It is desired to determine whether or not this instrument is better than the old standard, which has known variance 16. A test of $H_0$ (that this new instrument is the same in quality) versus $H_1$ (that it is of better quality) is to be made, where better quality means smaller variance. Find the U.M.P. size $\alpha = .01$ such test and plot its power function.

2. Suppose that $\mu$ in Exercise 1 is fixed but unknown. Show that even here there is a test of $\sigma = \sigma_0 = 4$ versus $\sigma < \sigma_0$ that is almost as good as the U.M.P. test of Exercise 1.
Hint: Estimate $\mu$ by

$$\bar{X}_{(n)} = \frac{1}{n} \sum_{i=1}^{n} X_i$$

and note that $\sum_{i=1}^{n} (X_i - \bar{X}_{(n)})^2/\sigma^2$ has a Chi square distribution with $n - 1$ degrees of freedom (see 1 in Chapter 8). Hence, not knowing $\mu$ seems equivalent to "losing" one observation.

3.° It is desired to test whether a given coin is fair or biased in favor of heads. Develop the U.M.P. size $\alpha$ test based on $n$ tosses.

4. It is desired to determine whether or not a given radioactive source is up to a given standard, or whether it is below some minimum standard. For this purpose, assume that $X_1, \ldots, X_n$ are independent Poisson with mean $\theta$. Set up the U.M.P. test of size $\alpha$ of $H_0: \theta = \theta_0$ versus $H_1: \theta \leq \theta_1$, where $\theta_1$ and $\theta_0$ are given with $\theta_1 < \theta_0$.

5.° Prior to the Salk vaccine, the incidence of polio in the United States was 19 cases per 100,000 people in any given year. In order to test the Salk, $n$ people are selected at random from the population. We let $p_0 = 19/100{,}000$ and want to test $p = p_0$ versus $p < p_0$. Determine the test (sample size and procedure) which is a U.M.P. test of $H_0$: $p = p_0$ of level .001 and such that the probability of type II error at $p = 10/100{,}000$ is .001. Use Central Limit Theorem approximations (see what Uspensky's normal approximation to the binomial yields). We mention that if the incidence of the disease varies from year to year, then it is necessary to have a control group from which the probability that an unvaccinated person will contract the disease in that year is estimated. The required sample size for the vaccinated group would then be the size needed for the lowest incidence that one expects to observe in the population.

6.° Show that there is no U.M.P. size $\alpha$ test that a coin is fair, assuming that $\alpha$ is a possible size value of an L.R. test of $p = 1/2$.

7. Same as the previous problem, but to test fairness of a die.

The following result is an almost immediate consequence of Theorem 4.5, the U.M.P. test of simple versus composite.

### 4.11 *Theorem: U.M.P. Tests of Composite versus Composite*

Suppose that there is a value $\theta_0 \in \Theta_0$ yielding a U.M.P. size $\alpha$ test of $\theta = \theta_0$ versus $\theta \in \Theta_1$ of the type specified in Theorem 4.5. Let $S_0$ be the acceptance region for the hypothesis $\theta = \theta_0$, and let $T$ be the test of $H_0$: $\theta \in \Theta_0$ versus $H_1$: $\theta \in \Theta_1$, whose acceptance region for $H_0$ is $S_0$. If $T$ is also a level $\alpha$ test of $H_0$, then it is a U.M.P. level $\alpha$ test of $H_0$ versus $H_1$. The distribution $P_{\theta_0}$ is called the *least favorable distribution associated with the null hypothesis* $H_0$.

*Proof:* By hypothesis, no region can yield a more powerful test of size $\alpha$ of $\theta = \theta_0$ versus $\theta \in \Theta_1$ than $S_0$. But any test of $\theta \in \Theta_0$ can be used as a test of $\theta = \theta_0$ when $\theta_0 \in \Theta_0$; simply use the acceptance region for $\theta \in \Theta_0$ as the acceptance region for $\theta = \theta_0$. Hence, no test of $\theta \in \Theta_0$ vs. $\theta \in \Theta_1$ of level $\alpha$ can have more power than one using $S_0$ as the acceptance region for $\theta \in \Theta_0$. Since this region yields a level $\alpha$ test of $\theta \in \Theta_0$, it is a U.M.P. level $\alpha$ test of $H_0$: $\theta \in \Theta_0$ vs. $H_1$: $\theta \in \Theta_1$, concluding the proof.

The following example illustrates the theorem nicely.

### 4.12 *Example: U.M.P. Level $\alpha$ Test of $\theta \leq \theta_0$ versus $\theta > \theta_0$*

Consider the test of $\theta = \theta_0$ versus $\theta = \theta_1$, treated in Example 4.6, and let $\Theta_0 = (-\infty, \theta_0]$, $\Theta_1 = (\theta_0, \infty)$. Recall that the test procedure of Example 4.6 was to reject the hypothesis $\theta = \theta_0$ (accept $\theta > \theta_0$) if $\sum_{i=1}^{n} x_i > d_\alpha$, where $d_\alpha$ was chosen so that

$$P_{\theta_0}\left\{\sum_{i=1}^{n} X_i > d_\alpha\right\} = \alpha.$$

We claim:

(a) that the procedure to reject the hypothesis $H_0$ that

$$\theta \leq \theta_0 \quad \text{if and only if} \quad \sum_{i=1}^{n} x_i > d_\alpha$$

is a level $\alpha$ test of $H_0$: $\theta \leq \theta_0$ against $H_1$: $\theta > \theta_0$, and

(b) that $P_{\theta_0}$ is the least favorable distribution associated with $H_0$.

We see that it is a level $\alpha$ test as follows: Let $Y_i$ represent independent normal random variables with mean 0 and variance 4. Let $P_\theta\{\ \}$ represent a probability in which the random variables $X_i$ inside the braces are independent normal with mean $\theta$ and variance 4. Now, for $\theta < \theta_0$,

$$\begin{aligned}
P_\theta\left\{\sum_{i=1}^{n} X_i > d_\alpha\right\} &= P\left\{\sum_{i=1}^{n} (Y_i + \theta) > d_\alpha\right\} \\
&= P\left\{\sum_{i=1}^{n} Y_i > d_\alpha - n\theta\right\} \\
&= P\left\{\sum_{i=1}^{n} (Y_i + \theta_0) > d_\alpha - n(\theta - \theta_0)\right\} \\
&= P_{\theta_0}\left\{\sum_{i=1}^{n} X_i > d_\alpha - n(\theta - \theta_0)\right\} \\
&< P_{\theta_0}\left\{\sum_{i=1}^{n} X_i > d_\alpha\right\} \\
&= \alpha.
\end{aligned}$$

Thus, we see that the test procedure of rejecting $H_0$: $\theta \leq \theta_0$ if and only if $\sum_{i=1}^{n} x_i > d_\alpha$ is a level $\alpha$ test of $H_0$. The last equality shows $P_{\theta_0}$ to be least favorable in $H_0$. Hence, the above procedure, which is essentially that of Example 4.6, is a U.M.P. level $\alpha$ test of $\theta \leq \theta_0$ versus $\theta > \theta_0$. △

---

### 4.13 EXERCISES

1. Show that the test whose rejection region is the same as that derived in the answer to Exercise 4.10(1) is a U.M.P. level $\alpha$ test of $H_0{:}\sigma^2 \geq \sigma^2 = 16$ versus $H_1{:}\sigma^2 < 16$. Plot the power function over the larger set of $\sigma$'s.

2.° Find the U.M.P. level $\alpha$ test of $H_0{:}p \leq p_0$ versus $H_1{:}p > p_0$, where $X_1, \ldots, X_n$ are independent Bernoulli random variables, that is, $P\{X_i = 1\} = p = 1 - P\{X_i = 0\}$.

3. Radiation measurements $X_1, \ldots, X_{100}$ are made. They are assumed to be independent Poisson with mean $\lambda$. Find the U.M.P. level $\alpha$ test of $\lambda \geq \lambda_0$ versus $\lambda < \lambda_0$.

4. Suppose that $T_1, \ldots, T_n$ are random variables whose values are the lifetimes of electric light bulbs of a new type. The old type of light bulb had an expected lifetime of 1000 hours. Suppose that $T_1, \ldots, T_{100}$ are independent, each with density $(1/\theta)e^{-t/\theta}$ for $t \geq 0$. We therefore set up a .05 level U.M.P. test of $H_0: \theta \leq 1000$ versus $H_1: \theta > 1000$. Find this test and plot its power function (you will have to use the Central Limit Theorem for approximate computations).

We now give another example of a case in which no U.M.P. test exists, followed by a reasonable test of the hypothesis in question.

### 4.14 *Example: Televised versus Classroom Instruction*

It is desired to determine whether the effect of televised instruction is different from that of normal classroom lectures. A set of $n$ students receiving televised instruction are given an examination. The score $x_j$ of the $j$th student is the value of a random variable $X_j$. It is assumed that $X_1, \ldots, X_n$ are independent normal random variables with Var $X_i = 9$. Normal classroom instruction yields $E(X_i) = 75$. If $E(X_i) = \theta$ for televised instruction, we desire to test $H_0$: $\theta = 75$ versus $H_1$: $\theta \neq 75$. (Actually, we are really interested in testing $H_0^*$: $|\theta - 75| \leq 6$ versus $H_1^*$: $|\theta - 75| > 6$, but the problem as posed is simpler to treat.) It is easy to see that there is no U.M.P. test of $H_0$ versus $H_1$ of size $\alpha$, for the U.M.P. test of size $\alpha$ of $\theta = 75$ versus $\theta > 75$ is of the form

$$\text{reject } H_0 \text{ if and only if } \sum_{i=1}^{n} x_i > d_\alpha,$$

whereas the U.M.P. size $\alpha$ test of $\theta = 75$ versus $\theta < 75$ is of the form

$$\text{reject } H_0 \text{ if and only if } \sum_{i=1}^{n} x_i < D_\alpha.$$

It follows from Theorem 4.5 that there is no U.M.P. test of $H_0$ vs. $H_1$. △

The fact that there is no U.M.P. test of $H_0$ versus $H_1$ in the last example does not mean that there is no "good" test of $H_0$ against $H_1$.

### 4.15 *Example: Reasonable Test of Televised versus Classroom Instruction*

We claim that a reasonable size $\alpha$ test of $H_0$: $\theta = 75$ versus $H_1$: $\theta \neq 75$ of the previous example is to reject $H_0$ if and only if $\left|\sum_{i=1}^{n} (x_i - 75)\right| > k_\alpha$, where $k_\alpha$ is chosen so that, under $H_0$,

$$P\left\{\left|\sum_{i=1}^{n} (X_i - 75)\right| > k_\alpha\right\} = \alpha.$$

We can compare the performance of this test with that of the best (likelihood ratio) size $\alpha$ test of $\theta = \theta_0$ versus $\theta = \theta_1$. For the suggested test, the probability of type II error at $\theta = \theta_1$ is

**4.16**
$$P_{\theta_1}\left\{\left|\sum_{i=1}^{n}(X_i - 75)\right| \leq k_\alpha\right\}$$

$$= P_{\theta_1}\left\{\left|\frac{1}{\sqrt{n}}\sum_{i=1}^{n}\frac{X_i - \theta_1}{3} + \sqrt{n}\left(\frac{\theta_1 - 75}{3}\right)\right| \leq \frac{k_\alpha}{3\sqrt{n}}\right\}$$

$$= P_{\theta_1}\left\{\left|\frac{1}{\sqrt{n}}\sum_{i=1}^{n}\frac{X_i - \theta_1}{3} + \sqrt{n}\left(\frac{\theta_1 - 75}{3}\right)\right| \leq \varphi_{1-\alpha/2}\right\}$$

(where $\varphi_{\alpha/2}$ is the upper $1 - \alpha/2$ point of the standard normal distribution, needed to obtain probability $\alpha$ of type I error)

$$= P_{\theta_1}\left\{-\varphi_{\alpha/2} - \sqrt{n}\left(\frac{\theta_1 - 75}{3}\right) \leq \frac{1}{\sqrt{n}}\sum_{i=1}^{n}\frac{X_i - \theta_1}{3} \leq \varphi_{\alpha/2} - \sqrt{n}\left(\frac{\theta_1 - 75}{3}\right)\right\}.$$

Now assume, for example, that $\theta_1 > 75$. The L.R. test of $\theta = 75$ versus $\theta = \theta_1$ of size $\alpha$ is of the form

$$\text{reject } \theta = 75 \text{ if and only if} \sum_{i=1}^{n}(x_i - 75) > c_\alpha.$$

It is not difficult to see that the probability of type II error for this test is

**4.17**
$$P_{\theta_1}\left\{\frac{1}{\sqrt{n}}\sum_{i=1}^{n}\frac{X_i - \theta_1}{3} \leq \varphi_\alpha - \sqrt{n}\left(\frac{\theta_1 - 75}{3}\right)\right\}.$$

There are now various ways to make the desired comparison.

(a) For some fixed $\theta_1$, determine how much larger $n$ has to be in Equation 4.16 to achieve the same result as in 4.17. For example, choosing $\alpha = .05$, $n = 9$, and $\theta_1 = 78$ in 4.17, we could (neglecting the left side of the event in 4.16) choose $n$ in 4.16 so that

$$1.96 - \sqrt{n} = 1.645 - 3 \qquad \text{or} \qquad \sqrt{n} = 3.315, \qquad n = 11.$$

We see that at $\theta = 78$, the cost of achieving the same probability of type II error as the best test is two observations. Note that at least as good a probability of type II error is achieved for all $\theta > 78$.

(b) It may be that sample size is determined in advance and cannot be increased (as in the case of aircraft collision warning). Here it is more reasonable to have $n$ the same in 4.16 and 4.17. In this case, for a given value of $\theta_1$ in 4.17, we might determine what value of $\theta_1$ to choose in 4.16 to achieve at least the same probability of type II error. If $\alpha = .05$, $n = 9$, and $\theta_1 = 78$ in 4.17, then choosing $\theta_1$ in 4.16 so that

$$1.645 - 3 = 1.96 - 3\left(\frac{\theta_1 - 75}{3}\right)$$

yields $\theta_1 = 78.315$. That is, at least as good a probability of type II error is achieved by the suggested test at 78.315 as is achieved by the best test at $\theta = 78$.

This type of comparison may have no useful interpretation in some problems. In a problem such as the one of collison warning, it is a very reasonable one, since it tells at what miss distance $d_1^*$ this size $\alpha$ test discriminates as well as the best size $\alpha$ test of $d = 0$ versus $d = d_1$.

(c) If the sample size must remain the same, we might choose $\theta_1$ and compare the probability of type II error of the best test of $\theta = \theta_0$ versus $\theta = \theta_1$ with the probability of type II error of the suggested test at $\theta = \theta_1$.

Such reasoning leads us to conclude that a *reasonable* test of $H_0$: $|\theta - 75| \leq 6$ versus: $|\theta - 75| > 6$ is

$$\text{reject } H_0 \text{ if and only if } \left|\sum_{i=1}^{n} (x_i - 75)\right| > q_\alpha,$$

where $q_\alpha$ is chosen so that $P_{81}\{\text{reject } H_0\} = \alpha = P_{69}\{\text{reject } H_0\}$.

The symmetry of the normal density permits us to satisfy both equalities with one choice of $q_\alpha$. For $\theta > 81$, this test behaves much like the U.M.P. test of $\theta = 81$ versus $\theta > 81$.

We want to urge students who want to use statistical procedures of the type being presented to take great care to ensure that the examination administered really measures what it's intended to. Examinations that measure the wrong attributes can do incalculable harm, as attested by the devastating impact of the "look-see" method once used as the only allowed method to teach reading. Similar remarks hold for tests of drugs, since short-term tests generally cannot be depended on to measure long-term effects and heavy dosage may tell little about the effects of long-term light dosage. △

We note an important principle, namely that ***the Neyman-Pearson tests given by Theorem 2.9 are useful as comparison yardsticks to measure the merit of tests of more complicated hypotheses.***

---

## *4.18 PROBLEMS*

1. Let $X_1, \ldots, X_{16}$ be independent normal random variables with $E(X_i) = 6.3$, Var $X_i = \sigma^2$, representing measurements made by a new type of device. The old device had $\sigma_0^2 = 9$ and an expectation of 6.3. Construct a level $\alpha$ test of $H_0: \sigma^2 = 9$ versus $H_1: \sigma^2 \neq 9$. Determine the merit of this test, as was done in Example 4.15, at $\sigma^2 = 16$.

2. Repeat Problem 1, with $H_0$ as $|\sigma^2 - 9| \leq 5$ and $H_1$ as $|\sigma^2 - 9| > 5$, and the comparison is with the U.M.P. size $\alpha$ test of $\sigma^2 = 14$ versus $\sigma^2 > 14$ in general.

3.° Construct a reasonable test of the hypothesis that a coin is fair, based on 100 observations, and graph an approximation to the power function when $\alpha = .05$.

4.° Construct a reasonable test of the hypothesis that a roulette wheel is fair, based on $n$ observations, and try to justify your claim that this test is reasonable.

## SECTION 5
## Generalized Likelihood Ratio Tests

The reader may have noticed that the situations treated in previous sections usually involved only one unknown parameter. In many cases, it happens that although we want to determine something about a parameter $\theta$, other parameters are a factor in determining the distribution of the observed random variables. Thus, we see that in many cases the methods developed up until now are of limited applicability. Although no really satisfactory theory has been developed to handle these situations, the ad hoc method of test construction we are about to present often yields quite useful and reasonable results.

Suppose $\mathscr{P}_0$ and $\mathscr{P}_1$ are two disjoint parametric families of distributions, where $\Theta_i$ is the set of parameter values associated with $\mathscr{P}_i$. Let $x_1, \ldots, x_n$ be the observed values of the random variables $X_1, \ldots, X_n$. Suppose that $f(x_1, \ldots, x_n; \theta)$ is the joint density of $X_1, \ldots, X_n$ at $(x_1, \ldots, x_n)$ when the parameter is $\theta$.

**5.1 Definition: Generalized Likelihood Ratio Test.** A test of $H_0$: $\theta \in \Theta_0$ versus $H_1$: $\theta \in \Theta_1$ of the form

reject $H_0$ (the hypothesis that $\theta \in \Theta_0$) if and only if

$$\mathscr{L} = \frac{\sup\limits_{\theta \in \Theta_0} f(x_1, \ldots, x_n; \theta)}{\sup\limits_{\theta \in \Theta_0} f(x_1, \ldots, x_n; \theta)} > c,$$

where $c$ is some constant, is called a *generalized likelihood ratio test*.†

A similar definition is given in the case in which a discrete probability function replaces the density. Intuitively, this test chooses $\Theta_1$ over $\Theta_0$ if the "best" possible explanation of the data, assuming $\theta \in \Theta_1$, is better than the "best" possible explanation assuming $\theta \in \Theta_0$.

Although the concept just introduced is not difficult to understand, the actual manipulations needed to compute the sup and to reduce the test to a simple form can be extremely tedious.

### 5.2 *Example: Two-Sided t-Test*

There are many cases in which random variables $X_1, X_2, \ldots, X_n$ are assumed independent normal with $E(X_i) = \mu$, $\text{Var } X_i = \sigma^2 > 0$, with both $\mu$ and $\sigma^2$ unknown, in which we want to know whether or not $\mu$ is equal to some given value $\mu_0$. Such a model is used when it is desired to test a new type of fertilizer for its effect

---

† By $\sup\limits_{\theta \in \Theta_i} f(x_1, \ldots, x_n; \theta)$, we mean a number $b$ such that:

(a) $f(x_1, \ldots, x_n; \theta) \leq b$ for each $\theta$ in $\Theta_i$, and

(b) no smaller number satisfies the previous condition.

The usual techniques of calculus are often helpful for finding $\sup\limits_{\theta \in \Theta} f(x_1, \ldots, x_n; \theta)$.

on average plant yield, in case the fertilizer may also affect variability. Here $\mu_0$ is the yield of plants not treated with this fertilizer. Cloud seeding tests might fall in this category, as well as testing the output of a new brand of magnetic tape, or the effects of a new drug. We derive the generalized likelihood ratio test of $H_0$: $\mu = \mu_0$ versus $H_1$: $\mu \neq \mu_0$. In this case, $\theta = (\mu, \sigma^2)$ and

**5.3**
$$f(x_1, \ldots, x_n; \theta) = \frac{1}{(2\pi\sigma^2)^{n/2}} \exp\left(-\frac{1}{2\sigma^2}\sum_{i=1}^{n}(x_i - \mu)^2\right).$$

Here $\Theta_0$ is the set of all pairs $(\mu_0, \sigma^2)$ for $\sigma^2 > 0$, and $\Theta_1$ is the set of all pairs $(\mu, \sigma^2)$ with $\mu \neq \mu_0$ and $\sigma^2 > 0$.

For the denominator of the generalized likelihood ratio test, $\sup_{\theta \in \Theta_1} f(x_1, \ldots, x_n; \theta)$, we maximize 5.3 over $\sigma^2$ with $\mu$ set equal to $\mu_0$. We take logarithms, then differentiate with respect to $\sigma^2$, and set the derivative equal to 0. It follows that if a maximum is achieved, it must occur at

$$\sigma^2 = \frac{1}{n}\sum_{i=1}^{n}(x_i - \mu_0)^2 = \tilde{\sigma}^2.$$

Using the criteria summarized in the Appendix, it can be shown that the maximum is achieved (the minimum is *not* achieved). Hence,

$$\sup_{\theta \in \Theta_0} f(x_1, \ldots, x_n; \theta) = \frac{1}{\left(\frac{2\pi}{n}\right)\left(\sum_{i=1}^{n}[x_i - \mu_0]^2\right)^{n/2}} e^{-n/2}.$$

For determination of $\sup_{\theta \in \Theta_1} f(x_1, \ldots, x_n; \theta)$, we carry out a similar procedure, but with partial derivatives, yielding two equations in two unknowns for the "maximizing" values of $\mu$, $\sigma^2$. Here we find

$$g(\mu, \sigma^2) = \log f(x_1, \ldots, x_n; \theta) = -\frac{n}{2}\log(2\pi\sigma^2) - \frac{1}{2\sigma^2}\sum_{i=1}^{n}(x_i - \mu)^2$$

$$\frac{\partial g}{\partial \mu} = \frac{1}{\sigma^2}\sum(x_i - \mu) \qquad \frac{\partial g}{\partial \sigma^2} = \frac{-n}{2\sigma^2} + \frac{1}{2\sigma^4}\sum_{i=1}^{n}(x_i - \mu)^2.$$

Setting

$$\frac{\partial g}{\partial \mu} = 0, \qquad \frac{\partial g}{\partial \sigma^2} = 0$$

we find

**5.4**
$$\mu = \frac{1}{n}\sum_{i=1}^{n} x_i, \qquad \sigma^2 = \frac{1}{n}\sum_{i=1}^{n}\left[x_i - \frac{1}{n}\sum_{j=1}^{n} x_j\right]^2$$

Note that for $\sigma^2 > 0$, the probability that

$$\frac{1}{n}\sum_{i=1}^{n} X_i$$

will assume the value $\mu_0$ is 0. Hence, there is no danger in using the unrestricted solution of

$$\frac{\partial g}{\partial \mu}(x_1, \ldots, x_n; \theta) = 0, \quad \frac{\partial g}{\partial \sigma^2}(x_1, \ldots, x_n; \theta) = 0$$

to determine $\sup_{\theta \in \Theta_1} f(x_1, \ldots, x_n; \theta)$. Substituting the values from 5.4 into $f$ and using the reasoning concerning achievement of the maximum given in the Appendix, we find that

$$\sup_{\theta \in \Theta_1} f(x_1, \ldots, x_n; \theta) = \frac{1}{\left(\frac{2\pi}{n}\sum_{i=1}^{n}\left[x_i - \frac{1}{n}\sum_{j=1}^{n} x_j\right]^2\right)^{n/2}} e^{-n/2}.$$

Hence,

$$\mathscr{L} = \frac{\sup_{\theta \in \Theta_1} f(x_1, \ldots, x_n; \theta)}{\sup_{\theta \in \Theta_0} f(x_1, \ldots, x_n; \theta)} = \left[\frac{\sum_{i=1}^{n}(x_i - \mu_0)^2}{\sum_{i=1}^{n}\left(x_i - \frac{1}{n}\sum_{j=1}^{n} x_j\right)^2}\right]^{n/2}.$$

For simplicity, let

$$\frac{1}{n}\sum_{j=1}^{n} x_j = \bar{x}_{(n)} \quad \text{and} \quad s_n = \sqrt{\frac{1}{n-1}\sum_{i=1}^{n}(x_i - \bar{x}_{(n)})^2}.$$

Then

$$\mathscr{L} = \left[\frac{\sum_{i=1}^{n}(x_i - \bar{x}_{(n)} + \bar{x}_{(n)} - \mu_0)^2}{(n-1)s_n^2}\right]^{n/2}.$$

But

$$\begin{aligned}
&\sum_{i=1}^{n}(x_i - \bar{x}_{(n)} + \bar{x}_{(n)} - \mu_0)^2 \\
&\quad = \sum_{i=1}^{n}(x_i - \bar{x}_{(n)})^2 + 2\sum_{i=1}^{n}(x_i - \bar{x}_{(n)})(\bar{x}_{(n)} - \mu_0) + \sum_{i=1}^{n}(\bar{x}_{(n)} - \mu_0)^2 \\
&\quad = (n-1)s_n^2 + 2(\bar{x}_{(n)} - \mu_0)\sum_{i=1}^{n}(x_i - \bar{x}_{(n)}) + n(\bar{x}_{(n)} - \mu_0)^2 \\
&\quad = (n-1)s_n^2 + n(\bar{x}_{(n)} - \mu_0)^2,
\end{aligned}$$

since

$$\sum_{i=1}^{n}(x_i - \bar{x}_{(n)}) = \sum_{i=1}^{n} x_i - n\bar{x}_{(n)} = n\bar{x}_{(n)} - n\bar{x}_{(n)} = 0.$$

Hence

$$\begin{aligned}
\mathscr{L} &= \left[\frac{(n-1)s_n^2 + n(\bar{x}_{(n)} - \mu_0)^2}{(n-1)s_n^2}\right]^{n/2} \\
&= \left[1 + \frac{1}{n-1}\frac{[\sqrt{n}(\bar{x}_{(n)} - \mu_0)]^2}{s_n^2}\right]^{n/2}.
\end{aligned}$$

We can therefore see that the generalized likelihood ratio test of $\mu \neq \mu_0$ against $\mu \neq \mu_0$,

$$\text{reject } \mu = \mu_0 \quad \text{if} \quad \mathscr{L} > c,$$

is equivalent to the procedure

$$\text{reject } \mu = \mu_0 \quad \text{if} \quad \left|\frac{\sqrt{n}(\bar{x}_{(n)} - \mu_0)}{s_n}\right|^2 > d^2,$$

where $d$ is some positive constant. That is, accept $\mu = \mu_0$ if and only if

**5.5**
$$\left|\frac{\sqrt{n}(\bar{x}_{(n)} - \mu_0)}{s_n}\right| \leq d.$$

We can make this a size $\alpha$ test of $\mu = \mu_0$ by noticing that

$$\frac{\sqrt{n}(\bar{x}_{(n)} - \mu_0)}{s_n}$$

is the value of a random variable having the $t$ distribution with $n - 1$ degrees of freedom under the hypothesis $\mu = \mu_0$ (see 8 in Chapter 8). Thus, to accomplish the desired result, we need only choose the constant $d$ to be the upper $1 - \alpha/2$ point of the $t$ distribution with $n - 1$ degrees of freedom.

It can be shown that the power of this test depends on $\mu$ and $\sigma$ only through $|(\mu - \mu_0)/\sigma|$. This fact should come as no great surprise when looking at 5.5.

Thus, if nothing is known about the value of $\sigma$, then there is no way to assure good power with this test at a pre-assigned value $\mu = \mu_1$. It may be, of course, that the cost of failing to reject $H_0$ depends on $|(\mu - \mu_0)/\sigma|$ in the same way that the power does. In such a case, the dependence of the power on $|(\mu - \mu_0)/\sigma|$ causes no difficulty. In the majority of practical situations in which this test is used, the experimenter will have some idea of the value of $\sigma$, based on past experience. He can then determine $n$ to discriminate as he desires against $\mu = \mu_1$ by appropriate power computations on the distribution that governs $\sqrt{n}(\bar{X}_{(n)} - \mu_0)/S_n$ when $E(X_i) = \mu_1$. This distribution is the non-central $t$ distribution with $n - 1$ degrees of freedom and non-centrality parameter $(\mu_1 - \mu_0)/\sigma$ (see 10, Chapter 8). △

### 5.6 *Example: One Sided t-Test*

In the applications that generated the last example, we might know from theoretical considerations that $\mu \geq \mu_0$. By building on the previous example, we can construct the generalized likelihood ratio test of $H_0$: $\mu = \mu_0$ versus $H_1$: $\mu > \mu_0$, where $X_1, \ldots, X_n$ are independent normal random variables with unknown mean $\mu$ and unknown variance $\sigma^2$. Here, $\Theta_0$ is the set of all pairs $(\mu_0, \sigma^2)$, where $\sigma^2 > 0$. However, $\Theta_1$ is the set of all pairs $(\mu, \sigma^2)$ with $\mu > \mu_0$ and $\sigma^2 > 0$. From the

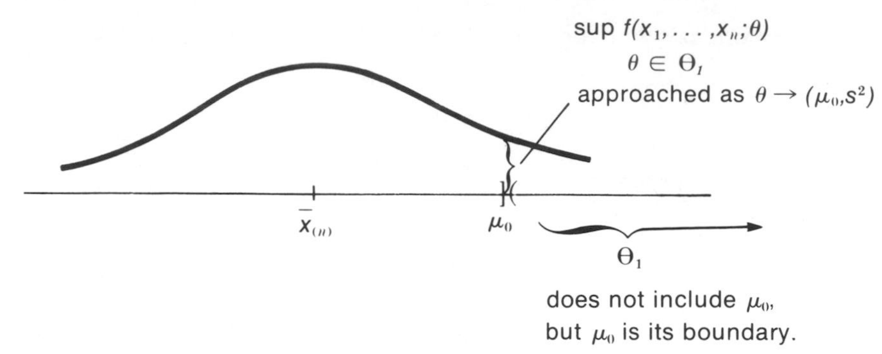


*FIGURE* 11–2.

previous example, we see that the generalized likelihood ratio is

$$\mathscr{L} = \frac{\sup\limits_{\theta\in\Theta_1} f(x_1, \ldots, x_n; \theta)}{\sup\limits_{\theta\in\Theta_0} f(x_1, \ldots, x_n; \theta)}$$

$$= \begin{cases} \left[1 + \dfrac{1}{n-1}\left(\dfrac{\sqrt{n}(\bar{x}_{(n)} - \mu_0)}{s_n}\right)^2\right]^{n/2} & \text{for} \quad \bar{x}_{(n)} > \mu_0 \\ 1 & \text{for} \quad \bar{x}_{(n)} \leq \mu_0. \end{cases}$$

This follows from the fact that when $\bar{x}_{(n)} \leq \mu_0$, we choose the boundary point of $\Theta_1$ closest to $\bar{x}_{(n)}$ (that is, $\mu_0$) in computing $\sup\limits_{\theta\in\Theta_1} f(x_1, \ldots, x_n; \theta)$. We can see the reasoning by looking at Figure 11.2 and using the fact that for unrestricted $\theta$ there is one maximum. If $\mu_0$ were in $\Theta_1$, the maximum over $\mu_1$ would be achieved at $\mu_0$. From this, it is not hard to see that when $c_\alpha > 1$, we have $\mathscr{L} > c_\alpha$ if and only if

$$\frac{\sqrt{n}(\bar{x}_{(n)} - \mu_0)}{s_n} > d_\alpha$$

where $d_\alpha$ is a constant. To obtain a size $\alpha$ test of $H_0$, we choose $d_\alpha$ as the upper $1 - \alpha$ point of the $t$ distribution, with $n - 1$ degrees of freedom. △

The next example is made optional not because of the conceptual difficulties, but because of the purely technical difficulties of determining the behavior of the likelihood ratio.

### 5.7 *Example: F Test*

We are given random variables $X_1, \ldots, X_m$ whose values are the measurements on one instrument, and random variables $Y_1, \ldots, Y_n$ whose values are measurements on another instrument. We would like to compare these instruments for quality. We assume that $X_1, \ldots, X_m, Y_1, \ldots, Y_n$ are independent normal with mean 0, Var $X_i = \sigma_x^2$, Var $Y_i = \sigma_y^2$. In order to carry out a comparison, we set up the generalized likelihood ratio test of the hypothesis $H_0: \sigma_x^2 = \sigma_y^2$ versus $H_1: \sigma_x^2 \neq \sigma_y^2$. Here $\theta = (\sigma_x^2, \sigma_y^2)$, while $\Theta_0$ is the collection of all $\theta$ where $\sigma_x^2 = \sigma_y^2$, and $\Theta_1$ is the collection of all $\theta$ with $\sigma_x^2 \neq \sigma_y^2$. Now

$$f(x_1, \ldots, x_m, y_1, \ldots, y_n; \theta) = \frac{1}{(2\pi\sigma_x^2)^{m/2}(2\pi\sigma_y^2)^{n/2}} \times \exp\left(-\frac{1}{2\sigma_x^2}\sum_{i=1}^{m} x_i^2 - \frac{1}{2\sigma_y^2}\sum_{i=1}^{n} y_i^2\right).$$

To find $\sup_{\theta \in \Theta_0} \cdots$, we first set $\sigma_x^2 = \sigma_y^2 = \sigma^2$ in the foregoing expression. We then maximize by taking logarithms, and setting

$$\frac{\partial \log f}{\partial \sigma^2} = 0.$$

This yields

$$\sup_{\theta \in \Theta_0} f(x_1, \ldots, x_m, y_1, \ldots, y_n; \theta) = \frac{1}{\left(\dfrac{2\pi}{m+n}\left[\sum_{i=1}^{m} x_i^2 + \sum_{j=1}^{n} y_j^2\right]\right)^{(m+n)/2}} \times \exp\left(-\frac{m+n}{2}\right).$$

To find $\sup_{\theta \in \Theta_1} f(x_1, \ldots, x_m, y_1, \ldots, y_n; \theta)$ we perform similar calculations, from which we find that the maximizing values of $\sigma_x^2$ and $\sigma_y^2$ are given by

$$\tilde{\sigma}_x^2 = \frac{1}{m}\sum_{i=1}^{m} x_i^2 \qquad \tilde{\sigma}_y^2 = \frac{1}{n}\sum_{i=1}^{n} y_i^2.$$

From this, we obtain

$$\sup_{\theta \in \Theta_1} f(x_1, \ldots, x_m, y_1, \ldots, y_n; \theta) = \frac{1}{\left(\dfrac{2\pi}{m}\sum_{i=1}^{m} x_i^2\right)^{m/2}\left(\dfrac{2\pi}{n}\sum_{i=1}^{n} y_i^2\right)^{n/2}} e^{-(m+n)/2}.$$

The generalized likelihood ratio

$$\mathscr{L} = \frac{\left(\dfrac{2\pi}{m+n}\left[\sum_{i=1}^{m} x_i^2 + \sum_{j=1}^{n} y_j^2\right]\right)^{(m+n)/2}}{\left(\dfrac{2\pi}{m}\sum_{i=1}^{m} x_i^2\right)^{m/2}\left(\dfrac{2\pi}{n}\sum_{j=1}^{n} y_j^2\right)^{n/2}}$$

$$= \frac{m^{m/2} n^{n/2}}{(m+n)^{(m+n)/2}}\left[1 + \frac{\sum_{j=1}^{n} y_j^2}{\sum_{i=1}^{m} x_i^2}\right]^{m/2}\left[1 + \frac{\sum_{i=1}^{m} x_i^2}{\sum_{j=1}^{n} y_j^2}\right]^{n/2}.$$

In order to see how $\mathscr{L}$ behaves, we examine

$$g(y) = [1 + y]^{m/2}\left[1 + \frac{1}{y}\right]^{n/2} = [1 + y]^{m/2}[1 + y]^{n/2} y^{-n/2}.$$

Let

$$h(y) = \log g(y) = \frac{m+n}{2}\log(1 + y) - \frac{n}{2}\log y.$$

Then

$$h'(y) = \frac{m+n}{2(1+y)} - \frac{n}{2y},$$

and

$$h'(y) < 0 \quad \text{for} \quad 0 < y < n/m, \quad h'(n/m) = 0, \quad h'(y) > 0 \quad \text{for} \quad y > n/m.$$

Therefore, $h(y)$ decreases for $0 < y < n/m$. Since $g$ is an increasing function of $h$ ($g = e^h$), it behaves in the same way. Thus, $g$ looks as pictured in Figure 11.3. From this, we see that the condition $\mathscr{L} > c$, which is the same as

$$\frac{m^{m/2}n^{n/2}}{(m+n)^{(m+n)/2}} g\left(\frac{\sum_{j=1}^{n} Y_j^2}{\sum_{i=1}^{m} X_i^2}\right) > c,$$

is equivalent to the condition that

$$\frac{\sum_{j=1}^{n} Y_j^2}{\sum_{i=1}^{m} X_i^2} \quad \text{lies outside} \quad [a, b].$$

If $m = n$, it is easy to see that the test is to reject $H_0$: $\sigma_x^2 = \sigma_y^2$ if

**5.8**
$$\frac{\sum_{j=1}^{m} Y_j^2}{\sum_{i=1}^{m} X_i^2} \quad \text{lies outside} \quad \left[\frac{1}{a}, a\right]$$

where $a$ is chosen so as to make this a level $\alpha$ test. To find $a$, we use the fact that under the null hypothesis, the statistic involved in expression 5.8 has an $F$ distribution with $(m, m)$ degrees of freedom. The power of the test is not hard to compute, since when Var $X_i = \sigma_x^2$ and Var $Y_i = \sigma_y^2$, we know that $X_i/\sigma_x$ and $Y_i/\sigma_y$ are standard

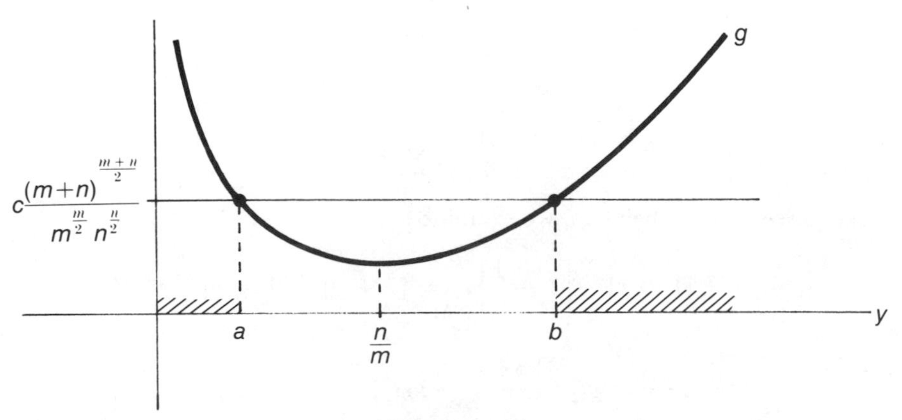


*FIGURE* 11–3.

normal (see Example 2.25, Chapter 4). Thus, from 11 in Chapter 8, we see that

$$\frac{\sigma_x^2 \sum_{j=1}^{m} Y_j^2}{\sigma_y^2 \sum_{i=1}^{m} X_i^2}$$

has an $F$ distribution with $(m, m)$ degrees of freedom. Hence, the probability of accepting the hypothesis $H_1$: $\sigma_x^2 \neq \sigma_y^2$ is

**5.9**
$$1 - P\left\{\frac{1}{a}\frac{\sigma_x^2}{\sigma_y^2} \leq Z_{m,m} \leq a\frac{\sigma_x^2}{\sigma_y^2}\right\}$$

where $Z_{m,m}$ has an $F$ distribution with $(m, m)$ degrees of freedom.

For $m \neq n$, a computation of a numerical nature is required. This arises from the fact that if we are given $a$, the left end point of the acceptance interval, we must find $b > a$ such that

$$[1 + b]^{m/2}\left[1 + \frac{1}{b}\right]^{n/2} = [1 + a]^{m/2}\left[1 + \frac{1}{a}\right]^{n/2}.$$

This is a non-trivial task, requiring techniques such as Newton's method. Then we compute the probability of type I error, and if it is too large or too small we adjust $a$ and then $b$ accordingly. Alternatively, we could allow a larger class of tests by choosing $a$ and $b$ as we pleased, satisfying only the condition that the probability of type I error be $\alpha$, with $a < 1$, $b > 1$. All of this will make use of the fact that

$$\frac{m}{n}\frac{\sigma_x^2}{\sigma_y^2}\frac{\sum_{j=1}^{n} Y_j^2}{\sum_{i=1}^{m} X_i^2}$$

has an $F$ distribution with $(n, m)$ degrees of freedom. We can see that the foregoing random variable can be used to generate confidence intervals for $\sigma_x^2/\sigma_y^2$. We can also obtain confidence intervals for $\sigma_x^2$ using the fact that

$$\sum_{i=1}^{m} \frac{X_i^2}{\sigma_x^2}$$

has a Chi square distribution with $m$ degrees of freedom. Similar remarks hold for $\sigma_y^2$. If $E(X_i) = \mu_x$ and $E(Y_j) = \mu_y$ are unknown, it can be shown that the G.L.R. test of the hypothesis $H_0$ that $\sigma_x^2 = \sigma_y^2$ is to reject $H_0$ if

$$\frac{\sum_{j=1}^{n} (y_j - \bar{y}_{(n)})^2}{\sum_{i=1}^{m} (x_i - \bar{x}_{(m)})^2}$$

lies outside some interval $[a, b]$. Details are given in the solution to Exercise 5.10(9). △

## 5.10 EXERCISES

1.° Arrivals at a service counter are assumed to be governed by a Poisson process. That is, the number of arrivals in a time period of duration $t$ has a Poisson distribution with mean $\lambda t$, and the numbers of arrivals in disjoint time intervals are independent of each other. Here $\lambda$ represents the average number of arrivals per hour. Suppose that $X_1, \ldots, X_n$ represent observations on the number of arrivals in disjoint time periods of one hour's duration and the service mechanism at the counter has been set up to handle the case $\lambda = \lambda_0$. In order to check on the validity of this assumption and enable the proper action to be taken:

(a) determine the generalized likelihood ratio test of $\lambda = \lambda_0$ versus $\lambda \neq \lambda_0$ (if you accept $\lambda \neq \lambda_0$, then it is presumed that some change is to be made in the service mechanism).

(b) determine the G.L.R. test of $\lambda = \lambda_0$ versus $\lambda > \lambda_0$ (here it is assumed that the service mechanism's rate of service is at a minimum when set for $\lambda = \lambda_0$, so that we don't care if $\lambda < \lambda_0$).

(c) determine the G.L.R. test of $|\lambda - \lambda_0| \leq c$ versus $|\lambda - \lambda_0| > c$ (here we just don't want to take action unless conditions have changed enough to warrant it).

2. Tests are made on the quality of units from a production line to determine whether or not the probability of a defective has changed from its usual value $p_0$. Set up the G.L.R. tests for testing:

(a) $p = p_0$ versus $p \neq p_0$,
(b) $p = p_0$ versus $p > p_0$, and
(c) $|p - p_0| \leq c$ versus $|p - p_0| > c$

where $X_1, \ldots, X_n$ are independent Bernoulli random variables.

3.° Suppose that the normal level of a particular component in the blood is the value of a random variable $X$, which is normal with mean $\mu = \mu_0$ and variance 1. It is suspected that a new drug alters this level. Find the G.L.R. test of $\mu = \mu_0$ versus $\mu \neq \mu_0$, based on independent measurements $X_1, \ldots, X_n$ of this component in persons who have taken the drug.

4. Suppose that the normal level of a certain antibody in the blood is the value of a random variable $X$, which is normal with mean $\mu = \mu_0$ and variance $\sigma_0^2$. A new vaccine is supposed to raise this antibody level. Develop the G.L.R. test of $\mu = \mu_0$ versus $\mu > \mu_0$ based on independent measurements on people who have been given the vaccine. The assumed variance is always $\sigma_0^2$.

5. Derive the G.L.R. test for the problem treated in Example 5.7, where $E(X_i) = \mu_x$, $E(Y_i) = \mu_y$, both means unknown.

6. Derive the G.L.R. test of $\mu_x = \mu_y$ versus $\mu_x \neq \mu_y$ when $X_1, \ldots, X_m$ are normal random variables with $E(X_i) = \mu_x$, Var $X_i = 1$, $Y_1, \ldots, Y_n$ are normal random variables with $E(Y_i) = \mu_y$, Var $Y_i = 1$, and $X_1, \ldots, X_m, Y_1, \ldots, Y_n$ are independent.

7. Let $X_1, \ldots, X_{n_x}, Y_1, \ldots, Y_{n_y}$ be independent normal random variables with $E(X_i) = \mu_x$, $E(Y_i) = \mu_y$, and common unknown variance $\sigma^2$.

(a) Show that the G.L.R. test of $H_0$: $\mu_x = \mu_y$ versus $H_1$: $\mu_x \neq \mu_y$ is of the form reject $H_0$ if and only if

$$\frac{(\bar{x}_{(n_x)} - \bar{y}_{(n_y)})^2}{\left(\frac{1}{n_x} + \frac{1}{n_y}\right)\frac{1}{n_x + n_y - 2}(n_x s_x^2 + n_y s_y^2)} > c,$$

where

$$\bar{x}_{(n_x)} = \frac{1}{n_x}\sum_{i=1}^{n_x} \mathsf{x}_i, \qquad \bar{y}_{(n_y)} = \frac{1}{n_y}\sum_{i=1}^{n_y} \mathsf{y}_i$$

$$s_x^2 = \frac{1}{n_x}\sum_{i=1}^{n_x} (\mathsf{x}_i - \bar{x}_{(n_x)})^2, \qquad s_y^2 = \frac{1}{n_y}\sum_{i=1}^{n_y} (\mathsf{y}_i - \bar{y}_{(n_y)})^2.$$

and $\mathsf{x}_1, \ldots, \mathsf{x}_{n_x}, \mathsf{y}_1, \ldots, \mathsf{y}_{n_y}$ are the observations.

*Remarks*: We know that

$$\text{(i)} \quad (\bar{x}_{(n_x)} - \bar{y}_{(n_y)} - [\mu_x - \mu_y])\Big/\left[\sigma\sqrt{\frac{1}{n_x} + \frac{1}{n_y}}\right]$$

is the value of a random variable with a standard normal distribution. Furthermore,

$$(\sqrt{n_x s_x^2 + n_y s_y^2}/\sigma)^2$$

has a Chi square distribution with $n_x + n_y - 2$ degrees of freedom, and can be shown to be independent of the variable whose value is (i). Thus,

$$\frac{(\bar{x}_{(n_x)} - \bar{y}_{(n_y)} - [\mu_x - \mu_y])}{\sqrt{\left(\frac{1}{n_x} + \frac{1}{n_y}\right)\frac{1}{n_x + n_y - 2}(n_x s_x^2 + n_y s_y^2)}}$$

is the value of a random variable with the $t$ distribution with $n_x + n_y - 2$ degrees of freedom. This enables us:

(ii) to choose $c$ so the test of $\mu_x = \mu_y$ versus $\mu_x \neq \mu_y$ has probability $\alpha$ of type I error and

(iii) to obtain a confidence interval for $\mu_x - \mu_y$.

(b) Show that the G.L.R. test of $H_0$: $\mu_x = \mu_y$ versus $H_1$: $\mu_x > \mu_y$ rejects $H_0$ if and only if

$$\frac{\bar{x}_{(n_x)} - \bar{y}_{(n_y)}}{\sqrt{\left(\frac{1}{n_x} + \frac{1}{n_y}\right)\frac{1}{n_x + n_y - 2}(n_x s_x^2 + n_y s_y^2)}} > d.$$

*Remarks.* If we want to determine whether two different normal populations have the same mean, the method of this question will fail if their variances are different. It may be very inefficient even if the random variables have the same variance. For example, we might want to test turbine blades manufactured under two different processes for use in aircraft engines. To avoid the difficulties just mentioned in making a test based on two samples, we might put a pair of blades, one from each process, on each of $n$ aircraft, and measure the values of $X_i - Y_i$, $i = 1, \ldots, n$. Here the values of $X_i$ and $Y_i$ are the respective lifetimes of the two types of blades on the $i$th aircraft. The advantage of this is that Var $(X_i - Y_i)$ is likely to be much smaller than

Var $X_i$ or Var $Y_i$, because both blades in a pair are subject to more nearly identical conditions. However, it is not unreasonable to believe that $X_i - Y_i$ would still be normal, with mean $\mu_x - \mu_y$ and unknown variance $\tau^2$ (which as we mentioned is likely to be smaller than the $\sigma^2$ of this problem). Testing the hypotheses $H_0$ that $\mu_x = \mu_y$ (that is, that $\mu_x - \mu_y = 0$) is a problem we already treated in Example 5.2.

8. Show that the G.L.R. of $\theta \in \Theta_0$ versus $\theta \in \Theta_1$ is a function of every sufficient statistic for $\theta$, and hence that the G.L.R. test is determined by the value of any sufficient statistic for $\theta$.

## SECTION 6

### Concluding Remarks

There are several techniques for testing composite hypotheses other than the ones we have treated here.

If we want to test $E(X_i) = 0$ versus $E(X_i) = 1$ where $\sigma^2 = \text{Var } X_i$ is unknown, we often derive a test by acting as if $\sigma^2$ were known. We then replace this test by one that uses an estimate of $\sigma^2$ in place of $\sigma^2$ itself. This is one way of handling the presence of the *nuisance parameter* $\sigma^2$. Note that in Example 5.2 (the $t$ test), this is essentially what is done by the G.L.R. test. The G.L.R. test in general estimates all parameters, using (restricted) maximum likelihood estimates, so the $t$ test result is not surprising. This method obviously extends to handling other nuisance parameters.

The following illustrates another way of handling composite hypotheses. Suppose we want to test $H_0$: $p = 1/2$ versus $H_1$: $p = 1/4$ or $p = 3/4$. One way to do this is to reduce $H_1$ to a simple hypothesis by testing $H_0$, the hypothesis that $X_1, \ldots, X_n$ have probability function $p_{1/2}$, against the hypothesis $H_1'$, that $X_1, \ldots, X_n$ have probability function $\frac{1}{2}(p_{1/4} + p_{3/4})$, where

$$p_q(x_1, \ldots, x_n) = q^{x_1+\cdots+x_n}(1 - q)^{n-(x_1+\cdots+x_n)}.$$

This technique is called the *method of mixtures*. Here, the method of mixtures acts as if $p = 1/4$ and $p = 3/4$ were equally likely when $H_1$ is true. This method also extends to rather large composite hypotheses.

In general, we should keep in mind that in testing $H_0$ against $H_1$, we are really asking which furnishes the better description, rather than which is true. If possible, we should attempt to determine the power function over a reasonably exhaustive set of possible descriptions in order to see whether our test procedure is performing as we desire.

We have mentioned earlier that confidence intervals may be used to construct tests of given level. Suppose, for example, that we have independent normal random variables $X_1, \ldots, X_n$ with mean $\mu$ and variance 1. We could construct a size .05 test of the hypothesis $H_0$ that $\mu = 0$ against $H_1$: $\mu = 1$ as follows. Construct a .95 level confidence interval for $\mu$ of the form

$$I = \left[\bar{X}_{(n)} - \frac{1.645}{\sqrt{n}}, \infty\right).$$

Reject $H_0$ if and only if $I$ fails to cover 0. Certainly, $P\{0 \in I\} = .95$ when $\mu = 0$. Hence, the probability of type I error is .05. If $\sigma^2$ were unknown, we could use a confidence interval of the form

$$J = \left[\bar{X}_{(n)} \pm t_\alpha \frac{S_{(n)}}{\sqrt{n}}\right], \qquad \text{(see 6.13 of Chapter 9)}$$

where

$$S_{(n)}^2 = \frac{1}{n-1} \sum_{i=1}^{n} (X_i - \bar{X}_{(n)})^2$$

and where $t_\alpha$ is the upper $1 - \alpha$ point of the $t$ distribution (see 8, Chapter 8, and Definition 6.2, Chapter 10).

It is common practice to use the tests developed for normal random variables on populations that are known to depart rather strongly from the normality assumption. This may simply cause a loss in efficiency over the best that could be done. This is what would happen if the data were really uniform on $[\theta - 1/2, \theta + 1/2]$ and we assumed it normal with mean $\theta$ and variance 1/12. It is not difficult to see that we can obtain much better discrimination for testing $\theta = .5$ versus $\theta = .6$ if we use $\frac{1}{2}[\max X_i + \min X_i]$ than if we use the test based on

$$\frac{1}{n} \sum_{i=1}^{n} X_i,$$

which is best if the data is normal. Alternatively, if the random variables $X_i$ were really Cauchy with median $\theta$, use of

$$\frac{1}{n} \sum_{i=1}^{n} X_i$$

would be absurd (see Problem 9.9(3), Chapter 6).

Generally, you must examine the performance of any test procedure under the various individual distributions that you have reason to believe may be good descriptions. This can be, of course, a formidable task.

It is tempting to take a peek at the data before deciding which tests look promising. This is a dangerous procedure, since it makes the power computations of the chosen tests invalid. Such peeks are worthwhile for suggesting future experiments. In fact, just recently, a look at some data suggested that Hodgkins' disease, an often fatal illness that resembles leukemia, may be infectious and may have a long incubation period. However, the data that suggested this conjecture may not be used in its verification, and new experiments will have to be run to find the answer.

Generally, we favor the construction of "short" confidence intervals to help draw conclusions about data when we are unsure of what to expect. For instance, a short confidence interval for a parameter $\mu$ is of use if we decide to ask, "Is $\mu$ near 0?", "Is $\mu > 10$?", and so forth.

Finally, it should be stressed that it is not wise to draw many conclusions from a given set of data by performing more than one test on this data. It is preferable to make use of separate confidence intervals, using the inequality $P(A \cap B) \geq 1 - P(A^c) - P(B^c)$. For instance, if you obtain a .95 level confidence interval for $\theta_1$ and a .97 level confidence interval for $\theta_2$, then the probability of simultaneous coverage is at least .92. Then the probability of simultaneous correctness of all statements that would result from these intervals would be at least .92. With confidence intervals of this type, you could validly draw simultaneous conclusions about $\theta_1 + \theta_2$, $\theta_1 \cdot \theta_2$, and so on.

# 12

# Non-Parametric Methods

## SECTION 1
## Introduction

Non-parametric methods arise naturally when the statistical model is not well enough developed to allow us to settle on a particular parametric family. They are useful when the assumption of a specific parametric family is unjustified, when certain questions are not readily formulated in terms of a parametric family, and when there is not sufficient data to justify the use of a Central Limit Theorem approximation. The non-parametric techniques that we shall study depend on four main concepts:

(a) the probability transformation (Theorem 3.6, Chapter 4), which tells us that if $F_X$ is continuous, the random variable $Y = F_X(X)$ is uniformly distributed on $[0, 1]$;

(b) the easily verified fact that the number of observations of $X_1, \ldots, X_n$ not exceeding the $p$-quantile $\theta_p$ has a binomial distribution with parameters $n$ and $p$ when $X_1, \ldots, X_n$ are independent and have a common continuous distribution;

(c) the fact that if $X_1, \ldots, X_n$ are independent variables with the same density, the conditional distribution of the sample given the order statistics is discrete uniform over the set of possible permutations of the given order statistics; and

(d) the Central Limit Theorem (Theorem 3.1, Chapter 7).

It is probably impossible to define non-parametric methods in any precise fashion. Generally, however, the tests being considered apply to large (non-parametric) classes of distributions. Furthermore, some of the following *distribution-free* properties are frequently exhibited.

(i) If $H_0$ is composite, but not "too" large, then the distribution of the test statistic varies only a little, if at all, as the true distribution $F$ ranges over the set of null hypothesis distributions (for example, the Chi square test of fit with estimated parameters).

(ii) In many cases, the distribution of the test statistic as the true distribution ranges over $H_0$ varies but little as $H_0$ ranges over some wider class (for example, the Chi square test of fit with a fixed number of intervals and the Wilcoxon test with continuous distributions).

Generally, the power of such tests will vary widely as the true distribution $F$ ranges over $H_1$.

Confidence estimates may still be generated for certain parameters when the class of possible governing distributions is non-parametric. Here we have the confidence intervals for quantiles in mind.

The tests will usually have less discriminating power than tests that ask the same question in a smaller class. For example, if we want to make a test on the median of normal random variables with mean $\mu$ and variance 1, then the usual non-parametric median test requires about $\pi/2$ times the number of observations required by the best test for normals to obtain the same discriminating power. However, the non-parametric median test applies to a wider class of distributions. Thus, the price paid for greater applicability is a loss of efficiency. Similar remarks hold for non-parametric confidence intervals.

## SECTION 2
## Tests of Fit and Inference about Distribution Functions

In most probability problems that we have investigated, we postulate a particular mathematical model that has a reasonable basis in theory. However, every such model must stand the test of practical utility. If we are not certain that our model is an adequate description of the phenomenon being studied, we may want to devise a direct test of its adequacy. Let $X_1, \ldots, X_n$ be independent random variables with $F_{X_i} = F$ where $F$ is unknown. Tests of hypotheses of the type $H_0$: $F \in \mathscr{F}_0$ against the alternative $F \notin \mathscr{F}_0$, where $\mathscr{F}_0$ is a given class of distribution functions, are called *tests of goodness-of-fit*. These tests play an important role in the early stages of scientific investigations.

### 2.1 Pearson's Chi Square Test of Fit

We shall first investigate the classic Chi square test of fit developed by Karl Pearson. Suppose first that $X$ is a discrete random variable that can take on the values $b_1, \ldots, b_q$. The simplest of Pearson's Chi square tests is a procedure to test the hypothesis $H_0$ that $P\{X = b_j\} = p_j, j = 1, \ldots, q$, where $p_j$ are specified probabilities with $\sum_{j=1}^{q} p_j = 1$. Suppose that $n$ independent observations $X_1, \ldots, X_n$ with the distribution of $X$ are observed, and that we let $\tilde{p}_j$ be the proportion of observations in the sample assuming the value $b_j$. The test statistic is then defined by

$$\chi^2 = \sum_{i=1}^{q} \frac{[\sqrt{n}(\tilde{p}_i - p_i)]^2}{p_i}.$$

It can be shown that for large $n$, the statistic $\chi^2$ has approximately a Chi square distribution with $q - 1$ degrees of freedom **when $H_0$ is true.** This is rather easy to see for the case $q = 2$. Here we have

$$p_1 = p, \qquad p_2 = 1 - p_1, \qquad \tilde{p}_1 = \tilde{p}, \qquad \tilde{p}_2 = 1 - \tilde{p}_1.$$

Hence

$$\sum_{i=1}^{q} \frac{(\sqrt{n}[\tilde{p}_i - p_i])^2}{p_i} = \frac{(\sqrt{n}[\tilde{p}_1 - p_1])^2}{p_1} + \frac{(\sqrt{n}[-\tilde{p}_1 + p_1])^2}{1 - p_1}$$

$$= \frac{(\sqrt{n}[\tilde{p}_1 - p_1])^2}{p_1(1 - p_1)}$$

$$= \left(\frac{\sqrt{n}(\tilde{p}_1 - p_1)}{\sqrt{p_1(1 - p_1)}}\right)^2.$$

Under $H_0$, we know that $n\tilde{p}_1$ is the value of a binomial random variable with expectation $np$ and variance $np_1(1 - p_1)$. From the Central Limit Theorem (Theorem 3.1, Chapter 7), we see that in this case the random variable whose value is

$$\frac{\sqrt{n}(\tilde{p}_1 - p_1)}{\sqrt{p_1(1 - p_1)}}$$

has a limiting standard normal distribution as $n$ gets large. From this, it follows that the limiting distribution of the random variable whose value is

$$\left(\frac{\sqrt{n}(\tilde{p}_1 - p_1)}{\sqrt{p_1(1 - p_1)}}\right)^2$$

when $H_0$ is true is Chi square with one degree of freedom (see 1 in Chapter 8). Now, we examine the random variable whose value is

$$\sum_{i=1}^{q} \frac{[\sqrt{n}(\tilde{p}_i - p_i)]^2}{p_i};$$

we notice that this expression should be larger when $H_0$ is false than when $H_0$ is true. In fact, it should grow at least at the rate $n(p_j^* - p_j)^2$, where $p_j^*$ is the true probability that $X = b_j$ and $p_j$ is the hypothesized probability that $X = b_j$. This should give a reasonable idea of the sample size needed to achieve good discrimination power against a given alternative in the test we now present. **The Chi square test of $H_0$: $P\{X = b_j\} = p_j$ is to reject $H_0$ if**

**2.2**
$$\sum_{i=1}^{q} \frac{n(\tilde{p}_i - p_i)^2}{p_i} > \chi^2_{q-1}(1 - \alpha)$$

where $\chi^2_{q-1}(1 - \alpha)$ is the upper $1 - \alpha$ point (see Definition 6.2, Chapter 10) of the Chi square distribution with $q - 1$ degrees of freedom, and $\tilde{p}_i$ is the proportion of the observed independent random variables $X_1, \ldots, X_n$ assuming the value $b_i$. A rule of thumb based on experimental evidence is to try to keep all of the $np_i$ no smaller than 10.

### 2.3 Modification of Chi Square Test for Densities

We can utilize the Chi square test for testing $H_0$: $f_X = f$ where $f$ is a given density by dividing up the real line into $q$ intervals $I_1, \ldots, I_q$, letting $p_j = P_f\{X \in I_j\}$ and letting $\tilde{p}_j$ be the proportion of observations falling in $I_j$. [We attempt to choose $I_1, \ldots, I_q$ so that $p_j \cong 1/q$ for all $j$, in order to satisfy $np_j \geq 10$ for all $j$ most easily.]

---

### *2.4 EXERCISES*

1.° Suppose that we are given the following data:

| | | |
|---|---|---|
| $x_1 = -.7751$ | $x_5 = -.9955$ | $x_9 = -.5673$ |
| $x_2 = -.1950$ | $x_6 = -.3371$ | $x_{10} = -.6513$ |
| $x_3 = -.2181$ | $x_7 = -.8527$ | $x_{11} = -.3800$ |
| $x_4 = -.5147$ | $x_8 = -.8513$ | $x_{12} = -.3740$ |

The random variables $X_1, \ldots, X_{12}$ represent the error made by discarding the digits to the right of the decimal point in some empirical data. Assuming these random variables to be independent, with common distribution function $F$, test the hypothesis that $F$ is uniform on $[-1, 0]$ at the .1 level. Use three intervals $[-1, -2/3]$, $(-2/3, -1/3]$, $(-1/3, 0]$. (The data was taken from a table of random numbers that were *supposed* to be uniformly distributed on $[-1, 0]$.)

2. Twelve independent measurements are made by a measuring instrument on a government standard. The errors $x_i$ of measurement are as follows:

| | | |
|---|---|---|
| $x_1 = .408$ | $x_5 = -1.798$ | $x_9 = -1.084$ |
| $x_2 = 2.865$ | $x_6 = -1.434$ | $x_{10} = .722$ |
| $x_3 = 3.666$ | $x_7 = -3.688$ | $x_{11} = -1.809$ |
| $x_4 = -.433$ | $x_8 = 2.261$ | $x_{12} = -2.890$ |

(a) Test $H_0$: that the $X_i$ have a normal distribution with mean 0, variance 1 (use four intervals, $(-\infty, -1]$, $(-1, 0]$, $(0, 1]$, and $(1, \infty)$) at the .1 level.

(b) Test $H_0$: that the $X_i$ have a normal distribution with mean 0 and variance 4, at the .1 level. (The data was actually drawn from a normal distribution with mean 0, variance 4.) Use the same intervals as in part (a).

It would be desirable to extend the Chi square test to handle composite null hypotheses, for example to test the hypothesis that $X$ is normally distributed. Suppose that we want to test the hypothesis that $F_X \in \mathscr{F}_0$, where $\mathscr{F}_0$ is some parametric family whose members are specified by the parameter $\theta = (\theta_1, \ldots, \theta_r)$. In this case, the $p_i$ in the Chi square test become $p_i(\theta)$, where $p_i(\theta)$ is the probability of the $i$th interval when $\theta$ is the governing parameter. Now let $\tilde{\theta}$ be the maximum likelihood estimate of $\theta$ based on the observed interval frequencies. That is, $\tilde{\theta}$ is the value of $\theta$ maximizing

$$[p_1(\theta)]^{n\tilde{p}_1}[p_2(\theta)]^{n\tilde{p}_2} \cdots [p_q(\theta)]^{n\tilde{p}_q}. \tag{2.5}$$

Define

**2.6**
$$\chi^2(\tilde{\theta}) = \sum_{i=1}^{q} n \frac{[\tilde{p}_i - p_i(\tilde{\theta})]^2}{p_i(\tilde{\theta})}.$$

Then it can be shown that under some restrictions,† $\chi^2(\tilde{\theta})$ is the value of a random variable whose limiting distribution as $n \to \infty$ is Chi square with $q - 1 - r$ degrees of freedom, *when $H_0$ is true.*

When we are using the modification of Chi square for densities in conjunction with the maximum likelihood estimate of $\theta$, it is generally very difficult to choose $\theta$ so as to maximize 2.5. In such a case, it is tempting to use the maximum likelihood estimate $\tilde{\theta}$ of $\theta$, based on the original data $x_1, \ldots, x_n$. Suppose we then reject $H_0$ for

**2.7**
$$\chi^2(\approx\theta) = \sum_{i=1}^{q} n \frac{[\tilde{p}_i - p_i(\approx\theta)]^2}{p_i(\approx\theta)} > c.$$

Chernoff and Lehman have shown that it is safer, for achievement of a probability $\leq \alpha$ of type I error, to act as if $\chi^2(\approx\theta)$ is the value of a random variable with the Chi square distribution with $q - 1$ degrees of freedom. That is, to achieve approximately a probability of type I error not exceeding $\alpha$, reject $H_0$ if

**2.8**
$$\chi^2(\approx\theta) > \chi^2_{q-1}(\alpha).$$

This is not too surprising, since use of the original data is closer to "knowing $\theta$." That is, use of the original data is closer to testing a simple hypothesis.

We illustrate this modification in the following example.

### 2.9 *Example: Testing Whether X is Normal with Variance 1*

Suppose that we are given the following data, assumed to be values of independent random variables with a common distribution function $F$, representing measurements on a battery of unknown voltage.

| | | |
|---|---|---|
| $x_1 = 1.707$ | $x_{11} = 3.627$ | $x_{21} = 1.269$ |
| $x_2 = 3.078$ | $x_{12} = 1.515$ | $x_{22} = 2.375$ |
| $x_3 = 1.376$ | $x_{13} = 2.578$ | $x_{23} = 3.197$ |
| $x_4 = 2.612$ | $x_{14} = -.771$ | $x_{24} = .319$ |
| $x_5 = 2.421$ | $x_{15} = 1.435$ | $x_{25} = .853$ |
| $x_6 = 1.014$ | $x_{16} = 2.045$ | $x_{26} = 2.774$ |
| $x_7 = 1.041$ | $x_{17} = 2.028$ | $x_{27} = .929$ |
| $x_8 = .765$ | $x_{18} = 1.949$ | $x_{28} = 1.765$ |
| $x_9 = 1.370$ | $x_{19} = 1.804$ | $x_{29} = 4.059$ |
| $x_{10} = 2.097$ | $x_{20} = 3.029$ | $x_{30} = 1.088$ |

† See Cramér [6].

We want to test $H_0$, that $F$ is a normal distribution function with unit variance, at the .1 level.

In this case, $\mathscr{F}_0$ is the family of normal distributions with unit variance, and we must estimate the mean $\theta$. We know the maximum likelihood estimator has the value

$$\bar{x}_{(n)} = \frac{1}{n} \sum_{i=1}^{n} x_i.$$

Here $n = 30$ and $\bar{x}_{(n)} = 1.845$. We use four intervals, $(-\infty, 1.145]$, $(1.145, 1.845]$, $(1.845, 2.545]$, and $(2.545, \infty)$. Since $\tilde{\theta} = 1.845$, we find

$$p_1(\tilde{\theta}) = .242$$
$$p_2(\tilde{\theta}) = .258$$
$$p_3(\tilde{\theta}) = .258$$
$$p_4(\tilde{\theta}) = .242$$
$$\tilde{p}_1 = 8/30,\ \tilde{p}_2 = 8/30,\ \tilde{p}_3 = 6/30,\ \tilde{p}_4 = 8/30,$$
$$\tilde{p}_1 = .267,\ \tilde{p}_2 = .267,\ \tilde{p}_3 = .2,\ \tilde{p}_4 = .267.$$

The Chi square statistic has the value

$$\begin{aligned}\sum_{i=1}^{4} \frac{30(\tilde{p}_i - p_i(\tilde{\theta}))^2}{p_i(\tilde{\theta})} &= 30\left[\frac{(.267 - .242)^2}{.242} + \frac{(.267 - .258)^2}{.258}\right. \\ &\quad \left. + \frac{(.2 - .258)^2}{.258} + \frac{(.267 - .242)^2}{.242}\right] \\ &= 30\left[\frac{.002704}{.242} + \frac{.000081}{.258} + \frac{.003364}{.258} + \frac{.002704}{.242}\right] \\ &\cong 1.07.\end{aligned}$$

As we've already mentioned, to be conservative we use $\chi^2_{q-1}(1 - \alpha) = \chi^2_3(1 - \alpha)$ in 2.8 (rather than $\chi^2_{q-1-r}(1 - \alpha)$). For $1 - \alpha = .9$, we find that $\chi^2_3(.9) = 6.25$. Therefore, we accept the hypothesis of normality with unit variance. (This data actually came from such a distribution with $\mu = 2$.) △

## 2.10 EXERCISES

1. Test the hypothesis that the data in Example 2.9 came from a normal distribution with variance 2 at approximately a level not exceeding .1, using the given four intervals.

2.° Test the hypothesis that the data of Example 2.9 came from a normal population with mean 1.5 at a level approximately not exceeding .1, using the given four intervals.

3. Test the hypothesis that the data of Example 2.9 came from a normal population at a level approximately not exceeding .1, using the given four intervals. Note that in this problem you must estimate both $\mu$ and $\sigma^2$.

## 2.11 The Empirical Distribution Function and Its Uses

Let $X_1, \ldots, X_n$ be independent random variables with a common unknown distribution function $F$. For each real $x$, let $F_n(x)$ be the proportion of $X_1, \ldots, X_n$ satisfying $X_i \leq x$. We call $F_n$ the *empirical distribution function based on* $X_1, \ldots, X_n$. The value of $F_n$ is pictured in Figure 12.1 for $n = 3$ when $X_i$ assumes the value $x_i$. Note that $F_n$ is what is known as a *random function*. Certainly the value of $F_n(x)$ (proportion of $X_i \leq x$) should be close to $F(x) = P\{X \leq x\}$ for $n$ large. In fact, since

$$nF_n(x) = \sum_{i=1}^{n} Y_i, \quad \text{where} \quad Y_i = \begin{cases} 1 & \text{if } X_i \leq x \\ 0 & \text{otherwise,} \end{cases}$$

we see that $nF_n(x)$ is a binomial random variable with parameters $n$ and $F(x)$. Thus, by the Chebyshev inequality, we see that

$$P\left\{|F_n(x) - F(x)| \leq \frac{t\sqrt{F(x)(1 - F(x))}}{\sqrt{n}}\right\} \geq 1 - 1/t^2.$$

Hence

$$P\{|F_n(x) - F(x)| \leq t/(2\sqrt{n})\} \geq 1 - 1/t^2,$$

which shows that for large $n$ we have $F_n(x)$ close to $F(x)$ with high probability. Suppose that we wish to test $H_0$, $F = F_0$ where $F_0$ is a specified distribution function. It seems reasonable to use some measure of the "distance" between $F_0$ and $F_n$ as a test statistic. The statistic we shall introduce here is essentially the maximum vertical distance, $d(F_n, F_0)$, between $F_n$ and $F_0$.

**2.12 Definition: $d(F,G)$.** Let $F$ and $G$ be two distribution functions. We define $d(F, G)$ to be the maximum of all numbers

$$\lim_{\substack{h \to 0 \\ h > 0}} |F(x + h) - G(x + h)|,$$

$$\lim_{\substack{h \to 0 \\ h < 0}} |F(x + h) - G(x + h)|$$

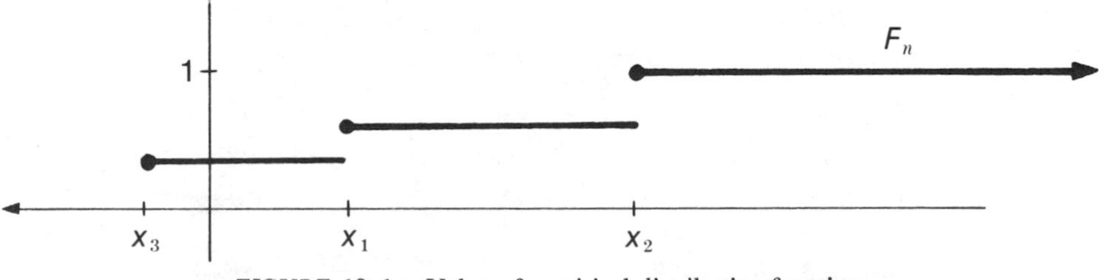


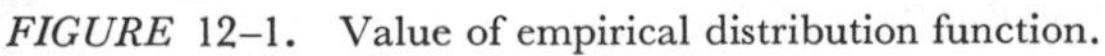
*FIGURE* 12–1. Value of empirical distribution function.

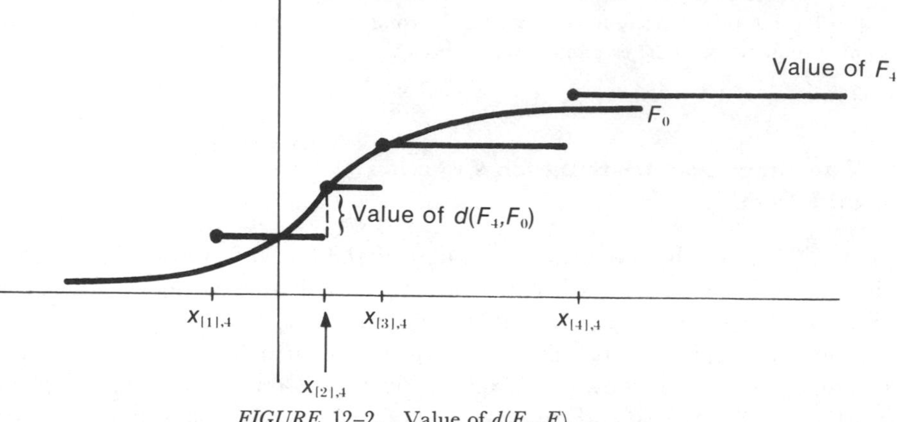


*FIGURE* 12–2. Value of $d(F_n, F)$.

as $x$ ranges over the set of real numbers. We illustrate $d(F_n, F)$ in Figure 12.2.

Because of the constancy of $F_n$ on intervals between the order statistics, $x_{[i],\,n}$, we see from Figure 12.2 that $d(F_n, F_0)$ can be determined by examining $F_0$ and $F_n$ at the $x_{[i],\,n}$ only. If we let $\max\limits_{1\le i\le n}\,(y_i, z_i)$ be the largest of the numbers $y_1, \ldots, y_n, z_1, \ldots, z_n$, then it is not difficult to show that the following lemma is true.

### *2.13 Lemma: Computation of d($F_n$, $F_0$)*

Let $x_{[1],\,n}, \ldots, x_{[n],\,n}$ be the observed values of $X_1, \ldots, X_n$, arranged in order of magnitude. Then the value of $d(F_n, F_0)$ is given by

$$\max_{1\le i\le n}\left(i/n - F_0(x_{[i],\,n}),\ F_0(x_{[i],\,n}) - \frac{i-1}{n}\right).$$

### *2.14 Example: Evaluation of d($F_4$, $F_0$)*

Suppose that

$$x_1 = 2.9, \qquad x_2 = 3.1, \qquad x_3 = 2.3, \qquad x_4 = 0$$

and $F_0$ is the standard normal distribution function. Then

$$x_{[1],\,4} = 0, \qquad x_{[2],\,4} = 2.3, \qquad x_{[3],\,4} = 2.9, \qquad x_{[4],\,4} = 3.1$$

$$F_0(x_{[1],\,4}) = .5, \qquad F_0(x_{[2],\,4}) = .989,$$

$$F_0(x_{[3],\,4}) = .998, \qquad F_0(x_{[4],\,4}) = .999$$

$$\tfrac{1}{4} - F_0(x_{[1],\,4}) = -.25, \qquad F_0(x_{[1],\,4}) - \tfrac{0}{4} = .5$$

$$\tfrac{1}{2} - F_0(x_{[2],\,4}) = -.489, \qquad F_0(x_{[2],\,4}) - \tfrac{1}{4} = .739$$

$$\tfrac{3}{4} - F_0(x_{[3],\,4}) = -.248, \qquad F_0(x_{[3],\,4}) - \tfrac{1}{2} = .498$$

$$1 - F_0(x_{[4],\,4}) = .001, \qquad F_0(x_{[4],\,4}) - \tfrac{3}{4} = -.249.$$

Hence, the value of $d(F_n, F_0)$ is .739. △

---

### 2.15 EXERCISE

Let $F_0$ be the standard normal distribution function and suppose that

$$x_1 = .42, \quad x_2 = -.53, \quad x_3 = -.79, \quad x_4 = -2.43, \quad \text{and} \quad x_5 = -.29.$$

Find the value of $d(F_5, F_0)$.

We shall now show that the distribution of $d(F_n, F_0)$ does not depend on which $F_0$ is chosen when the governing distribution $F = F_0$, provided $F_0$ is continuous. We see this as follows. Let $X_1, \ldots, X_n$ be independent random variables with common distribution function $F$, where $F$ is continuous. Then the random variables $Y_1, \ldots, Y_n$ given by $Y_i = F(X_i)$ are independent with the uniform distribution on $[0, 1]$ (see Theorem 3.6, Chapter 4). Now let us form the empirical distribution function $U_n$, based on $Y_1, \ldots, Y_n$. Let $U$ denote the uniform distribution function on $[0, 1]$, $U(y) = y, 0 \leq y \leq 1$. Then it is easy to see that $d(U_n, U) = d(F_n, F)$. Let $y = F(x)$; then $Y_j \leq y$ if and only if $X_j \leq x$. Hence, $U_n(y) = F_n(x)$ and $U_n(y) - U(y) = F_n(x) - F(x)$. From this, the asserted result follows. It follows, then, that we can compute rejection points for tests based on $d(F_n, F_0)$ by restricting attention to order statistics from the uniform distribution. The following theorem can be shown to hold.

### 2.16 *Kolmogorov-Smirnov Theorem*

Let $X_1, \ldots, X_n$ be independent random variables with a continuous distribution function $F$. Then for each $\alpha$ in $(0, 1)$, there exists a value $h_\alpha$ such that

$$\lim_{n \to \infty} P\{\sqrt{n}\, d(F_n, F) \leq h_\alpha\} = 1 - \alpha.$$

In fact,

$$\lim_{n \to \infty} P\left\{\sqrt{n}\, d(F_n, F) \leq c\right\} = 1 - 2\sum_{j=1}^{\infty} (-1)^{j+1} e^{-2j^2c^2}.$$

From this expansion we can numerically determine $h_\alpha$. Tables of $h_\alpha$ for various values of $\alpha$ are available in Owen [15].

### 2.17 *Kolmogorov-Smirnov Test of Fit*

To test the hypothesis $H_0$, that the distribution function $F$ of the independent random variables $X_1, \ldots, X_n$ is a specified distribution function $F_0$ at approximately the $\alpha$ level, reject $H_0$ if the value of $\sqrt{n}\, d(F_n, F_0)$ exceeds $h_\alpha$. *To test the hypothesis $H_0'$ that $d(F, F_0) \leq k$ at approximately the $\alpha$ level, reject $H_0'$ if the value of $\sqrt{n}\, d(F_n, F_0) > \sqrt{n}\, k + h_\alpha$.*

The Kolmogorov-Smirnov test possesses several advantages over the simplest Chi square test of fit (2.2). The first is that it can be applied directly to any data with no need to discretize. But if $F_0$ is the distribution function of a discrete random variable, the Kolmogorov-Smirnov test can still be applied, with the added advantage of a reduction in the probability of type I error when using $h_\alpha$ as the "rejection constant." Secondly, the Kolmogorov-Smirnov test, unlike the Chi square test, admits a simple and meaningful approximate power computation. Suppose that we use the Kolmogorov-Smirnov statistic $d(F_n, F_0)$ to test the hypothesis $H_0$, that $F = F_0$. If, in fact, $|F(x) - F_0(x)| \geq \ell$ for some $x$, then for large $n$ we have the approximation that the power (probability of rejecting $H_0$) is at least

**2.18**
$$\int_{2[h_\alpha - \sqrt{n}\,\ell]} \frac{1}{\sqrt{2\pi}} e^{-t^2/2}\, dt.$$

Note that the condition $|F(x) - F_0(x)| \geq \ell$ means that the probability assigned by $F_0$ to the interval $(-\infty, x]$ differs from the true probability by at least $\ell$.

The result on power given by 2.18 is an easily derived consequence of the fact that $nF_n(x)$ has a binomial distribution with parameters $n$ and $p = F(x)$, together with the normal approximation to the binomial distribution. Similarly, in the test of $H_0'$: $d(F, F_0) \leq k$, if for some $x$ we have $|F(x) - F_0(x)| \geq k + \ell$, an approximate lower bound for the power is also given by 2.18.

### *2.19 Example: Test that X is Standard Normal*

Test at the .1 level that $X$ is standard normal where

$$x_1 = 3.381, \qquad x_2 = 1.616, \qquad x_3 = -5.827,$$

and

$$x_4 = -.202.$$

It can be shown that

$$h_\alpha \cong \sqrt{\frac{1}{2} \log_e \frac{2}{\alpha}}.$$

Hence, $h_{.1} \cong \sqrt{(1/2) \log 20} \cong 1.225$. Here

$$\begin{aligned} x_{[1],\,4} &= -5.827 & F_0(x_{[1],\,4}) &\cong 0 \\ x_{[2],\,4} &= -.202 & F_0(x_{[2],\,4}) &\cong .421 \\ x_{[3],\,4} &= 1.616 & F_0(x_{[3],\,4}) &\cong .947 \\ x_{[4],\,4} &= 3.381 & F_0(x_{[4],\,4}) &\cong 1. \end{aligned}$$

From this, it follows that $d(F_n, F_0)$ has the value .447 and that $\sqrt{n}\, d(F_n, F_0)$ has the value .894. Thus we do not reject $H_0$ with the given data. This data was generated from a normal random variable with mean 0 and variance 4. *However, four observations cannot in general give much discriminating power.* △

Because it turns out to be somewhat difficult to modify the Kolmogorov-Smirnov test to handle composite hypotheses, we shall omit this topic.

---

## 2.20 EXERCISES

1.° Test the hypothesis that $F$ is standard normal on the following data, using the Kolmogorov-Smirnov test of fit (2.17):

| | | |
|---|---|---|
| $x_1 = 3.778$ | $x_4 = 1.228$ | $x_7 = 1.012$ |
| $x_2 = 1.924$ | $x_5 = -.847$ | $x_8 = .599$ |
| $x_3 = -2.104$ | $x_6 = -.968$ | $x_9 = 1.036$ |

The data came from a normal distribution with mean 0 and variance 4.

2. Test the hypothesis that $F$ is standard normal on the data of Exercise 2.4(2), using 2.17.

3. Test the hypothesis that $F$ is normal with mean 0 and variance 4 on the data of Exercise 2.4(2), using 2.17.

In medical diagnosis, we would like to obtain simultaneous confidence intervals for the probabilities of various symptom sets associated with a given disease. A disease might exhibit only two possible symptom sets, {fever, high white count, nausea}, and {nausea, abdominal tenderness}. In the experiment of determining the symptoms of a person with the disease, assume that there are $m$ possible *sets of symptoms*, $1, 2, \ldots, m$, with probabilities $p_1, p_2, \ldots, p_m$, respectively. Suppose that we observe $n$ patients and let the random variable $X_i$ assume the value $j$ when the $i$th person exhibits symptom set $j$. It is possible to use the Kolmogorov-Smirnov Theorem (2.16) to obtain simultaneous confidence intervals for the probabilities $p_j$. However, somewhat better results can be obtained as follows.

Let $I$ denote an interval, and for any distribution function $Q$, let $P_Q(I)$ denote the probability of $I$ when $Q$ is the true distribution function. We let $D(F, G)$ stand for the largest of the values

$$|P_F(I) - P_G(I)|$$

as $I$ ranges over the set of all intervals. That is,

**2.21**
$$D(F, G) = \max_{\text{all intervals } I} |P_F(I) - P_G(I)|.$$

Next, suppose that $X_1, \ldots, X_n$ are independent random variables with distribution function $F$. Form the empirical distribution function $F_n$ (2.11) based on $X_1, \ldots, X_n$. As in the Kolmogorov-Smirnov Theorem (2.16), it can be shown that for each $\alpha$ in (0, 1), there is a value $H_\alpha$ such that

**2.22**
$$\lim_{n\to\infty} P\{\sqrt{n}\, D(F_n, F) \le H_\alpha\} \ge 1 - \alpha.$$

A table of $H_\alpha$ is available in Owen [15], who calls it $h_{2,\alpha}$. Now, the distribution function $F$ of the symptom set random variables $X_i$ is pictured in Figure 12.3. From 2.21, we see that

$$D(F_n, F) \leq H_\alpha/\sqrt{n} \text{ implies that for all intervals } I,$$
$$|P_{F_n}(I) - P_F(I)| \leq H_\alpha/\sqrt{n}.$$

In particular, if $D(F_n, F) \leq H_\alpha/\sqrt{n}$, we see that

**2.23** $$|P_{F_n}(I_j) - P_F(I_j)| \leq H_\alpha/\sqrt{n}$$

for all $j = 1, 2, \ldots, m$. But from Figure 12.3 we know that $P_F(I_j) = p_j$. Furthermore, $P_{F_n}(I_j) = \hat{p}_j$, the random variable whose value is the proportion of patients exhibiting the $j$th symptom set. Thus, we may write 2.23 as

$$|\hat{p}_j - p_j| \leq H_\alpha/\sqrt{n} \qquad \text{for all} \qquad j = 1, \ldots, m.$$

Hence, we see that if $D(F_n, F) \leq H_\alpha/\sqrt{n}$, then

$$p_j \in [\hat{p}_j - H_\alpha/\sqrt{n}, \hat{p}_j + H_\alpha/\sqrt{n}]$$

for all $j = 1, \ldots, m$. But, from 2.22, we know that

$$P\{D(F_n, F) \leq H_\alpha/\sqrt{n}\} \geq 1 - \alpha$$

approximately for large $n$. Thus, we see that

$P\{\text{simultaneously for all } j = 1, \ldots, m, p_j \in [\hat{p}_j - H_\alpha/\sqrt{n}, \hat{p}_j + H_\alpha/\sqrt{n}]\} \geq 1 - \alpha$

approximately for large $n$.

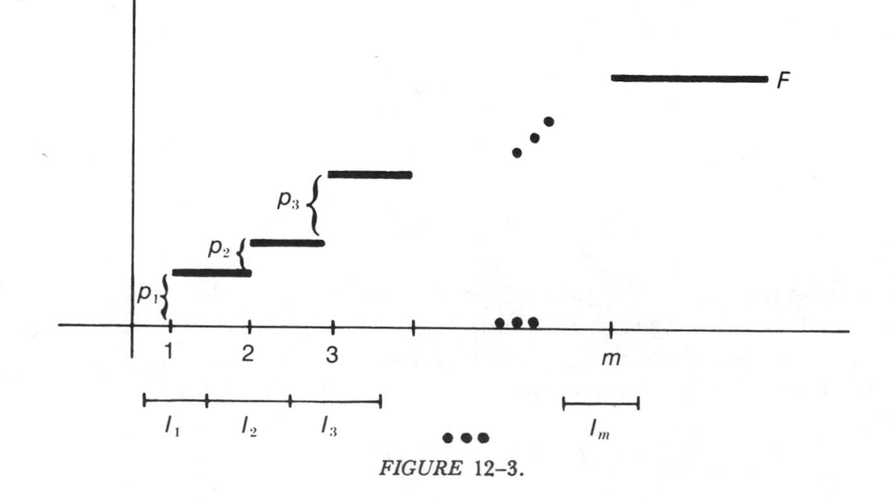


*FIGURE* 12–3.

It can be shown that

**2.4** $$\alpha \cong (2 - 8H_\alpha^2)e^{-2H\alpha^2}$$

when $\alpha$ is small. We summarize our results in the following theorem.

### *2.25 Theorem: Simultaneous Confidence Intervals for Probabilities*

Suppose that an experiment having $m$ possible outcomes $1, 2, \ldots, m$ with probabilities $p_1, \ldots, p_m$ is repeated $n$ times. The values of confidence intervals yielding probability approximately at least $1 - \alpha$ of coverage of all $p_j$ simultaneously are

$$[\tilde{p}_j - H_\alpha/\sqrt{n}, \tilde{p}_j + H_\alpha/\sqrt{n}], \qquad j = 1, \ldots, m.$$

Here, $\tilde{p}_j$ is the observed proportion of experiments resulting in outcome $j$.

The value $H_\alpha$ can be computed approximately from Equation 2.24 and is tabled in Owen [15], p. 442.

We repeat that this result is especially useful for efficient accurate determination of the probabilities of various symptom sets associated with some disease.

### *2.26 Example: Dice*

A die is tossed 60 times. The various observed proportions are:

$$1: \ 11/60 = .183$$
$$2: \ 14/60 = .233$$
$$3: \ 11/60 = .183$$
$$4: \ 10/60 = .167$$
$$5: \ 5/60 = .083$$
$$6: \ 9/60 = .15$$

From Equation 2.24 or Owen's tables, we find that $H_{.05} \cong 1.75$, and hence

$$H_{.05}/\sqrt{n} = H_{.05}/\sqrt{60} \cong .226.$$

Thus, our confidence intervals have the following values, respectively:

$p_1$: [0, .409]  $p_4$: [0, .393]

$p_2$: [.007, .459]  $p_5$: [0, .309]

$p_3$: [0, .409]  $p_6$: [0, .376].

Note that for obvious reasons we have chosen the lower limit to be 0 in those cases in which $\tilde{p}_j - H_{.05}/\sqrt{n} < 0$.

The data here were actually generated by tossing a die. Note that the evidence is insufficient to enable us to assert that the die is close to being fair, but note also that we surely are unable to assert that it is unfair, since these intervals all include $1/6 \cong .167$. To obtain a length as short as .0175, we would require about 10,000 observations. △

---

*2.27 EXERCISES*

1.° Suppose that any one of three symptoms may each be present or absent in a given disease: fever, swelling, and low red blood cell count. Let the actual outcome be an ordered triple $(s_1, s_2, s_3)$, where

$$s_1 = \begin{cases} 0 \text{ if no fever,} \\ 1 \text{ if fever} \end{cases}$$

and so on. There are eight possible outcomes:

$$(0\ 0\ 0)\ (1\ 0\ 0)\ (0\ 1\ 0)\ (0\ 0\ 1)\ (1\ 1\ 0)\ (1\ 0\ 1)\ (0\ 1\ 1) \text{ and } (1\ 1\ 1),$$

having probabilities $p_1, \ldots, p_8$ respectively. Suppose that in 10,000 cases, $\tilde{p}_1 = .01$, $\tilde{p}_2 = .05$, $\tilde{p}_3 = .04$, $\tilde{p}_4 = .07$, $\tilde{p}_5 = .21$, $\tilde{p}_6 = .17$, $\tilde{p}_7 = .32$, and $\tilde{p}_8 = .13$.

(a) Compute the value of confidence intervals of level approximately .95 for the $p_i$.

(b) What is an approximate lower bound for the probability that a person with the given disease will exhibit at least two symptoms, if we want confidence of at least .95 that our bound is correct?

2. Repeat question 1 with 100,000 rather than 10,000 cases and $\tilde{p}_1 = .02$, $\tilde{p}_2 = .06$, $\tilde{p}_3 = .06$, $\tilde{p}_4 = .08$, $\tilde{p}_5 = .20$, $\tilde{p}_6 = .10$, $\tilde{p}_7 = .28$, and $\tilde{p}_8 = .20$.

3. How many observations are needed in the previous question to:
(a)° yield approximately .99 level confidence intervals of length .01?
(b) yield approximately .99 level confidence intervals of length .005?

## SECTION 3
## Inference Concerning Quantiles

It may happen that in our model the variance does not exist, or that the sample size is not large enough to allow use of the Central Limit Theorem for inference concerning the mean. In such cases, we often find it satisfactory to turn our attention to the quantiles of the distribution. The median may also be preferred over the mean as a measure of the "center" when the distribution is highly unsymmetric, as in income distribution.

The basis for the theory we are about to develop lies in the fact that if $\theta_p$ is the $p$ quantile (see Definition 9.1, Chapter 6), then

$$P\{X \leq \theta_p\} \geq p \quad \text{and} \quad P\{X \geq \theta_p\} \geq 1 - p.$$

Hence, the number of observed values of the independent identically distributed random variables $X_1, \ldots, X_n$ not exceeding $\theta_p$ has a binomial distribution with parameters $n$ and $p^* \geq p$. Similarly, the number of these values greater than or equal to $\theta_p$ has a binomial distribution with parameters $n$ and $1 - p^{**} \geq 1 - p$. But if $X_{[j],\, n}$ is the $j$th order statistic from $X_1, \ldots, X_n$ (see Definition 9.6, Chapter 6) then $X_{[j],\, n} \leq \theta_p$ if and only if at least $j$ of the $X_1, \ldots, X_n$ take on values not exceeding $\theta_p$. Thus

$$P\{X_{[j],\, n} \leq \theta_p\} = \sum_{i=j}^{n} \binom{n}{i} p^{*i}(1 - p^*)^{n-i} \geq \sum_{i=j}^{n} \binom{n}{i} p^i(1 - p)^{n-i}.$$

This inequality holds, since the probability of at least $j$ successes is greater for $p^* > p$. Similarly, $X_{[k],\, n} \geq \theta_p$ if and only if at least $n - k + 1$ of the $X_1, \ldots, X_n$ take on values greater than or equal to $\theta_p$. Hence

$$\begin{aligned} P\{X_{[k],\, n} \geq \theta_p\} &= \sum_{u=n-k+1}^{n} \binom{n}{u} (1 - p^{**})^u p^{**n-u} \\ &= \sum_{u=n-k+1}^{n} \binom{n}{n-u} (1 - p^{**})^u p^{**n-u} \\ &\geq \sum_{u=n-k+1}^{n} \binom{n}{n-u} (1 - p)^u p^{n-u}. \end{aligned}$$

Now let $i = n - u$. The last sum becomes

$$\sum_{i=k-1}^{0} \binom{n}{i} p^i(1 - p)^{n-i} = \sum_{i=0}^{k-1} \binom{n}{i} p^i(1 - p)^{n-i}.$$

Thus, we have the following theorem.

### 3.1 *Theorem: Order Statistics and Quantiles*

Let $X_1, \ldots, X_n$ be independent random variables with a common distribution having $p$ quantile $\theta_p$. Let $X_{[i],\, n}$ denote the $i$th order statistic. Then

$$P\{X_{[j],\, n} \leq \theta_p\} \geq \sum_{i=j}^{n} \binom{n}{i} p^i(1 - p)^{n-i}$$

$$P\{X_{[k],\, n} \geq \theta_p\} \geq \sum_{i=0}^{k-1} \binom{n}{i} p^i(1 - p)^{n-i}.$$

Furthermore,

$$\begin{aligned}P\{X_{[j],n} \le \theta_p \le X_{[k],n}\} &= 1 - P\{X_{[j],n} > \theta_p \quad \text{or} \quad X_{[k],n} < \theta_p\}\\ &= 1 - P\{X_{[j],n} > \theta_p\} - P\{X_{[k],n} < \theta_p\}\\ &= P\{X_{[j],n} \le \theta_p\} + P\{X_{[k],n} \ge \theta_p\} - 1\\ &\ge \sum_{i=j}^{k-1} \binom{n}{i} p^i(1-p)^{n-i}.\end{aligned}$$

### *3.2 Corollary: Confidence Limits for Quantiles*

If $n$ is chosen so that

$$\sum_{i=j}^{n} \binom{n}{i} p^i(1-p)^{n-i} \ge 1 - \alpha,$$

then $X_{[j],n}$ is a lower $1 - \alpha$ level confidence limit for $\theta_p$. If $n$ is chosen so that

$$\sum_{i=0}^{k-1} \binom{n}{i} p^i(1-p)^{n-i} \ge 1 - \alpha,$$

then $X_{[k],n}$ is an upper $1 - \alpha$ level confidence limit for $\theta_p$. If $n$ is chosen so that for $j < k$

$$\sum_{i=j}^{k-1} \binom{n}{i} p^i(1-p)^{n-i} \ge 1 - \alpha,$$

then $[X_{[j],n}, X_{[k],n}]$ is a $1 - \alpha$ level confidence interval for $\theta_p$.

### *3.3 Example: Confidence Interval for Median*

Suppose that we want to determine something about the median income in the United States. How much information can we obtain from a representative sample consisting of the income of five people chosen from the population (of wage earners) with replacement? If we denote these five observations by $x_1, \ldots, x_5$, then

$$P\{X_{[1],5} \le \theta_{1/2} \le X_{[5],5}\} \ge \sum_{i=1}^{4} \binom{5}{i} (1/2)^i(1/2)^{5-i} = 1 - (1/2)^4 = 15/16.$$

Hence, the interval $[x_{[1],5}, x_{[5],5}]$ is the value of a 15/16 level confidence interval for the median income.

The real difficulty here lies in obtaining a representative sample rather than in the mathematical manipulations.

Note that it is very simple to use Corollary 3.2 (on confidence bounds for quantiles) to test hypotheses concerning quantiles. We will omit the details. △

### *3.4 EXERCISES*

1.° How many observations are required in order that $[X_{[1],\,n}, X_{[n],\,n}]$ be a .99 level confidence interval for the median?

2. Find the confidence level of the interval $[X_{[1],\,10}, X_{[10],\,10}]$ for the median.

3.° A confidence interval of confidence .9 is desired for the quantile $\theta_{1/5}$ of the distribution of income. If the interval is of the form $[X_{[1],\,n}, X_{[2n/5],\,n}]$, what should $n$ be?

4. Using the Central Limit Theorem, find approximately the level of the confidence interval $[X_{[25],\,100}, X_{[75],\,100}]$ for the median.

5. Same as Exercise 4, with $\theta_{.6}$ replacing the median.

## SECTION 4
## Two Sample Tests and General Tests of Homogeneity

There are many situations in which we observe two independent random samples, $Y_1, \ldots, Y_m$ and $Z_1, \ldots, Z_n$, with continuous distribution functions $F_Y$ and $F_Z$ respectively, and wish to test the hypothesis that $F_Y = F_Z$. The observations $Y_i$ may represent the effect of one type of drug on a given disease, with $Z_j$ the effect of another type of drug. Or, the $Y_i$ may be the measured hardness of a given product under one production process and the $Z_j$ the hardness under another; or the $Y_i$ may be scores of one group on an intelligence test and $Z_j$ the scores of another group. A variety of non-parametric tests of $H_0$: $F_Y = F_Z$ have been developed, each possessing different discriminating power characteristics. We shall treat only the most popular of these tests—the Wilcoxon, or Mann-Whitney, test.

To start our investigation, we first examine the meaning of the inequality $F_Z > F_Y$ where $F_Z$ and $F_Y$ are two distribution functions. We see from Figure 12.4

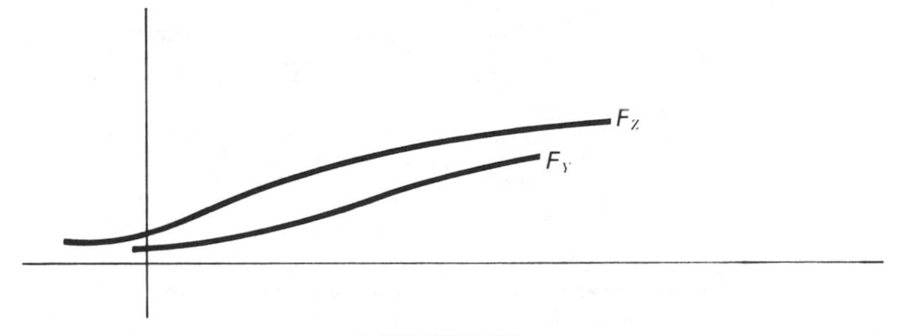


*FIGURE* 12–4.

that in this case, $F_Z$ concentrates its mass *to the left* of that of $F_Y$. That is, if $F_Z > F_Y$, then $Z$ "tends to be" less than $Y$, which we will write $Z$ "$<$" $Y$. Note that if $F_Y$ is the normal distribution function with mean $\mu_Y$ and variance $\sigma^2$, and $F_Z$ is the normal distribution function with mean $\mu_Z$ and variance $\sigma^2$, then $F_Y = F_Z$ corresponds to $\mu_Y = \mu_Z$, and $F_Z > F_Y$ corresponds to $\mu_Y > \mu_Z$.

We already have the $t$ test of $\mu_Y = \mu_Z$ versus $\mu_Y > \mu_Z$ (Exercise 5.10(7)(b), Chapter 11). The Wilcoxon test that we are developing now can also be used for this purpose. It will have less discriminating power than the $t$ test, but it will test $Y$ "$>$" $Z$ properly over a much wider class of distributions.

The Wilcoxon test of the hypothesis $H_0$, that $F_Y = F_Z$, against the alternative $H_1$, that $F_Z > F_Y$, is developed as follows: Combine the two sets of observations $y_1, \ldots, y_m$, and $z_1, \ldots, z_n$ into one sample of size $m + n$, and let the random variables whose values are the observations arranged in order of magnitude be denoted by $W_{[1]} < W_{[2]} < \cdots < W_{[m+n]}$. That is, the value of $W_{[1]}$ is the smallest of the values $y_1, \ldots, y_m, z_1, \ldots, z_n$, the value of $W_{[2]}$ is the next smallest of the values $y_1, \ldots, y_m, z_1, \ldots, z_n$, and so on. (For simplicity, we assume that there are no ties.) We define $r_1$ as the position (rank) of the smallest $y$ value in the "ordered sample," $r_2$ as the position of the next smallest $y$ value, and so on. For example, the positioning of $z$'s and $y$'s illustrated in Figure 12.5 leads to $r_1 = 3$, $r_2 = 5$, $r_3 = 6$.

Now, note that if the $y$'s tend to exceed the $z$'s, then the $y$'s will tend to have larger ranks than they should in the case in which the $z$'s and $y$'s have the same distribution. Thus, it seems reasonable to

$$\text{assert that } F_Z > F_Y (Y \text{“}>\text{”} Z) \quad \text{if} \quad \sum_{j=1}^{m} r_j \geq c$$

and to

$$\text{assert that } F_Z = F_Y \qquad \text{if} \quad \sum_{j=1}^{m} r_j < c.$$

Since we are assuming continuous distribution functions, whenever $F_Y = F_Z$, the random vector whose value is the observed rankings $(r_1, \ldots, r_m, \rho_1, \ldots, \rho_n)$ of the full sample is equally likely to be any permutation of $1, 2, \ldots, m + n$. Let $R_i$ be the random variable whose observed value is $r_i$. From the observation just made, we see that when $F_Y = F_Z$, then

$$\textbf{4.1} \quad P\{R_1 = r_1, \ldots, R_m = r_m\} = \frac{\begin{array}{c}\text{number of permutations of } 1, 2, \ldots, m + n \\ \text{with 1 in position } r_1, \ldots, m \text{ in position } r_m\end{array}}{\text{total number of permutations of } 1, 2, \ldots, m + n}$$

$$= \frac{n!}{(m + n)!},$$

z z y z y y

*FIGURE* 12–5.

since the numerator is the number of possible assignments of the $z_1, \ldots, z_n$ to the $n$ available positions. From 4.1, we see that we can choose $c$ so that when $F_Y = F_Z$, under $H_0$,

**4.2**
$$P\left\{\sum_{j=1}^{m} R_j \geq c\right\} = \alpha.$$

Now we are ready to present the Wilcoxon tests.

### 4.3 Wilcoxon Tests of $F_Y = F_Z$

To test the hypothesis $H_0$: $F_Y = F_Z$ against the alternative $H_1$: $F_Z > F_Y$, we observe the values $y_1, \ldots, y_m, z_1, \ldots, z_n$ of two independent random samples governed by $F_Y$ and $F_Z$ respectively. In the combined ordered sample, let $r_i$ be the observed rank of $y_i$.

Accept $H_0$ if

$$\sum_{j=1}^{m} r_j < c.$$

Otherwise, accept $H_1$. We can choose $c$ so that, under $H_0$,

$$P\left\{\sum_{j=1}^{m} R_j \geq c\right\} = \alpha$$

for the allowable $\alpha$ values (by means of Equation 4.1).

To test $H_0$ against $H_1'$: $F_Z < F_Y$ or $F_Z > F_Y$, we accept $H_0$ if $c_1 \leq \sum_{j=1}^{m} r_j \leq c_2$. We see that $c_1$ and $c_2$ can be chosen to yield desired values of probability of type I error.

For the practical purpose of choosing the critical values in the Wilcoxon test, the following theorem can be shown.

### *4.4 Theorem: Approximate Distribution of* $\sum_{j=1}^{m} R_j$

Under $H_0$ of the Wilcoxon tests (4.3) for $m \geq 10$, the random variable $\sum_{j=1}^{m} R_j$ has approximately a normal distribution function with

$$E\left(\sum_{j=1}^{m} R_j\right) = \frac{m(m+n+1)}{2} \quad \text{and} \quad \text{Var}\left(\sum_{j=1}^{m} R_j\right) = \frac{mn(m+n+1)}{12}.$$

We give one illustration.

### 4.5 *Example: Test of $H_0$ against $H_1'$*

Suppose that $m = n = 10$. Then, under $H_0$,

$$E\left(\sum_{j=1}^{10} R_j\right) = 105, \quad \text{Var}\left(\sum_{j=1}^{10} R_j\right) = 175.$$

Using Theorem 4.4, we see that the test of $H_0$ versus $H_1'$ of approximate .95 level is to accept $H_0$ if $\sum_{j=1}^{m} r_j$ falls in the interval $[105 - 1.96\sqrt{175},\ 105 + 1.96\sqrt{175}]$, which would be the interval [79, 131]. △

It is not hard to see that the discriminating power of the Wilcoxon tests when $H_0$ fails to hold depends very much on which are the true distributions. For instance, if $F_Z$ is uniform on [0, 1] and $F_Y$ is uniform on $[\varepsilon, 1 + \varepsilon]$ where $\varepsilon$ is small, then it is easy to see that there is very little discriminating power. But just the opposite holds if $F_Z$ is uniform on [0, 1] and $F_Y$ is uniform on [2, 3].

Although we do not investigate it here, it is worth mentioning that the Kolmogorov-Smirnov test can be extended to testing $F_Z = F_Y$ by means of the statistic $d(F_m^{(y)}, F_n^{(z)})$, where $F_m^{(y)}$ and $F_n^{(z)}$ are the empirical distribution functions formed from $Y_1, \ldots, Y_m$ and $Z_1, \ldots, Z_n$ respectively.

---

### 4.6 *EXERCISES*

Suppose that $m = n = 15$ and

| | | | |
|---|---|---|---|
| $z_1 = .787$ | $z_2 = .858$ | $z_3 = 1.286$ | $z_4 = 1.121$ |
| $z_5 = .752$ | $z_6 = .047$ | $z_7 = .418$ | $z_8 = .705$ |
| $z_9 = 1.537$ | $z_{10} = 1.434$ | $z_{11} = 1.202$ | $z_{12} = -.626$ |
| $z_{13} = .086$ | $z_{14} = .696$ | $z_{15} = -1.445$ | |

| | | |
|---|---|---|
| $y_1 = 1.9416$ | $y_2 = .875$ | $y_3 = 2.659$ |
| $y_4 = 1.719$ | $y_5 = 1.037$ | $y_6 = 1.699$ |
| $y_7 = 1.320$ | $y_8 = .201$ | $y_9 = 1.373$ |
| $y_{10} = .352$ | $y_{11} = .073$ | $y_{12} = 2.572$ |
| $y_{13} = .681$ | $y_{14} = 3.049$ | $y_{15} = -2.077$ |

1.° Test $H_0$: $F_Z = F_Y$ against $H_1$: $F_Z > F_Y$ at approximately the .1 level. The data were generated as follows: $Z$ is normal with $E(Z) = 0$ and Var $Z = 1$; $Y$ is normal with $E(Y) = 1$ and Var $Y = 1$.

2. Test $H_0$: $F_Z = F_Y$ against $H_1'$: $F_Z > F_Y$ or $F_Z < F_Y$ at approximately the .05 level.

## 4.7 The Chi Square Test for Homogeneity

Suppose that we are given $n$ populations and want to test the hypothesis $H_0$ that they all are describable by the same distribution. In the previous parts of this section, we have treated the case $n = 2$, but these tests do not extend conveniently to the case $n > 2$. When the populations are discrete, the following is usable. Suppose that the $j$th population is describable by probabilities $p_{1j}, p_{2j}, \ldots, p_{Ij}$ where $j = 1, 2, \ldots, J$. We want to test the hypothesis $H_0$, that

$$p_{i1} = p_{i2} = \cdots = p_{iJ} \qquad \text{for all} \qquad i = 1, 2, \ldots, I.$$

Let us draw samples of size $n_{.1}$ from the first population

.

.

.

size $n_{.J}$ from the $J$th population.

Denote by $n_{ij}$ the number of elements of the sample from the $j$th population assuming the value $i$; let $N_{ij}$ be the random variable whose value is $n_{ij}$. Define the random variable

$$\sum_{i=1}^{I} \sum_{j=1}^{J} \frac{\left(N_{ij} - \dfrac{N_{.j}N_{i.}}{N}\right)^2}{\left(\dfrac{N_{.j}N_{i.}}{N}\right)} = \chi^2$$

where

$$N_{.j} = \sum_{i-1}^{I} N_{ij} = n_{.j}, \quad N_{i.} = \sum_{j=1}^{J} N_{ij}, \quad \text{and} \quad N = \sum_{i=1}^{I} \sum_{j=1}^{J} N_{ij}.$$

It can be shown that if all $n_{.j}$ are large enough, the random variable $\chi^2$ has approximately a Chi square distribution with $(I - 1)(J - 1)$ degrees of freedom when $H_0$ is true. The statistic $\chi^2$ tends to be much larger when $H_0$ is not true than when it is. Hence, it seems reasonable to reject $H_0$ if $\chi^2 > c_{1-\alpha}$, where $c_{1-\alpha}$ is the upper $1 - \alpha$-point of the Chi square distribution with $(I - 1)(J - 1)$ degrees of freedom.

## 4.8 *Example: Mortality Rates*

It is desired to see whether or not mortality rates are the same in cities, suburbs, and rural areas. By the mortality rate, we mean the probability that a person selected randomly from the given population will die in a given period of time (often three years). Here $J = 3$ (three populations), $I = 2$ (live or die). Here are the actual figures over five years:

| | Lived | Died |
|---|---|---|
| Center City | 272,913,536 | 2,642,230 |
| Suburbs | 64,714,732 | 497,096 |
| Rural | 195,267,581 | 1,934,350 |

$\chi^2$ turns out to be quite large, so these populations apparently cannot be described by the same probabilities.

A confidence interval approach, yielding confidence intervals for the $p_{ij}$ would seem more informative. △

---

*4.9 EXERCISE*

Obtain .99 Level Confidence Intervals for the $p_{ij}$ just described.

## SECTION 5 Tests of Independence

In case the random variables $X$ and $Y$ are jointly normal, then they are independent precisely when the correlation coefficient $\rho = 0$. Hence, in this parametric case, to test independence we would test $H_0$: $\rho = 0$. Such a test could be based on Fisher's $Z$ (see Problem 6.17(1), Chapter 10).

If we are unwilling to assume normality, then a more basic type of test, linked more directly to the definition of independence may be desired.

### 5.1 Chi Square Test of Independence

Let $X$ be a random variable that can take on values $x_1, \ldots, x_m$ and $Y$ be a random variable which can take on values $y_1, \ldots, y_q$. Observe the values of $n$ independent random vectors, each having distribution $P_{X,Y}$. Let $\tilde{p}_{ij}$ be the observed proportion of pairs with values $(x_i, y_j)$. Let $\tilde{p}_{i.}$ be the observed proportion of pairs whose first coordinate is $x_i$, and let $\tilde{p}_{.j}$ be the observed proportion of pairs whose second coordinate is $y_j$. Suppose that

$$P\{X = x_i, Y = y_j\} = p_{ij} \qquad P\{X = x_i\} = p_{i.} \qquad P\{Y = y_j\} = p_{.j}.$$

If $X$ and $Y$ are independent, then $p_{ij} = p_{i.}p_{.j}$. Also, we expect

$\tilde{p}_{ij}$ to be close to $p_{ij}$

$\tilde{p}_{i.}$ to be close to $p_{i.}$

$\tilde{p}_{.j}$ to be close to $p_{.j}$

for large $n$. Hence, a reasonable test of independence of $X$ and $Y$ is based on the statistic whose value is

$$c^2 = n \sum_{\substack{i=1,\ldots,m \\ j=1,\ldots,q}} \frac{(\tilde{p}_{ij} - \tilde{p}_{i.}\tilde{p}_{.j})^2}{\tilde{p}_{i.}\tilde{p}_{.j}}.$$

Let $H_0$ be the hypothesis that $X$ and $Y$ are independent; in this case it can be shown by reasoning similar to that used for the Chi square test of fit that $c^2$ is the value of a statistic whose distribution is approximately Chi square with $(m - 1)(q - 1)$ degrees of freedom for large $n$. When $H_0$ is not true, $c^2$ should tend to be larger than when $H_0$ is true. Thus, a reasonable test of $H_0$ is to reject $H_0$ when

$$c^2 > \chi^2_{(m-1)(q-1)}(1 - \alpha)$$

where $\chi^2_{(m-1)(q-1)}(1 - \alpha)$ is the upper $1 - \alpha$ point on the Chi square distribution with $(m - 1)(q - 1)$ degrees of freedom. It is covenient to list the various observed and computed frequencies in a table, called a *contingency table*, as shown in Table 12.1.

**TABLE 12.1. Contingency Table**

| $\tilde{p}_{11}$ | $\tilde{p}_{21}$ | · | ... | $\tilde{p}_{.1}$ |
|---|---|---|---|---|
| $\tilde{p}_{12}$ | ⋮ | · | · | $\tilde{p}_{.2}$ |
| ⋮ | ⋮ | | | ⋮ |
| $\tilde{p}_{1.}$ | $\tilde{p}_{2.}$ | | | |

This test for independence can be adapted to random variables with a density as follows: Divide the sample space for $X$ (the real line) into $m$ intervals. You should attempt to choose these intervals so that each one has a probability as close to $1/m$ as possible. Similarly, divide the sample space for $Y$ into $q$ intervals, each of which you attempt to choose so as to obtain probability as close to $1/q$ as possible. As a rule of thumb, experimental evidence indicates that it is desirable to have $n/q$ and $n/m$ at least equal to 5.

Notice that the values of $X$ and $Y$ are irrelevant. All we need to compute are the observed frequencies. $\tilde{p}_{ij}$ is the proportion of the observed vectors for which the value of the first coordinate is in the $i$th $X$ interval and the value of the second coordinate is in the $j$th $Y$ interval; $\tilde{p}_{i.}$ is the proportion of observed vectors for which the value of the first coordinate is in the $i$th $X$ interval, and $\tilde{p}_{.j}$ is the proportion of these vectors whose second coordinate value is in the $j$th $Y$ interval.

### 5.2 *Example*

| | | |
|---|---|---|
| $\tilde{p}_{11} = {}^{14}\!/_{60}$ | $\tilde{p}_{12} = {}^{12}\!/_{60}$ | ${}^{26}\!/_{60} = \tilde{p}_{1.}$ |
| $\tilde{p}_{21} = {}^{22}\!/_{60}$ | $\tilde{p}_{22} = {}^{12}\!/_{60}$ | ${}^{34}\!/_{60} = \tilde{p}_{2.}$ |
| ${}^{36}\!/_{60} = \tilde{p}_{.1}$ | ${}^{24}\!/_{60} = \tilde{p}_{.2}$ | |

Among a group of 60 persons chosen at random, two attributes were determined about each person: whether or not the person wore glasses, and whether or not the person had been involved in any traffic accident in the past three years. Let $\tilde{p}_{11}$ be the proportion of people who wore glasses and who were involved in a traffic accident. Do the data indicate that these attributes are dependent? We shall test this at the .1 level. We form

$$60\sum_{i=1}^{2}\sum_{j=1}^{2}\frac{(\tilde{p}_{ij}-\tilde{p}_{i.}\tilde{p}_{.j})^2}{\tilde{p}_{i.}\tilde{p}_{.j}} = 60\sum_{i=1}^{2}\left[\frac{(\tilde{p}_{i1}-\tilde{p}_{i.}\tilde{p}_{.1})^2}{\tilde{p}_{i.}\tilde{p}_{.1}}+\frac{(\tilde{p}_{i2}-\tilde{p}_{i.}\tilde{p}_{.2})^2}{\tilde{p}_{i.}\tilde{p}_{.2}}\right]$$

$$= 60\left[\frac{(\tilde{p}_{11}-\tilde{p}_{1.}\tilde{p}_{.1})^2}{\tilde{p}_{1.}\tilde{p}_{.1}}+\frac{(\tilde{p}_{21}-\tilde{p}_{2.}\tilde{p}_{.1})^2}{\tilde{p}_{2.}\tilde{p}_{.1}}\right.$$

$$\left.+\frac{(\tilde{p}_{12}-\tilde{p}_{1.}\tilde{p}_{.2})^2}{\tilde{p}_{1.}\tilde{p}_{.2}}+\frac{(\tilde{p}_{22}-\tilde{p}_{2.}\tilde{p}_{.2})^2}{\tilde{p}_{2.}\tilde{p}_{.2}}\right]$$

$$= 60\left[\frac{({}^{14}\!/_{60}-{}^{26}\!/_{60}\cdot{}^{36}\!/_{60})^2}{{}^{26}\!/_{60}\cdot{}^{36}\!/_{60}}+\frac{({}^{22}\!/_{60}-{}^{34}\!/_{60}\cdot{}^{36}\!/_{60})^2}{{}^{34}\!/_{60}\cdot{}^{36}\!/_{60}}\right.$$

$$\left.+\frac{({}^{12}\!/_{60}-{}^{26}\!/_{60}\cdot{}^{24}\!/_{60})^2}{{}^{26}\!/_{60}\cdot{}^{24}\!/_{60}}+\frac{({}^{12}\!/_{60}-{}^{34}\!/_{60}\cdot{}^{23}\!/_{60})^2}{{}^{34}\!/_{60}\cdot{}^{24}\!/_{60}}\right]$$

$$= 60\left[\frac{(.234-2.6)^2}{.26}+\frac{(.367-.34)^2}{.34}\right.$$

$$\left.+\frac{(.2-.174)^2}{.174}+\frac{(.2-.227)^2}{.227}\right]$$

$$= 60[.0026+.0021+.0039+.0032]$$

$$= .728.$$

In this case, under the hypothesis $H_0$ of independence, the random variable whose value is .728 has a Chi square distribution with $(m-1)(q-1) = (2-1)(2-1) = 1$ degree of freedom. From tables, $\chi_1^2(.1) = 2.71$. Hence, the data do not indicate dependence. △

### 5.3 *EXERCISES*

A pair of dice is tossed 60 times. The following contingency table results.

**TABLE 12.3**

| Die 1 \ Die 2 | 1 | 2 | 3 | 4 | 5 | 6 | |
|---|---|---|---|---|---|---|---|
| 1 | 0 | 2/60 | 0 | 4/60 | 4/60 | 0 | 10/60 |
| 2 | 2/60 | 1/60 | 1/60 | 0 | 1/60 | 2/60 | 7/60 |
| 3 | 0 | 2/60 | 2/60 | 4/60 | 2/60 | 3/60 | 13/60 |
| 4 | 3/60 | 2/60 | 2/60 | 5/60 | 2/60 | 2/60 | 16/60 |
| 5 | 0 | 0 | 2/60 | 2/60 | 2/60 | 5/60 | 11/60 |
| 6 | 1/60 | 1/60 | 0 | 1/60 | 0 | 0 | 3/60 |
| | 6/60 | 8/60 | 7/60 | 16/60 | 11/60 | 12/60 | |

Test the hypothesis that the dice are independent at the .05 level.

A slightly different approach to the problem of independence is indicated in important cases in which we would like to be sure of the meaning of our data. For instance, when we desire to determine if there is a relation between cigarette smoking and lung cancer, the authors feel that a confidence interval approach is warranted, since the power of the most frequently used tests is either not known or often not taken into account in the design of the experiment.

### 5.4 *Example: Smoking and Lung Cancer*

In the experiment of selecting one person at random from the population, let $C$ be the event that this person is a cigarette smoker and let $\mathscr{L}$ be the event that the person contracts lung cancer. Suppose we select a large number $n$ of people at random from the population. If enough care is taken to follow up all cases, in time we can compute

$\tilde{p}_{C\mathscr{L}}$, the proportion of the sample who smoke cigarettes and who ultimately contract lung cancer,

$\tilde{p}_C$, the proportion of the sample who smoke cigarettes, and

$\tilde{p}_{\mathscr{L}}$, the proportion of the sample who contract lung cancer.

Now $n\tilde{p}_C$, $n\tilde{p}_{\mathscr{L}}$, and $n\tilde{p}_{C\mathscr{L}}$ are the values of binomial random variables with parameters $(n, p_C)$, $(n, p_{\mathscr{L}})$, and $(n, p_{C\mathscr{L}})$, respectively. Hence, we can obtain $1 - \alpha/3$ confidence intervals for $p_{\mathscr{L}}$, $p_C$, and $p_{C\mathscr{L}}$ either directly from binomial tables (which may require the aid of a large scale computer), or with the help of Uspensky's normal approximation to the binomial. We stress this because in important studies we cannot afford to rely on simple approximations without error terms. The probability that *all* three intervals cover the parameters is at least $1 - \alpha$, since $P(A \cap B \cap C) \geq 1 - P(A^c) - P(B^c) - P(C^c)$. Let the computed values of the confidence intervals be denoted by

$$[p_{\mathscr{L}L}, p_{\mathscr{L}U}], [p_{CL}, p_{CU}], \text{ and } [p_{C\mathscr{L}L}, p_{C\mathscr{L}U}],$$

respectively. Then, if $p_{\mathscr{L}|C} = p_{C\mathscr{L}}/p_C$ denotes the conditional probability that a cigarette smoker contracts lung cancer, we will be correct a proportion $1 - \alpha$ of the time in asserting that

**5.5**
$$p_{C\mathscr{L}L}/p_{CU} - p_{\mathscr{L}U} \leq p_{\mathscr{L}|C} - p_{\mathscr{L}} \leq p_{C\mathscr{L}U}/p_{CL} - p_{\mathscr{L}L}.$$

This assertion is quite meaningful. However, even if we concluded that $p_{\mathscr{L}|C} - p_{\mathscr{L}}$ was positive, this *alone* would not be convincing evidence that smoking *causes* lung cancer. With only this evidence, an equally satisfactory explanation would be that the type of person who smokes is also the type prone to lung cancer. Our insistence on more evidence can be further justified by pointing out that if we used these statistics *by themselves* to conclude that smoking causes lung cancer, then we would also conclude that mortuaries cause death, airplane crashes cause thunderstorms, hospitals cause diseases, and so on. (However, the authors feel that the statistical evidence, together with experimentation on the effects of smoke inhalation in mice, the effects of nicotine on blood pressure, and so on, *is quite convincing evidence of the danger of smoking*.)

Let us suppose that $n = 10{,}000$ and that $\alpha = .01$. Let us assume that $\tilde{p}_C = .3765$ and construct the $1 - \alpha/3$ confidence interval for $p_C$ based on the Uspensky normal approximation. We construct a $1 - \alpha/6$ lower confidence limit and a $1 - \alpha/6$ upper confidence limit. Recall from Theorem 6.25 in Chapter 10 that to find a lower $1 - \alpha/6$ confidence limit $\theta_L$ for $\theta$, we must choose $\theta_L$ to satisfy

$$P_{\theta_L}\{\hat{\theta} \leq \tilde{\theta} - 1\} \geq 1 - \alpha/6, \quad \text{i.e.,}$$

$$P_{\theta_L}\{\hat{\theta} \geq \tilde{\theta}\} \leq \frac{\alpha}{6},$$

that is, for this case,

$$P_{p_{CL}}\{\hat{p}_C \geq \tilde{p}_C\} \leq \frac{\alpha}{6}.$$

Specifically,

$$P_{p_{CL}}\left\{\frac{\sum_{i=1}^{10,000} X_i}{10{,}000} \geq .3765\right\} \leq \frac{.01}{6}$$

where $\sum_{i=1}^{10,000} X_i$ is the number of cigarette smokers in our sample. From the Uspensky result (Theorem 3.16, Chapter 7), we know that

**5.6**
$$P_p\left\{\sum_{i=1}^{10,000} X_i \geq 3765\right\} = \int_w^\infty \frac{1}{\sqrt{2\pi}} e^{-u^2/2}\, du - \frac{1 - 2p}{600\sqrt{2\pi p(1-p)}}(1 - w^2)e^{-w^2/2} + \varepsilon$$

where

$$|\varepsilon| \le \frac{.065 + .09\,|1 - 2p|}{10^4 p(1 - p)} + e^{-(300/2)\sqrt{p(1-p)}}$$

and

$$w = \frac{3764.5 - 10^4 p}{100\sqrt{p(1 - p)}}.$$

Now, since

$$\int_{2.935}^{\infty} \frac{1}{\sqrt{2\pi}} e^{-u^2/2}\,du \cong .00166\ldots = .01/6,$$

we know that we want to choose $p$ so that approximately

$$\frac{3764.5 - 10^4 p}{100\sqrt{p(1 - p)}} = 2.935.$$

We may find a number $p$ satisfying this by direct solution of a quadratic equation or by some numerical method. One quick way is to realize that $100\sqrt{p(1 - p)}$ should be somewhere around $100\sqrt{.38(1 - .38)} \cong 48.4$. Hence,

$$3764.5 - 10^4 p \cong 2.935 \times 48.4 \cong 142.$$

Then $10^4 p \cong 3764.5 - 142 = 3622.5$. So, we try $p = .36$ in the right side of Equation 5.6, to see if this side is then less than .01/6. If $p = .36$ furnishes a lower confidence limit, we try $p = .361$. If $p = .361$ does not furnish such a limit, we try .362, and so on, achieving as much accuracy as Equation 5.6 permits. Here $\varepsilon$ is taken to be positive.

The method presented here is quite suitable for a digital computer, but a more sophisticated numerical technique is probably preferable when the computation is done by hand. In any case, we find that $p = .36$ is a lower confidence limit, that the terms following the normal integral are approximately $1.2 \times 10^{-5}$, and that

$p = .361$ is a lower $1 - .01/6$ confidence limit,

$p = .362$ is a lower $1 - .01/6$ confidence limit, and

$p = .363$ is not a lower $1 - .01/6$ confidence limit.

A similar computation for the upper confidence limit yields

$p = .391$ is an upper $1 - .01/6$ confidence limit while

$p = .390$ is not an upper $1 - .01/6$ confidence limit.

Hence, a $1 - .01/3$ confidence interval for $p_C$ when $\tilde{p}_C = .3765$ and $n = 10^4$ is

$$[.362, .391]. \qquad \triangle$$

## *5.7 PROBLEMS*

1.° Suppose that in our sample of 10,000 people, of whom 3765 are smokers (Example 5.4), 654 smokers and 737 non-smokers developed lung cancer. Obtain $1 - .01/3$ confidence intervals for $p_{\mathscr{L}}$ and $p_{C\mathscr{L}}$, and from this apply Equation 5.5 to determine something about $p_{\mathscr{L}|C} - p_{\mathscr{L}}$.

2. Repeat Problem 1 for the case in which 538 smokers (approximately 1/7 of them) and 779 (approximately 1/8) of the non-smokers developed lung cancer.

3. Given a sample of 10,000 people of which 3765 are smokers, suppose that 500 smokers developed lung cancer. What is the largest number of non-smokers that could develop lung cancer for which we would still conclude that $p_{\mathscr{L}|C} - p_{\mathscr{L}} > 0$ at the .99 level?

# 13

# Simple Linear Regression

## SECTION 1
## The Basic Model

One of the most important aspects of all scientific work is the determination of relations between pairs of measured variables. For instance, in nutritional studies we may want to determine the relation between vitamin D intake and the incidence of rickets. In automobile design, we may try to determine the relation between octane and miles per gallon.

In this chapter, we fix our attention on the case in which we want to determine the relation between a scalar variable $x$ and a scalar variable $y$. We assume that the value of $x$ determines a corresponding value of $y$ so that their relationship is a functional one, as shown in Figure 13.1. This might very well be the relation between octane ($x$) and miles per gallon ($y$). Notice that any small portion of this functional relation looks very much like a straight line. Thus, if we are mainly interested in only a small range of $x$ values, we may postulate a relation of the form

**1.1** $$y = \alpha x + \beta$$

for $x$ in some small interval $[a, b]$. (Of course, such a relationship may be adequate even for large intervals in some cases, as in the number of miles versus time while driving at constant speed.)

Now, if we could make perfect measurements, then any two points $(x_1, y_1)$ and $(x_2, y_2)$ (with $x_1 \neq x_2$) obtained from measurements would completely determine the coefficients $\alpha$ and $\beta$ in 1.1. Unfortunately, as we know, perfect measurements are virtually unattainable. A somewhat more realistic description of our actual measurements is the model

**1.2** $$Y = \alpha x + \beta + \varepsilon,$$

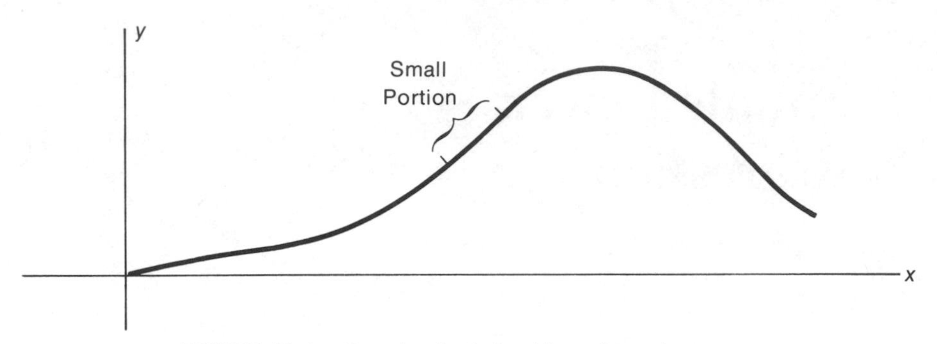


*FIGURE* 13–1. Functional relationship, $x$ determines $y$.

where $Y$ is a random variable whose value is the measurement of the second variable, $x$ is the controlled value of the first variable, and $\varepsilon$ is a random variable arising from measurement error. In this so-called *regression* model, $\varepsilon$ is usually assumed to be normal with $E(\varepsilon) = 0$, Var $\varepsilon = \sigma^2$, where $\sigma^2$ may be known or unknown, but with $\sigma^2$ not dependent on $x$, $\alpha$, or $\beta$.

Now, suppose that we obtain $n$ pairs of measurements

$$(x_1, y_1), (x_2, y_2), \ldots, (x_n, y_n),$$

where $x_1, \ldots, x_n$ are determined by us in advance, with at least two of the $x_i$'s distinct, and where $y_1, \ldots, y_n$ are the values of independent random variables $Y_1, \ldots, Y_n$, which are normal with $E(Y_i) = \alpha x_i + \beta$ and Var $Y_i = \sigma^2$. We would still like to obtain some information about $\alpha$ and $\beta$ even in the presence of error. Looking at Figure 13.2, we examine a typical candidate for the ideal line relating $Y$ and $x$.

If there were no error, then we would expect the vertical distances $d_i$ to be 0. This suggests that a convenient measure of the overall fit of any candidate for the line we seek is $\sum_{i=1}^{n} d_i^2$. Another possibility is $\sum_{i=1}^{n} |d_i|$, but this is less convenient mathematically.

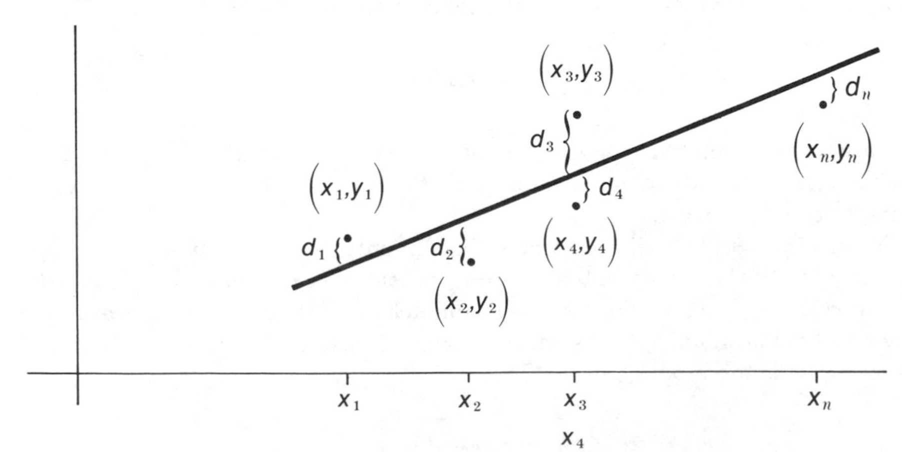


*FIGURE* 13–2. Data points and candidate for ideal line.

## SECTION 2
## The Least Squares Algorithm

Because of its mathematical convenience and its wide usage, we shall concentrate on the following principle for determining the *regression line* $\alpha x + \beta$ of 1.2.

### 2.1 Least Squares Algorithm

We choose $\alpha$ and $\beta$ so as to minimize

$$\sum_{i=1}^{n} d_i^2 = \sum_{i=1}^{n} (y_i - [\alpha x_i + \beta])^2.$$

The resulting line, whose equation is $y = \alpha x + \beta$, is called the *least squares line* or the *line of "best" fit*. For fixed data $(x_i, y_i)$, we see that $\sum_{i=1}^{n} d_i^2$ is determined by $\alpha$, $\beta$. Hence, we write

$$G(\alpha, \beta) = \sum_{i=1}^{n} (y_i - [\alpha x_i + \beta])^2.$$

Our object is to determine $\alpha$, $\beta$ so as to minimize $G(\alpha, \beta)$. It can be shown that in this case the values of $\alpha$ and $\beta$ that minimize $G(\alpha, \beta)$, say $\tilde{\alpha}$ and $\tilde{\beta}$, are the solutions of the equations

**2.2**
$$\frac{\partial}{\partial \alpha} G(\alpha, \beta) = 0$$
$$\frac{\partial}{\partial \beta} G(\alpha, \beta) = 0.$$

(Those who are unfamiliar with partial differentiation should consult the Appendix.) Now,

$$\frac{\partial}{\partial \alpha} G(\alpha, \beta) = \sum_{i=1}^{n} 2(y_i - [\alpha x_i + \beta])(-x_i)$$
$$\frac{\partial}{\partial \beta} G(\alpha, \beta) = \sum_{i=1}^{n} 2(y_i - [\alpha x_i + \beta])(-1).$$

Hence, 2.2 reduces to the two equations

$$\sum_{i=1}^{n} y_i x_i - \alpha \sum_{i=1}^{n} x_i^2 - \beta \sum_{i=1}^{n} x_i = 0$$
$$\sum_{i=1}^{n} y_i - \alpha \sum_{i=1}^{n} x_i - \beta n = 0,$$

which we rewrite as

**2.3**
$$\left(\sum_{i=1}^{n} x_i^2\right)\alpha + \left(\sum_{i=1}^{n} x_i\right)\beta = \sum_{i=1}^{n} y_i x_i$$
$$\left(\sum_{i=1}^{n} x_i\right)\alpha + (n)\beta = \sum_{i=1}^{n} y_i.$$

Notice that 2.3 consists of two linear equations in two unknowns. These equations are called the *normal* equations for the unknowns $\tilde{\alpha}$, $\tilde{\beta}$.

Before attempting to solve 2.3, it is convenient to introduce some simplifying notation. Let

**2.4**
$$\sum_{i=1}^{n} x_i^2 = n\overline{x^2_{(n)}}; \quad \sum_{i=1}^{n} x_i = n\bar{x}_{(n)}; \quad \sum_{i=1}^{n} y_i x_i = n(\overline{yx})_{(n)}; \quad \sum_{i=1}^{n} y_i = n\bar{y}_{(n)},$$
$$\sum_{i=1}^{n} Y_i x_i = n(\overline{Yx})_{(n)}, \quad \text{and so on.}$$

Then 2.3 becomes (after cancelling $n$)

**2.5**
$$\overline{x^2_{(n)}}\alpha + \bar{x}_{(n)}\beta = (\overline{yx})_{(n)}$$
$$\bar{x}_{(n)}\alpha + \beta = \bar{y}_{(n)}.$$

We solve the first of these equations for $\alpha$, obtaining

**2.6**
$$\alpha = \frac{(\overline{yx})_{(n)}}{\overline{x^2_{(n)}}} - \frac{\bar{x}_{(n)}}{\overline{x^2_{(n)}}}\beta.$$

We substitute this into the second equation in 2.5 to find

$$\frac{\bar{x}_{(n)}(\overline{yx})_{(n)}}{\overline{x^2_{(n)}}} - \frac{(\bar{x}_{(n)})^2}{\overline{x^2_{(n)}}}\beta + \beta = \bar{y}_{(n)}.$$

Solving this for $\beta$, we have

**2.7**
$$\beta = \frac{\bar{y}_{(n)}\overline{x^2_{(n)}} - \bar{x}_{(n)}(\overline{yx})_{(n)}}{\overline{x^2_{(n)}} - (\bar{x}_{(n)})^2}.$$

Substituting 2.7 into 2.6, we have

**2.8**
$$\alpha = \frac{(\overline{yx})_{(n)}}{\overline{x^2_{(n)}}} - \frac{\bar{x}_{(n)}}{\overline{x^2_{(n)}}}\,\frac{\bar{y}_{(n)}\overline{x^2_{(n)}} - \bar{x}_{(n)}(\overline{yx})_{(n)}}{\overline{x^2_{(n)}} - (\bar{x}_{(n)})^2}$$
$$= \frac{(\overline{yx})_{(n)} - \bar{x}_{(n)}\bar{y}_{(n)}}{\overline{x^2_{(n)}} - (\bar{x}_{(n)})^2}.$$

That is, the solutions $\tilde{\alpha}$, $\tilde{\beta}$ of Equations 2.2 are

**2.9**
$$\tilde{\alpha} = \frac{\overline{(yx)}_{(n)} - \bar{x}_{(n)}\bar{y}_{(n)}}{\overline{x^2_{(n)}} - (\bar{x}_{(n)})^2}; \qquad \tilde{\beta} = \frac{\bar{y}_{(n)}\overline{x^2_{(n)}} - \bar{x}_{(n)}\overline{(yx)}_{(n)}}{\overline{x^2_{(n)}} - (\bar{x}_{(n)})^2}.$$

We note here that $\tilde{\alpha}$, $\tilde{\beta}$ are the values of the *least squares (point) estimators.* From these we will also be able to derive confidence intervals for $\alpha$, $\beta$ as well as to test hypotheses concerning them.

### *2.10 Example: Mileage*

Suppose that we are given the following data, where the first element of each ordered pair is the number of gallons put in the tank at each stop and the second is the corresponding number of miles.

$$(5, 45.33), \quad (8, 71.53), \quad (7, 65.05), \quad (10, 91.13), \quad (5, 43.82).$$

We want to find the values $\tilde{\alpha}$, $\tilde{\beta}$ of the least squares estimators. To do so, using the notation of 2.4, we compute

$$\overline{x^2_{(n)}} = 52.6, \quad \bar{x}_{(n)} = 7, \quad \overline{(yx)}_{(n)} = 476.928, \quad \bar{y}_{(n)} = 63.372$$

Thus

$$\tilde{\alpha} = \frac{476.928 - 7 \times 63.372}{52.6 - 49} = \frac{33.324}{3.6} \cong 9.25$$

$$\tilde{\beta} = \frac{63.372 \times 52.6 - 7 \times 476.928}{3.6} = \frac{-5.1288}{3.6} \cong -1.42.$$

(Actually, the data here were generated from the line $y = 9x$, or 9 miles per gallon.) △

---

### *2.11 EXERCISES*

1. Suppose that we are given the following data, generated from the regression line $y = 3x + 7$ and a table of normal deviates.

$$(2, 14), (3, 17.5), (3, 15.5), (5, 22.7), (5, 18.6), (5, 21.5),$$
$$(6, 25.2), (6, 25.6), (7, 27), (7, 29.1).$$

(a) Plot the data on graph paper.
(b) Find $\tilde{\alpha}$, $\tilde{\beta}$.
(c) Compare these values with values obtained from measurements on a line drawn through the data by eye.

2. Repeat Exercise 1, but with these data, generated from the line $y = 7x - 4$ and a table of normal deviates.

(1, 2.98), (1, 4.39), (2, 8.97), (2, 10.99), (3, 18.50),
(3, 17.29), (4, 21.82), (4, 24.41), (6, 37.37), (6, 37.32).

3.° (a) Solve the normal equations for $\tilde{\alpha}$, $\tilde{\beta}$ where the data are:

(1, 5.52), (1, 5.53), (2, 6.17), (2, 7.27), (3, 4.89),
(3, 5.12), (4, 1.53), (4, 2.65), (4, 3.80).

(b) Extend the method of least squares by finding $\approx\!\alpha$, $\approx\!\beta$, and $\approx\!\gamma$ that minimize $\sum_i (y_i - [\alpha x_i + \beta + \gamma x_i^2])^2$ for the foregoing data.

(c) Compare $\sum_i (y_i - [\alpha x_i + \tilde{\beta}])^2$ obtained in (a) with $\sum_i (y_i - [\approx\!\alpha x_i + \approx\!\beta + \approx\!\gamma x_i^2])^2$ obtained in (b). The results should not be surprising, since the curve from which these data were generated is $y = 2 + 4x - x^2$.

(d) Plot the data on graph paper.

The results of (c) give an indication of the error made under a linear and a quadratic model. The quantity

$$\frac{1}{9}\sum_{i=1}^{9} (y_i - [\tilde{\alpha} x_i + \tilde{\beta}])^2 \cong 1.1605$$

is an estimate of the expectation of the square of the error under the linear model, $y = \alpha x + \beta$, and the quantity

$$\frac{1}{9}\sum_{i=1}^{9} (y_i - [\approx\!\alpha x_i + \approx\!\beta + \approx\!\gamma x_i^2])^2 \cong 0.422$$

is an estimate of the expectation of the square of the error under the quadratic model, $y = \alpha x + \beta + \gamma x^2$. If a mean square error of 1.16 is tolerable, then the linear model is useful.

2.12 *Remarks on Simplifying Computation.* If we need to manipulate data a lot, sometimes we can simplify the computation by first making certain changes in the data. It is often useful to move the decimal point, or to subtract some fixed number from each value and then do the manipulation. But we must know how to "undo" the transformations to get the answer we want. For instance, in computing

$$\frac{1}{n}\sum_{i=1}^{n} x_i,$$

if the values $x_i$ were all very small, such as .0073, .0052, and .0084, we might work with values 73, 52, and 84 (just multiply each value by $10^4$). The result of averaging these transformed values is too large by a factor of $10^4$. So, to obtain

$$\frac{1}{n}\sum_{i=1}^{n} x_i,$$

we take the average of the transformed data and divide by $10^4$. Similarly, if the $x_i$ were 1.1103, 1.1101, 1.1126, and so on, we might first subtract 1.11 from each, yielding .0003, .0001, and .0026, and then multiply by $10^4$, yielding 3, 1, and 26. If we average this result, divide this average by $10^4$, and add back 1.11, we obtain

$$\frac{1}{n}\sum_{i=1}^{n} x_i.$$

We will not pursue this topic further; we suggest that the student look at each computation to see whether or not there are simpler ways to handle data than doing the indicated computation directly.

## SECTION 3
## Estimation and Testing in Simple Linear Regression

In this section, we assume the model of Equation 1.2. We are given $n$ pairs of measurements

$$(x_1, y_1), \ldots, (x_n, y_n).$$

Here the $x_i$ are under our control and at least two of them are distinct. The numbers $y_1, \ldots, y_n$ are assumed to be the values of independent normal random variables $Y_1, \ldots, Y_n$ with

**3.1** $$E(Y_i) = \alpha x_i + \beta \quad \text{and} \quad \operatorname{Var} Y_i = \sigma^2.$$

The *regression coefficients* $\alpha$ and $\beta$ are presumed to be unknown to us, and $\sigma^2$ may be known or unknown.

This model is usually adequate over only a limited range of $x$ values. Any attempt to use it over too large a range is likely to show up in the form of an apparently large variance $\sigma^2$. We will treat the problem of choosing and analyzing a more appropriate model for such situations in the next chapter.

The values $\tilde{\alpha}$ and $\tilde{\beta}$ of the least squares estimates $\hat{\alpha}$, $\hat{\beta}$ are given in 2.9. It can be shown that, under the given model, the estimates $\hat{\alpha}$, $\hat{\beta}$ are the "best linear unbiased estimates" (see Example 3.18, chapter 10) of $\alpha$ and $\beta$. This result, known as the Gauss-Markov theorem, provides further justification for using the least squares estimates.

Our objectives now are to obtain confidence intervals for $\alpha$, $\beta$ and, when $\sigma^2$ is unknown, for $\sigma^2$, and to construct tests of hypotheses about these parameters. The confidence intervals so obtained are useful for determining the accuracy of predicted values based on the least squares estimates. The tests are useful for answering questions concerning the regression coefficients and $\sigma^2$.

The expressions for the least squares estimates given in 2.9 look rather formidable, so we shall attempt to simplify them before proceeding with our analysis.

From 3.1, using the notation in 2.4, we find

**3.2**

$$E(\overline{Y}_{(n)}) = \frac{1}{n}\sum_{i=1}^{n} (\alpha x_i + \beta) = \alpha\bar{x}_{(n)} + \beta$$

$$E((\overline{Yx})_{(n)}) = \frac{1}{n}\sum_{i=1}^{n} x_i(\alpha x_i + \beta) = \alpha\overline{x^2}_{(n)} + \beta\bar{x}_{(n)}.$$

Substituting these expressions into 2.9 shows that $\hat{\beta}$ and $\hat{\alpha}$ are unbiased;

**3.3**

$$E(\hat{\beta}) = \beta \quad \text{and} \quad E(\hat{\alpha}) = \alpha.$$

The fact that $E(\hat{\beta} - \beta) = 0$ suggests that we use 3.2 to write $\hat{\beta} - \beta$ in the form

$$\hat{\beta} - \beta = \frac{\overline{x^2_{(n)}}(\overline{Y}_{(n)} - \alpha\bar{x}_{(n)} - \beta) - \bar{x}_{(n)}((\overline{Yx})_{(n)} - \alpha\overline{x^2_{(n)}} - \beta\bar{x}_{(n)})}{\overline{x^2_{(n)}} - (\bar{x}_{(n)})^2}.$$

Now let

**3.4**

$$\varepsilon_i = Y_i - \alpha x_i - \beta, \qquad i = 1, \ldots, n.$$

Notice that $\varepsilon_1, \ldots, \varepsilon_n$ are independent normal random variables with mean 0 and variance $\sigma^2$. We see from 3.4 that

$$\overline{Y}_{(n)} = \alpha\bar{x}_{(n)} + \beta + \frac{1}{n}\sum_{i=1}^{n} \varepsilon_i$$

and

$$(\overline{Yx})_{(n)} = \alpha\overline{x^2_{(n)}} + \beta\bar{x}_{(n)} + \frac{1}{n}\sum_{i=1}^{n} x_i\varepsilon_i.$$

Hence, we see that

**3.5**

$$\hat{\beta} - \beta = \frac{\sum_{i=1}^{n} [\overline{x^2_{(n)}} - \bar{x}_{(n)}x_i]\varepsilon_i}{n(\overline{x^2_{(n)}} - (\bar{x}_{(n)})^2)}, \quad \text{and similarly}$$

$$\hat{\alpha} - \alpha = \frac{\sum_{i=1}^{n} (x_i - \bar{x}_{(n)})\varepsilon_i}{n(\overline{x^2_{(n)}} - (\bar{x}_{(n)})^2)}.$$

Let us now use 3.5 and the independence of $\varepsilon_1, \ldots, \varepsilon_n$ to determine the variance of $\hat{\alpha}$. Recall that

$$\text{Var} \sum_{i=1}^{n} q_i\varepsilon_i = \sum_{i=1}^{n} q_i^2 \text{ Var } \varepsilon_i = \sigma^2 \sum_{i=1}^{n} q_i^2,$$

since Var $\varepsilon_i = \sigma^2$ by Theorem 4.11 of Chapter 6. Hence,

$$\operatorname{Var}(\hat{\alpha}) = \frac{\sigma^2 \sum_{i=1}^{n} (x_i - \bar{x}_{(n)})^2}{n^2(\overline{x^2_{(n)}} - (\bar{x}_{(n)})^2)^2}$$

$$= \frac{\sigma^2 \left[ \sum_{i=1}^{n} x_i^2 - 2\bar{x}_{(n)} \sum_{i=1}^{n} x_i + n(\bar{x}_{(n)}{}^2) \right]}{n^2(\overline{x^2_{(n)}} - (\bar{x}_{(n)})^2)^2}$$

$$= \frac{\sigma^2[n\overline{x^2_{(n)}} - n(\bar{x}_{(n)})^2]}{n^2(\overline{x^2_{(n)}} - (\bar{x}_{(n)})^2)^2}$$

$$= \frac{\sigma^2}{n(\overline{x^2_{(n)}} - (\bar{x}_{(n)})^2)}.$$

In a similar fashion, we can find Var $\hat{\beta}$. Finally, using

$$E(\varepsilon_i \varepsilon_j) = \begin{cases} 0 & \text{if } i \neq j \\ \sigma^2 & \text{if } i = j, \end{cases} \qquad \text{(see Theorem 4.10 in Chapter 6)}$$

we see that

$$\operatorname{cov}(\hat{\alpha}, \hat{\beta}) = E[(\hat{\alpha} - \alpha)(\hat{\beta} - \beta)]$$

$$= E\left[ \frac{\left( \sum_{i=1}^{n} [\overline{x^2_{(n)}} - \bar{x}_{(n)} x_i] \varepsilon_i \right) \left( \sum_{j=1}^{n} [x_j - \bar{x}_{(n)}] \varepsilon_j \right)}{n^2(\overline{x^2_{(n)}} - (\bar{x}_{(n)})^2)^2} \right]$$

$$= \frac{\sigma^2 \sum_{i=1}^{n} [\overline{x^2_{(n)}} - \bar{x}_{(n)} x_i][x_i - \bar{x}_{(n)}]}{n^2(\overline{x^2_{(n)}} - (\bar{x}_{(n)})^2)^2}$$

$$= \frac{\sigma^2 \left[ \overline{x^2_{(n)}} \sum_{i=1}^{n} x_i - n\overline{x^2_{(n)}}\bar{x}_{(n)} - \bar{x}_{(n)} \sum_{i=1}^{n} x_i^2 + (\bar{x}_{(n)})^2 \sum_{i=1}^{n} x_i \right]}{n^2(\overline{x^2_{(n)}} - (\bar{x}_{(n)})^2)^2}$$

$$= \frac{\sigma^2[n\overline{x^2_{(n)}}\bar{x}_{(n)} - n\overline{x^2_{(n)}}\bar{x}_{(n)} - n\bar{x}_{(n)}\overline{x^2_{(n)}} + n(\bar{x}_{(n)})^2 \bar{x}_{(n)}]}{n^2(\overline{x^2_{(n)}} - (\bar{x}_{(n)})^2)^2}$$

$$= \frac{\sigma^2 \bar{x}_{(n)}[(\bar{x}_{(n)})^2 - \overline{x^2_{(n)}}]}{n(\overline{x^2_{(n)}} - (\bar{x}_{(n)})^2)^2} = \frac{-\sigma^2 \bar{x}_{(n)}}{n(\overline{x^2_{(n)}} - (\bar{x}_{(n)})^2)}.$$

We see from 3.5 and Theorem 7.16, Chapter 6, that for any constants $a$ and $b$, the statistic $a\hat{\alpha} + b\hat{\beta}$ is a normal variable. By 3.3 and Corollary 4.3, Chapter 6, we know that

$$E(a\hat{\alpha} + b\hat{\beta}) = a\alpha + b\beta.$$

Summarizing, we have $E(a\hat{\alpha} + b\hat{\beta}) = a\alpha + b\beta$ and $a\hat{\alpha} + b\hat{\beta}$ is a normal random variable with

**3.6**

$$\text{Var } \hat{\alpha} = \frac{\sigma^2}{n(\overline{x^2_{(n)}} - (\bar{x}_{(n)})^2)}$$

$$\text{Var } \hat{\beta} = \frac{\sigma^2 \overline{x^2_{(n)}}}{n(\overline{x^2_{(n)}} - (\bar{x}_{(n)})^2)}$$

$$\text{Cov}(\hat{\alpha}, \hat{\beta}) = \frac{-\sigma^2 \bar{x}_{(n)}}{n(\overline{x^2_{(n)}} - (\bar{x}_{(n)})^2)}.$$

Using 3.6 and Theorem 4.8, Chapter 6, we can compute $\text{Var}(a\hat{\alpha} + b\hat{\beta})$. Hence, when $\sigma^2$ is known we may find confidence intervals for any sum of the form $a\alpha + b\beta$ based on $a\hat{\alpha} + b\hat{\beta}$ (see Example 6.11, Chapter 10).

### 3.7 *Example: Confidence Interval for* $2\alpha + \beta$

We might desire to know something about the value of a measurement at some $x$ value that might be of interest later. For instance, if $x$ represents the number of years from the present and $Y$ is a measurement of residential electric power consumption, then $2\alpha + \beta$ would be the expected residential power consumption two years hence. If $x = 2$, then we have $Y = 2\alpha + \beta + \varepsilon$. Its expectation, the quantity of interest, is given by $E(Y) = 2\alpha + \beta$. Hence, it is reasonable to try to find a confidence interval for $2\alpha + \beta$. We know that $2\hat{\alpha} + \hat{\beta}$ is normal, with expectation $2\alpha + \beta$ and variance as obtained from 3.6 and Theorem 4.8, Chapter 6, given by

(*)
$$\text{Var}(2\hat{\alpha} + \hat{\beta}) = \frac{4 + 4\overline{x^2_{(n)}} - 4\bar{x}_{(n)}}{n(\overline{x^2_{(n)}}) - (\bar{x}_{(n)})^2}\sigma^2.$$

Let $\sigma_{2\hat{\alpha}+\hat{\beta}}$ denote the standard deviation of $2\hat{\alpha} + \hat{\beta}$. Then, for known $\sigma^2$, we have

$$P\left\{-1.96 \leq \frac{2\hat{\alpha} + \hat{\beta} - (2\alpha + \beta)}{\sigma_{2\hat{\alpha}+\hat{\beta}}} \leq 1.96\right\} \cong .95,$$

from which we find

$$P\{2\hat{\alpha} + \hat{\beta} - 1.96\sigma_{2\hat{\alpha}+\hat{\beta}} \leq 2\alpha + \beta \leq 2\hat{\alpha} + \hat{\beta} + 1.96\sigma_{2\hat{\alpha}+\hat{\beta}}\} \cong .95.$$

That is, the desired interval is

$$[2\hat{\alpha} + \hat{\beta} - 1.96\sigma_{2\hat{\alpha}+\hat{\beta}}, 2\hat{\alpha} + \hat{\beta} + 1.96\sigma_{2\hat{\alpha}+\hat{\beta}}].$$

Note that we can compute $\sigma_{2\hat{\alpha}+\hat{\beta}}$ in terms of $\sigma$ from (*).

We cannot help but point out that such computations can be quite misleading and dangerous in the situation being considered. It is very possible that electric power consumption grows faster than linearly (Figure 13.3). Thus, a line fitted to

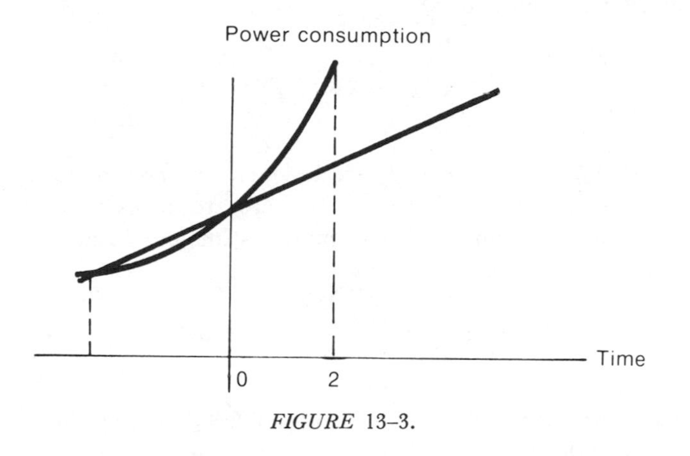


*FIGURE* 13–3.

data would seriously underestimate power requirements for two years hence. Such naive extrapolation could lead to overloads and power failures (like the great black-out in the northeast states on November 9, 1965). We should try to guard against extrapolation beyond the reasonable limits of a model. △

---

### 3.8 EXERCISES

1.[0] Given that $\sigma^2 = 1$, use the data of Exercise 2.11(1) to find:

(a) a .95 level confidence interval for $\alpha$.

(b) a pair of intervals, one for $\alpha$ and one for $\beta$, such that both parameters are covered with probability at least .98.

(Hint: $P(A \cap B) \geq 1 - P(A^c) - P(B^c)$).

(c) a .95 level confidence interval for $2\alpha + \beta$.

(d) a .95 level confidence interval for $4\alpha + 3\beta$.

(e) a .99 level confidence interval for $4\alpha + 3\beta$.

2. Repeat Exercise 1, using the data of Exercise 2.11(2).

If $\sigma^2$ is unknown, then we must first estimate $\sigma^2$ in order to obtain confidence intervals of the kind just described.

Notice that if $\sigma^2$ is large, then the statistic

$$\frac{1}{n-2}\sum_{i=1}^{n}(Y_i - [\hat{\alpha}x_i + \hat{\beta}])^2 \equiv \hat{\sigma}^2$$

will tend to be large. This suggests using this statistic to help estimate $\sigma^2$ when $\sigma^2$ is unknown. For this purpose, we quote the following result, which will be proved in Chapter 14.

### *3.9 Theorem: Distribution of*

$$\sum_{i=1}^{n}(Y_i - \hat{\alpha}x_i - \hat{\beta})^2$$

Suppose that $Y_1, \ldots, Y_n$, $n > 2$, are independent normal random variables, with $E(Y_i) = \alpha x_i + \beta$ and Var $Y_i = \sigma^2$. If there are at least two distinct values among $x_1, \ldots, x_n$, and $\hat{\alpha}$, $\hat{\beta}$ are the least squares estimates of $\alpha$ and $\beta$ (see Equation 2.9), then

$$(n-2)\frac{\hat{\sigma}^2}{\sigma^2} = \sum_{i=1}^{n}(Y_i - \hat{\alpha}x_i - \hat{\beta})^2/\sigma^2$$

has a Chi square distribution with $n - 2$ degrees of freedom, and this (unobservable) random variable is independent of the random variables $\hat{\alpha}$ and $\hat{\beta}$.

Let us see how we can apply this theorem.

### *3.10 Example: Confidence Interval for $\sigma^2$, $\alpha$*

Consider the following data, assumed to be deviations from average yield as a function of the amount of a new type of fertilizer.

| $i$ | $x_i$ | $y_i$ |
|---|---|---|
| 1 | 1 | −3.84 |
| 2 | 1 | −1.54 |
| 3 | 3 | 3.45 |
| 4 | 3 | 2.05 |

The data were obtained from the line $y = 2x - 4$ with $\sigma^2 = 1$. We find

$$\bar{x}_{(4)} = 2,\ \overline{x^2}_{(4)} = 5,\ \bar{y}_{(4)} = .0175,\ (\overline{yx})_{(4)} = 2.78.$$

From this, we may compute the values of $\hat{\alpha}$ and $\hat{\beta}$ to obtain the observed values

$$\tilde{\alpha} = 2.78 - .035 = 2.745$$

$$\tilde{\beta} = .0875 - 5.56 = -5.4725.$$

Hence,

$$\sum_{i=1}^{4}(y_i - \tilde{\alpha}x_i - \tilde{\beta}_i)^2 = 3.66.$$

Let $\chi_2^2$ represent a random variable having the Chi square distribution with two degrees of freedom. Then, from tables of this distribution, we find that

$$P\{.206 \le \chi_2^2 \le 10.6\} \cong .9.$$

Hence

$$P\left\{.206 \le \sum_{i=1}^{4}(Y_i - \hat{\alpha}x_i - \hat{\beta})^2/\sigma^2 \le 10.6\right\} \cong .9.$$

We rewrite this as

$$P\left\{\frac{\sum_{i=1}^{4}(Y_i - \hat{\alpha}x_i - \hat{\beta})^2}{10.6} \le \sigma^2 \le \frac{\sum_{i=1}^{4}(Y_i - \hat{\alpha}x_i - \hat{\beta})^2}{.206}\right\} \cong .9.$$

Thus, a .9 level confidence interval for $\sigma^2$ has the value

$$\left[\frac{3.66}{10.6}, \frac{3.66}{.206}\right] \cong [.346, 17.8].$$

Therefore, the value of a .9 level confidence interval for $\sigma$ is [.58, 4.22]. (Note that $\sigma$ rather than $\sigma^2$ is the parameter with a direct operational meaning, because of the Chebyshev inequality and the Central Limit Theorem. It is $\sigma$ that is measured in the same units as our data.) We know from Theorem 3.9 that $\hat{\alpha}$ and

$$\frac{1}{\sigma^2}\sum_{i=1}^{4}(Y_i - \hat{\alpha}x_i - \hat{\beta})^2$$

are independent. Hence, we know from 8 in Chapter 8 that

$$\frac{(\hat{\alpha} - \alpha)/\sigma_{\hat{\alpha}}}{\sqrt{\frac{1}{2}\sum_{i=1}^{4}(Y_i - \hat{\alpha}x_i - \hat{\beta})^2/\sigma^2}}$$

has a $t$ distribution with two degrees of freedom.

Using the equations in 3.6, we find that

$$\sigma_{\hat{\alpha}}^2 = \frac{\sigma^2}{4},$$

and therefore

$$\frac{\hat{\alpha} - \alpha}{\frac{1}{2}\sqrt{\frac{1}{2}\sum_{i=1}^{4}(Y_i - \hat{\alpha}x_i - \hat{\beta})^2}}$$

has a $t$ distribution with two degrees of freedom. Using computations similar to those in Example 6.13, Chapter 10, with

$$t_{\ell,.1} = -2.92 \qquad \text{and} \qquad t_{u,.1} = 2.92,$$

we see that the value of the desired .9 level confidence interval for $\alpha$ is given by $[2.74 - 2.92\sqrt{.458}, 2.74 + 2.92\sqrt{.458}] \cong 2.74 \pm 1.97 = [.97, 4.71]$. △

## 3.11 EXERCISES

1.° The gross national product of Krancovia for four years looked as follows.

| Year | GNP in millions of Kranques |
|---|---|
| 1950 | 18.699 |
| 1951 | 17.737 |
| 1952 | 16.897 |
| 1953 | 16.339 |

(The generating line is actually $y = -.8(x - 1950) + 18.6$ with $\sigma = .1$.) Assume a model of the form $Y = \alpha x + \beta + \varepsilon$.

(a) Compute the value of a .9 level confidence interval for the rate of decline $-\alpha$ of the Krancovian economy:

(i) assuming $\sigma = .1$.

(ii) assuming $\sigma$ unknown.

(b) What is your point estimate for $-\alpha$?

(c) Compute the value of a .9 level confidence interval for $\sigma$ of the smallest length you can obtain from available tables.

(d) Compute the value of a .95 level lower confidence limit and a .95 level upper confidence limit for $1954\,\alpha + \beta$:

(i) assuming $\sigma = .1$.

(ii) assuming $\sigma$ unknown.

For these cases, combine the limits to compute the value of a .9 level confidence interval for $1954\,\alpha + \beta$.

(e) Estimate the year Krancovia goes down the drain.

(f) Find the values of a pair of confidence intervals for $\alpha$ and $\beta$ whose probability of simultaneous coverage is at least .98.

(Hint: $P(A \cap B) \geq 1 - P(A^c) - P(B^c)$.)

2. Using the data of Exercise 2.11(3), find the value of a .95 level upper and lower confidence limits for $\sigma$ under the model $Y = \alpha x + \beta + \varepsilon$. Because this model is inadequate, the values of these limits should be much larger than 1, the true $\sigma$ of the random variables used in generating the data, and much larger than you would expect, knowing that $\sigma = 1$ and assuming the simple linear model is valid.

The final topic we will take up in this chapter is that of testing hypotheses concerning the regression coefficients. We continue to assume that for some $i \neq j$, we have $x_i \neq x_j$.

We can easily test hypotheses about any weighted sum $a\alpha + b\beta$. For, using the equations in 3.6, we see that the random variable

$$\frac{[a\hat{\alpha} + b\hat{\beta} - (a\alpha + b\beta)]\sqrt{n(\overline{x^2_{(n)}} - (\bar{x}_{(n)})^2)}}{\sqrt{a^2 + b^2\overline{x^2_{(n)}} - 2ab\bar{x}_{(n)}}\;\sigma}$$

has a standard normal distribution, whereas

$$\frac{[a\hat{\alpha} + b\hat{\beta} - (a\alpha + b\beta)]\sqrt{n(\overline{x^2_{(n)}} - (\bar{x}_{(n)})^2)}}{\sqrt{a^2 + b^2\overline{x^2_{(n)}} - 2ab\bar{x}_{(n)}}\sqrt{\dfrac{1}{n-2}\sum_{i=1}^{n}(Y_i - \hat{\alpha}x_i - \hat{\beta})^2}}$$

has a $t$ distribution with $n - 2$ degrees of freedom. This enables us to perform tests both when $\sigma^2$ is known and when it is unknown.

### 3.12 *Example: Test of $H_0$: $\alpha \leq 2$*

Suppose that we are given the following data as pairs of measurements, where $x_i$ is the speed in miles per hour of an automobile and $y_i$ is the corresponding measurement of the cooling system's water temperature in degrees Fahrenheit. We postulate $Y = \alpha x + \beta$. If $\alpha > 2$, then a larger cooling system than the standard one may be required. The data were generated from the line $y = 2x + 60$ with $\sigma = 4$.

| $x_i$ | $y_i$ |
|---|---|
| 20 mph | 94.684 F° |
| 30 | 125.136 |
| 40 | 142.476 |
| 50 | 162.796 |
| 60 | 180.404 |

To test $H_0$: $\alpha \leq 2$, we assume that $\alpha = 2$ in order to obtain a test with the proper level. Under $H_0$, if $\sigma$ is unknown, we see from Theorem 3.9 and from 8 in Chapter 8 that

**3.13**
$$\frac{(\hat{\alpha} - 2)\sqrt{5(\overline{x^2_{(5)}}) - (\bar{x}_{(5)})^2}}{\sqrt{\dfrac{1}{3}\sum_{i=1}^{5}(Y_i - \hat{\alpha}x_i - \hat{\beta})^2}}$$

has a $t$ distribution with three degrees of freedom.

The test of $H_0$ is to reject $H_0$ if the value of the statistic in 3.13 exceeds the upper $1 - \gamma$ point on the $t$ distribution with three degrees of freedom. From the given data,

$$\bar{x}_{(5)} = 40, \qquad \overline{x^2_{(5)}} = 1800, \qquad \overline{x^2_{(5)}} - (\bar{x}_{(5)})^2 = 200,$$
$$\bar{y}_{(5)} \cong 141.1,$$
$$\overline{(yx)}_{(5)} = 6059.$$

Thus, the values $\tilde{\alpha}$, $\tilde{\beta}$ of $\hat{\alpha}$ and $\hat{\beta}$ are $\tilde{\alpha} = 2.075$, $\tilde{\beta} = 58.05$. The estimate of $\sigma^2$ has the value

$$\frac{1}{3}\sum_{i=1}^{5}(y_i - \tilde{\alpha}x_i - \tilde{\beta})^2 = 17.1.$$

The statistic in 3.13 has the value

$$\frac{.075\sqrt{1000}}{\sqrt{17.1}} \cong .57.$$

The upper .9 point of the $t$ distribution with three degrees of freedom is 1.64, so if we desire the probability of rejecting $H_0$ when it is true to be less than or equal to .1, we would in this case accept $H_0$: $\alpha \leq 2$, based on the value of our $t$ statistic.

*Remarks.* Rather than worry about the power of this test, it is easier to form a confidence interval for $\alpha$, so that we can see the discriminating ability of procedures based on $\hat{\alpha}$. We note here that

$$\frac{(\hat{\alpha} - \alpha)\sqrt{1000}}{\sqrt{\frac{1}{3}\sum_{i=1}^{5}(Y_i - \hat{\alpha}x_i - \hat{\beta})^2}}$$

has a $t$ distribution with three degrees of freedom. This enables us to find a .95 level confidence interval for $\alpha$. We rewrite the event

$$\left\{-t_{.975} \leq \frac{(\hat{\alpha} - \alpha)\sqrt{1000}}{\sqrt{\frac{1}{3}\sum_{i=1}^{5}(Y_i - \hat{\alpha}x_i - \hat{\beta})^2}} \leq t_{.975}\right\}$$

as

$$\left\{\hat{\alpha} - \frac{t_{.975}\sqrt{\frac{1}{3}\sum_{i=1}^{5}(Y_i - \hat{\alpha}x_i - \hat{\beta})^2}}{\sqrt{1000}} \leq \alpha \leq \hat{\alpha} + \frac{t_{.975}\sqrt{\frac{1}{3}\sum_{i=1}^{5}(Y_i - \hat{\alpha}x_i - \hat{\beta})^2}}{\sqrt{1000}}\right\}.$$

In this case, the confidence interval has the value $[2.075 - .416, 2.075 + .416]$. Note that this shows that if $\alpha$ were bigger than 2.5, we would reject $H_0$ with high probability. △

---

## *3.14 EXERCISE*

An insurance company might assume a model of the form $Y_x = \alpha x + \beta + \varepsilon$, where $Y_x$ is a random variable whose value is the proportion of the population of age

$x$ which dies at age $x$. Suppose that the following measurements are made:

| $x$ = age in years | $y_x$ – proportion of this age group dying at this age |
|---|---|
| 20 | .02122 |
| 25 | .03639 |
| 30 | .02848 |
| 35 | .04689 |
| 40 | .05375 |
| 45 | .03990 |
| 50 | .05831 |

The data were obtained from the line $y_x = .001x + .007$ with $\sigma = .01$. Note that $\beta$ measures mortality from causes not associated with age.

(a)° Test $H_0$: $\beta = 0$ at the .05 level.
(b) Obtain a .9 level confidence interval for $\beta$.
(c)° Test $H_0$: $\alpha \leq .0005$
(d) Obtain a .9 level confidence interval for $\alpha$.

*Concluding Remarks.* This chapter constitutes only the simplest introduction to regression analysis. A more complicated topic in regression, which we have not treated, is the subject of *mixed models*. Here some of the regression coefficients may be thought of as random variables. For example, we might have as a model $Y_i = a + b_i x_i + \varepsilon_i$, with $a$ an unknown constant, and $b_1, \ldots, b_n$ independent normal random variables with mean 0 and variance $\sigma_x^2$ (usually unknown). $x_1, \ldots, x_n$ are given constants, and $\varepsilon_1, \ldots, \varepsilon_n$ are independent normal random variables with $E(\varepsilon_i) = 0$ and Var $\varepsilon_i = \sigma^2$ (known or unknown).

In contrast to the model $Y = a + bx + \varepsilon$, where we try to learn something about the constant $b$, we now want to determine something about the population from which the $b_i$ were drawn. If we are testing typists on various typewriters to learn characteristics of these specific typists, we might use the original model. If we wanted to draw conclusions about the population of all typists, and assume that these particular typists are a random sample from this population, we might use a mixed model. Many of the standard tests and estimation procedures of the mixed model arise from likelihood ratio tests.

# 14

# General Linear Models*

## SECTION 1
## Introduction

In the previous chapter, we investigated the very restricted situation in which two measured variables are linearly related. Now we will treat more general problems of determining relationships among measured quantities. For instance, we might be interested in determining the relationship between such quantities as:

(a) the amplification $y$ of some electronic device
and
the electrical values $x_1, \ldots, x_m$ of some of its components;

(b) the demand $y$ for some product
and
its cost $x_1$, the amount of advertising $x_2$, the number of salesmen $x_3$, the season $x_4$, the state of the economy $x_5$, the quality of the product $x_6$, and the amount of competition $x_7$;

(c) the yield $y$ of some given crop
and
the amount $x_i$ of fertilizer $i$, $i = 1, \ldots, m$.

Under certain circumstances, it may be reasonable to assume that $y$ is functionally related to $x_1, x_2, \ldots, x_m$. That is to say, we assume that there is a function $G$ such that

$$y = G(x_1, \ldots, x_m).$$

Taylor's Theorem for functions of one variable (Appendix) states that under rather general conditions a function of one variable can be approximated rather well by a polynomial function. This theorem can be rather easily extended to functions

* We are not using Gill Sans e.g., x, y, to indicate data. We will use lower case letters in boldface to denote the observed values of random vectors, i.e., to indicate a vector of data.

of more than one variable. With this multidimensional version of Taylor's Theorem as our guide, we frequently obtain a useful model by assuming that $G$ is a polynomial:

$$G(x_1, \ldots, x_m) = \sum_{0 \le i_1 + i_2 + \cdots + i_m \le n} \beta_{i_1, i_2, \ldots, i_m} x_1^{i_1} x_2^{i_2} \cdots x_m^{i_m}.$$

We can always rename our variables so as to make our notation more convenient. For instance, if

$$y = G(x_1, x_2) = \beta_{10}x_1 + \beta_{01}x_2 + \beta_{20}x_1^2 + \beta_{02}x_2^2 + \beta_{11}x_1x_2,$$

we could let

$$\begin{aligned} \beta_{10} &= \beta_1 & & \\ \beta_{01} &= \beta_2 & & \\ \beta_{20} &= \beta_3 & x_1^2 &= x_3 \\ \beta_{02} &= \beta_4 & x_2^2 &= x_4 \\ \beta_{11} &= \beta_5 & x_1x_2 &= x_5 \end{aligned}$$

to obtain

$$y = \sum_{r=1}^{5} \beta_r x_r.$$

Hence, we will restrict consideration to models of the form

**1.1** $$y = \sum_{r=1}^{m} \beta_r x_r.$$

Here $x_r$ are variables whose values generally are known and essentially under the experimenter's control, whereas the $\beta_r$ are unknown parameters. In particular, $\beta_r$ represents the effect on $y$ of a unit dose of the $r$th variable.

Given the model of 1.1, we can try to determine the parameters $\beta_r$ by observing the values of $y$ for $m$ distinct choices $[x_{11}, x_{12}, \ldots, x_{1m}]$, $[x_{21}, x_{22}, \ldots, x_{2m}]$, $\ldots$, $[x_{m1}, x_{m2}, \ldots, x_{mm}]$. That is, we would observe

$$y_1 = \sum_{r=1}^{m} \beta_r x_{1r}$$

$$y_2 = \sum_{r=1}^{m} \beta_r x_{2r}$$

$$\vdots$$

$$y_m = \sum_{r=1}^{m} \beta_r x_{mr}$$

and try to solve this system of equations for the unknowns $\beta_1, \ldots, \beta_m$ in terms of the known $x$ and $y$ values.

We now introduce a very usual complication, namely that we cannot usually observe $\sum_{r=1}^{m} \beta_r x_r$ perfectly. Rather, we can measure only

**1.2**
$$y = \sum_{r=1}^{m} \beta_r x_r + \varepsilon,$$

where $\varepsilon$ represents measurement error. This error may arise from the imperfection of our model, from uncontrollable variation of the data, or from an inherent variability characterizing our measuring instruments. The random variable $\varepsilon$ is most often taken to be normal with expectation 0 and unknown variance $\sigma^2$. Although this last assumption can furnish a poor description, it is mathematically tractable and usually yields reasonable results.

We are now in a position to state in full generality the model that we will be investigating.

### 1.3 The General Linear Model

We can observe the values $y_j$ of independent normal random variables $Y_j$, for $j = 1, \ldots, n$. We assume that for each $j$, we have

**1.4**
$$E(Y_j) = \sum_{i=1}^{m} \beta_i x_{ji}$$

and Var $Y_j = \sigma^2$, unknown. By the proper choice of $x_{ji}$ we can, of course, repeat measurements under identical conditions as often as we desire. The values $x_{ji}$ are known in advance of observing the data and the parameters $\beta_i$, representing the effect of a unit amount of the $i$th variable, are unknown. We want to use the data $y_j, j = 1, \ldots, n$ to learn something about the unknown parameters $\beta_1, \ldots, \beta_m, \sigma^2$.

There is no simple way to choose the $x_{ji}$. Even the most elementary case treated in Chapter 13 is full of hidden difficulties. If you believe that the simple linear model is a very good description, then it seems proper to take observations at only two $x$-points, spaced as far away as possible (Figure 14.1). It appears that the most accurate determination of the least squares line would be accomplished in this way. But what if the relationship between $x$ and $y$ were as pictured in Figure 14.2? Then use of the two indicated $x$ values is likely to overestimate $y$ for $y$'s in $[x_1, x_2]$ and to

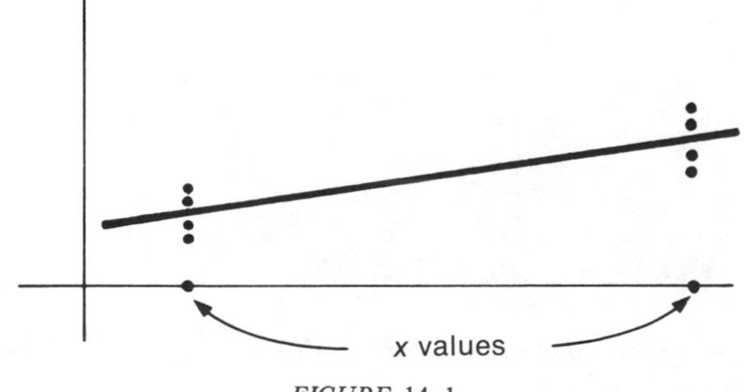


*FIGURE* 14–1.

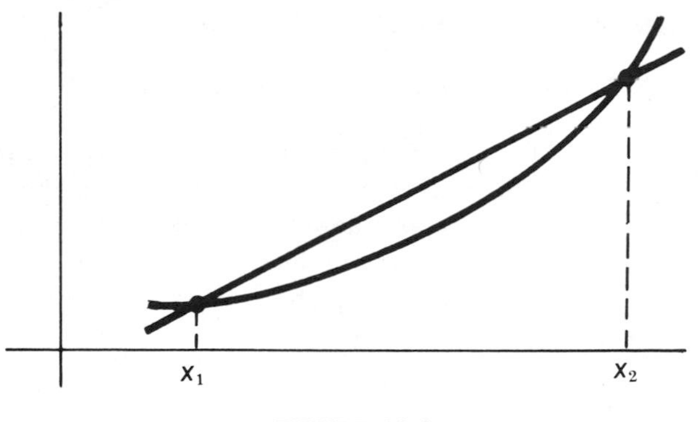


*FIGURE* 14–2.

underestimate $y$ outside $[x_1, x_2]$. With only two $x$ values, the estimate of $\sigma^2$ is likely to be fairly small. This seriously affects future predictions. More $x$ values between the indicated ones would likely lower the least squares line, thus lowering even further our point estimates of $y$ outside $[x_1, x_2]$. However, the raised estimate of $\sigma^2$ may give our confidence intervals a better real chance of coverage. About the only guidelines that can legitimately be given are to try to determine the likely departures of the model from reality, and if possible, to guard against those departures that will adversely affect the uses to which you will apply these results.

In order to facilitate our analysis, it is most convenient to view the set of random variables $Y_j$, $j = 1, \ldots, n$ as a random vector in $n$ dimensional Euclidean space. Here we write

$$\mathbf{Y} = \begin{bmatrix} Y_1 \\ \cdot \\ \cdot \\ Y_j \\ \cdot \\ \cdot \\ Y_n \end{bmatrix} = \begin{bmatrix} \beta_1 x_{11} + \beta_2 x_{12} + \cdots + \beta_m x_{1m} \\ \cdot \\ \cdot \\ \beta_1 x_{j1} + \beta_2 x_{j2} + \cdots + \beta_m x_{jm} \\ \cdot \\ \cdot \\ \beta_1 x_{n1} + \beta_2 x_{n2} + \cdots + \beta_m x_{nm} \end{bmatrix} + \begin{bmatrix} \varepsilon_1 \\ \cdot \\ \cdot \\ \varepsilon_j \\ \cdot \\ \cdot \\ \varepsilon_n \end{bmatrix},$$

which is just a symbolic way of writing

$$Y_j = \sum_{r=1}^{m} \beta_r x_{jr} + \varepsilon_j$$

for all $j = 1, \ldots, n$. This may also be written:

$$\mathbf{Y} = \begin{bmatrix} Y_1 \\ \cdot \\ \cdot \\ \cdot \\ Y_j \\ \cdot \\ \cdot \\ \cdot \\ Y_n \end{bmatrix} = \begin{bmatrix} x_{11} \\ \cdot \\ \cdot \\ \cdot \\ x_{j1} \\ \cdot \\ \cdot \\ \cdot \\ x_{n1} \end{bmatrix} \beta_1 + \begin{bmatrix} x_{12} \\ \cdot \\ \cdot \\ \cdot \\ x_{j2} \\ \cdot \\ \cdot \\ \cdot \\ x_{n2} \end{bmatrix} \beta_2 + \cdots + \begin{bmatrix} x_{1m} \\ \cdot \\ \cdot \\ \cdot \\ x_{jm} \\ \cdot \\ \cdot \\ \cdot \\ x_{nm} \end{bmatrix} \beta_m + \begin{bmatrix} \varepsilon_1 \\ \cdot \\ \cdot \\ \cdot \\ \varepsilon_j \\ \cdot \\ \cdot \\ \cdot \\ \varepsilon_n \end{bmatrix},$$

or better yet, the **general linear model** is given by

**1.5**
$$\mathbf{Y} = \mathbf{x}_1\beta_1 + \mathbf{x}_2\beta_2 + \cdots + \mathbf{x}_m\beta_m + \boldsymbol{\epsilon}.$$

Here $\mathbf{Y}$ is the vector of observable random variables, $\mathbf{x}_1, \ldots, \mathbf{x}_m$ are known column vectors, the regression coefficients $\beta_1, \ldots, \beta_m$ are unknown parameters, and $\boldsymbol{\epsilon}$ is a vector of independent normal random variables with mean 0 and unknown variance $\sigma^2$, all in $n$-dimensional Euclidean space.

Thus, our model is one in which the observation vector $\mathbf{Y}$ can assume any value in $n$-dimensional space. However, *its expectation*

$$E(\mathbf{Y}) = \begin{bmatrix} E(Y_1) \\ \cdot \\ \cdot \\ \cdot \\ E(Y_j) \\ \cdot \\ \cdot \\ \cdot \\ E(Y_n) \end{bmatrix}$$

must be of the form

$$E(\mathbf{Y}) = \mathbf{x}_1\beta_1 + \cdots + \mathbf{x}_m\beta_m.$$

That is to say, $E(\mathbf{Y})$ must be a *linear combination* of the known vectors $\mathbf{x}_1, \ldots, \mathbf{x}_m$.

Let us explore the significance of this observation in the following example.

### *1.6 Example: Mileage*

Suppose that we want to learn something about the dependence of our automobile's performance on both the octane rating of its fuel and the outside temperature. We can surely control the octane rating, and we can control the outside temperature by waiting until the temperature we want is achieved. Let us assume that we can make three observations, and that

$$Y_1 = 95\beta_1 + 70\beta_2 + \varepsilon_1$$
$$Y_2 = 95\beta_1 + 70\beta_2 + \varepsilon_2$$
$$Y_3 = 100\beta_1 + 80\beta_2 + \varepsilon_3.$$

Our model corresponds to making two measurements at 95 octane and 70 degrees temperature and one measurement at 100 octane and 80 degrees. We may write

$$\mathbf{Y} = \begin{bmatrix} 95 \\ 95 \\ 100 \end{bmatrix}\beta_1 + \begin{bmatrix} 70 \\ 70 \\ 80 \end{bmatrix}\beta_2 + \boldsymbol{\epsilon}$$

$$E(\mathbf{Y}) = \begin{bmatrix} 95 \\ 95 \\ 100 \end{bmatrix}\beta_1 + \begin{bmatrix} 70 \\ 70 \\ 80 \end{bmatrix}\beta_2.$$

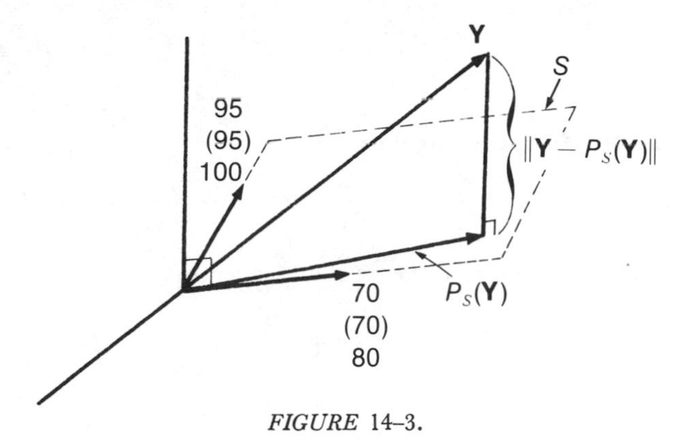


*FIGURE* 14–3.

Written in this fashion, our model has an interesting and enlightening geometrical interpretation. The observation vector $\mathbf{Y}$ may be any point in three-dimensional Euclidean space. Its expectation $E(\mathbf{Y})$, however, must be a linear combination of the vectors

$$\begin{bmatrix} 95 \\ 95 \\ 100 \end{bmatrix} \quad \text{and} \quad \begin{bmatrix} 70 \\ 70 \\ 80 \end{bmatrix}.$$

Hence, $E(\mathbf{Y})$ must lie in the plane $S$ formed by these vectors, as illustrated in Figure 14.3. The "best" estimate of $E(\mathbf{Y})$ is the least squares estimate. Geometrically, this is the point, $P_S(\mathbf{Y})$, in the plane $S$, that is closest to the vector $\mathbf{Y}$. We call $P_S(\mathbf{Y})$ the projection of $\mathbf{Y}$ on $S$. If we let

$$\|\mathbf{v}\| = \left\| \begin{bmatrix} v_1 \\ v_2 \\ v_3 \end{bmatrix} \right\| = \sqrt{v_1^2 + v_2^2 + v_3^2}$$

denote the length of the vector $\mathbf{v}$, as motivated by the Pythagorean Theorem, then $P_S(\mathbf{Y})$ is that vector in the plane $S$ that minimizes

$$\|\mathbf{Y} - P_S(\mathbf{Y})\|.$$

We shall show later that the statistic $\|\mathbf{Y} - P_S(\mathbf{Y})\|^2$ is used in estimating $\sigma^2$. △

In order to develop a workable theory of testing and estimation for the parameters $\beta_i$, $\sigma^2$, we must first develop some linear algebra techniques of finding projections and of determining the dimensions of the "planes" that arise naturally in our analysis. The next section is devoted to this development.

## SECTION 2
## Some Fundamentals of Linear Algebra

### 2.1 Basic Definitions

*Euclidean n-dimensional space* $\mathscr{E}_n$ is the set of all sequences

$$\mathbf{v} = \begin{bmatrix} v_1 \\ v_2 \\ \cdot \\ \cdot \\ \cdot \\ v_n \end{bmatrix}$$

of $n$ elements whose *length* $\|\mathbf{v}\|$ is defined by

$$\|\mathbf{v}\| = \sqrt{\sum_{i=1}^{n} v_i^2}.$$

Elements of $\mathscr{E}_n$ are called *n-dimensional vectors*. If $\mathbf{v} \in \mathscr{E}_n$, $\mathbf{w} \in \mathscr{E}_n$ and $a$ and $b$ are real numbers, then the *linear combination*

$$a\mathbf{v} + b\mathbf{w} = a\begin{bmatrix} v_1 \\ \cdot \\ \cdot \\ \cdot \\ v_n \end{bmatrix} + b\begin{bmatrix} w_1 \\ \cdot \\ \cdot \\ \cdot \\ w_n \end{bmatrix} \equiv \begin{bmatrix} av_1 + bw_1 \\ \cdot \\ \cdot \\ \cdot \\ av_n + bw_n \end{bmatrix}.$$

If $\mathbf{x}_1, \mathbf{x}_2, \ldots, \mathbf{x}_m$ are in $\mathscr{E}_n$ and $\beta_1, \ldots, \beta_m$ are real numbers, then the *linear combination* $\sum_{r=1}^{m} \mathbf{x}_r\beta_r$ is defined in the natural fashion

$$\sum_{r=1}^{m} \mathbf{x}_r\beta_r = \sum_{r=1}^{m} \begin{bmatrix} x_{1r} \\ \cdot \\ \cdot \\ \cdot \\ x_{nr} \end{bmatrix} \beta_r = \begin{bmatrix} \sum_{r=1}^{m} x_{1r}\beta_r \\ \cdot \\ \cdot \\ \cdot \\ \sum_{r=1}^{m} x_{nr}\beta_r \end{bmatrix}.$$

When it is convenient, we will write $\beta\mathbf{x}$ in place of $\mathbf{x}\beta$.

If $\mathbf{x}_1, \ldots, \mathbf{x}_m$ are $n$-dimensional vectors, then we denote *the set of all linear combinations of* $\mathbf{x}_1, \ldots, \mathbf{x}_m$ by the symbol $[\![\mathbf{x}_1, \ldots, \mathbf{x}_m]\!]$ and refer to it as the *subspace of* $\mathscr{E}_n$ *spanned by* $\mathbf{x}_1, \ldots, \mathbf{x}_m$. If $S$ is some subspace of $\mathscr{E}_n$ and $\mathbf{y} \in \mathscr{E}_n$, then the element $\mathbf{s} \in S$ that minimizes $\|\mathbf{s} - \mathbf{y}\|$ is called the *projection of* $\mathbf{y}$ *on* $S$ and is denoted by $P_S(\mathbf{y})$.

If $S$ is the subspace spanned by the single nonzero vector $\mathbf{x}$,

$$S = [\![\mathbf{x}]\!], \; \|\mathbf{x}\| \neq 0,$$

so that $S$ consists of all scalar multiples of $\mathbf{x}$, then $P_S(\mathbf{y}) = P_{[\![\mathbf{x}]\!]}(\mathbf{y})$ is found as follows: Every element $\mathbf{s}$ of $[\![\mathbf{x}]\!]$ is of the form $\mathbf{s} = b\mathbf{x}$ where $b$ is a real number. Hence, the problem of finding the $\mathbf{s} \in [\![\mathbf{x}]\!]$ minimizing $\|\mathbf{s} - \mathbf{y}\|$ is the same as that of finding the real number $b$ minimizing $\|b\mathbf{x} - \mathbf{y}\|$, or equivalently minimizing $g(b) = \|b\mathbf{x} - \mathbf{y}\|^2 = \sum_{i=1}^{n} (bx_i - y_i)^2$. Using the usual calculus technique, we find

$$b = \frac{\sum_{i=1}^{n} x_i y_i}{\sum_{i=1}^{n} x_i^2}.$$

To facilitate matters, we give the following definition.

**2.2 Definition: Dot Product.** We call $\sum_{i=1}^{n} x_i y_i$ the *dot product* (or inner product or scalar product) of the vectors $\mathbf{x}$ and $\mathbf{y}$. We denote the dot product of $\mathbf{x}$ and $\mathbf{y}$ by the symbol

$$\mathbf{x} \cdot \mathbf{y}.$$

The following easily established properties of the dot product are needed in our later work.

For all real $a$ and $b$, and all $\mathbf{x}$, $\mathbf{y}$, and $\mathbf{z}$ in $\mathscr{E}_n$,

$$\textbf{2.3} \qquad \begin{cases} (a\mathbf{x} + b\mathbf{y}) \cdot \mathbf{z} = a(\mathbf{x} \cdot \mathbf{z}) + b(\mathbf{y} \cdot \mathbf{z}) \\ \mathbf{x} \cdot \mathbf{y} = \mathbf{y} \cdot \mathbf{x} \\ \mathbf{x} \cdot \mathbf{x} \geq 0, \text{ equality holding if and only if } \mathbf{x} = \begin{bmatrix} 0 \\ \cdot \\ \cdot \\ \cdot \\ 0 \end{bmatrix} \\ \|\mathbf{x}\| = \sqrt{\mathbf{x} \cdot \mathbf{x}}. \end{cases}$$

In terms of dot products, we have shown that in determining the projection $b\mathbf{x}$ of $\mathbf{y}$ on the nonzero vector $\mathbf{x}$, we have

$$b = \frac{\mathbf{x} \cdot \mathbf{y}}{\|\mathbf{x}\|^2}.$$

We state our result formally in the following theorem.

***2.4 Theorem:*** $P_{[\![\mathbf{x}]\!]}(y)$

The projection of $\mathbf{y}$ on the subspace $[\![\mathbf{x}]\!]$ of all scalar multiples of $\mathbf{x}$ is given by

$$P_{[\![\mathbf{x}]\!]}(\mathbf{y}) = \frac{\mathbf{x} \cdot \mathbf{y}}{\|\mathbf{x}\|^2} \mathbf{x}$$

whenever $\|\mathbf{x}\| \neq 0$. Clearly,

$$P_{[\![\mathbf{x}]\!]}(\mathbf{y}) = \begin{bmatrix} 0 \\ \cdot \\ \cdot \\ \cdot \\ 0 \end{bmatrix} \text{ if } \mathbf{x} = \begin{bmatrix} 0 \\ \cdot \\ \cdot \\ \cdot \\ 0 \end{bmatrix}.$$

That is, the closest vector of the form $b\mathbf{x}$ to $\mathbf{y}$ is given by $P_{[\![\mathbf{x}]\!]}(\mathbf{y})$.

If

$$P_{[\![\mathbf{x}]\!]}(\mathbf{y}) = \begin{bmatrix} 0 \\ \cdot \\ \cdot \\ \cdot \\ 0 \end{bmatrix},$$

then we say that $\mathbf{x}$ and $\mathbf{y}$ are *perpendicular* or *orthogonal* and write $\mathbf{y} \perp \mathbf{x}$. We illustrate orthogonality and its absence in Figure 14.4.

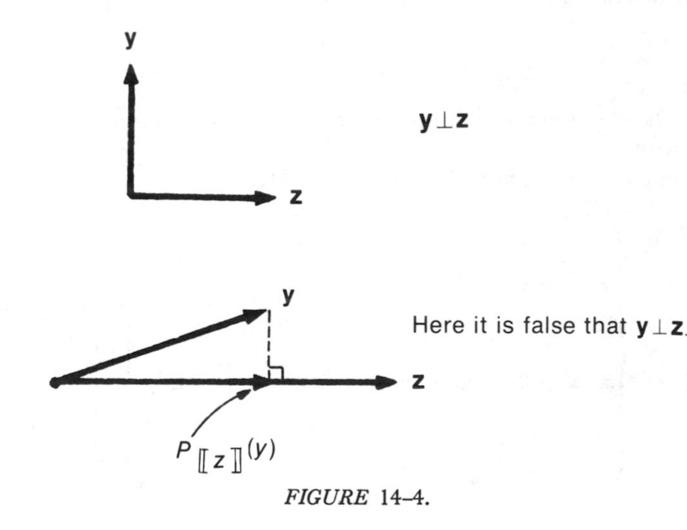


*FIGURE* 14–4.

***2.5 Corollary:*** $y \perp x$

$\mathbf{y} \perp \mathbf{x}$ if and only if $\mathbf{x} \cdot \mathbf{y} = 0$.

In order to be able to develop the distribution theory needed in regression analysis, we must investigate the notion of the *dimension of a subspace*. For this, we need the following definition.

**2.6 Definition: Linear Dependence and Independence.** A vector $\mathbf{v}$ is said to be *linearly dependent* on a set of vectors $\{\mathbf{w}_1, \ldots, \mathbf{w}_m\}$ if $\mathbf{v}$ can be expressed as a linear combination of the $\mathbf{w}_i$, that is, if there are real numbers $a_1, \ldots, a_m$ such that

$$\mathbf{v} = \sum_{i=1}^{m} \mathbf{w}_i a_i.$$

A sequence of vectors $\mathbf{v}_1, \ldots, \mathbf{v}_q$ is said to be *linearly dependent* if at least one of its elements is linearly dependent on the remaining ones. Hence, $\mathbf{v}_1, \ldots, \mathbf{v}_q$ is linearly dependent if there are real numbers $a_i$ such that for some $i$ we have

$$\mathbf{v}_i = \sum_{\substack{j=1 \\ j \neq i}}^{q} \mathbf{v}_j a_j.$$

The sequence of vectors $\mathbf{v}_1, \ldots, \mathbf{v}_q$ is said to be *linearly independent* if it is not linearly dependent.

**2.7 Definition: Dimension of a Subspace.** The *dimension* dim $S$ of the subspace $S = [\![\mathbf{x}_1, \ldots, \mathbf{x}_m]\!]$ is the minimum number of vectors needed to span $S$. That is, dim $S$ is the minimum number $q$ such that

$$S = [\![\mathbf{s}_1, \mathbf{s}_2, \ldots, \mathbf{s}_q]\!].$$

We now claim that if $S = [\![\mathbf{s}_1, \ldots, \mathbf{s}_q]\!]$, where $\mathbf{s}_1, \ldots, \mathbf{s}_q$ is a linearly independent sequence, then dim $S = q$. We prove this by means of the following theorem.

***2.8 Theorem:***

If $[\![\mathbf{v}_1, \ldots, \mathbf{v}_q]\!] = [\![\mathbf{w}_1, \ldots, \mathbf{w}_p]\!] = S$, and both sequences $\mathbf{v}_1, \ldots, \mathbf{v}_q$ and $\mathbf{w}_1, \ldots, \mathbf{w}_p$ are linearly independent, then $q = p$.

*Proof.* We first note that if the sequence $\mathbf{z}_1, \ldots, \mathbf{z}_m$ is linearly dependent, with $\|\mathbf{z}_i\| \neq 0$ for each $i$, then for some $j > 1$,

(*) $\qquad \mathbf{z}_j$ is linearly dependent on $\mathbf{z}_1, \ldots, \mathbf{z}_{j-1}$.

This follows from the fact that, because of the linear dependence, there exists an $i$ with

$$\mathbf{z}_i = \sum_{\substack{k=1 \\ k \neq i}}^{m} b_k \mathbf{z}_k$$

where not all of the $b_k$ are 0. Thus

$$b_1\mathbf{z}_1 + \cdots + b_{i-1}\mathbf{z}_{i-1} - \mathbf{z}_i + b_{i+1}\mathbf{z}_{i+1} + \cdots + b_m\mathbf{z}_m = \begin{bmatrix} 0 \\ \cdot \\ \cdot \\ \cdot \\ 0 \end{bmatrix}.$$

Now, if $i \neq 1$, then either

(a) $b_k = 0$ for $k > i$,

in which case $\mathbf{z}_i = b_1\mathbf{z}_1 + \cdots + b_{i-1}\mathbf{z}_{i-1}$, or

(b) there is a largest $k > i$ for which $b_k \neq 0$.

Then

$$\mathbf{z}_k = -\frac{b_{k-1}}{b_k}\mathbf{z}_{k-1} - \cdots - \frac{b_{i+1}}{b_k}\mathbf{z}_{i+1} + \frac{1}{b_k}\mathbf{z}_i - \frac{b_{i-1}}{b_k}\mathbf{z}_{i-1} - \cdots - \frac{b_1}{b_k}\mathbf{z}_1.$$

If $i = 1$, then the argument of part (b) still holds, yielding

$$\mathbf{z}_k = -\frac{b_{k-1}}{b_k}\mathbf{z}_{k-1} - \cdots - \frac{b_2}{b_k}\mathbf{z}_2 + \frac{1}{b_k}\mathbf{z}_1,$$

because if $b_k = 0$ for all $k > 1$, then we would have

$$\mathbf{z}_1 = \begin{bmatrix} 0 \\ \cdot \\ \cdot \\ \cdot \\ 0 \end{bmatrix},$$

which is prohibited.

We have thus shown (*).

Now, to prove Theorem 2.8, suppose that $q \neq p$. We lose no generality supposing $q > p$. Since $\mathbf{v}_p \in S = [\![\mathbf{w}_1, \ldots, \mathbf{w}_p]\!]$, we know that $\mathbf{v}_p, \mathbf{w}_1, \ldots, \mathbf{w}_p$ is a linearly dependent sequence. Hence, for some $i$, $1 \leq i \leq p$, we may eliminate $\mathbf{w}_i$, that is,

$$[\![\mathbf{v}_p, \mathbf{w}_1, \ldots, \mathbf{w}_{i-1}, \mathbf{w}_{i+1}, \ldots, \mathbf{w}_p]\!] = [\![\mathbf{v}_1, \ldots, \mathbf{v}_q]\!].$$

Here we used relation (*). Now, because $\mathbf{v}_{p-1} \in S$, we see that $\mathbf{v}_{p-1}, \mathbf{v}_p, \mathbf{w}_1, \ldots, \mathbf{w}_{i-1}, \mathbf{w}_{i+1}, \ldots, \mathbf{w}_p$ is also a linearly dependent sequence. Now $\mathbf{v}_{p-1}$ and $\mathbf{v}_p$ are linearly independent. Therefore, again by (*), we see that

$$[\![\mathbf{v}_{p-1}, \mathbf{v}_p, \mathbf{w}_1, \ldots, \mathbf{w}_{i-1}, \mathbf{w}_{i+1}, \ldots, \mathbf{w}_p]\!]$$
$$= [\![\mathbf{v}_{p-1}, \mathbf{v}_p, \mathbf{w}_1, \ldots, \mathbf{w}_{j-1}, \mathbf{w}_{j+1}, \ldots, \mathbf{w}_{i-1}, \mathbf{w}_{i+1}, \ldots, \mathbf{w}_p]\!].$$

That is, another of the **w**'s can be eliminated. Proceeding in this way (rigorously by induction) we finally obtain

$$[\![\mathbf{v}_1, \ldots, \mathbf{v}_p]\!] = [\![\mathbf{v}_1, \ldots, \mathbf{v}_q]\!],$$

that is,

$$[\![\mathbf{v}_1, \ldots, \mathbf{v}_p]\!] = [\![\mathbf{v}_1, \ldots, \mathbf{v}_p, \ldots, \mathbf{v}_q]\!].$$

But if $p < q$, then we see that $\mathbf{v}_q$ is linearly dependent on $\mathbf{v}_1, \ldots, \mathbf{v}_p$. This contradicts the hypothesis that $\mathbf{v}_1, \ldots, \mathbf{v}_q$ is a linearly independent sequence. Hence, we must have that $q = p$.

Note that $p$ depends only on the subspace $S$ and not on the particular sequence $\mathbf{v}_1, \ldots, \mathbf{v}_p$ which generates $S$.

### *2.9 Theorem: Dimension,* dim $\mathscr{E}_n$

If the subspace $S = [\![\mathbf{v}_1, \ldots, \mathbf{v}_p]\!]$, where $\mathbf{v}_1, \ldots, \mathbf{v}_p$ is a linearly independent sequence, then $\dim S = p$. In particular, $\dim \mathscr{E}_n = n$.

*Proof:* If $\dim S = q$, then $S = [\![\mathbf{w}_1, \ldots, \mathbf{w}_q]\!]$. Surely $\mathbf{w}_1, \ldots, \mathbf{w}_q$ must be linearly independent (or $S = [\![\mathbf{w}_1, \ldots, \mathbf{w}_{j-1}, \mathbf{w}_{j+1}, \ldots, \mathbf{w}_q]\!]$, showing that $\dim S < q$). But then, by the previous theorem, $q = p$.

Let $\mathbf{e}_i$ be the vector in $\mathscr{E}_n$ with 1 in the $i$th coordinate and 0 in all other coordinates. Then it is easily seen that $\mathbf{e}_1, \ldots, \mathbf{e}_n$ are linearly independent and $\mathscr{E}_n = [\![\mathbf{e}_1, \ldots, \mathbf{e}_n]\!]$. Hence, $\dim \mathscr{E}_n = n$.

We now develop a simple method for choosing a linearly independent sequence of orthogonal vectors that span a subspace $S$ and that will therefore yield us the dimension of $S$.

## 2.10 Gram-Schmidt Orthogonalization Procedure

With this method, we will be able to compute $P_S(\mathbf{y})$ quite easily. Suppose that $S = [\![\mathbf{v}_1, \ldots, \mathbf{v}_p]\!]$, where for all $i$, $\|\mathbf{v}_i\| \neq 0$, but $\mathbf{v}_1, \ldots, \mathbf{v}_p$ do not necessarily form a linearly independent sequence. Let

$$\left.\begin{aligned} \mathbf{w}_1' &= \mathbf{v}_1 \\ \mathbf{w}_2' &= \mathbf{v}_2 - P_{[\![\mathbf{w}_1']\!]}(\mathbf{v}_2) \end{aligned}\right\} \text{illustrated in Figure 14.5}$$

$$\mathbf{w}_3' = \mathbf{v}_3 - P_{[\![\mathbf{w}_1']\!]}(\mathbf{v}_3) - P_{[\![\mathbf{w}_2']\!]}(\mathbf{v}_3)$$

$$\vdots$$

$$\mathbf{w}_p' = \mathbf{v}_p = \sum_{i=1}^{p-1} P_{[\![\mathbf{w}_i']\!]}(\mathbf{v}_p) \,.$$

(Recall that

$$P_{[\![\mathbf{w}]\!]}(\mathbf{v}) = \begin{bmatrix} 0 \\ \cdot \\ \cdot \\ \cdot \\ 0 \end{bmatrix}$$

if $\|\mathbf{w}\| = 0$.) Now discard all $\mathbf{w}_i'$ that are the zero vector

$$\begin{bmatrix} 0 \\ \cdot \\ \cdot \\ \cdot \\ 0 \end{bmatrix}$$

and rename the remaining $\mathbf{w}_i'$ as

$$\mathbf{w}_1, \mathbf{w}_2, \ldots, \mathbf{w}_q \qquad q \leq p.$$

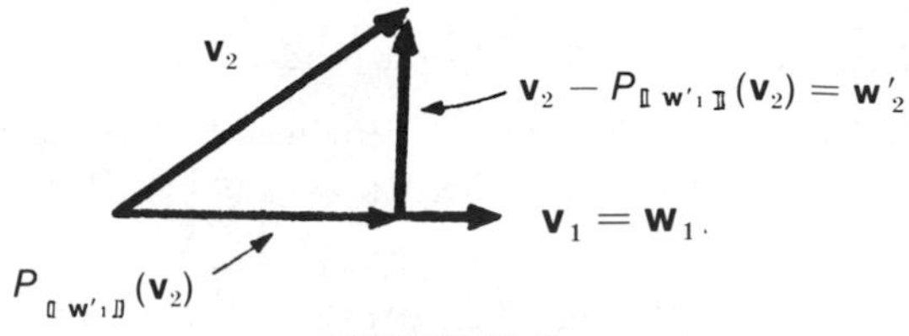


*FIGURE* 14-5.

### 2.11 *Theorem:* dim $S$

When the Gram Schmidt procedure is used on the sequence of vectors $\mathbf{v}_1, \ldots, \mathbf{v}_p$ and $S = [\![\mathbf{v}_1, \ldots, \mathbf{v}_p]\!]$, then

$$\dim S = q.$$

*Proof:* It is easy to verify that $\mathbf{w}_i \cdot \mathbf{w}_j = 0$ if $i \neq j$. For instance,

$$\mathbf{w}_1 \cdot \mathbf{w}_2 = \mathbf{v}_1 \cdot [\mathbf{v}_2 - P_{[\![\mathbf{v}_1]\!]}(\mathbf{v}_2)]$$

$$= \mathbf{v}_1 \cdot \left[\mathbf{v}_2 - \frac{\mathbf{v}_1 \cdot \mathbf{v}_2}{\mathbf{v}_1 \cdot \mathbf{v}_1}\mathbf{v}_1\right]$$

$$= \mathbf{v}_1 \cdot \mathbf{v}_2 - \mathbf{v}_1 \cdot \mathbf{v}_2 = 0.$$

Now, from their construction, we see that

$$\mathbf{w}_i \in [\![\mathbf{v}_1, \ldots, \mathbf{v}_p]\!].$$

For instance,

$$\mathbf{w}_1 = \mathbf{v}_1, \qquad \mathbf{w}_2' = \mathbf{v}_2 - k\mathbf{w}_1' = \mathbf{v}_2 - k\mathbf{v}_1, \text{ and so on.}$$

Hence, $[\![\mathbf{w}_1, \ldots, \mathbf{w}_q]\!] \subseteq [\![\mathbf{v}_1, \ldots, \mathbf{v}_p]\!]$. But $\mathbf{v}_i \in [\![\mathbf{w}_1, \ldots, \mathbf{w}_q]\!]$, since

$$\mathbf{v}_1 = \mathbf{w}_1, \qquad \mathbf{v}_2 = \mathbf{w}_2' + k\mathbf{w}_1' \cdots, \text{ and so on.}$$

Hence, $[\![\mathbf{w}_1, \ldots, \mathbf{w}_q]\!] = [\![\mathbf{v}_1, \ldots, \mathbf{v}_p]\!] \equiv S$. Now we need only show that $\mathbf{w}_1, \ldots, \mathbf{w}_q$ are linearly independent. Because of its importance, we state the needed result formally as the following lemma.

### 2.12 *Lemma: Linear Independence of Mutually $\perp$ Non-Zero Vectors*

If $\mathbf{w}_1, \ldots, \mathbf{w}_q$ satisfy

(a) $\mathbf{w}_i \perp \mathbf{w}_j$ for $i \neq j$ and

(b) $\|\mathbf{w}_j\| \neq 0$ for each $j$ $\left(\text{that is, } \mathbf{w}_j \neq \begin{bmatrix} 0 \\ \cdot \\ \cdot \\ \cdot \\ 0 \end{bmatrix}\right)$,

then the sequence $\mathbf{w}_1, \ldots \mathbf{w}_q$ is linearly independent.

*Proof of the Lemma:* If this sequence were linearly dependent, then for some $j$ we would have $\mathbf{w}_j = \sum_{i \neq j} a_i \mathbf{w}_i$. Taking the dot product of $\mathbf{w}_j$ with both sides to obtain

$$\mathbf{w}_j \cdot \mathbf{w}_j = \sum_{i \neq j} a_i \mathbf{w}_i \cdot \mathbf{w}_j = 0,$$

we would have $\|\mathbf{w}_j\|^2 = 0$, which is impossible. Since the assumption of linear dependence has led to a contradiction, it is untenable. Thus, $\mathbf{w}_1, \ldots, \mathbf{w}_q$ must be linearly independent, as asserted.

Since we discarded all $\mathbf{0}$ vectors, this lemma shows that the Gram-Schmidt orthogonalization procedure yields a sequence of orthogonal linearly independent vectors $\mathbf{w}_1, \ldots, \mathbf{w}_q$ such that

$$[\![\mathbf{v}_1, \ldots, \mathbf{v}_p]\!] = S = [\![\mathbf{w}_1, \ldots, \mathbf{w}_q]\!].$$

It now follows from 2.12 applied to the Gram-Schmidt orthogonalization procedure that $q = \dim S$.

### 2.13 *Theorem:* $P_S(\mathbf{y})$

Suppose that $S = [\![\mathbf{w}_1, \ldots, \mathbf{w}_q]\!]$, where the sequence $\mathbf{w}_1, \ldots, \mathbf{w}_q$ consists of mutually orthogonal non-zero vectors. Then

**2.14**
$$P_S(\mathbf{y}) = P_{[\![\mathbf{w}_1]\!]}(\mathbf{y}) + \cdots + P_{[\![\mathbf{w}_q]\!]}(\mathbf{y}).$$

*Proof:* $P_S(\mathbf{y})$ is that linear combination $\sum_{i=1}^{q} b_i\mathbf{w}_i$ that minimizes

$$\begin{aligned} H(b_1, \ldots, b_q) &= \left\|\mathbf{y} - \sum_{i=1}^{q} b_i\mathbf{w}_i\right\|^2 \\ &= \left(\mathbf{y} - \sum_{i=1}^{q} b_i\mathbf{w}_i\right)\cdot\left(\mathbf{y} - \sum_{i=1}^{q} b_i\mathbf{w}_i\right) \\ &= \mathbf{y}\cdot\mathbf{y} - 2\sum_{i=1}^{q} b_i\mathbf{w}_i\cdot\mathbf{y} + \sum_{i=1}^{q} b_i^2\mathbf{w}_i\cdot\mathbf{w}_i \end{aligned}$$

(where we utilized the fact that $\mathbf{w}_i \cdot \mathbf{w}_j = 0$ if $j \neq i$). To minimize $H$, we set

$$\frac{\partial H}{\partial b_i} = 0,$$

obtaining

$$-2\mathbf{w}_i \cdot \mathbf{y} + 2b_i\mathbf{w}_i \cdot \mathbf{w}_i = 0$$

or

$$b_i = \frac{\mathbf{w}_i \cdot \mathbf{y}}{\mathbf{w}_i \cdot \mathbf{w}_i}.$$

Thus,

$$b_i\mathbf{w}_i = \frac{\mathbf{w}_i \cdot \mathbf{y}}{\mathbf{w}_i \cdot \mathbf{w}_i}\mathbf{w}_i = P_{[\![\mathbf{w}_i]\!]}(\mathbf{y}),$$

establishing the desired result.

The following corollary to 2.13 is used quite a bit later on.

### 2.15 *Corollary to 2.13:* $P_S(a\mathbf{y} + b\mathbf{z})$

If $S$ is a subspace of $\mathscr{E}_n$, $\mathbf{y}$ and $\mathbf{z}$ are any elements of $\mathscr{E}_n$, and $a$ and $b$ are real numbers, then

$$P_S(a\mathbf{y} + b\mathbf{z}) = aP_S(\mathbf{y}) + bP_S(\mathbf{z}).$$

We prove it by direct substitution, using Equation 2.14, Theorem 2.4, and the dot product (Definition 2.2).

When we have chosen a sequence $\mathbf{w}_1, \ldots, \mathbf{w}_q$ of orthogonal vectors to span a subspace $S$, we often find it convenient to extend this sequence by adding $n - q$ orthogonal vectors so that $\mathbf{w}_1, \ldots, \mathbf{w}_q, \mathbf{w}_{q+1}, \ldots, \mathbf{w}_n$ spans $\mathscr{E}_n$. This is not at all difficult to do, since

$$\mathbf{e}_1 = \begin{bmatrix} 1 \\ 0 \\ \cdot \\ \cdot \\ \cdot \\ 0 \end{bmatrix}, \quad \mathbf{e}_2 = \begin{bmatrix} 0 \\ 1 \\ 0 \\ \cdot \\ \cdot \\ \cdot \\ 0 \end{bmatrix}, \ldots, \mathbf{e}_n = \begin{bmatrix} 0 \\ \cdot \\ \cdot \\ \cdot \\ 0 \\ 1 \end{bmatrix}$$

span $\mathscr{E}_n$. Obtain $\mathbf{w}_{q+1}$ as

$$\mathbf{w}_{q+1} = \mathbf{e}_1 - P_{[\![\mathbf{w}_1]\!]}(\mathbf{e}_1) - \cdots - P_{[\![\mathbf{w}_q]\!]}(\mathbf{e}_1)$$

(provided this is not the zero vector), discarding any zero vectors so generated.

If $S$ is a subspace of $\mathscr{E}_n$, say $S = [\![\mathbf{v}_1, \ldots, \mathbf{v}_p]\!]$ and $S'$ is a subspace of $S$, that is, $S' = [\![\mathbf{z}_1, \ldots, \mathbf{z}_k]\!]$ with $S' \subseteq S$, then it should be clear how to choose an orthogonal linearly independent sequence $\mathbf{w}_1, \ldots, \mathbf{w}_n$ so that

$$S' = [\![\mathbf{w}_1, \ldots, \mathbf{w}_r]\!], \qquad S = [\![\mathbf{w}_1, \ldots, \mathbf{w}_r, \mathbf{w}_{r+1}, \ldots, \mathbf{w}_q]\!]$$

$$\mathscr{E}_n = [\![\mathbf{w}_1, \ldots, \mathbf{w}_n]\!].$$

We first use the Gram-Schmidt procedure (2.10) on $[\![\mathbf{z}_1, \ldots, \mathbf{z}_k]\!]$, obtaining $\mathbf{w}_1, \ldots, \mathbf{w}_r$. We continue with the Gram-Schmidt procedure, using $[\![\mathbf{v}_1, \ldots, \mathbf{v}_p]\!]$, generating the additional vectors $\mathbf{w}_{r+1}, \ldots, \mathbf{w}_q$, and finally continue with $\mathbf{e}_1, \ldots, \mathbf{e}_n$ to generate $\mathbf{w}_{q+1}, \ldots, \mathbf{w}_n$. We state this result formally as follows.

### *2.16 Theorem: Orthogonal Bases for Subspaces*

If $S$ is a subspace of $\mathscr{E}_n$ and $S'$ is a subspace of $S$, then there is an orthogonal sequence of non-zero vectors, $\mathbf{w}_1, \ldots, \mathbf{w}_n$, such that

$$S' = [\![\mathbf{w}_1, \ldots, \mathbf{w}_r]\!], \qquad S = [\![\mathbf{w}_1, \ldots, \mathbf{w}_q]\!] \qquad \text{and} \qquad \mathscr{E}_n = [\![\mathbf{w}_1, \ldots, \mathbf{w}_n]\!].$$

We are now able to show the following corollary.

### *2.17 Corollary to 2.13 and 2.16:* $P_{S'}(P_S(\mathbf{y}))$

If $S$ is a subspace of $\mathscr{E}_n$, and $S'$ is a subspace of $S$, then

$$P_{S'}(P_S(\mathbf{y})) = P_{S'}(\mathbf{y})$$

for each $\mathbf{y}$ in $\mathscr{E}_n$.

*Proof:* Use Theorem 2.16 to choose orthogonal vectors $\mathbf{w}_1, \ldots, \mathbf{w}_n$ with

$$S' = [\![\mathbf{w}_1, \ldots, \mathbf{w}_r]\!], \qquad S = [\![\mathbf{w}_1, \ldots, \mathbf{w}_r, \ldots, \mathbf{w}_q]\!].$$

Now, by Theorem 2.13,

$$(*) \qquad P_{S'}(\mathbf{y}) = \sum_{i=1}^{r} P_{[\![\mathbf{w}_i]\!]}(\mathbf{y})$$

$$P_S(\mathbf{y}) = \sum_{m=1}^{q} P_{[\![\mathbf{w}_m]\!]}(\mathbf{y}),$$

and therefore

$$P_{S'}(P_S(\mathbf{y})) = P_{S'}\left(\sum_{m=1}^{q} P_{[\![\mathbf{w}_m]\!]}(\mathbf{y})\right)$$

$$= \sum_{m=1}^{q} P_{S'}(P_{[\![\mathbf{w}_m]\!]}(\mathbf{y})) \qquad \text{by 2.15}$$

$$= \sum_{m=1}^{q} \sum_{i=1}^{r} P_{[\![\mathbf{w}_i]\!]}(P_{[\![\mathbf{w}_m]\!]}(\mathbf{y})) \qquad \text{using (*) above with } \mathbf{y} \text{ replaced by } P_{[\![\mathbf{w}_m]\!]}(\mathbf{y}).$$

But, using the orthogonality of the sequence $\mathbf{w}_1, \ldots, \mathbf{w}_q$ together with Theorem 2.4 shows that

$$P_{[\![\mathbf{w}_i]\!]}(P_{[\![\mathbf{w}_m]\!]}(\mathbf{y})) = \begin{cases} P_{[\![\mathbf{w}_i]\!]}(\mathbf{y}) & \text{if } m = i \\ \begin{bmatrix} 0 \\ \cdot \\ \cdot \\ 0 \end{bmatrix} & \text{if } m \neq i. \end{cases}$$

Hence,

$$P_{S'}(P_S(\mathbf{y})) = \sum_{i=1}^{r} P_{[\![\mathbf{w}_i]\!]}(\mathbf{y}) = P_{S'}(\mathbf{y}) \qquad \text{by } (*),$$

proving 2.17.

### *2.18 Theorem: Basis*

If dim $S = q$ and the sequence $\mathbf{v}_1, \ldots, \mathbf{v}_q$ of elements of $S$ is linearly independent, then $S = [\![\mathbf{v}_1, \ldots, \mathbf{v}_q]\!]$.

*Proof:* Let $S' = [\![\mathbf{v}_1, \ldots, \mathbf{v}_q]\!]$. Using the Gram-Schmidt procedure, construct a sequence of non-zero orthogonal vectors $\mathbf{w}_1, \ldots, \mathbf{w}_q$ such that

$$[\![\mathbf{w}_1, \ldots, \mathbf{w}_q]\!] = S'.$$

(Note from Lemma 2.12 that $\mathbf{w}_1, \ldots, \mathbf{w}_q$ are independent and from Theorem 2.8 that $q$ such vectors are required.) Now construct a basis for $S$, as in Theorem 2.16. If $S' \neq S$, then $S = [\![\mathbf{w}_1, \ldots, \mathbf{w}_q, \mathbf{w}_{q+1}, \ldots, \mathbf{w}_{q+j}]\!]$, contradicting the hypothesis that dim $S = q$. Hence $S' = S$, that is, $[\![\mathbf{v}_1, \ldots, \mathbf{v}_q]\!] = S$.

We know from Lemma 2.12 that any sequence $\mathbf{w}_1, \ldots, \mathbf{w}_n$ of non-zero mutually orthogonal vectors is a linearly independent sequence. Since dim $\mathscr{E}_n = n$ (2.9), we see from Theorem 2.18 that we may think of $\mathbf{w}_1, \ldots, \mathbf{w}_n$ as the basis vectors for a new, and possibly more convenient, coordinate system for $\mathscr{E}_n$. That is, if we are given a vector

$$\mathbf{y} = \begin{bmatrix} y_1 \\ \cdot \\ \cdot \\ \cdot \\ y_n \end{bmatrix} = \mathbf{e}_1 y_1 + \mathbf{e}_2 y_2 + \cdots + \mathbf{e}_n y_n,$$

we might want to express $\mathbf{y}$ as a linear combination of the $\mathbf{w}$'s

**2.19**
$$\mathbf{y} = \mathbf{w}_1 z_1 + \mathbf{w}_2 z_2 + \cdots + \mathbf{w}_n z_n.$$

We may write 2.19 in the abbreviated symbolic form,

**2.20**
$$\mathbf{y} = W\mathbf{z},$$

where

$$\mathbf{z} = \begin{bmatrix} z_1 \\ \cdot \\ \cdot \\ \cdot \\ z_n \end{bmatrix}, \; W = \mathbf{w}_1, \ldots, \mathbf{w}_n \quad \text{or} \quad W = \begin{bmatrix} w_{11} \\ \cdot \\ \cdot \\ \cdot \\ w_{n1} \end{bmatrix}, \ldots, \begin{bmatrix} w_{1n} \\ \cdot \\ \cdot \\ \cdot \\ w_{nn} \end{bmatrix}.$$

Note that 2.20 means that $\mathbf{y}$ is a linear combination of the columns of $W$, where the (scalar) coefficients are the elements of $\mathbf{z}$. Usually, the inner brackets are omitted and we write

$$W = \begin{bmatrix} w_{11} & \cdots & w_{1n} \\ \cdot & & \cdot \\ \cdot & & \cdot \\ \cdot & & \cdot \\ w_{n1} & \cdots & w_{nn} \end{bmatrix}$$

and refer to $W$ as a *matrix*, or as an $n \times n$ matrix.

A vector $\mathbf{z}$ for which Equation 2.20 is a true statement is called a solution of 2.20. In fact, we recognize 2.20 as a system of $n$ linear equations in $n$ unknowns $z_1, \ldots, z_n$. The solution of Equation 2.20 furnishes the coordinates of $\mathbf{y}$ corresponding to the basis vectors $\mathbf{w}_1, \ldots, \mathbf{w}_n$.

When the $\mathbf{w}_i$ are orthogonal (as we are assuming now), solution of 2.19 is quite easy. Simply take the dot product of both sides of 2.19 with $\mathbf{w}_i$ to obtain

$$\mathbf{y} \cdot \mathbf{w}_i = (\mathbf{w}_i \cdot \mathbf{w}_i) z_i,$$

or,

**2.21 Solution of 2.19 for Orthogonal Basis**

$$z_i = \frac{\mathbf{y} \cdot \mathbf{w}_i}{\mathbf{w}_i \cdot \mathbf{w}_i}.$$

If we write the above in matrix-vector form, we obtain

$$\mathbf{z} = \begin{bmatrix} \dfrac{\mathbf{y} \cdot \mathbf{w}_1}{\mathbf{w}_1 \cdot \mathbf{w}_1} \\ \cdot \\ \cdot \\ \cdot \\ \dfrac{\mathbf{y} \cdot \mathbf{w}_n}{\mathbf{w}_n \cdot \mathbf{w}_n} \end{bmatrix} = \begin{bmatrix} \dfrac{\sum_j y_j w_{j1}}{\sum_k w_{k1}^2} \\ \cdot \\ \cdot \\ \cdot \\ \dfrac{\sum_j y_j w_{jn}}{\sum_k w_{kn}^2} \end{bmatrix} = \begin{bmatrix} \dfrac{w_{11}}{\sum_k w_{k1}^2} \\ \cdot \\ \cdot \\ \cdot \\ \dfrac{w_{1n}}{\sum_k w_{kn}^2} \end{bmatrix} y_1 + \cdots + \begin{bmatrix} \dfrac{w_{n1}}{\sum_k w_{k1}^2} \\ \cdot \\ \cdot \\ \cdot \\ \dfrac{w_{nn}}{\sum_k w_{kn}^2} \end{bmatrix} y_n$$

$$\equiv W^{-1}\mathbf{y},$$

where $W^{-1}$ is the matrix whose columns appear on the line above.

In order to simplify the notation and to present a slightly simpler explanation, *we find it convenient to assume that the orthogonal basis vectors* $\mathbf{w}_1, \ldots, \mathbf{w}_n$ *have unit length.* This only requires replacing $\mathbf{w}_i$ by

$$\mathbf{w}_i \frac{1}{\|\mathbf{w}_i\|}.$$

We have then derived the following result.

***2.22 Theorem: New Basis***

Suppose that $\mathbf{w}_1, \ldots, \mathbf{w}_n$ is a sequence of orthogonal vectors of unit length in $\mathscr{E}_n$. The coordinates of

$$\mathbf{y} = \begin{bmatrix} y_1 \\ \cdot \\ \cdot \\ \cdot \\ y_n \end{bmatrix}$$

with respect to $\mathbf{w}_1, \ldots, \mathbf{w}_n$ are defined to be the coefficients $z_1, \ldots, z_n$ such that

$$\mathbf{y} = W\mathbf{z} \equiv \mathbf{w}_1 z_1 + \cdots + \mathbf{w}_n z_n \equiv \begin{bmatrix} w_{11} \\ \cdot \\ \cdot \\ \cdot \\ w_{n1} \end{bmatrix} z_1 + \cdots + \begin{bmatrix} w_{1n} \\ \cdot \\ \cdot \\ \cdot \\ w_{nn} \end{bmatrix} z_n.$$

We have just shown that

**2.23**
$$\mathbf{z} = W^{-1}\mathbf{y} = \begin{bmatrix} w_{11} \\ \cdot \\ \cdot \\ \cdot \\ w_{1n} \end{bmatrix} y_1 + \cdots + \begin{bmatrix} w_{n1} \\ \cdot \\ \cdot \\ \cdot \\ w_{nn} \end{bmatrix} y_n.$$

There are several ways to interpret the matrices $W$ and $W^{-1}$. *We may think of $W$ as a function* whose value at $\mathbf{z}$ is $\mathbf{y} = W\mathbf{z}$. In this case, the matrix $W^{-1}$ *is interpreted as the inverse function* whose value at $\mathbf{y}$ is $\mathbf{z} = W^{-1}\mathbf{y}$. Now the columns of $W$ are orthogonal and of unit length. Square matrices (ones with $n$ rows and $n$ columns) whose columns are of unit length and orthogonal to each other are called *orthonormal* matrices.

### 2.24 Orthonormal or O.N. Matrices

We now claim that an orthonormal matrix interpreted as a function essentially represents a rotation† about the origin. We see this from the fact that

$$\mathbf{z} = \begin{bmatrix} z_1 \\ \cdot \\ \cdot \\ \cdot \\ z_n \end{bmatrix} = z_1 \begin{bmatrix} 1 \\ 0 \\ \cdot \\ \cdot \\ \cdot \\ 0 \end{bmatrix} + z_2 \begin{bmatrix} 0 \\ 1 \\ 0 \\ \cdot \\ \cdot \\ \cdot \\ 0 \end{bmatrix} + \cdots + z_n \begin{bmatrix} 0 \\ \cdot \\ \cdot \\ \cdot \\ 0 \\ 1 \end{bmatrix} = \sum z_i \mathbf{e}_i$$

is the same linear combination of the "old" coordinate orthogonal unit vectors $\mathbf{e}_1, \ldots, \mathbf{e}_n$ as is

$$W\mathbf{z} = \sum z_i \mathbf{w}_i$$

of the new coordinate orthogonal unit vectors $\mathbf{w}_1, \ldots, \mathbf{w}_n$. But, since the new coordinate vectors $\mathbf{w}_1, \ldots, \mathbf{w}_n$ are orthogonal and of unit length, they were obtained by rotating the old orthogonal unit vectors $\mathbf{e}_1, \ldots, \mathbf{e}_n$ about the origin. This rotation of the coordinate vectors rotates all other vectors correspondingly, as shown in Figure 14.6.

The matrix $W^{-1}$ reverses the action of the rotation $W$, and hence it is itself a rotation. If we view the columns of $W$ as the basis of a new coordinate system, then $W^{-1}$ *is the transformation that relates the old coordinates* $\mathbf{y}$ *to the new ones* $\mathbf{z} = W^{-1}\mathbf{y}$.

No matter which viewpoint we take regarding O.N. matrices, it certainly follows that distance between vectors should be preserved under an O.N. transformation. We establish this necessary result in Theorem 2.25.

† Reflections of the coordinate axes are also allowed.

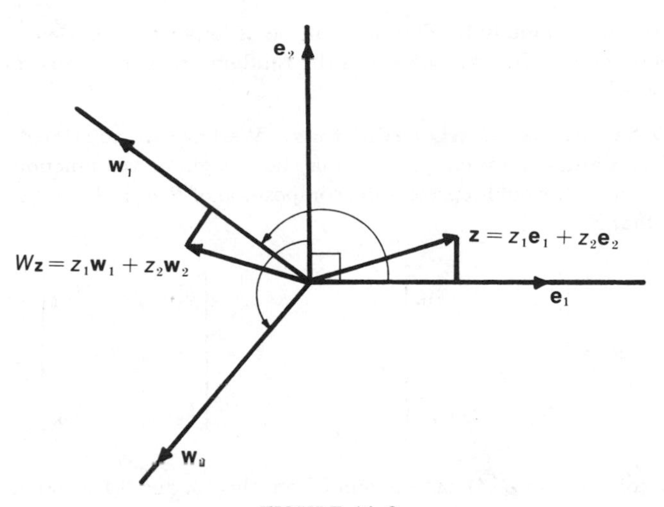


*FIGURE* 14–6.

### 2.25 *Theorem: Distance-Preserving Property of O.N. Matrices*

Let **u** and **v** be $n$-dimensional vectors and let $W$ be an $n \times n$ orthonormal matrix. Then

$$\|\mathbf{u} - \mathbf{v}\| = \|W\mathbf{u} - W\mathbf{v}\|.$$

*Proof.* Since it is easily verified that $W\mathbf{u} - W\mathbf{v} = W(\mathbf{u} - \mathbf{v})$, we need only show that for each $n$-dimensional vector **z**,

$$\|\mathbf{z}\| = \|W\mathbf{z}\|.$$

But
$$\begin{aligned}
\|W\mathbf{z}\|^2 &= (W\mathbf{z}) \cdot (W\mathbf{z}) && \text{by 2.3}\\
&= \left(\sum_{i=1}^{n} z_i \mathbf{w}_i\right) \cdot \left(\sum_{j=1}^{n} z_j \mathbf{w}_j\right) && \text{by 2.20}\\
&= \sum_{i=1}^{n} \sum_{j=1}^{n} z_i z_j \mathbf{w}_i \cdot \mathbf{w}_j && \text{by 2.3}\\
&= \sum_{i=1}^{n} z_i^2 && \text{using the orthonormality of } W\\
&= \|\mathbf{z}\|^2.
\end{aligned}$$

This proves $\|\mathbf{z}\| = \|W\mathbf{z}\|$, as asserted.

One more fact about orthonormal matrices is needed, namely, if $W$ is an orthonormal matrix, then its determinant, $\|W\|$, satisfies

**2.26** $$\big|\|W\|\big| = 1.$$

This follows from the fact that the magnitude of the determinant of a matrix represents geometrically the volume of the parallelepiped formed by the column vectors of $W$ (see the Appendix). But the columns of $W$ are orthogonal and of unit length, so this volume is 1. This fact will be of importance in determining the distribution of $\mathbf{Z} = W^{-1}\mathbf{Y}$, where $\mathbf{Y}$ is the random vector of observations in our linear model.

**2.27 Some Final Matrix Definitions.** We have already stated that when $A$ is a matrix and $\mathbf{z}$ is a vector, then $A$ may be thought of as a function. If $B$ is a matrix, then $B(A)$ would represent the composition of $B$ with $A$. It is not difficult to show that if

$$A = \begin{bmatrix} a_{11} & \cdots & a_{1n} \\ \cdot & & \\ \cdot & & \\ \cdot & & \\ a_{n1} & \cdots & a_{nn} \end{bmatrix} \quad \text{and} \quad B = \begin{bmatrix} b_{11} & \cdots & b_{1n} \\ \cdot & & \\ \cdot & & \\ \cdot & & \\ b_{n1} & \cdots & b_{nn} \end{bmatrix},$$

then the composition $B(A)$ can be found from the matrix $BA$ whose $i$th row $j$th column element is $\sum_{k=1}^{n} b_{ik}a_{kj}$.

If $C$ is a matrix of $n$ column vectors of $k$ coordinates each, it is useful to think of $C$ as *a function* whose domain consists of $n$-dimensional vectors, and whose range consists of $k$-dimensional vectors, defined by

$$C\mathbf{z} = \begin{bmatrix} c_{11} & c_{12} & \cdots & c_{1n} \\ \cdot & & & \\ \cdot & & & \\ \cdot & & & \\ c_{k1} & c_{k2} & & c_{kn} \end{bmatrix} \mathbf{z} = \begin{bmatrix} c_{11} \\ \cdot \\ \cdot \\ \cdot \\ c_{k1} \end{bmatrix} z_1 + \cdots + \begin{bmatrix} c_{1n} \\ \cdot \\ \cdot \\ \cdot \\ c_{kn} \end{bmatrix} z_n$$

$$= \mathbf{c}_1 z_1 + \cdots + \mathbf{c}_n z_n.$$

In the same way, we may extend the notion of composition. If

$$C = \begin{bmatrix} c_{11} & \cdots & c_{1n} \\ \cdot & & \\ \cdot & & \\ \cdot & & \\ c_{k1} & & c_{kn} \end{bmatrix} \quad \text{and} \quad D = \begin{bmatrix} d_{11} & \cdots & d_{1q} \\ \cdot & & \\ \cdot & & \\ \cdot & & \\ d_{n1} & & d_{nq} \end{bmatrix},$$

then it is natural to let the *product* $CD$ of $C$ and $D$ be the matrix whose $i$th row $j$th column element is $\sum_{k=1}^{n} c_{ik}\, d_{kj}$. We may then think of every vector as a matrix. Then we have

$$\mathbf{z} \cdot \mathbf{y} = [z_1, \ldots, z_n] \begin{bmatrix} y_1 \\ \cdot \\ \cdot \\ \cdot \\ y_n \end{bmatrix}.$$

In order to be able to write the dot product in matrix notation, it is convenient to define the *transpose* $D^T$ of the matrix

$$D = \begin{bmatrix} d_{11} & \cdots & d_{1q} \\ \cdot & & \\ \cdot & & \\ \cdot & & \\ d_{n1} & \cdots & d_{nq} \end{bmatrix} \quad \text{as} \quad D^T = \begin{bmatrix} d_{11} & \cdots & d_{n1} \\ \cdot & & \\ \cdot & & \\ \cdot & & \\ d_{1q} & \cdots & d_{nq} \end{bmatrix}.$$

That is, the rows and columns have been interchanged to form $D^T$ from $D$. Then it is easy to show that

$$(CD)^T = D^T C^T.$$

These facts will make it easier to manipulate multivariate normal densities. We note that in terms of the transpose notation, the result of 2.23 for an O.N. matrix $W$ may be written

**2.28** $$W^{-1} = W^T.$$

## SECTION 3
## Distribution Theory for Linear Models

Suppose that $\mathbf{Y}$ is a vector of independent normal random variables with variance $\sigma^2$ and that $E(\mathbf{Y})$ lies in $S = [\![\mathbf{x}_1, \ldots, \mathbf{x}_m]\!]$, dim $S = q$. Let $W$ be an orthonormal matrix whose first $q$ columns span $S$.

### 3.1 *Theorem: Distribution of* $\mathbf{Z} = W^{-1}\mathbf{Y}$

Under the foregoing conditions, $\mathbf{Z} = W^{-1}\mathbf{Y}$ is a vector of independent normal random variables with variance $\sigma^2$. The mean vector $E(\mathbf{Z}) = W^{-1}E(\mathbf{Y})$, and the last $n - q$ coordinates of $E(\mathbf{Z})$ are 0.

*Proof:* Let $\boldsymbol{\mu} = E(\mathbf{Y})$, a column matrix. Then, in terms of matrices,

$$f_{Y_1,\ldots,Y_n}(y_1, \ldots, y_n) = \frac{1}{(2\pi\sigma^2)^{n/2}} e^{-(1/2\sigma^2)\sum_{i=1}^{n}(y_i-\mu_i)^2}$$

$$= \frac{1}{(2\pi\sigma^2)^{n/2}} e^{-(1/2\sigma^2)(\mathbf{y}-\boldsymbol{\mu})^T(\mathbf{y}-\boldsymbol{\mu})}.$$

Then, according to the theory on change of variable (see the Appendix),

$$f_{\mathbf{Z}}(\mathbf{z}) = f_{\mathbf{Y}}(W\mathbf{z})\,|J|$$

where $J$ is the Jacobian determinant. Here $J = \|W\|$,† since $\mathbf{y} = W\mathbf{z}$ means $y_i = \sum_{j=1}^{n} w_{ij} z_j$, and thus

$$\frac{\partial y_i}{\partial z_j} = w_{ij}.$$

But $|\|W\|| = 1$, since $W$ is orthonormal, by 2.26. Thus

$$f_{\mathbf{Z}}(\mathbf{z}) = \frac{1}{(2\pi\sigma^2)^{n/2}} \exp -\left[\left(\frac{1}{2\sigma^2}\right)(W\mathbf{z} - \boldsymbol{\mu})^T(W\mathbf{z} - \boldsymbol{\mu})\right]$$

$$= \frac{1}{(2\pi\sigma^2)^{n/2}} \exp -\left\{\left(\frac{1}{2\sigma^2}\right)(W[\mathbf{z} - W^{-1}\boldsymbol{\mu}])^T(W[\mathbf{z} - W^{-1}\boldsymbol{\mu}])\right\}$$

$$= \frac{1}{(2\pi\sigma^2)^{n/2}} \exp -\left[\left(\frac{1}{2\sigma^2}\right)(\mathbf{z} - W^{-1}\boldsymbol{\mu})^T W^T W(\mathbf{z} - W^{-1}\boldsymbol{\mu})\right]$$

$$= \frac{1}{(2\pi\sigma^2)^{n/2}} \exp -\left[\left(\frac{1}{2\sigma^2}\right)(\mathbf{z} - W^{-1}\boldsymbol{\mu})^T(\mathbf{z} - W^{-1}\boldsymbol{\mu})\right],$$

since we know that $W^T = W^{-1}$ for an orthonormal matrix (see 2.28), and hence $W^T W$ represents the identity function.

Thus, we see that

$$f_{\mathbf{Z}}(\mathbf{z}) = \frac{1}{(2\pi\sigma^2)^{n/2}} \exp -\left\{\left(\frac{1}{2\sigma^2}\right) \sum_{i=1}^{n} (\mathbf{z}_i - [W^{-1}\boldsymbol{\mu}]_i)^2\right\},$$

where $[W^{-1}\boldsymbol{\mu}]_i$ is the $i$th coordinate of $W^{-1}\boldsymbol{\mu}$. This shows that $Z_1, \ldots, Z_n$ are independent and normal with variance $\sigma^2$. Now $E(\mathbf{Z}) = W^{-1}\boldsymbol{\mu}$. Let $\boldsymbol{\nu} = E(\mathbf{Z})$. Then $W\boldsymbol{\nu} = \boldsymbol{\mu}$, and because $\mu$ lies in $S$, and the first $q$ columns of $W$ span $S$, only the first $q$ coordinates of $\boldsymbol{\nu}$ can be nonzero (because of the independence of the columns of $W$). Hence, the last $n - q$ coordinates of $\boldsymbol{\nu} = E(\mathbf{Z})$ must be 0. Q.E.D.

### 3.2 *Corollary: Distribution of* $\|\mathbf{Y} - P_S(\mathbf{Y})\|^2/\sigma^2$, $P_S(\mathbf{Y})$

Suppose that $\dim S = q$ and that $\mathbf{Y}$ is an $n$-dimensional random vector whose components are independent normal with variance $\sigma^2$. If $E(\mathbf{Y})$ is known only to lie in $S$, then $\|\mathbf{Y} - P_S(\mathbf{Y})\|^2/\sigma^2$ has a Chi square distribution with $n - q$ degrees of freedom. Furthermore, $\mathbf{Y} - P_S(\mathbf{Y})$ is independent of $P_S(\mathbf{Y})$.

*Proof:* The intuitive idea behind the proof is to convert to a coordinate system whose first $q$ axes span $S$. In this coordinate system, computations are greatly simplified. The details are as follows: Use Theorem 2.16 to construct an orthonormal matrix $W$ whose first $q$ columns span $S$. Let $\mathbf{Z} = W^{-1}\mathbf{Y}$. Then

$$\mathbf{Y} - P_S(\mathbf{Y}) = W\mathbf{Z} - P_S(W\mathbf{Z})$$

$$= \sum_{i=1}^{n} Z_i \mathbf{w}_i - P_S\left(\sum_{i=1}^{n} Z_i \mathbf{w}_i\right) \qquad \text{by 2.20}$$

$$= \sum_{i=1}^{n} Z_i \mathbf{w}_i - \sum_{j=1}^{q} P_{[\![\mathbf{w}_j]\!]}\left(\sum_{i=1}^{n} Z_i \mathbf{w}_i\right) \qquad \text{by Theorem 2.13.}$$

† We use $\|W\|$ to denote the determinant of $W$ when $W$ is a matrix.

Now, using the orthonormality of $\mathbf{w}_1, \ldots, \mathbf{w}_n$ and Theorem 2.4, we know that

$$P_{[\![\mathbf{w}_j]\!]}(\mathbf{w}_i) = \begin{cases} \mathbf{w}_j & \text{if } i = j \\ \begin{bmatrix} 0 \\ \cdot \\ \cdot \\ \cdot \\ 0 \end{bmatrix} & \text{if } i \neq j. \end{cases}$$

Hence, from Corollary 2.15, we conclude that

**3.3** $$\mathbf{Y} - P_S(\mathbf{Y}) = \sum_{i=q+1}^{n} Z_i \mathbf{w}_i.$$

Thus, since $Y = \sum_{i=1}^{n} Z_i \mathbf{w}_i$, we also have

**3.4** $$P_S(\mathbf{Y}) = \sum_{i=1}^{q} Z_i \mathbf{w}_i.$$

The independence of $\mathbf{Y} - P_S(\mathbf{Y})$ and $P_S(\mathbf{Y})$ follows from Theorem 3.1. Now, from 3.3, we compute

$$\|\mathbf{Y} - P_S(\mathbf{Y})\|^2/\sigma^2 = \left\|\sum_{i=q+1}^{n} Z_i \mathbf{w}_i\right\|^2/\sigma^2$$

$$= \frac{1}{\sigma^2}\left\| W \begin{bmatrix} 0 \\ \cdot \\ \cdot \\ 0 \\ Z_{q+1} \\ \cdot \\ \cdot \\ Z_n \end{bmatrix} \right\|^2 = \frac{1}{\sigma^2}\left\| \begin{bmatrix} 0 \\ \cdot \\ \cdot \\ 0 \\ Z_{q+1} \\ \cdot \\ \cdot \\ Z_n \end{bmatrix} \right\|^2 \quad \text{by Theorem 2.25}$$

$$= \sum_{i=q+1}^{n} Z_i^2/\sigma^2$$

$$= \sum_{i=q+1}^{n} (Z_i/\sigma)^2.$$

From Theorem 3.1, we know that $Z_{q+1}/\sigma, \ldots, Z_n/\sigma$ are independent normal random variables with mean 0 and unit variance. Thus, by 1 in Chapter 8, $\|\mathbf{Y} - P_S(\mathbf{Y})\|^2/\sigma^2$ has a Chi square distribution with $n - q$ degrees of freedom, proving the theorem.

### 3.5 *Example: t Statistic*

Suppose that $Y_1, \ldots, Y_n$ are independent normal $E(Y_i) = \mu$, Var $Y_i = \sigma^2$. Then $E(\mathbf{Y}) \in S = [\![\mathbf{v}]\!]$, where

$$\mathbf{v} = \begin{bmatrix} 1 \\ 1 \\ \cdot \\ \cdot \\ \cdot \\ 1 \end{bmatrix}.$$

So

$$P_S(\mathbf{Y}) = P_{[\![\mathbf{v}]\!]}(\mathbf{Y}) = \frac{\sum_{i=1}^{n} Y_i}{n} \begin{bmatrix} 1 \\ 1 \\ \cdot \\ \cdot \\ \cdot \\ 1 \end{bmatrix} = \begin{bmatrix} \bar{Y}_{(n)} \\ \cdot \\ \cdot \\ \cdot \\ \bar{Y}_{(n)} \end{bmatrix}$$

(see 2.4). Now

$$\|\mathbf{Y} - P_S(\mathbf{Y})\|^2/\sigma^2 = \left\| \begin{bmatrix} Y_1 - \bar{Y}_{(n)} \\ \cdot \\ \cdot \\ \cdot \\ Y_n - \bar{Y}_{(n)} \end{bmatrix} \right\|^2 /\sigma^2 = \sum_{i=1}^{n} (Y_i - \bar{Y}_{(n)})^2/\sigma^2$$

is statistically independent of $\bar{Y}_{(n)}$, and, since dim $S = 1$, it has a Chi square distribution with $n - 1$ degrees of freedom. Thus, by 8 in Chapter 8,

$$\frac{\sqrt{n}[\bar{Y}_{(n)} - \mu]/\sigma}{\sqrt{\dfrac{1}{n-1}\sum_{i=1}^{n}(Y_i - \bar{Y}_{(n)})^2/\sigma^2}} = \frac{\sqrt{n}[\bar{Y}_{(n)} - \mu]}{\sqrt{\dfrac{1}{n-1}\sum_{i=1}^{n}(Y_i - \bar{Y}_{(n)})^2}}$$

has a $t$ distribution with $n - 1$ degrees of freedom. △

To conclude this section, we derive the formula for the least squares estimates of the regression coefficients $\beta_1, \ldots, \beta_n$ when $E(Y) = \beta_1\mathbf{x}_1 + \cdots + \beta_n\mathbf{x}_n$, and $\mathbf{x}_1, \ldots, \mathbf{x}_n$ are linearly independent. Recall that the least squares estimates of $\beta_1, \ldots, \beta_n$ are those values that minimize

$$\left\|\mathbf{y} - \sum_{i=1}^{n} \beta_i\mathbf{x}_i\right\|^2 = \left(\mathbf{y} - \sum_{i=1}^{n} \beta_i\mathbf{x}_i\right) \cdot \left(\mathbf{y} - \sum_{j=1}^{n} \beta_j\mathbf{x}_j\right)$$

$$= \mathbf{y} \cdot \mathbf{y} - 2\sum_{i=1}^{n} \beta_i\mathbf{x}_i \cdot \mathbf{y} + \sum_{i=1}^{n}\sum_{j=1}^{n} \beta_i\beta_j\mathbf{x}_i \cdot \mathbf{x}_j.$$

Taking partial derivatives with respect to $\beta_k$, $k = 1, \ldots, n$ and setting these derivatives equal to 0 yields the normal equations

$$-2\mathbf{x}_k \cdot \mathbf{y} + 2\sum_{j \neq k} \beta_j\mathbf{x}_k \cdot \mathbf{x}_j + 2\beta_k\mathbf{x}_k \cdot \mathbf{x}_k = 0,\ k = 1, \ldots, n$$

or

$$\begin{gathered} \mathbf{x}_1 \cdot \mathbf{x}_1\beta_1 + \mathbf{x}_2 \cdot \mathbf{x}_1\beta_2 + \cdots + \mathbf{x}_n \cdot \mathbf{x}_1\beta_n = \mathbf{x}_1 \cdot \mathbf{y} \\ \cdot \\ \cdot \\ \cdot \\ \mathbf{x}_1 \cdot \mathbf{x}_n\beta_1 + \mathbf{x}_1 \cdot \mathbf{x}_n\beta_2 + \cdots + \mathbf{x}_n \cdot \mathbf{x}_n\beta_n = \mathbf{x}_n \cdot \mathbf{y}. \end{gathered}$$

We write this in matrix form as

$$\begin{bmatrix} \mathbf{x}_1^T \\ \cdot \\ \cdot \\ \cdot \\ \mathbf{x}_n^T \end{bmatrix} [\mathbf{x}_1, \ldots, \mathbf{x}_n]\boldsymbol{\beta} = \begin{bmatrix} \mathbf{x}_1^T \\ \cdot \\ \cdot \\ \cdot \\ \mathbf{x}_n^T \end{bmatrix} \mathbf{y}$$

or

$$X^T X \boldsymbol{\beta} = X^T \mathbf{y}.$$

We know the solution to be unique, since the solution is $P_S(\mathbf{y})$ where $S = [\![\mathbf{x}_1, \ldots, \mathbf{x}_n]\!]$. But by 2.13, $P_S(\mathbf{y})$ is unique, and because of the linear independence of $\mathbf{x}_1, \ldots, \mathbf{x}_n$, it is a unique linear combination of $\mathbf{x}_1, \ldots, \mathbf{x}_n$. Hence we know that the matrix function $X^T X$ has an inverse. Thus, the *least squares estimates are given by*

**3.6** $$\tilde{\boldsymbol{\beta}} = (X^T X)^{-1} X^T \mathbf{y}.$$

It can be shown that under the model of 1.3, when the columns $\mathbf{x}_1, \ldots, \mathbf{x}_n$ of $X$ are linearly independent, then the least squares estimates are the best linear unbiased estimates of the regression coefficients. (This is the *Gauss-Markov Theorem.*)

## SECTION 4
## Testing Hypotheses in Linear Models

Consider a model for the effect on yield of two additives:

**4.1** $$\mathbf{Y} = \mathbf{x}_1\beta_1 + \mathbf{x}_2\beta_2 + \boldsymbol{\epsilon}$$

where $\beta_i$ is the average effect of a unit of additive $i$ on the yield. Here, $x_{ji}$ (the $j$th element of the column vector $\mathbf{x}_i$) gives the amount of additive $i$ in the $j$th measurement. We might want to determine whether or not the two additives have the same average effect. That is to say, we might want to test the hypothesis that $\beta_1 = \beta_2$. If $\beta_1 = \beta_2 \equiv \beta$, then

$$E(\mathbf{Y}) = (\mathbf{x}_1 + \mathbf{x}_2)\beta.$$

That is, the set of all possible mean vectors $E(\mathbf{Y})$ consists of all multiples of the vector $\mathbf{x}_1 + \mathbf{x}_2 \in S = [\![\mathbf{x}_1, \mathbf{x}_2]\!]$. Furthermore, if the set of all possible mean vectors $E(\mathbf{Y})$ consists of multiples of $\mathbf{x}_1 + \mathbf{x}_2$, then we may conclude that $\beta_1 = \beta_2$, provided $\mathbf{x}_1$ and $\mathbf{x}_2$ are linearly independent. This follows from the fact that if

$$\mathbf{x}_1\beta_1 + \mathbf{x}_2\beta_2 = \mathbf{x}_1\beta + \mathbf{x}_2\beta,$$

then $\mathbf{x}_1(\beta_1 - \beta) = \mathbf{x}_2(\beta - \beta_2)$. But when $\mathbf{x}_1$ and $\mathbf{x}_2$ are linearly independent, the preceding equation is impossible unless $\beta_1 - \beta = 0$ and $\beta - \beta_2 = 0$.

Now notice that the hypothesis of interest here corresponds to the condition that the mean actually lies in some given subspace $S' = [\![\mathbf{x}_1 + \mathbf{x}_2]\!]$ of $S$.

***4.2 In general, under the general linear model 1.5, we shall see that the hypotheses that can be tested easily are those of the form***

$$H_0\colon E(\mathbf{Y}) \in S'$$

where $S'$ is a subspace of $S$. That is,

$$S' = [\![\mathbf{w}_1, \ldots, \mathbf{w}_r]\!] \qquad \mathbf{w}_i \in S = [\![\mathbf{x}_1, \ldots, \mathbf{x}_m]\!].$$

### *4.3 Example: Test of Equal Effect of Additives*

Suppose that in our model 4.1,

$$\mathbf{x}_1 = \begin{bmatrix} 1 \\ 2 \\ 1 \\ 2 \end{bmatrix} \quad \text{and} \quad \mathbf{x}_2 = \begin{bmatrix} 1 \\ 1 \\ 2 \\ 2 \end{bmatrix}$$

and we want to test $H_0$: $\beta_1 = \beta_2$. As we saw, the hypothesis $H_0$ corresponds to the subspace $S' = [\![\mathbf{x}_1 + \mathbf{x}_2]\!]$, and since $\mathbf{x}_1$, $\mathbf{x}_2$ are linearly independent, $E(\mathbf{Y}) \in S'$ implies that $\beta_1 = \beta_2$.

A reasonable test of $H_0$ is to reject $H_0$ if and only if

$$\frac{\dfrac{1}{q-r}\|P_S(\mathbf{y}) - P_{S'}(\mathbf{y})\|^2}{\dfrac{1}{n-q}\|\mathbf{y} - P_S(\mathbf{y})\|^2} \equiv \mathscr{F}_{q-r,n-q} > c$$

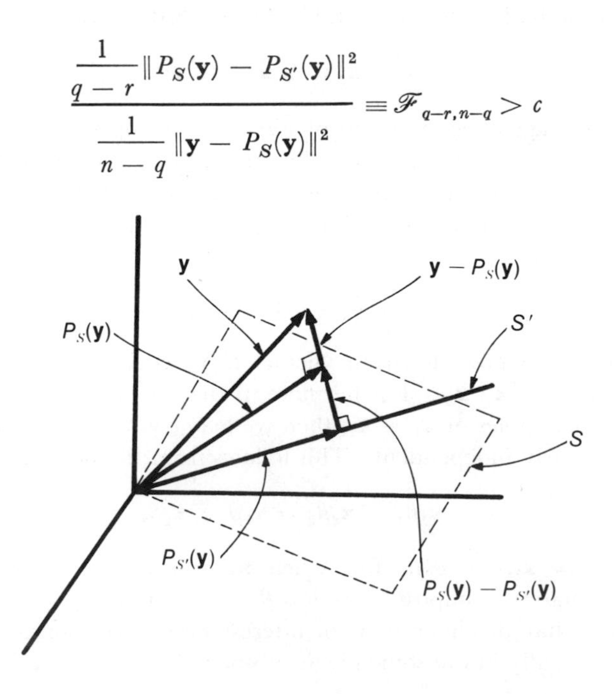

where $q = \dim S = 2$, $r = \dim S' = 1$, $n = 4$, the number of coordinates of the vector $\mathbf{y}$ of observed data. The denominator is a normalizing factor, and we know that $\|\mathbf{Y} - P_S(\mathbf{Y})\|^2/\sigma^2$ has a $\chi^2_{n-q}$ distribution. The numerator is a measure of how much the null hypothesis is violated. The numerator and denominator can be seen to be values of statistically independent random variables, using arguments exactly like those in the proof of Corollary 3.2 (and which we will give in generality at the end of this section). Furthermore, via these same arguments,

$$\|P_S(\mathbf{Y}) - P_{S'}(\mathbf{Y})\|^2/\sigma^2$$

is a $\chi^2_{q-r}$ random variable under $H_0$. Hence, under $H_0$, the test statistic $\mathscr{F}_{q-r,\, n-q}$ is the value of a random variable having an $F$ distribution with $(q - r, n - q)$ degrees of freedom (see 11, Chapter 8).

In order to carry out the test, we must compute the various projections. Now

$$\mathbf{x}_1 + \mathbf{x}_2 = \begin{bmatrix} 2 \\ 3 \\ 3 \\ 4 \end{bmatrix}.$$

So we let

$$\mathbf{w}_1 = \frac{1}{\sqrt{38}} \begin{bmatrix} 2 \\ 3 \\ 3 \\ 4 \end{bmatrix},$$

noting that $[\![\mathbf{w}_1]\!] = S'$. To find $\mathbf{w}_2 \perp \mathbf{w}_1$ such that $[\![\mathbf{w}_1, \mathbf{w}_2]\!] = S$, we use the Gram-Schmidt procedure (2.10), trying $\mathbf{x}_1$ first (see Theorem 2.16). That is to say, we let $\mathbf{w}_2' = \mathbf{x}_1 - P_{[\![\mathbf{w}_1]\!]}(\mathbf{x}_1)$ (hoping $\|\mathbf{w}_2'\| \neq 0$). Then

$$\begin{aligned} \mathbf{w}_2' &= \mathbf{x}_1 - (\mathbf{w}_1 \cdot \mathbf{x}_1)\mathbf{w}_1 \\ &= \begin{bmatrix} 1 \\ 2 \\ 1 \\ 2 \end{bmatrix} - \frac{1}{\sqrt{38}} \left( \begin{bmatrix} 2 \\ 3 \\ 3 \\ 4 \end{bmatrix} \cdot \begin{bmatrix} 1 \\ 2 \\ 1 \\ 2 \end{bmatrix} \right) \frac{1}{\sqrt{38}} \begin{bmatrix} 2 \\ 3 \\ 3 \\ 4 \end{bmatrix} \\ &= \begin{bmatrix} 1 \\ 2 \\ 1 \\ 2 \end{bmatrix} - \frac{19}{38} \begin{bmatrix} 2 \\ 3 \\ 3 \\ 4 \end{bmatrix} = \begin{bmatrix} 0 \\ .5 \\ -.5 \\ 0 \end{bmatrix}. \end{aligned}$$

Thus

$$\mathbf{w}_2 = \sqrt{2} \begin{bmatrix} 0 \\ .5 \\ -.5 \\ 0 \end{bmatrix}.$$

Hence,

$$\mathbf{w}_1 = \frac{1}{\sqrt{38}}\begin{bmatrix}2\\3\\3\\4\end{bmatrix}, \qquad \mathbf{w}_2 = \sqrt{2}\begin{bmatrix}0\\.5\\-.5\\0\end{bmatrix}.$$

Now

$$P_S(\mathbf{y}) = P_{[\![\mathbf{w}_1]\!]}(\mathbf{y}) + P_{[\![\mathbf{w}_2]\!]}(\mathbf{y}) = (\mathbf{y}\cdot\mathbf{w}_1)\mathbf{w}_1 + (\mathbf{y}\cdot\mathbf{w}_2)\mathbf{w}_2$$

and

$$P_{S'}(\mathbf{y}) = (\mathbf{y}\cdot\mathbf{w}_1)\mathbf{w}_1.$$

Thus, our test statistic is

$$\mathscr{F}_{2-1,\,4-2} = \frac{(1/1)\ \|(\mathbf{y}\cdot\mathbf{w}_2)\mathbf{w}_2\|^2}{(1/2)\ \|\mathbf{y} - (\mathbf{y}\cdot\mathbf{w}_1)\mathbf{w}_1 - (\mathbf{y}\cdot\mathbf{w}_2)\mathbf{w}_2\|^2}$$

$$= \frac{2[\mathbf{y}\cdot\mathbf{w}_2]^2}{\|\mathbf{y} - (\mathbf{y}\cdot\mathbf{w}_1)\mathbf{w}_1 - (\mathbf{y}\cdot\mathbf{w}_2)\mathbf{w}_2\|^2}. \qquad \triangle$$

We now proceed to carry out an investigation of the nature of the hypotheses, which can be easily tested in the general linear model, 1.5. We start by noticing that in the particular model of the previous example, the hypothesis $\beta_1 = \beta_2$ may also be written as $\beta_1 - \beta_2 = 0$. That is, this hypothesis is that some given linear combination of the $\beta_i$ is 0.

In the case in which $\beta_1$ and $\beta_2$ are the unit effects of two additives of a certain type, and $\beta_3$, $\beta_4$, and $\beta_5$ are the unit effects of three additives of some other type, we might well want to determine whether

$$\beta_1 = \beta_2 \qquad \text{and} \qquad \beta_3 = \beta_4 = \beta_5,$$

that is, whether

$$\beta_1 - \beta_2 = 0 \qquad \text{and} \qquad \beta_3 - \beta_4 = 0 \qquad \text{and} \qquad \beta_3 - \beta_5 = 0.$$

***This leads us to investigate the question of whether all of some given set of linear combinations of the $\beta_i$ are 0, in the general linear model, 1.5.***

To be precise, let $\alpha_{ji}$ be given numbers for $1 \le i \le m$, $1 \le j \le \omega \le m$. We want to find out the conditions under which it is possible to test the hypothesis that

**4.4** $$\sum_{i=1}^{m} \alpha_{ji}\beta_i = 0 \quad \text{for all } j = 1, 2, \ldots, \omega.$$

We note that if we let $\alpha$ be the matrix with $\alpha_{ji}$ in its $j$th row and $i$th column, and let

$$\boldsymbol{\beta} = \begin{bmatrix}\beta_1\\ \cdot\\ \cdot\\ \cdot\\ \beta_m\end{bmatrix},$$

then 4.4 may be written as

**4.4′** $$\alpha\boldsymbol{\beta} = \mathbf{0}_\omega,$$

where

$$\mathbf{0}_\omega = \begin{bmatrix} 0 \\ \cdot \\ \cdot \\ \cdot \\ 0 \end{bmatrix},$$

the vector with all 0's in its $\omega$ coordinates. Then we can see that 4.4 geometrically represents the condition that $\boldsymbol{\beta}$ is $\perp$ to (the transpose of each of) the row vectors of $\alpha$.

We can see that this condition is equivalent to the condition that $\boldsymbol{\beta}$ belongs to a subspace of $m$-dimensional Euclidean space, which we shall denote by $\alpha^{\perp}_{\text{row}}$. Using the Gram-Schmidt procedure as in Theorem 2.16, we construct an orthogonal sequence of unit vectors $\mathbf{c}_1, \mathbf{c}_2, \ldots, \mathbf{c}_\omega, \ldots, \mathbf{c}_m$ in Euclidean $m$-dimensional space (that is, an orthonormal basis for $\mathscr{E}_m$) whose first $\omega$ elements span the subspace determined by the subspace $[\![\boldsymbol{\alpha}_1^T, \ldots, \boldsymbol{\alpha}_\omega^T]\!]$, where $\boldsymbol{\alpha}_i$ is the $i$th row vector of $\alpha$. (We may assume this subspace is $\omega$-dimensional, since otherwise some of the equations in 4.4 would be redundant. This is in any case a minor point.) We claim that $\boldsymbol{\beta}$ satisfies 4.4 if and only if

**4.5** $$\boldsymbol{\beta} \in [\![\mathbf{c}_{\omega+1}, \ldots, \mathbf{c}_m]\!].$$

We see this as follows: Surely if $\boldsymbol{\beta} \in [\![\mathbf{c}_{\omega+1}, \ldots, \mathbf{c}_m]\!]$, then $\boldsymbol{\beta} \perp \boldsymbol{\alpha}_i^T$ for each $i = 1, \ldots, \omega$, since $\mathbf{c}_{\omega+j} \perp \mathbf{c}_k$ for $k \leq \omega$ and $j \geq 1$, and every $\boldsymbol{\alpha}_i^T \in [\![\mathbf{c}_1, \ldots, \mathbf{c}_\omega]\!]$.

Similarly, if 4.4 holds, then $\boldsymbol{\beta} \perp \boldsymbol{\alpha}_i^T$ for each $i$. Hence, $\boldsymbol{\beta} \perp \mathbf{c}_i$ for $i = 1, \ldots, \omega$. But then, since $\mathbf{c}_1, \ldots, \mathbf{c}_m$ span $\mathscr{E}_m$, we know that

(*) $$\boldsymbol{\beta} = \sum_{i=1}^{m} a_i \mathbf{c}_i,$$

and since $\boldsymbol{\beta} \perp \mathbf{c}_i$ for $i = 1, \ldots, \omega$, it follows that $a_i = 0$ for $i = 1, \ldots, \omega$. (Just take the dot product of (*) with $\mathbf{c}_i$.) Hence, $\boldsymbol{\beta} \in [\![\mathbf{c}_{\omega+1}, \ldots, \mathbf{c}_m]\!]$. We have shown the following theorem.

### *4.6 Theorem: Hypothesis as Subspace of $\mathscr{E}_m$*

Equations 4.4 hold if and only if $\boldsymbol{\beta} \in \alpha^{\perp}_{\text{row}}$, where $\alpha^{\perp}_{\text{row}}$ *is the subspace of* $\mathscr{E}_m$ $\perp$ *to the row space of* $\alpha$—the space spanned by the orthogonal vectors $\mathbf{c}_{\omega+1}, \ldots, \mathbf{c}_m$, where $\mathbf{c}_1, \ldots, \mathbf{c}_m$ are orthogonal vectors spanning $\mathscr{E}_m$, $\boldsymbol{\alpha}_j = [\alpha_{j1}, \ldots, \alpha_{jm}]$ and $[\![\mathbf{c}_1, \ldots, \mathbf{c}_\omega]\!] = [\![\boldsymbol{\alpha}_1^T, \ldots, \boldsymbol{\alpha}_\omega^T]\!]$.

We will now show that if $\boldsymbol{\beta} \in \alpha^{\perp}_{\text{row}}$, then $E(\mathbf{Y})$ belongs to a particular subspace $S'$ of $S$ that can be determined from $\alpha^{\perp}_{\text{row}}$. The condition $\boldsymbol{\beta} \in \alpha^{\perp}_{\text{row}}$ is equivalent to the assertion

$$\boldsymbol{\beta} = \sum_{i=\omega+1}^{m} a_i \mathbf{c}_i$$

for some $a_i$. But then

$$E(\mathbf{Y}) = \sum_{i=1}^{m} \beta_i \mathbf{x}_i = X\boldsymbol{\beta},$$

where $X$ is the matrix of column vectors $\mathbf{x}_1, \ldots, \mathbf{x}_m$ of the general linear model. Hence,

$$E(\mathbf{Y}) = X\boldsymbol{\beta} = X \sum_{i=\omega+1}^{m} a_i \mathbf{c}_i = \sum_{i=\omega+1}^{m} a_i X \mathbf{c}_i$$

for some $a_i$. Thus,

$$E(\mathbf{Y}) \in [\![X\mathbf{c}_{\omega+1}, \ldots, X\mathbf{c}_m]\!] \equiv S'.$$

We have thus established the following lemma.

### 4.7 *Lemma: Significance of the Condition* $\boldsymbol{\beta} \in \boldsymbol{\alpha}_{\text{row}}^{\perp}$

If $\boldsymbol{\beta} \in \alpha_{\text{row}}^{\perp}$, where $\alpha$ is the matrix given in 4.4, and the general linear model (1.5) holds, then

$$E(\mathbf{Y}) \in S' \equiv [\![X\mathbf{c}_{\omega+1}, \ldots, X\mathbf{c}_m]\!],$$

derived previously. Here, $X$ is the matrix of column vectors $\mathbf{x}_1, \ldots, \mathbf{x}_m$ of the general linear model (1.5).

Now we would like to determine the conditions under which

**4.8**
$$\begin{cases} E(\mathbf{Y}) \in [\![X\mathbf{c}_{\omega+1}, \ldots, X\mathbf{c}_m]\!] \text{ implies that} \\ \boldsymbol{\beta} \in [\![\mathbf{c}_{\omega+1}, \ldots, \mathbf{c}_m]\!] \end{cases}$$

where $E(\mathbf{Y}) = \sum_{i=1}^{m} \beta_i \mathbf{x}_i$. Note that 4.8 must be satisfied if testing $H_0$ that $E(\mathbf{Y}) \in S'$ really tests what we want to test—namely, 4.4.

In order to accomplish our aim, let us see what it means if 4.8 does not hold. Then there would exist a vector

**4.9**
$$\boldsymbol{\beta} \notin [\![\mathbf{c}_{\omega+1}, \ldots, \mathbf{c}_m]\!]$$

such that $X\boldsymbol{\beta} \in [\![X\mathbf{c}_{\omega+1}, \ldots, X\mathbf{c}_m]\!]$. Since we may write $\boldsymbol{\beta} = \sum_{i=1}^{m} a_i \mathbf{c}_i$ with $a_i \neq 0$ for at least one $i \geq \omega$ (because of 4.9), we see that there must exist a *non-zero* vector $\mathbf{C} \in [\![\mathbf{c}_1, \ldots, \mathbf{c}_\omega]\!]$ such that $X\mathbf{C} \in [\![X\mathbf{c}_{\omega+1}, \ldots, X\mathbf{c}_m]\!]$. Hence, there must be numbers $\delta_j$, $\omega + 1 \leq j \leq m$ such that

$$X\left(\mathbf{C} - \sum_{j=\omega+1}^{m} \delta_j \mathbf{c}_j\right) = \mathbf{0}_n \qquad \text{(the zero vector in } \mathscr{E}_n\text{).}$$

This can surely be prevented by insisting that the set of all $m$-dimensional vectors mapped into $\mathbf{0}_n$ by $X$ be a subset of $[\![\mathbf{c}_{\omega+1}, \ldots, \mathbf{c}_m]\!]$, that is, that

**4.10**
$$\begin{cases} X^{-1}\mathbf{0}_n \subseteq [\![\mathbf{c}_{\omega+1}, \ldots, \mathbf{c}_m]\!] \\ \text{where} \\ X^{-1}\mathbf{0}_n = \{\boldsymbol{\beta} \in \mathscr{E}_m \colon X\boldsymbol{\beta} = \mathbf{0}_n\}. \end{cases}$$

Condition 4.10 is also necessary to accomplish 4.8. Suppose that $X^{-1}\mathbf{0}_n$ were not contained in $[\![\mathbf{c}_{\omega+1}, \ldots, \mathbf{c}_m]\!]$. Then there would exist $\boldsymbol{\beta}^* = \sum_{i=1}^{m} d_i \mathbf{c}_i$, where $d_i \neq 0$ for some $i \leq \omega$, with $X\boldsymbol{\beta}^* = \mathbf{0}_n$. Then clearly, for $\boldsymbol{\beta}' \in [\![\mathbf{c}_{\omega+1}, \ldots, \mathbf{c}_m]\!]$, we would have $X\boldsymbol{\beta}' = X(\boldsymbol{\beta}' + \boldsymbol{\beta}^*)$. Hence, we see that $\boldsymbol{\beta} \in [\![X\mathbf{c}_{\omega+1}, \ldots, X\mathbf{c}_m]\!]$ would not imply that $\boldsymbol{\beta} \in [\![\mathbf{c}_{\omega+1}, \ldots, \mathbf{c}_m]\!]$, that is, 4.8 would be violated. We have proved the following theorem.

### *4.11 Theorem: Rephrasal of* $H_0: \sum \boldsymbol{\alpha}_{ji}\boldsymbol{\beta}_i = \mathbf{0}$

In order that the hypothesis $H_0$ that $\sum_{i=1}^{m} \alpha_{ji}\beta_i = 0$ for all $j = 1, 2, \ldots, \omega$ be *equivalent* to the condition $E(\mathbf{Y}) \in [\![X\mathbf{c}_{\omega+1}, \ldots, X\mathbf{c}_m]\!]$ of 4.8, it is necessary and sufficient that

$$(*) \qquad X^{-1}\mathbf{0}_n \equiv \{\boldsymbol{\beta} \in \mathscr{E}_m : X\boldsymbol{\beta} = \mathbf{0}_n\} \subseteq [\![\mathbf{c}_{\omega+1}, \ldots, \mathbf{c}_m]\!].$$

It is convenient to rephrase this result.

Recall that $\alpha_{\text{row}} = [\![\mathbf{c}_1, \ldots, \mathbf{c}_\omega]\!] = [\![\boldsymbol{\alpha}_1^T, \ldots, \boldsymbol{\alpha}_\omega^T]\!]$ and that the set $\alpha_{\text{row}}^{\perp} = [\![\mathbf{c}_{\omega+1}, \ldots, \mathbf{c}_m]\!]$. Now notice that

$$4.12 \qquad X^{-1}\mathbf{0}_n = X_{\text{row}}^{\perp},$$

since a vector $\boldsymbol{\beta} \in \mathscr{E}_m$ gets mapped into $\mathbf{0}_n$ by $X$ if and only if $\boldsymbol{\beta}$ is $\perp$ to every (transposed) row of $X$.

Hence, condition (*) of 4.10 may be written as

$$4.13 \qquad \begin{cases} \text{(a) } X_{\text{row}}^{\perp} \subseteq \alpha_{\text{row}}^{\perp}, \text{ or equivalently,} \\ \text{(b) } X_{\text{row}} \supseteq \alpha_{\text{row}}. \end{cases}$$

In order to verify 4.13, we need only do the following:

**4.14** Form an orthonormal sequence $\mathbf{D}_1, \ldots, \mathbf{D}_\gamma, \mathbf{D}_{\gamma+1}, \ldots, \mathbf{D}_m$, where $\mathbf{D}_1, \ldots, \mathbf{D}_\gamma$ are obtained by using the Gram-Schmidt procedure (2.10) on the (transposes of the) rows of $X$ ( it can be shown that $\gamma = q = \dim S$). Then 4.13 is satisfied if and only if $\boldsymbol{\alpha}_i^T \cdot \mathbf{D}_j = 0$ for each $j = \gamma + 1, \ldots, m$. This follows from the fact that $\boldsymbol{\alpha}_{\text{row}} \subseteq X_{\text{row}}$ if and only if $\mathbf{F} \in \alpha_{\text{row}}$ implies $\mathbf{F} \perp \mathbf{G}$ for every $\mathbf{G}$ in $X_{\text{row}}^{\perp}$.

We have thus established our vital practical result, the following theorem.

### *4.15 Theorem: Testability of* $H_0: \sum \boldsymbol{\alpha}_{ji}\boldsymbol{\beta}_i = \mathbf{0}$

We repeat the foregoing procedure to see whether we can "properly" test the hypothesis $H_0$ that $\sum_{i=1}^{m} \alpha_{ji}\beta_i = 0$ for $j = 1, \ldots, \omega$ via testing

$$E(\mathbf{Y}) \in [\![X\mathbf{c}_{\omega+1}, \ldots, X\mathbf{c}_m]\!],$$

where $[\![\mathbf{c}_{\omega+1}, \ldots, \mathbf{c}_m]\!] = \alpha^{\perp}_{\text{row}}$, the space of $m$-dimensional vectors $\perp$ to the (transposes of the) row vectors $\boldsymbol{\alpha}_i$ of

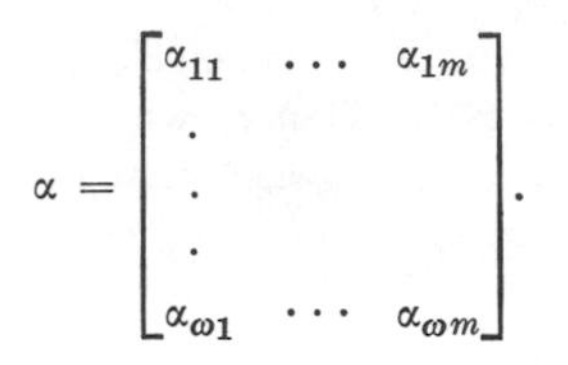

$$\alpha = \begin{bmatrix} \alpha_{11} & \cdots & \alpha_{1m} \\ \cdot & & \\ \cdot & & \\ \cdot & & \\ \alpha_{\omega 1} & \cdots & \alpha_{\omega m} \end{bmatrix}.$$

The following corollary, which follows from 4.15, is quite useful.

### *4.16 Corollary: When $\mathbf{x}_1, \ldots, \mathbf{x}_m$ are Linearly Independent*

When the column vectors $\mathbf{x}_1, \ldots, \mathbf{x}_m$ are a linearly independent sequence, it is always possible to test $H_0$ properly.

*Proof:* By the linear independence of $\mathbf{x}_1, \ldots, \mathbf{x}_m$, we know that the equation

$$\mathbf{x}_1\beta_1 + \mathbf{x}_2\beta_2 + \cdots + \mathbf{x}_m\beta_m \equiv X\boldsymbol{\beta} = \mathbf{0}_n$$

has only the trivial solution $\boldsymbol{\beta} = \mathbf{0}_m$. That is, $X^{\perp}_{\text{row}} = \{\mathbf{0}_m\}$. But then, 4.13($a$) is trivially satisfied.

This corollary is particularly useful when we can see by inspection that $\mathbf{x}_1, \ldots, \mathbf{x}_m$ are linearly independent.

As an aside, we mention that this shows that if $X$ is any matrix with $n$ rows and $n$ columns, then the column vectors are linearly independent if and only if the row vectors are linearly independent too.

The general form of the test of $H_0$ that we proposed in Example 4.3 was to reject the hypothesis $H_0$ that $E(\mathbf{Y}) \in S'$ if and only if

**4.17**
$$\frac{n-q}{q-r}\,\frac{\|P_S(\mathbf{y}) - P_{S'}(\mathbf{y})\|^2}{\|\mathbf{y} - P_S(\mathbf{y})\|^2} \equiv \mathscr{F}_{q-r,\,n-q} \geq c,$$

where

$$\mathbf{y} = \begin{bmatrix} y_1 \\ \cdot \\ \cdot \\ \cdot \\ y_n \end{bmatrix}$$

is the vector of actual observed data. It is the observed value of the random vector $\mathbf{Y}$ whose mean $E(\mathbf{Y})$ is known to lie in $S$, with $\dim S = q$ and $\dim S' = r < q$.

We will now show the following theorem.

### *4.18 Theorem: Distribution of F Statistic under $H_0$*

Under the general linear model, 1.5, suppose that $H_0$: $E(\mathbf{Y}) \in S'$ holds, where

$$\mathbf{y} = \begin{bmatrix} y_1 \\ \cdot \\ \cdot \\ \cdot \\ y_n \end{bmatrix},$$

is the observed value of $\mathbf{Y}$, $S = [\![\mathbf{x}_1, \ldots, \mathbf{x}_m]\!]$, $\dim S = q$, and $\dim S' = r < q$. Then the quantity $\mathscr{F}_{q-r,\, n-q}$ of 4.17 is the value of a random variable having the $F$ distribution with $(q - r, n - q)$ degrees of freedom.

*Proof:* Use Theorem 2.16 to choose an O.N. matrix $W$ with the subspace $S' = [\![\mathbf{w}_1, \ldots, \mathbf{w}_r]\!]$, $S = [\![\mathbf{w}_1, \ldots, \mathbf{w}_q]\!]$, and $\mathscr{E}_n = [\![\mathbf{w}_1, \ldots, \mathbf{w}_n]\!]$. As in the proof of Corollary 3.2, if we let $\mathbf{Z} = W^{-1}\mathbf{Y}$, we find that

$$\mathbf{Y} - P_S(\mathbf{Y}) = \sum_{i=q+1}^{n} Z_i \mathbf{w}_i$$

$$P_S(\mathbf{Y}) - P_{S'}(\mathbf{Y}) = \sum_{i=r+1}^{q} Z_i \mathbf{w}_i.$$

Just as in Corollary 3.2, we find that

$$\|\mathbf{Y} - P_S(\mathbf{Y})\|^2/\sigma^2 = \sum_{i=q+1}^{n} (Z_i/\sigma)^2$$

$$\|P_S(\mathbf{Y}) - P_{S'}(\mathbf{Y})\|/\sigma^2 = \sum_{i=r+1}^{q} (Z_i/\sigma)^2.$$

But, by applying Theorem 3.1 with $S$ replaced by $S'$, we conclude that when $H_0$ holds, $Z_{r+1}/\sigma, \ldots, Z_n/\sigma$ are independent standard normal random variables. Hence, by 1 in Chapter 8,

$$\|\mathbf{Y} - P_S(\mathbf{Y})\|^2/\sigma^2 \quad \text{and} \quad \|P_S(\mathbf{Y}) - P_{S'}(\mathbf{Y})\|^2/\sigma^2$$

are independent random variables with Chi square distribution of $n - q$ and $q - r$ degrees of freedom, respectively. The conclusion we desire now follows from 11, Chapter 8, which defines the $F$ distribution.

Since a great deal of ground has been covered in this section, we will summarize briefly.

### 4.19 Summary of Testing Results

Consider the general linear model, 1.5, and let $X$ be the matrix of column vectors $\mathbf{x}_1, \ldots, \mathbf{x}_m$ in 1.5.

In order to properly test the hypothesis $H_0$, that

$$\sum_{i=1}^{m} \alpha_{ji}\beta_i = 0,$$

for all $j = 1, \ldots, \omega$, we form an O.N. sequence $\mathbf{D}_1, \ldots, \mathbf{D}_\gamma, \mathbf{D}_{\gamma+1}, \ldots, \mathbf{D}_m$, where $\mathbf{D}_1, \ldots, \mathbf{D}_\gamma$ are obtained from the Gram-Schmidt procedure (2.10) on the transposes of the rows of $X$.

We find that $H_0$ can be properly tested if and only if

**4.20** $$\boldsymbol{\alpha}_i^T \cdot \mathbf{D}_j = 0, \quad \text{where} \quad \boldsymbol{\alpha}_i \text{ is the } i\text{th } \textit{row} \text{ of } \alpha$$

for each $j = \gamma + 1, \ldots, m$. (It can be shown that $\gamma = q = \dim S$.) Conditions 4.20 are always satisfied when the columns of $X$ are linearly independent. Provided 4.20 is satisfied we reject $H_0$ if and only if

$$\frac{n-q}{q-r} \frac{\|P_S(\mathbf{y}) - P_{S'}(\mathbf{y})\|^2}{\|\mathbf{y} - P_S(\mathbf{y})\|^2} \equiv \mathscr{F}_{q-r,n-q} \geq c$$

where

**4.21** $$S = [\![\mathbf{x}_1, \ldots, \mathbf{x}_m]\!], \qquad S' = [\![X\mathbf{c}_{\omega+1}, \ldots, X\mathbf{c}_m]\!],$$

and $\mathbf{c}_1, \ldots, \mathbf{c}_\omega, \mathbf{c}_{\omega+1}, \ldots, \mathbf{c}_m$ is an O.N. sequence of vectors in $\mathscr{E}_m$ such that

$$[\![\boldsymbol{\alpha}_1^T, \ldots, \boldsymbol{\alpha}_\omega^T]\!] = [\![\mathbf{c}_1, \ldots, \mathbf{c}_\omega]\!].$$

**4.22** $$\dim S' = r \qquad \text{and} \qquad \dim S = q. \quad \text{The vector } \mathbf{y} = \begin{bmatrix} y_1 \\ \cdot \\ \cdot \\ \cdot \\ y_n \end{bmatrix}$$

is the vector of observed values.

In order that this test be of level $\lambda$, we choose $c$ to be the upper $1 - \lambda$ point on the $F$ distribution with $(q - r, n - q)$ degrees of freedom.

### *4.23 Example: Test of $H_0$: $\boldsymbol{\beta}_1 = \boldsymbol{\beta}_2$*

Suppose that in the general linear model, $n = 4$ and

$$\mathbf{x}_1 = \begin{bmatrix} 0 \\ 1 \\ 1 \\ 1 \end{bmatrix}, \quad \mathbf{x}_2 = \begin{bmatrix} 1 \\ 0 \\ 1 \\ 1 \end{bmatrix}, \quad \mathbf{x}_3 = \begin{bmatrix} 1 \\ 1 \\ 2 \\ 2 \end{bmatrix}.$$

We see that $\mathbf{x}_3 = \mathbf{x}_2 + \mathbf{x}_1$. Hence, we must check to see whether the hypothesis $H_0$ can be tested.† We first form an orthogonal sequence from the rows [0, 1, 1], [1, 0, 1], and [1, 1, 2].

$$\mathbf{D}_1 = [0, 1, 1]^T$$

$$\mathbf{D}_2 = [1, 0, 1]^T - P_{[\mathbf{D}_1]}[1, 0, 1]^T$$

$$= \begin{bmatrix} 1 \\ 0 \\ 1 \end{bmatrix} - \frac{\begin{bmatrix} 1 \\ 0 \\ 1 \end{bmatrix} \cdot \begin{bmatrix} 0 \\ 1 \\ 1 \end{bmatrix}}{\left\| \begin{bmatrix} 0 \\ 1 \\ 1 \end{bmatrix} \right\|^2} \begin{bmatrix} 0 \\ 1 \\ 1 \end{bmatrix} = \begin{bmatrix} 1 \\ -\frac{1}{2} \\ \frac{1}{2} \end{bmatrix}.$$

(1,0,1)$^T$

$\mathbf{D}_1$

To obtain the third orthogonal vector, $\mathbf{D}_3$, try

$$\mathbf{D}_3 = \begin{bmatrix} 1 \\ 0 \\ 0 \end{bmatrix} - P_{[\mathbf{D}_1]} \begin{bmatrix} 1 \\ 0 \\ 0 \end{bmatrix} - P_{[\mathbf{D}_2]} \begin{bmatrix} 1 \\ 0 \\ 0 \end{bmatrix} = \begin{bmatrix} \frac{1}{3} \\ \frac{1}{3} \\ -\frac{1}{3} \end{bmatrix}.$$

$\mathbf{D}_1$ and $\mathbf{D}_2$ span the rows of $X$, and

$$\mathbf{D}_3 = \begin{bmatrix} \frac{1}{3} \\ \frac{1}{3} \\ -\frac{1}{3} \end{bmatrix}$$

corresponds to $\mathbf{D}_{\gamma+1}, \ldots, \mathbf{D}_m$.

Now we check whether $\boldsymbol{\alpha}_i^T \cdot \mathbf{D}_3 = 0$. Here

$$\boldsymbol{\alpha}_1^T = \begin{bmatrix} 1 \\ -1 \\ 0 \end{bmatrix}.$$

Luckily,

$$\begin{bmatrix} 1 \\ -1 \\ 0 \end{bmatrix} \cdot \begin{bmatrix} \frac{1}{3} \\ \frac{1}{3} \\ -\frac{1}{3} \end{bmatrix} = 0.$$

† The linear dependence was introduced here solely for illustrative purposes and is somewhat artificial. However, such dependence does arise naturally in the analysis of variance.

(Note that we cannot properly test the hypothesis that $\beta_1 = 2\beta_2$.) Now we must phrase the test as a subspace of $\mathscr{E}_4$. To do this, we have

$$\mathbf{c}_1 = \begin{bmatrix} 1 \\ -1 \\ 0 \end{bmatrix}.$$

We must find two more non-zero vectors, $\mathbf{c}_2$, $\mathbf{c}_3$ such that $\mathbf{c}_1$, $\mathbf{c}_2$, $\mathbf{c}_3$ are orthogonal. By inspection, we see that

$$\mathbf{c}_2 = \begin{bmatrix} 1 \\ 1 \\ 0 \end{bmatrix} \quad \text{and} \quad \mathbf{c}_3 = \begin{bmatrix} 0 \\ 0 \\ 1 \end{bmatrix}$$

will do. (We could have used the Gram-Schmidt procedure.) Hence,

$$S' = \left[\!\!\left[ \begin{bmatrix} 0 & 1 & 1 \\ 1 & 0 & 1 \\ 1 & 1 & 2 \\ 1 & 1 & 2 \end{bmatrix} \begin{bmatrix} 1 \\ 1 \\ 0 \end{bmatrix}, \begin{bmatrix} 0 & 1 & 1 \\ 1 & 0 & 1 \\ 1 & 1 & 2 \\ 1 & 1 & 2 \end{bmatrix} \begin{bmatrix} 0 \\ 0 \\ 1 \end{bmatrix} \right]\!\!\right] = \left[\!\!\left[ \begin{bmatrix} 1 \\ 1 \\ 2 \\ 2 \end{bmatrix}, \begin{bmatrix} 1 \\ 1 \\ 2 \\ 2 \end{bmatrix} \right]\!\!\right] = \left[\!\!\left[ \begin{bmatrix} 1 \\ 1 \\ 2 \\ 2 \end{bmatrix} \right]\!\!\right].$$

Note that

$$E(\mathbf{Y}) \in \left[\!\!\left[ \begin{bmatrix} 1 \\ 1 \\ 2 \\ 2 \end{bmatrix} \right]\!\!\right]$$

does imply that $\beta_1 = \beta_2$, since

$$\beta_1 \begin{bmatrix} 0 \\ 1 \\ 1 \\ 1 \end{bmatrix} + \beta_2 \begin{bmatrix} 1 \\ 0 \\ 1 \\ 1 \end{bmatrix} + \beta_3 \begin{bmatrix} 1 \\ 1 \\ 2 \\ 2 \end{bmatrix} = c \begin{bmatrix} 1 \\ 1 \\ 2 \\ 2 \end{bmatrix}$$

for some $c$ if and only if

$$(\beta_1 - \beta_2) \begin{bmatrix} 0 \\ 1 \\ 1 \\ 1 \end{bmatrix} + \beta_2 \begin{bmatrix} 1 \\ 1 \\ 2 \\ 2 \end{bmatrix} + \beta_3 \begin{bmatrix} 1 \\ 1 \\ 2 \\ 2 \end{bmatrix} = c \begin{bmatrix} 1 \\ 1 \\ 2 \\ 2 \end{bmatrix},$$

which happens only if

$$(\beta_1 - \beta_2)\begin{bmatrix}0\\1\\1\\1\end{bmatrix} = (c - \beta_2 - \beta_3)\begin{bmatrix}1\\1\\2\\2\end{bmatrix}.$$

This can happen only if $\beta_1 - \beta_2 = 0$. The test of $H_0$ is now easy:

$$q = \dim S = 2, \qquad r = \dim S' = 1.$$

An orthogonal basis for $S$ is

$$\{\mathbf{v}_1, \mathbf{v}_2\} = \left\{\begin{bmatrix}1\\1\\2\\2\end{bmatrix}, \begin{bmatrix}1\\-1\\0\\0\end{bmatrix}\right\},$$

(obtained by inspection, but we could have used the Gram-Schmidt procedure) the first vector $\mathbf{v}_1 = \mathbf{x}_1 + \mathbf{x}_2$ spanning $S'$, the second $\mathbf{v}_2 = \mathbf{x}_2 - \mathbf{x}_1$ clearly orthogonal to $\mathbf{v}_1$:

$$P_S(\mathbf{y}) = P_{[\mathbf{v}_1]}(\mathbf{y}) + P_{[\mathbf{v}_2]}(\mathbf{y}) = \frac{\begin{bmatrix}1\\1\\2\\2\end{bmatrix}\cdot\begin{bmatrix}y_1\\y_2\\y_3\\y_4\end{bmatrix}}{10}\begin{bmatrix}1\\1\\2\\2\end{bmatrix} + \frac{\begin{bmatrix}1\\-1\\0\\0\end{bmatrix}\cdot\begin{bmatrix}y_1\\y_2\\y_3\\y_4\end{bmatrix}}{2}\begin{bmatrix}1\\-1\\0\\0\end{bmatrix}$$

$$= \frac{y_1 + y_2 + 2y_3 + 2y_4}{10}\begin{bmatrix}1\\1\\2\\2\end{bmatrix} + \frac{y_1 - y_2}{2}\begin{bmatrix}1\\-1\\0\\0\end{bmatrix}$$

$P_{S'}(\mathbf{y}) = P_{[\mathbf{v}_1]}(\mathbf{y})$. Computing

$$\frac{n-q}{q-r}\,\frac{\|P_S(\mathbf{y}) - P_{S'}(\mathbf{y})\|^2}{\|\mathbf{y} - P_S(\mathbf{y})\|^2}$$

is now routine. △

---

### 4.24 *EXERCISES*

1.° In the model of 4.3, set up the test of $H_0$: $\beta_1 = \beta_2 = 0$.
2. In the model of 4.3, set up the test of $H_0$: $\beta_1 = 3\beta_2$.

3. In the model of 4.23:
   (a) can we test the hypothesis $\beta_2 + \beta_3 = 0$?
   (b)° can we test the hypothesis $\beta_2 - \beta_3 = 0$?
   (c) can we test $\beta_1 + \beta_2 - 2\beta_3 = 0$?
   (d) set up the test of $\beta_1 + \beta_3 = 0$.

You may have wondered whether it were possible to conveniently test hypotheses of the form

**4.25** $$\sum_{i=1}^{m} \alpha_{ji}\beta_i = \delta_j, \qquad j = 1, \ldots, \omega,$$

where $\delta_1, \ldots, \delta_\omega$ are given numbers. We show how to do this as the final topic in this section.

There is nothing lost by assuming $\boldsymbol{\alpha}_1^T, \ldots, \boldsymbol{\alpha}_\omega^T$ to be linearly independent. Certainly

$$\boldsymbol{\delta} = \begin{bmatrix} \delta_1 \\ \cdot \\ \cdot \\ \cdot \\ \delta_\omega \end{bmatrix}$$

must be in the subspace spanned by the *column* vectors of $\alpha$, or 4.25 couldn't possibly be satisfied. Under these circumstances, there is a vector

$$\mathbf{A} = \begin{bmatrix} A_1 \\ \cdot \\ \cdot \\ \cdot \\ A_m \end{bmatrix}$$

such that

**4.26** $$\alpha\mathbf{A} = \boldsymbol{\delta}.$$

Then we see that 4.25 is equivalent to

**4.27** $$\sum_{i=1}^{m} \alpha_{ji}(\beta_i - A_i) = 0 \quad \text{for } j = 1, \ldots, \omega.$$

Thus, 4.25 is equivalent to $\boldsymbol{\beta} - \mathbf{A} \in \alpha_{\text{row}}^{\perp}$.

We now insist that 4.13 must hold, in order that 4.25 be equivalent to

**4.28** $$X(\boldsymbol{\beta} - \mathbf{A}) \in [\![X\mathbf{c}_{\omega+1}, \ldots, X\mathbf{c}_m]\!] \qquad \text{(see Theorem 4.11).}$$

If we just let $\mathbf{Y}^* = \mathbf{Y} - X\mathbf{A}$, thinking of $\mathbf{Y}^*$ as our observation vector, then $E(\mathbf{Y}^*) = E(\mathbf{Y}) - X\mathbf{A} = X(\boldsymbol{\beta} - \mathbf{A})$. Hence, the hypothesis 4.25, in which $E(\mathbf{Y}) = X\boldsymbol{\beta}$, is equivalent to 4.27, in which $\mathbf{Y}^*$ is our observation vector with $E(\mathbf{Y}^*) = X(\boldsymbol{\beta} - \mathbf{A})$, where $\mathbf{A}$ is any solution of 4.26. Thus, to test 4.25, simply solve 4.26 for $\mathbf{A}$, transform from $\mathbf{Y}$ to $\mathbf{Y}^* = \mathbf{Y} - X\mathbf{A}$, and test whether $E(\mathbf{Y}^*) \in [\![X\mathbf{c}_{\omega+1}, \ldots, X\mathbf{c}_m]\!]$. Note that $\mathbf{Y}^*$ still consists of independent normal random variables with variance $\sigma^2$, and that

$$[\![\mathbf{c}_{\omega+1}, \ldots, \mathbf{c}_m]\!] = \alpha_{\text{row}}^{\perp} \quad \text{where} \quad \alpha = \begin{bmatrix} \alpha_{11} & \cdots & \alpha_{1m} \\ \cdot & & \\ \cdot & & \\ \cdot & & \\ \alpha_{\omega 1} & \cdots & \alpha_{\omega m} \end{bmatrix}.$$

## SECTION 5
## Power of the $F$ Test and the Choice of Replications

All too often, the only reasonable conclusion that can be drawn from performing an $F$ test is that we failed to gather enough data. When we do not make enough observations, our test does not allow much discriminating power. If we are testing at the $\alpha$ level with small $\alpha$, this usually results in our being forced to accept the null hypothesis because we have no convincing evidence of a description preferable to the null hypothesis. Unless this is the experimenter's aim, attention must be paid to the power of the test to see to it that it is capable of achieving the desired discrimination.

With this goal in mind, we first determine the distribution of the $F$ statistic

$$\frac{n-q}{q-r}\,\frac{\|P_S(\mathbf{Y}) - P_{S'}(\mathbf{Y})\|^2}{\|\mathbf{Y} - P_S(\mathbf{Y})\|^2} = \frac{n-q}{q-r}\,\frac{\|P_S(\mathbf{Y}) - P_{S'}(\mathbf{Y})\|^2/\sigma^2}{\|\mathbf{Y} - P_S(\mathbf{Y})\|^2/\sigma^2}$$

when $\mathbf{Y}$ is a vector of independent normal random variables with common variance $\sigma^2$ and expectation $E(\mathbf{Y}) \in S$. We assume that $E(\mathbf{Y}) \notin S'$, since the case $E(\mathbf{Y}) \in S'$ has been handled by Theorem 4.18.

In order to carry out our aim we first examine the numerator, writing

**5.1** $\|P_S(\mathbf{Y}) - P_{S'}(\mathbf{Y})\|^2/\sigma^2$

$$= \|\underbrace{P_S(\mathbf{Y}) - E(\mathbf{Y})}_{\mathbf{0}\text{ mean}} - \underbrace{[P_{S'}(\mathbf{Y}) - P_{S'}(E(\mathbf{Y}))]}_{\mathbf{0}\text{ mean}} + E(\mathbf{Y}) - P_{S'}(E(\mathbf{Y}))\|^2/\sigma^2.$$

This suggests using the method of Theorem 2.16 to define an orthonormal matrix $W$ with column vectors $\mathbf{w}_1, \ldots, \mathbf{w}_n$ such that

$$[\![\mathbf{w}_1, \ldots, \mathbf{w}_r]\!] = S', \qquad \mathbf{w}_{r+1} \in [\![E(\mathbf{Y}) - P_{S'}(E(\mathbf{Y}))]\!].$$

$E(\mathbf{Y})$
$E(\mathbf{Y}) - P_{S'}(E(\mathbf{Y}))$
$S'$
$P_{S'}(E(\mathbf{Y}))$

Note that this can be done, since $\mathbf{w}_j \perp E(\mathbf{Y}) - P_{S'}(E(\mathbf{Y}))$ for $1 \le j \le r$. This is shown as follows:

$$\mathbf{w}_j \cdot (E(\mathbf{Y}) - P_{S'}(E(\mathbf{Y}))) = \mathbf{w}_j \cdot (E(\mathbf{Y}) - \sum_{i=1}^{r} P_{[\![\mathbf{w}_i]\!]}(E(\mathbf{Y})) \quad \text{by Theorem 2.13}$$

$$= \mathbf{w}_j \cdot (E(\mathbf{Y}) - \sum_{i=1}^{r} (E(\mathbf{Y}) \cdot \mathbf{w}_i)\mathbf{w}_i) \quad \text{by Theorem 2.4 and the fact that } W \text{ is orthonormal}$$

$$= E(\mathbf{Y}) \cdot \mathbf{w}_j - \sum_{i=1}^{r} (E(\mathbf{Y}) \cdot \mathbf{w}_i)(\mathbf{w}_i \cdot \mathbf{w}_j)$$

$$= E(\mathbf{Y}) \cdot \mathbf{w}_j - E(\mathbf{Y}) \cdot \mathbf{w}_j \quad \text{since } \mathbf{w}_1, \ldots, \mathbf{w}_r \text{ are orthogonal}$$

$$= 0.$$

Let $S^* = [\![\mathbf{w}_1, \ldots, \mathbf{w}_{r+1}]\!]$. It is clear that $E(\mathbf{Y}) \in S^*$. Let $\mathbf{Z} = W^{-1}\mathbf{Y}$ and apply Theorem 3.1 with $S$ replaced by $S^*$ to conclude that

**5.2** $Z_1, \ldots, Z_n$ are independent normal variables with variance $\sigma^2$

and

**5.3** $$E(Z_i) = 0 \quad \text{for} \quad i = r + 2, \ldots, n.$$

Then, as in the proof of Corollary 3.2,

$$\|P_S(\mathbf{Y}) - P_{S'}(\mathbf{Y})\|^2/\sigma^2 = \left\|\sum_{i=1}^{q} Z_i\mathbf{w}_i - \sum_{i=1}^{r} Z_i\mathbf{w}_i\right\|^2 \Big/ \sigma^2$$

$$= \sum_{i=r+1}^{q} (Z_i/\sigma)^2.$$

Now $E(Z_{r+2}) = E(Z_{r+3}) = \cdots = E(Z_q) = 0$. We claim that

$$|E(Z_{r+1})| = \|E(\mathbf{Y}) - P_{S'}(E(\mathbf{Y}))\|.$$

This follows easily from the computation

$$E(\mathbf{Y}) - P_{S'}(E(\mathbf{Y})) = E(W\mathbf{Z}) - P_{S'}(E(W\mathbf{Z}))$$

$$= E\left(\sum_{i=1}^{n} Z_i\mathbf{w}_i\right) - P_{S'}(WE(\mathbf{Z}))$$

$$= E\left(\sum_{i=1}^{n} Z_i\mathbf{w}_i\right) - P_{S'}\left(\sum_{i=1}^{n} E(Z_i)\mathbf{w}_i\right)$$

$$= \sum_{i=1}^{n} E(Z_i)\mathbf{w}_i - \sum_{i=1}^{n} E(Z_i)\mathbf{w}_i$$

since $S' = [\![\mathbf{w}_1, \ldots, \mathbf{w}_r]\!]$, using Theorem 2.13, Theorem 2.4, Corollary 2.15, and the orthonormality of the $\mathbf{w}_i$, as we have done earlier

$$= \sum_{i=r+1}^{n} E(Z_i)\mathbf{w}_i$$

$$= E(Z_{r+1})\mathbf{w}_{r+1}.$$

It is easy to verify, as in Corollary 3.2, that

$$\|\mathbf{Y} - P_S(\mathbf{Y})\|^2 = \sum_{i=q+1}^{n} Z_i^2.$$

Thus, we have shown that there are *independent* normal random variables $Z_1, \ldots, Z_n$ with variance $\sigma^2$,

$$|E(Z_{r+1})| = \|E(\mathbf{Y}) - P_{S'}(E(\mathbf{Y}))\|$$

$$E(Z_i) = 0, \qquad i = r+2, \ldots, n$$

such that

$$D = \|\mathbf{Y} - P_S(\mathbf{Y})\|^2/\sigma^2 = \sum_{i=q+1}^{n} (Z_i/\sigma)^2$$

and

$$N = \|P_S(\mathbf{Y}) - P_{S'}(\mathbf{Y})\|^2/\sigma^2 = (Z_{r+1}/\sigma)^2 + \sum_{i=r+2}^{q} (Z_i/\sigma)^2.$$

This shows that the two random variables $D$ and $N$ are independent. $D$ has the Chi square distribution with $n - q$ degrees of freedom, and $N$ has the non central Chi square distribution with $q - r$ degrees of freedom, with non-centrality parameter

$$\frac{\|E(\mathbf{Y}) - P_{S'}(E(\mathbf{Y}))\|^2}{2\sigma^2} \qquad \text{(see 6 in Chapter 8).}$$

Thus, from Definition 12 in Chapter 8 of the non-central $F$ distribution, we have proved

## 5.4 *Theorem: Distribution of the F Statistic*

The quantity $\mathscr{F}_{q-r,\, n-q}$ of Equation 4.17 is the value of a random variable having the non-central $F$ distribution with $(q - r, n - q)$ degrees of freedom and non-centrality parameter

$$\frac{\|E(\mathbf{Y}) - P_{S'}(E(\mathbf{Y}))\|^2}{2\sigma^2}.$$

The non-central $F$ distribution is tabled in several different ways. In order to compute the power of the $F$ test for a given $X$ at some fixed alternative, we may use the tables developed by Tang [in Graybill, 9]. These are provided in your *Solution Manual*, with directions for their use.

---

### 5.5 *EXERCISES*

1.° In Example 4.3, the test of equal effect of additives, suppose that the level $\lambda = .05$. Using the Tang tables, find the power when $\beta_1 = 1$, $\beta_2 = 2$, and $\sigma^2 = 1$.

2. In Example 4.3, suppose that the level $\lambda = .01$. Using the Tang tables, find the power when $\beta_1 = 2$, $\beta_2 = 1$, and $\sigma^2 = 4$.

The more relevant and useful computation involves finding the sample size that will yield the desired power at a specified alternative. We will only consider the special case in which we only have the option of choosing the number of repetitions of some basic design, such as the one in Example 4.3. First, we must introduce some notation. Suppose that $\mathbf{x}$ is a vector in $\mathscr{E}_n$. By $\mathbf{x}_{(k)}$ we will mean the vector in $\mathscr{E}_{kn}$ each of whose elements is repeated $k$ times. To illustrate, if

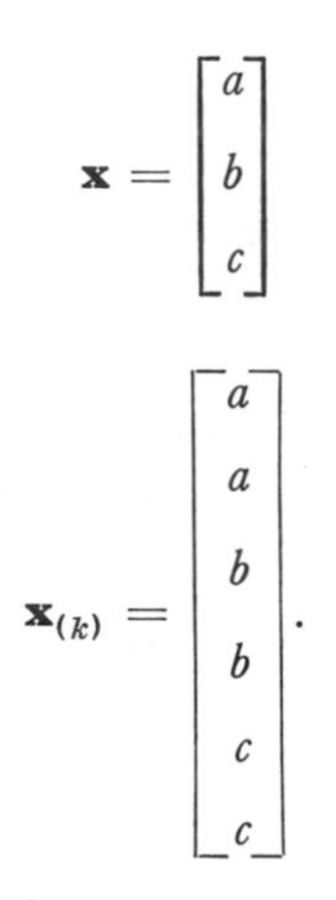

$$\mathbf{x} = \begin{bmatrix} a \\ b \\ c \end{bmatrix}$$

and $k = 2$, then

$$\mathbf{x}_{(k)} = \begin{bmatrix} a \\ a \\ b \\ b \\ c \\ c \end{bmatrix}.$$

Now suppose we have a linear model

**5.6**
$$\mathbf{Y} = \beta_1\mathbf{x}_1 + \cdots + \beta_m\mathbf{x}_m + \boldsymbol{\varepsilon},$$

where $\mathbf{Y}$ has $n$ coordinates and $\boldsymbol{\varepsilon}$ is a vector of $n$ independent normal random variables with mean 0 and variance $\sigma^2$. Let $\mathbf{Y}^{(k)}$ be a random vector with $nk$ coordinates and $\boldsymbol{\varepsilon}^{(k)}$ be a vector consisting of $nk$ independent normal random variables with mean 0 and variance $\sigma^2$.

**5.7 Definition: Replications.** We will say that $\mathbf{Y}^{(k)}$ consists of $k$ replications of the model in 5.6 if

$$\mathbf{Y}^{(k)} = \beta_1\mathbf{x}_{1,(k)} + \cdots + \beta_m\mathbf{x}_{m,(k)} + \boldsymbol{\varepsilon}^{(k)}.$$

Now, the subspace $S = [\![\mathbf{x}_1, \ldots, \mathbf{x}_m]\!]$ of the original model 5.6 induces a subspace $S^{(k)} = [\![\mathbf{x}_{1,(k)}, \ldots, \mathbf{x}_{m,(k)}]\!]$ in the model consisting of $k$ replications of 5.6. It is evident that

$$\dim S = \dim S^{(k)}$$

and that in the model of $k$ replications of 5.6, $E(\mathbf{Y}^{(k)}) \in S^{(k)}$. Similarly, any subspace $S' = [\![\mathbf{w}_1, \ldots, \mathbf{w}_r]\!]$ of $S$ corresponds to the subspace

$$S'^{(k)} = [\![\mathbf{w}_{1,(k)}, \ldots, \mathbf{w}_{r,(k)}]\!]$$

in the model of $k$ replications of 5.6 and $\dim S'^{(k)} = \dim S'$.

We see that any hypothesis concerning the regression coefficients that can be tested in 5.6 can be tested in 5.7, as follows. Suppose that $\dim S = q$,

$\dim S' = r$, and that the test of $H_0$: $E(\mathbf{Y}) \in S'$ in 5.6 is to reject $H_0$ if and only if

**5.8**
$$\frac{n-q}{q-r}\frac{\|P_S(\mathbf{y}) - P_{S'}(\mathbf{y})\|^2}{\|\mathbf{y} - P_S(\mathbf{y})\|^2} \geq f_{q-r,n-q}(1-\lambda)$$

where $f_{q-r,n-q}(1-\lambda)$ is the upper $1-\lambda$ point of the $F$ distribution with $(q-r, n-q)$ degrees of freedom. The corresponding test in 5.7 of level $\lambda$ of the hypothesis $H_0^{(k)}$: $E(\mathbf{Y}^{(k)}) \in S'^{(k)}$ is to reject $H_0^{(k)}$ if and only if

**5.9**
$$\frac{kn-q}{q-r}\frac{\|P_{S^{(k)}}(\mathbf{y}^{(k)}) - P_{S'^{(k)}}(\mathbf{y}^{(k)})\|^2}{\|\mathbf{y}^{(k)} - P_{S^{(k)}}(\mathbf{y}^{(k)})\|^2} \geq f_{q-r,kn-\lambda}(1-\lambda).$$

Note that the two tests ask the same question about the regression coefficients $\beta_1, \ldots, \beta_m$. They differ in their discriminating power against an alternative. The higher $k$ is, the greater the discriminating power is. We shall now see how to choose $k$ so as to achieve at least a given power $1-\beta$ against any specific simple alternative, that is, one with specified mean vector and $\sigma^2$.

Suppose that we want to test against a specific alternative such that the distribution of the test statistic in 5.8 was the non-central $F$ distribution with $(q-r, n-q)$ degrees of freedom and non-centrality parameter

$$\mathcal{N} = \frac{\|E(\mathbf{Y}) - P_{S'}(E(\mathbf{Y}))\|^2}{2\sigma^2}.$$

Then we will need the following theorem.

## 5.10 *Theorem: Distribution of Test Statistic with Replications under Alternative*

Under a specific alternative, 5.9 yields a statistic having the non-central $F$ distribution with $(q-r, kn-q)$ degrees of freedom and non-centrality parameter

$$\mathcal{N}^{(k)} = \frac{\|E(\mathbf{Y}^{(k)}) - P_{S'^{(k)}}(E(\mathbf{Y}^{(k)}))\|^2}{2\sigma^2} = k\mathcal{N} = k\frac{\|E(\mathbf{Y}) - P_{S'}(E(\mathbf{Y}))\|^2}{2\sigma^2}.$$

The only part that requires proving is that $\mathcal{N}^{(k)} = k\mathcal{N}$. The easy way to accomplish this is to convert *first* to the variables $Z_1, \ldots, Z_n$, as in the proof of Theorem 5.4, and to act as if these were the actual observed values. Recall that $Z_1, \ldots, Z_n$ are independent random variables with variance $\sigma^2$. The general linear model is $E(\mathbf{Y}) \in [\![\mathbf{w}_1, \ldots, \mathbf{w}_q]\!]$, and the hypothesis to be tested is that $E(\mathbf{Y}) \in [\![\mathbf{w}_1, \ldots, \mathbf{w}_r]\!]$. That is, the model is

$$E(WZ) \in [\![\mathbf{w}_1, \ldots, \mathbf{w}_q]\!]$$

and the hypothesis is that

$$E(WZ) \in [\![\mathbf{w}_1, \ldots, \mathbf{w}_r]\!].$$

Let $\mathbf{e}_i$ be the vector in $\mathscr{E}_n$ with a 1 in the $i$th coordinate and 0 in all others. Now use the fact that $E(WZ) = WE(Z)$ and $W^{-1} = W^T$ (see Equation 2.28), to see that the model is

$$E(Z) \in [\![\mathbf{e}_1, \ldots, \mathbf{e}_q]\!] \equiv S_Z$$

and the hypothesis to be tested is that

$$E(Z) \in [\![\mathbf{e}_1, \ldots, \mathbf{e}_r]\!] \equiv S'_Z.$$

In Theorem 5.4, we showed that by proper choice of $w_{r+1}$, the particular alternative with non-centrality parameter $\mathcal{N}$ corresponded to

$$|E(Z_{r+1})| = (\sqrt{2\mathcal{N}})\sigma$$
$$E(Z_{r+2}) = \cdots = E(Z_q) = 0.$$

In the model with $k$ replications, we note that $S_Z'^{(k)} = [\![\mathbf{e}_{1,(k)}, \ldots, \mathbf{e}_{r,(k)}]\!]$ and the $F$ ratio

$$\frac{kn - q}{q - r} \frac{\|P_{S_Z^{(k)}}(\mathbf{z}^{(k)}) - P_{S_Z'^{(k)}}(\mathbf{z}^{(k)})\|^2}{\|\mathbf{z}^{(k)} - P_{S_Z^{(k)}}(\mathbf{z}^{(k)})\|^2}$$

is the value of a random variable with the non-central $F$ distribution with $(q - r, kn - q)$ degrees of freedom and non-centrality parameter

$$\frac{\|E(\mathbf{Z}^{(k)}) - P_{S_Z'^{(k)}}(E(\mathbf{Z}^{(k)}))\|^2}{2\sigma^2}.$$

Now

$$E(\mathbf{Z}^{(k)}) = \sum_{i=1}^{r+1} E(Z_i)\mathbf{e}_{i,(k)}.$$

Using the orthogonality of the $\mathbf{e}_{i,(k)}$, Theorem 2.13, Theorem 2.4, and Corollary 2.15, we have

$$P_{S_Z'^{(k)}}(E(\mathbf{Z}^{(k)})) = \sum_{i=1}^{r} E(Z_i)P_{[\![\mathbf{e}_{i,(k)}]\!]}(\mathbf{e}_{i,(k)})$$
$$= \sum_{i=1}^{r} E(Z_i)\mathbf{e}_{i,(k)}.$$

Thus

$$\mathcal{N}^{(k)} = \|E(\mathbf{Z}^{(k)}) - P_{S_Z'^{(k)}}(E(\mathbf{Z}^{(k)}))\|^2/(2\sigma^2)$$
$$= \|E(Z_{r+1})\mathbf{e}_{i,(k)}\|^2/(2\sigma^2)$$
$$= k\ |E(Z_{r+1})\ |^2/(2\sigma^2)$$
$$= k\mathcal{N},$$

as asserted.

We can use the tables prepared by Martin Fox (in the *Solution Manual*) and this last result to determine the number of replications $k$ that will yield power† $\beta$ against a specific alternative when the test has level $\lambda$. We first compute the basic non-centrality parameter

$$\mathcal{N} = \frac{\|E(\mathbf{Y}) - P_{S'}(E(\mathbf{Y}))\|^2}{2\sigma^2}$$

† The Fox tables use $\beta$ rather than $1 - \beta$ for power.

corresponding to the alternative we have in mind. To use the Fox tables, set

$$f_1 = q - r \qquad \text{and} \qquad \varphi = \sqrt{\frac{2k\mathcal{N}}{f_1 + 1}}.$$

Choose the desired level $\lambda$ and power $\beta$ and go to the Fox table with these values of $\lambda$ and $\beta$. Choose a value of $k$, compute $f_1$ and $\varphi$, and on the curve labeled with the computed $\varphi$, read off the $f_2$ value on the second axis corresponding to the computed $f_1$ value. If this value exceeds $kn - q$, then $k$ must be raised. The smallest $k$ for which $f_2 \leq kn - q$ is the desired number of replications. We supply one elementary example.

### 5.11 *Example: Determination of Number of Replications*

Suppose that we have four treatments in the model

$$E(\mathbf{Y}) = \beta_1 \begin{bmatrix} 1 \\ 0 \\ 0 \\ 0 \end{bmatrix} + \beta_2 \begin{bmatrix} 0 \\ 1 \\ 0 \\ 0 \end{bmatrix} + \beta_3 \begin{bmatrix} 0 \\ 0 \\ 1 \\ 0 \end{bmatrix} + \beta_4 \begin{bmatrix} 0 \\ 0 \\ 0 \\ 1 \end{bmatrix}.$$

Let $H_0$ be $\beta_1 = \beta_2 = \beta_3 = \beta_4$. Then $H_0$ corresponds to

$$E(\mathbf{Y}) \in S' = \left[\begin{bmatrix} 1 \\ 1 \\ 1 \\ 1 \end{bmatrix}\right].$$

Suppose we want a level of $\lambda = .01$ with power $\beta = .9$ when $\beta_1 = \beta_2 = \beta_3 = 0$ and $\beta_4 = 1$, $\sigma^2 = 1$. At this given alternative, we have

$$E(\mathbf{Y}) = \begin{bmatrix} 0 \\ 0 \\ 0 \\ 1 \end{bmatrix}, \quad P_{S'}(E(\mathbf{Y})) = \frac{E(\mathbf{Y}) \cdot \begin{bmatrix} 1 \\ 1 \\ 1 \\ 1 \end{bmatrix}}{\left\| \begin{bmatrix} 1 \\ 1 \\ 1 \\ 1 \end{bmatrix} \right\|} \begin{bmatrix} 1 \\ 1 \\ 1 \\ 1 \end{bmatrix} = \begin{bmatrix} 1/4 \\ 1/4 \\ 1/4 \\ 1/4 \end{bmatrix}.$$

Hence $\|E(\mathbf{Y}) - P_{S'}(E(\mathbf{Y}))\|^2 = 3/4$. Thus $\mathcal{N} = 3/8$. Since $q = 4$, $r = 1$, $q - r = f_1 = 3$. Thus

$$\varphi = \frac{\sqrt{3}}{4}\sqrt{k} \cong .434\sqrt{k}.$$

For a trial $k$ of 16, $\varphi = 1.736$ we go off scale. Try $k = 36$; then $\varphi = 2.604$. Here we read from the chart that $f_2 = 18$. But $kn - q = 144 - 4 = 140$, so that 36 replications certainly are sufficient. We find $k = 25$ too small. Using bisection to search (i.e., trying $k = 30$ next), we find that $k = 27$ replications seems to be the minimum number needed. △

---

### *5.12 EXERCISES*

1.° In the model of Example 5.11, how many replications are needed to test $\beta_1 = \beta_2 = \beta_3 = \beta_4 = 0$ at the .01 level with power .9 when $\sigma^2 = 4$, $\beta_1 = \beta_2 = 1$ and $\beta_3 = \beta_4 = 0$?

2. Suppose we replace the model of 5.11 with one in which

$$\mathbf{x}_1 = \begin{bmatrix} 0 \\ 1 \\ 1 \\ 1 \end{bmatrix}, \quad \mathbf{x}_2 = \begin{bmatrix} 1 \\ 0 \\ 1 \\ 1 \end{bmatrix}, \quad \mathbf{x}_3 = \begin{bmatrix} 1 \\ 1 \\ 0 \\ 1 \end{bmatrix}, \quad \mathbf{x}_4 = \begin{bmatrix} 1 \\ 1 \\ 1 \\ 0 \end{bmatrix}.$$

(a)° Set up the test that $\beta_1 = \beta_2 = \beta_3 = \beta_4$ at the .05 level to achieve power .7 when $\beta_1 = \beta_2 = \beta_3 = 1$ and $\beta_4 = 2$, $\sigma^2 = .25$

(b) Set up the test that $\beta_1 = \beta_2 = \beta_3 = \beta_4 = 0$ at the .05 level to achieve power .7 when $\beta_1 = \beta_2 = \beta_3 = 0$ and $\beta_4 = 2$, $\sigma^2 = 25$

## SECTION 6
## Confidence Interval Estimation

If we consider the model of 4.1 and we want to predict the mileage corresponding to 5 units of additive one and 2 units of additive two, we would want to use $5\beta_1 + 2\beta_2$. Hence, it is natural to want to determine confidence intervals for linear combinations of the $\beta_i$'s.

It is an evident geometrical fact that a vector $\mathbf{v}$ lies in some given circle if and only if the projection $P_L(\mathbf{v})$ of $\mathbf{v}$ along each line $L$ through the origin lies inside the circle's projection on $L$, as indicated in the sketch.

The following was developed by Scheffé.

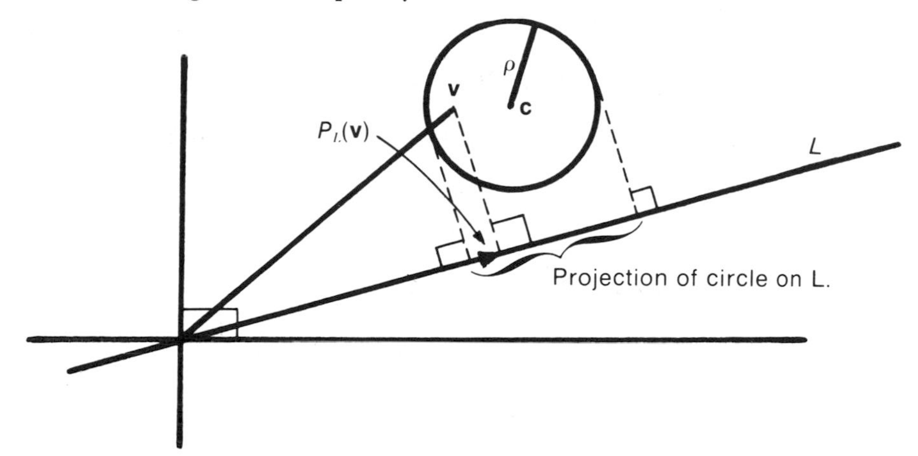


We start our investigation by noting that if $S'$ is any subspace of $S$ (say $S' = [\![\mathbf{w}_1, \ldots, \mathbf{w}_r]\!]$, $\mathbf{w}_i \in S$), then from our earlier results it follows that

$$\frac{(n-q)\,\|P_{S'}(\mathbf{y}) - P_{S'}(E(\mathbf{Y}))\|^2}{r\,\|\mathbf{y} - P_S(\mathbf{y})\|^2}$$

is the value of a random variable (generally unobservable) that has an $F$ distribution with $r$ and $n - q$ degrees of freedom, where $q = \dim S$, $r = \dim S'$. Hence, the probability

**6.1** $$P\left\{\frac{n-q}{r}\,\frac{\|P_{S'}(\mathbf{Y}) - P_{S'}(E(\mathbf{Y}))\|^2}{\|\mathbf{Y} - P_S(\mathbf{Y})\|^2} \leq f_{r,\,n-q}(1-\beta)\right\} = 1 - \beta$$

where $f_{r,n-q}(1 - \beta)$ is the upper $1 - \beta$ point on the $F$ distribution with $r$ and $n - q$ degrees of freedom.

We rewrite 6.1 as

**6.2** $$P\left\{\|P_{S'}(\mathbf{Y}) - P_{S'}(E(\mathbf{Y}))\|^2 \leq \frac{r}{n-q}\, f_{r,\,n-q}(1-\beta)\,\|\mathbf{Y} - P_S(\mathbf{Y})\|^2\right\} = 1 - \beta.$$

We will now prove the fundamental geometric lemma stated earlier.

### *6.3 Lemma*

Let $S'$ be any subspace of $S$. Then for given $\mathbf{c} \in S'$ (the "center" of our "circle") and $\rho \geq 0$ (radius of this circle), if $\mathbf{y}$ is any vector in $\mathscr{E}_n$, we have

$$\|P_{S'}(\mathbf{y}) - \mathbf{c}\|^2 \leq \rho^2$$

if and only if for each one-dimensional subspace $L$ of $S'$,

$$\|P_L(\mathbf{y}) - P_L(\mathbf{c})\|^2 \leq \rho^2.$$

*Proof:* First let $L$ be any given one-dimensional subspace of $S'$. Choose an orthonormal set of vectors

**6.4** $$\mathbf{w}_1, \ldots, \mathbf{w}_n \text{ spanning } \mathscr{E}_n \text{ such that}$$

$$L = [\![\mathbf{w}_1]\!], \qquad S' = [\![\mathbf{w}_1, \ldots, \mathbf{w}_r]\!], \qquad S = [\![\mathbf{w}_1, \ldots, \mathbf{w}_q]\!].$$

Then if

$$\mathbf{y} = \sum_{i=1}^{n} \mathbf{w}_i z_i \quad \text{and} \quad \mathbf{c} = \sum_{i=1}^{q} \mathbf{w}_i b_i,$$

we know from Theorem 2.13 and Theorem 2.4 that

$$P_{S'}(\mathbf{y}) = \sum_{i=1}^{r} \mathbf{w}_i z_i, \qquad P_L(\mathbf{y}) = \mathbf{w}_1 z_1, \quad \text{and} \quad P_L(\mathbf{c}) = \mathbf{w}_1 b_1.$$

Hence

$$\|P_L(\mathbf{y}) - P_L(\mathbf{c})\|^2 = (b_1 - z_1)^2$$

while

$$\|P_{S'}(\mathbf{y}) - \mathbf{c}\|^2 = \sum_{i=1}^{r} (b_i - z_i)^2.$$

(This follows from the relation $\|\mathbf{v}\|^2 = \mathbf{v} \cdot \mathbf{v}$ and the orthonormality of $\mathbf{w}_1, \ldots, \mathbf{w}_n$.) Hence,

$$\|P_L(\mathbf{y}) - P_L(\mathbf{c})\|^2 \leq \|P_{S'}(\mathbf{y}) - \mathbf{c}\|^2.$$

Thus, if $\|P_{S'}(\mathbf{y}) - \mathbf{c}\|^2 \leq \rho^2$, then surely $\|P_L(\mathbf{y}) - P_L(\mathbf{c})\|^2 \leq \rho^2$ for each one-dimensional subspace $L$ of $S'$.

We now show that

**6.5** $$\begin{cases} \|P_L(\mathbf{y}) - P_L(\mathbf{c})\|^2 \leq \rho^2 \text{ for each one-dimensional subspace } L \text{ of } S' \text{ implies} \\ \|P_{S'}(\mathbf{y}) - \mathbf{c}\|^2 \leq \rho^2. \end{cases}$$

Our analytic proof is suggested by the sketch.

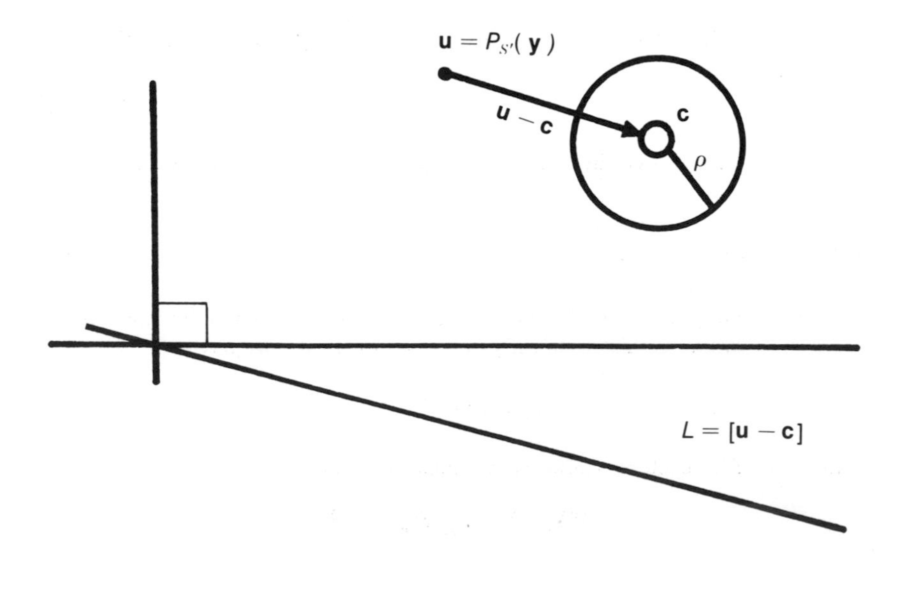

To establish the implication in 6.5, we shall show that there is a one-dimensional subspace $L$ of $S'$ such that $\|P_L(\mathbf{y}) - P_L(\mathbf{c})\|^2 = \|P_{S'}(\mathbf{y}) - \mathbf{c}\|^2$. From the sketch, we expect that $L = [\![\mathbf{u} - \mathbf{c}]\!]$ is such a subspace, where $\mathbf{u} = P_{S'}(\mathbf{y})$.

To see this analytically, we first note that $P_L(\mathbf{y}) = P_L(\mathbf{u})$ where $\mathbf{u} = P_{S'}(\mathbf{y})$. This can be seen by using the orthonormal system of 6.4 (or Corollary 2.17). We have $P_{S'}(\mathbf{y}) - \mathbf{c} = \mathbf{u} - \mathbf{c}$. Letting $L = [\![\mathbf{u} - \mathbf{c}]\!]$, we see that

$$
\begin{aligned}
P_L(\mathbf{y}) - P_L(\mathbf{c}) &= P_L(\mathbf{u}) - P_L(\mathbf{c}) && \text{by 2.17} \\
&= P_L(\mathbf{u} - \mathbf{c}) && \text{by 2.15} \\
&= P_{[\![\mathbf{u}-\mathbf{c}]\!]}(\mathbf{u} - \mathbf{c}) = \mathbf{u} - \mathbf{c}.
\end{aligned}
$$

Hence, for $L = [\![\mathbf{u} - \mathbf{c}]\!]$,

$$\|P_{S'}(\mathbf{y}) - \mathbf{c}\|^2 = \|P_L(\mathbf{y}) - P_L(\mathbf{c})\|^2,$$

establishing our desired result.

Let us now identify $\mathbf{c}$ with $P_{S'}(E(\mathbf{Y}))$ in Lemma 6.3. Using this lemma, we may rewrite 6.2 in the form

**6.6**
$$P\left\{\begin{matrix}\text{for all 1-dimensional}\\ \text{subspaces } L \text{ of } S'\end{matrix} \quad \|P_L(\mathbf{Y}) - P_L(E(\mathbf{Y}))\|^2 \leq \frac{r}{n-q} f_{r,\,n-q}(1-\beta)\, \|\mathbf{Y} - P_S(\mathbf{Y})\|^2\right\} = 1 - \beta.$$

Here $S'$ is an arbitrary subspace of the subspace $S$ in which $E(\mathbf{Y})$ is assumed to lie. In order to arrive at our goal, we must understand the meaning of

$$\|P_L(\mathbf{y}) - P_L(E(\mathbf{Y}))\|^2 \leq \rho^2.$$

Let us write $L = [\![\ell]\!]$ where $\|\ell\| = 1$. Then we know from 2.4 that

$$P_L(\mathbf{y}) = (\mathbf{y} \cdot \ell)\ell$$

and

$$P_L(E(\mathbf{Y})) = (E(\mathbf{Y}) \cdot \ell)\ell = \left[\left(\sum_{i=1}^{m} \beta_i \mathbf{x}_i\right) \cdot \ell\right]\ell = \ell \sum_{i=1}^{m} \beta_i (\mathbf{x}_i \cdot \ell).$$

Hence, using $\|\ell\| = 1$, we have

$$\|P_L(\mathbf{y}) - P_L(E(\mathbf{Y}))\|^2 = \left\|\left[\mathbf{y} \cdot \ell - \sum_{i=1}^{m} \beta_i (\mathbf{x}_i \cdot \ell)\right]\ell\right\|^2 = \left[\mathbf{y} \cdot \ell - \sum_{i=1}^{m} (\mathbf{x}_i \cdot \ell)\beta_i\right]^2.$$

Hence, for $L = [\![\ell]\!]$ with $\|\ell\| = 1$,

**6.7**
$$\begin{cases} \|P_L(\mathbf{y}) - P_L(E(\mathbf{Y}))\|^2 \leq \rho^2 \quad \text{if and only if} \\ \mathbf{y} \cdot \ell - \rho \leq \displaystyle\sum_{i=1}^{m} (\mathbf{x}_i \cdot \ell)\beta_i \leq \mathbf{y} \cdot \ell + \rho. \end{cases}$$

Combining 6.6 and 6.7, we see the following theorem.

### *6.8 Theorem: Simultaneous Confidence Intervals*

Let $S'$ be an arbitrary subspace of the subspace $S$ of $n$-dimensional space. Suppose that $\dim S' = r \leq q = \dim S$. Let

$$\rho = \sqrt{\frac{r}{n-q} f_{r,\,n-q}(1-\beta)}\ \|\mathbf{Y} - P_S(\mathbf{Y})\|,$$

where $f_{r,\,n-q}(1-\beta)$ is the upper $1-\beta$ point of the $F$ distribution with $(r, n-q)$ degrees of freedom. Then, under the general linear model, 1.5,

$$P\left\{\begin{matrix}\text{for all unit}\\ \text{vectors } \ell \in S'\end{matrix}\ \mathbf{Y}\cdot\ell - \rho \leq \sum_{i=1}^{m} (\mathbf{x}_i \cdot \ell)\beta_i \leq \mathbf{Y}\cdot\ell + \rho\right\} = 1-\beta.$$

If we let $\mathscr{L}$ be any positive multiple of $\ell$, then we may write this in the form

$$P\left\{\begin{matrix}\text{for all}\\ \mathscr{L} \in S'\end{matrix}\ \mathbf{Y}\cdot\mathscr{L} - \rho\,\|\mathscr{L}\| \leq \sum_{i=1}^{m} (\mathbf{x}_i \cdot \mathscr{L})\beta_i \leq \mathbf{Y}\cdot\mathscr{L} + \rho\,\|\mathscr{L}\|\right\} = 1-\beta.$$

This states that the probability of *simultaneous* coverage of the linear combinations $\sum_{i=1}^{m} (\mathbf{x}_i \cdot \mathscr{L})\beta_i$ by the respective random intervals $[\mathbf{Y}\cdot\mathscr{L} - \rho\|\mathscr{L}\|, \mathbf{Y}\cdot\mathscr{L} + \rho\|\mathscr{L}\|]$ is $1-\beta$.

The question that we must answer to conclude this section is, "Which linear combinations $\sum_{i=1}^{m} A_i\beta_i$ can be estimated in the manner just given?"

It is easy to see that if $A_1, \ldots, A_m$ are given, then we can so estimate $\sum_{i=1}^{m} A_i\beta_i$ if and only if we can find a vector $\mathscr{L} \in S$ such that

**6.9**
$$\mathbf{x}_i \cdot \mathscr{L} = A_i$$

for each $i = 1, \ldots, m$.

Expanding 6.9, we have

$$x_{1i}\mathscr{L}_1 + x_{2i}\mathscr{L}_2 + \cdots + x_{ni}\mathscr{L}_n = A_i$$

for each $i = 1, \ldots, m$, or

**6.10**
$$\begin{bmatrix} x_{11} \\ \cdot \\ \cdot \\ \cdot \\ x_{1m} \end{bmatrix} \mathscr{L}_1 + \cdots + \begin{bmatrix} x_{n1} \\ \cdot \\ \cdot \\ \cdot \\ x_{nm} \end{bmatrix} \mathscr{L}_n = \begin{bmatrix} A_1 \\ \cdot \\ \cdot \\ \cdot \\ A_m \end{bmatrix}.$$

But the vector

$$\begin{bmatrix} x_{i1} \\ \cdot \\ \cdot \\ \cdot \\ x_{im} \end{bmatrix}$$

is just the transpose of the $i$th row vector of the matrix $X$. Hence, the question of whether $\sum_{i=1}^{m} A_i\beta_i$ can be estimated reduces to the question of whether there is an $\mathscr{L} \in S$ such that 6.10 holds. Now surely, 6.10 alone has a solution, say $\mathscr{L}^*$, when $\mathbf{A} \in X_{\text{row}}$. If we write

$$\mathscr{L}^* = [\mathscr{L}^* - P_S(\mathscr{L}^*)] + P_S(\mathscr{L}^*)$$

and note that $\mathscr{L}^* - P_S(\mathscr{L}^*) \perp S$, we see that $P_S(\mathscr{L}^*)$ satisfies Equation 6.9. Hence $P_S(\mathscr{L}^*)$ satisfies Equation 6.10. Since $P_S(\mathscr{L}^*) \in S$, this is the $\mathscr{L}$ we are seeking. In fact, $\mathscr{L}$ is unique, since if $\mathscr{L}_1$ and $\mathscr{L}_2$ are in $S$ and both satisfy 6.9 (or 6.10), then $\mathscr{L}_1 - \mathscr{L}_2$ satisfies

$$\mathbf{x}_i \cdot (\mathscr{L}_1 - \mathscr{L}_2) = 0 \quad \text{for all} \quad i.$$

Hence, $\mathscr{L}_1 - \mathscr{L}_2 \in S^{\perp}$. But $\mathscr{L}_1 - \mathscr{L}_2 \in S$, and this is only possible if $\mathscr{L}_1 - \mathscr{L}_2 = \mathbf{0}_n$, since

$$\sum_{i=1}^{q} b_i\mathbf{w}_i = \sum_{i=q+1}^{n} b_i\mathbf{w}_i$$

for orthogonal $\mathbf{w}_1, \ldots, \mathbf{w}_n$ implies $b_i = 0$. Hence, we have proved the following theorem.

### 6.11 *Theorem: Estimability of* $\sum_{i=1}^{m} + A_i\boldsymbol{\beta}_i$

The linear combination $\sum_{i=1}^{m} A_i\beta_i$ may be estimated as in 6.8 if and only if

$$\mathbf{A} = \begin{bmatrix} A_1 \\ \cdot \\ \cdot \\ \cdot \\ A_m \end{bmatrix} \subset X_{\text{row}}$$

The desired estimate is unique. It may be easiest to determine whether $\mathbf{A}$ satisfies this condition by finding an orthonormal set $\mathbf{D}_1, \ldots, \mathbf{D}_\gamma, \mathbf{D}_{\gamma+1}, \ldots, \mathbf{D}_m$, as in 4.14, where $X_{\text{row}} = [\mathbf{D}_1, \ldots, \mathbf{D}_\gamma]$. Then

$$\mathbf{A} \in X_{\text{row}} \quad \text{if and only if} \quad \mathbf{A} \cdot \mathbf{D}_{\gamma+j} = 0 \quad \text{for} \quad j = 1, \ldots, m - \gamma.$$

Often it will be the case that $\gamma = m$. Then $\mathbf{A} \in X_{\text{row}}$ is surely satisfied. To actually find $\mathscr{L}$ when $\mathbf{A} \in X_{\text{row}}$, write $\mathscr{L}$ in the form

$$\mathscr{L} = \sum_{i=1}^{m} \eta_i\mathbf{x}_i$$

and substitute this into 6.10 to determine $\eta_i$.

To find the subspace $S'$ needed to estimate, say, $L_j(\beta) = \sum_{i=1}^{m} A_{ji}\beta_i$, $j = 1, \ldots, \omega$, check each $L_j$ for estimability and (assuming this is satisfied) find the corresponding $\mathscr{L}_j$'s. Then $S' = [\![\mathscr{L}_1, \ldots, \mathscr{L}_\omega]\!]$.

For example, if you want simultaneous intervals for all linear combinations of $\beta_1$ and $\beta_2$, you let

$$L_1(\boldsymbol{\beta}) = \beta_1, \quad \text{and find } \mathscr{L}_1,$$

$$L_2(\boldsymbol{\beta}) = \beta_2, \quad \text{and find } \mathscr{L}_2.$$

We now present a few simple illustrative examples.

### *6.12 Example: Two Additives*

Let us once again return to the model of 4.3,

$$\mathbf{Y} = \begin{bmatrix} 1 \\ 2 \\ 1 \\ 2 \end{bmatrix}\beta_1 + \begin{bmatrix} 1 \\ 1 \\ 2 \\ 2 \end{bmatrix}\beta_2 + \boldsymbol{\varepsilon} = \mathbf{x}_1\beta_1 + \mathbf{x}_2\beta_2 + \boldsymbol{\varepsilon},$$

with equal amounts of additives one and two. We may then want to estimate $\beta_1 + \beta_2$.

It is seen by inspection that $\mathbf{x}_1$ and $\mathbf{x}_2$ are linearly independent. Hence, $X_{\text{row}}^{\perp} = \{\mathbf{0}_2\}$, and therefore we can surely estimate $\beta_1 + \beta_2$ as in Theorem 6.8. If we also desire to estimate other really different linear combinations of the $\beta_i$, we would choose $S' = S = [\![\mathbf{x}_1, \mathbf{x}_2]\!]$. Here we shall assume that our only interest is in estimating $\beta_1 + \beta_2$.

We can obtain shorter confidence intervals than for $S' = S$ by choosing $S' = [\![\mathscr{L}]\!]$ where $\mathscr{L} \in S$ is chosen to satisfy Equation 6.9 (or 6.10). Since $\mathscr{L} \in S$, we write

$$\mathscr{L} = t\begin{bmatrix} 1 \\ 2 \\ 1 \\ 2 \end{bmatrix} + s\begin{bmatrix} 1 \\ 1 \\ 2 \\ 2 \end{bmatrix} = \begin{bmatrix} t + s \\ 2t + s \\ t + 2s \\ 2t + 2s \end{bmatrix}.$$

Here $\mathbf{A} = \begin{bmatrix} A_1 \\ A_2 \end{bmatrix} = \begin{bmatrix} 1 \\ 1 \end{bmatrix}$. Hence, Equation 6.10 becomes

$$1 \cdot (t + s) + 2(2t + s) + 1 \cdot (t + 2s) + 2(2t + 2s) = 1 = A_1$$

$$1 \cdot (t + s) + 1 \cdot (2t + s) + 2(t + 2s) + 2(2t + 2s) = 1 = A_2$$

or

$$10t + 9s = 1, \qquad 9t + 10s = 1,$$

yielding

$$t = \tfrac{1}{19} = s.$$

Hence

$$\mathscr{L} = \tfrac{1}{19}\begin{bmatrix} 2 \\ 3 \\ 3 \\ 4 \end{bmatrix}.$$

Thus, our confidence interval for $\beta_1 + \beta_2$ is, by Theorem 6.8,

$$[\mathbf{Y} \cdot \mathscr{L} - \rho \|\mathscr{L}\|, \mathbf{Y} \cdot \mathscr{L} + \rho \|\mathscr{L}\|]$$
$$= [\tfrac{1}{19}(2Y_1 + 3Y_2 + 3Y_3 + 4Y_4) - \rho\sqrt{2/19},$$
$$\tfrac{1}{19}(2Y_1 + 3Y_2 + 3Y_3 + 4Y_4) + \rho\sqrt{2/19}].$$

Here

$$\rho = \sqrt{\tfrac{1}{2}f_{1,2}(1-\beta)}\, \|\mathbf{Y} - P_S(\mathbf{Y})\|,$$

where $f_{1,2}(1 - \beta)$ is the upper $1 - \beta$ point of the $F$ distribution with (1, 2) degrees of freedom. Now, to find $P_S(\mathbf{Y})$, we form an orthonormal system from $\mathbf{x}_1$, $\mathbf{x}_2$. Let

$$\mathbf{w}_1 = \frac{\mathbf{x}_1}{\|\mathbf{x}_1\|} = \frac{1}{\sqrt{10}}\begin{bmatrix} 1 \\ 2 \\ 1 \\ 2 \end{bmatrix}$$

$$\mathbf{w}_2' = \mathbf{x}_2 - P_{[\mathbf{w}_1]}(\mathbf{x}_2) = \begin{bmatrix} 1 \\ 1 \\ 2 \\ 2 \end{bmatrix} - (\mathbf{x}_2 \cdot \mathbf{w}_1)\mathbf{w}_1$$

$$= \begin{bmatrix} 1 \\ 1 \\ 2 \\ 2 \end{bmatrix} - \frac{1}{\sqrt{10}}\, 9\, \frac{1}{\sqrt{10}}\begin{bmatrix} 1 \\ 2 \\ 1 \\ 2 \end{bmatrix}$$

$$= \begin{bmatrix} 1 \\ 1 \\ 2 \\ 2 \end{bmatrix} - \begin{bmatrix} .9 \\ 1.8 \\ .9 \\ 1.8 \end{bmatrix} = \begin{bmatrix} .1 \\ -.8 \\ 1.1 \\ .2 \end{bmatrix}$$

$$\mathbf{w}_2 = \frac{\mathbf{w}_2'}{\|\mathbf{w}_2'\|} = \frac{1}{\sqrt{190}}\begin{bmatrix} 1 \\ -8 \\ 11 \\ 2 \end{bmatrix}.$$

Thus

$$P_S(\mathbf{Y}) = (\mathbf{Y} \cdot \mathbf{w}_1)\mathbf{w}_1 + (\mathbf{Y} \cdot \mathbf{w}_2)\mathbf{w}_2,$$

that is,

$$P_S(\mathbf{Y}) = \tfrac{1}{10}(Y_1 + 2Y_2 + Y_3 + 2Y_4)\begin{bmatrix}1\\2\\1\\2\end{bmatrix} + \tfrac{1}{190}(Y_1 - 8Y_2 + 11Y_3 + 2Y_4)\begin{bmatrix}1\\1\\2\\2\end{bmatrix}.$$

It is now easy to determine $\|\mathbf{Y} - P_S(\mathbf{Y})\|$, and hence $\rho$. △

### *6.13 Example: Estimation in the Model of 4.23*

Recall that in 4.23 we had

$$\mathbf{x}_1 = \begin{bmatrix}0\\1\\1\\1\end{bmatrix}, \qquad \mathbf{x}_2 = \begin{bmatrix}1\\0\\1\\1\end{bmatrix},$$

and $\mathbf{x}_3 = \mathbf{x}_1 + \mathbf{x}_2$. Recall from 6.11 that $\sum_{i=1}^{m} A_i\beta_i$ can be estimated only if the dot products $\mathbf{A} \cdot \mathbf{D}_{\gamma+j} = 0$ for $j = 1, \ldots, m - \gamma$. We saw in 4.23 that this meant $\sum A_i\beta_i$ could be estimated if and only if

$$\mathbf{A} \cdot \mathbf{D}_3 = \mathbf{A} \cdot \begin{bmatrix}1/3\\1/3\\-1/3\end{bmatrix} = 0.$$

(That is, we can estimate any linear combinations that we can make tests on.) Let us first find a $1 - \beta$ level confidence interval for $\beta_1 - \beta_2$. We must find $\mathscr{L} \in S$ such that 6.10 is satisfied—here $A_1 = 1$, $A_2 = -1$, and $A_3 = 0$. So 6.10 becomes

**6.14**
$$\begin{bmatrix}0\\1\\1\end{bmatrix}\mathscr{L}_1 + \begin{bmatrix}1\\0\\1\end{bmatrix}\mathscr{L}_2 + \begin{bmatrix}1\\1\\2\end{bmatrix}\mathscr{L}_3 + \begin{bmatrix}1\\1\\2\end{bmatrix}\mathscr{L}_4 = \begin{bmatrix}1\\-1\\0\end{bmatrix}.$$

Since we must have $\mathscr{L} \in S$, we write

$$\mathscr{L} = t\begin{bmatrix}0\\1\\1\\1\end{bmatrix} + s\begin{bmatrix}1\\0\\1\\1\end{bmatrix} = \begin{bmatrix}s\\t\\t+s\\t+s\end{bmatrix} = \begin{bmatrix}\mathscr{L}_1\\\mathscr{L}_2\\\mathscr{L}_3\\\mathscr{L}_4\end{bmatrix} \quad \text{(no need to worry about } \mathbf{x}_3\text{)}.$$

Then 6.14 becomes

$$\begin{bmatrix}0\\1\\1\end{bmatrix}s + \begin{bmatrix}1\\0\\1\end{bmatrix}t + \begin{bmatrix}1\\1\\2\end{bmatrix}(t+s) + \begin{bmatrix}1\\1\\2\end{bmatrix}(t+s) = \begin{bmatrix}1\\-1\\0\end{bmatrix},$$

or

$$3t + 2s = 1$$

$$2t + 3s = -1$$

$$5t + 5s = 0,$$

or $t = 1$, $s = -1$. Hence

$$\mathscr{L} = \begin{bmatrix}-1\\1\\0\\0\end{bmatrix}.$$

Thus,

$$S' = \left[\!\left[\begin{pmatrix}-1\\1\\0\\0\end{pmatrix}\right]\!\right] = \|\mathbf{x}_1 - \mathbf{x}_2\|,$$

$r = \dim S' = 1$, $n = 4$, $q = 2$.

We can check that $\mathbf{x}_1 \cdot \mathscr{L} = 1$, $\mathbf{x}_2 \cdot \mathscr{L} = -1$, and $\mathbf{x}_3 \cdot \mathscr{L} = 0$. The desired $1 - \beta$ level confidence interval is

$$[\mathbf{Y} \cdot \mathscr{L} - \rho\|\mathscr{L}\|, \mathbf{Y} \cdot \mathscr{L} + \rho\|\mathscr{L}\|]$$
$$= [-Y_1 + Y_2 - \sqrt{2}\rho, -Y_1 + Y_2 + \sqrt{2}\rho]$$

where $\rho$ is computed in the answer to Exercise 6.15 (4).

If we had desired simultaneous $1 - \beta$ level confidence intervals for $\beta_1 - \beta_2$ and $\beta_1 + \beta_3$, we would have had to find, in addition to the $\mathscr{L}$ already found, an $\mathscr{L}' \in S$ such that

$$\begin{bmatrix}0\\1\\1\end{bmatrix}\mathscr{L}_1' + \begin{bmatrix}1\\0\\1\end{bmatrix}\mathscr{L}_2' + \begin{bmatrix}1\\1\\2\end{bmatrix}\mathscr{L}_3' + \begin{bmatrix}1\\1\\2\end{bmatrix}\mathscr{L}_4' = \begin{bmatrix}1\\0\\1\end{bmatrix}.$$

Since it would be independent of $\mathscr{L}$, we would have had to choose $S' = S$. Then $r = 2 = q$, $n = 4$. △

## 6.15 EXERCISES

1. In the model of 6.12, find a $1 - \beta$ level confidence interval for $\beta_1 - \beta_2$.

2. In the model of 6.12, find simultaneous $1 - \beta$ level confidence intervals for $\beta_1$ and $\beta_2$.

3.° In the model for 6.13, is it possible to find a confidence interval of level $1 - \beta$ for $\beta_1 + \beta_2 + \beta_3$?

4.° In the model used in 6.13, find a $1 - \beta$ level confidence interval for the linear combination $\beta_1 + \beta_2 + 2\beta_3$.

5.° Suppose that we have the following model:

$$\mathbf{x}_1 = \begin{bmatrix} 1 \\ 1 \\ 1 \\ 0 \end{bmatrix}, \quad \mathbf{x}_2 = \begin{bmatrix} 1 \\ 1 \\ 0 \\ 1 \end{bmatrix}, \quad \mathbf{x}_3 = \begin{bmatrix} 1 \\ 0 \\ 1 \\ 1 \end{bmatrix}.$$

Find $1 - \beta$ level simultaneous confidence intervals for $\beta_1 + \beta_2$ and $\beta_1 - \beta_3$.

*Some Final Comments on the Subject of Linear Models.* If we knew $\sigma^2$, we could replace the denominator

$$\frac{1}{n - q} \|\mathbf{Y} - P_S(\mathbf{Y})\|^2$$

by $\sigma^2$. We can easily use $\|\mathbf{Y} - P_S(\mathbf{Y})\|^2$ to obtain confidence limits for $\sigma^2$, since $\|\mathbf{Y} - P_S(\mathbf{Y})\|^2/\sigma^2$ has a Chi square distribution with $n - q$ degrees of freedom.

If we want confidence intervals of *fixed length* for the linear combinations $\sum_{i=1}^{m} A_i\beta_i$, we might use a two-stage scheme, using the first stage to obtain a high probability upper confidence limit for $\sigma^2$. We would then let the second stage sample size be determined so as to yield the desired result if $\sigma^2 = \sigma_u^2$, the upper limit.

The topics we have treated here constitute only a small part of the subject of general linear models; we have omitted efficient computational methods and other models, such as the case in which the elements of $\mathbf{Y}$ have some known type of statistical dependence, or the case in which the parameters $\beta_1, \ldots, \beta_m$ may be the values of random variables whose statistical properties we want to determine. Such a situation might arise if, say, $\beta_i$ is the actual effect of a given lot of an additive of type $i$, but we want to infer something about this type of additive rather than about the particular lot being tested.

# Appendix

## A.1 Introduction

The main purpose of this Appendix is to provide a convenient reference for the various theorems that we use from calculus. We hope to provide the reader who is unfamiliar with these results some insight into the reasons for expecting these results and maybe some intuition that could lead to proofs. We usually provide one or more worked out examples to help illustrate the theory.

## A.2 The Fundamental Theorem of Calculus

Suppose that $f$ is a continuous real valued function whose domain consists of the real line. Then for each real $x$, the integral $G(x) = \int_a^x f(t)\,dt$ exists, and $G$ is differentiable with $G'(x) = f(x)$.

The intuitive idea behind the last result is the following: $G(x+h) - G(x)$ is geometrically the area indicated in Figure A.1. But we see that

$$G(x+h) - G(x) = f(t)h \quad \text{where} \quad x \leq t \leq x + h,$$

and hence

$$\frac{G(x+h) - G(x)}{h} = f(t) \simeq f(x)$$

for $h$ small, by the continuity of $f$. Thus, letting $h \to 0$ yields

$$G'(x) = f(x).$$

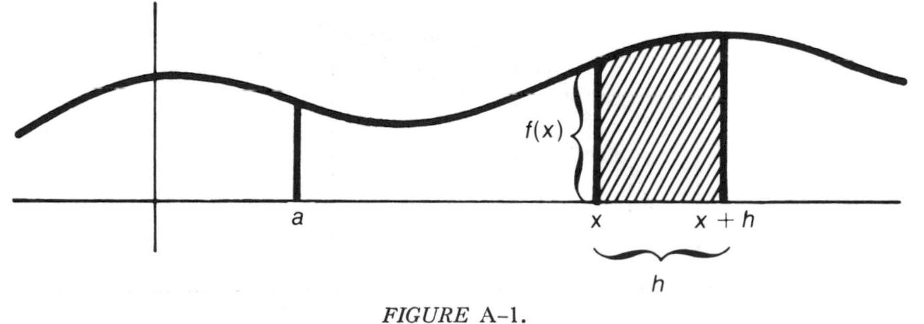


*FIGURE* A-1.

***2.1 Example:***

$$F(x) = \int_0^{x^2} \frac{1}{t^2 + t + 1}\, dt.$$

We want to find $F'(x)$. Let

$$f(t) = \frac{1}{t^2 + t + 1}.$$

Then for $G(x) = \int_0^x f(t)\, dt$, we see that $F(x) = G(x^2)$. Hence, by the chain rule, $F'(x) = G'(x^2) \cdot 2x$. But

$$G'(y) = f(y) = \frac{1}{y^2 + y + 1}$$

by the fundamental theorem. Therefore,

$$G'(x^2) = \frac{1}{x^4 + x^2 + 1},$$

and hence

$$F'(x) = \frac{1}{x^4 + x^2 + 1} \cdot 2x. \qquad \triangle$$

## A.3 The Mean Value Theorem

If $f$ is a real valued function with a derivative at every point of the closed bounded interval $[a, b]$, then there is a point $c$ with $a < c < b$ such that

$$\frac{f(b) - f(a)}{b - a} = f'(c).$$

Geometrically, this asserts the existence of a tangent line with the same slope as the segment connecting the point $(a, f(a))$ with the point $(b, f(b))$, as shown in Figure A.2, and this result is very easy to accept. It is often convenient to replace $f(b) - f(a)$

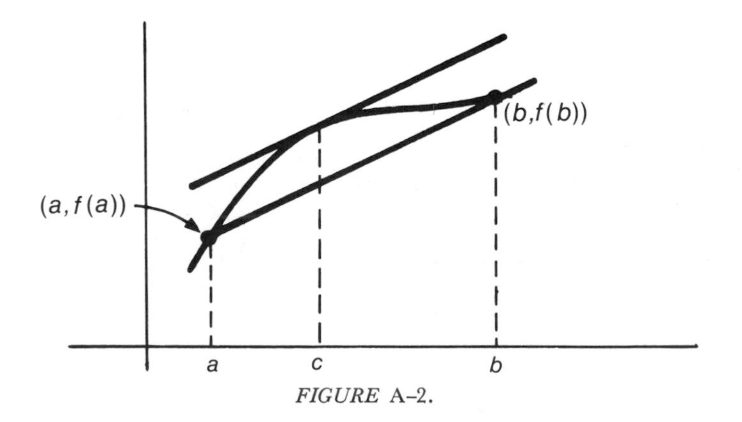


*FIGURE* A-2.

by $(b - a)f'(c)$ in computational work, although the location of $c$ may be unknown; this can prove useful if upper and lower bounds for $f'(c)$ can be found.

### 3.1 *Example: log 4/3*

Let $f(x) = \log x$. Then

$$\frac{\log 4 - \log 3}{4 - 3} = \frac{1}{c},$$

with $3 \le c \le 4$, by the mean value theorem. But note that

$$\frac{\log 4 - \log 3}{4 - 3} = \log \frac{4}{3},$$

and that

$$\frac{1}{4} \le \frac{1}{c} \le \frac{1}{3}.$$

Therefore,

$$1/4 \le \log 4/3 \le 1/3. \qquad \triangle$$

## A.4 L'Hospital's Rule

A result that can be derived from the mean value theorem and that is often of great practical use in computation is L'Hospital's rule, stated in the following theorem.

### 4.1 *Theorem*

Suppose that $f$ and $g$ are functions with $\lim_{x \to a} f(x) = 0 = \lim_{x \to a} g(x)$. If

$$\lim_{x \to a} \frac{f'(x)}{g'(x)}$$

exists, then so does

$$\lim_{x \to a} \frac{f(x)}{g(x)}, \qquad \text{and} \qquad \lim_{x \to a} \frac{f(x)}{g(x)} = \lim_{x \to a} \frac{f'(x)}{g'(x)}.$$

For $f$ and $g$ continuous at $a$, this result is believable because then

$$\frac{f(x)}{g(x)} = \frac{f(x) - f(a)}{g(x) - g(a)} = \frac{(x - a)f'(c)}{(x - a)g'(d)} = \frac{f'(c)}{g'(d)}$$

by the mean value theorem, where $c$ and $d$ lie between $a$ and $x$. If $g'$ is continuous and $g'(a)$ is not 0, then the desired result follows by letting $x$ tend to $a$. (Other cases require more effort to prove.)

***4.2 Example***

$$\lim_{x\to 0} \frac{e^x - 1 - x}{x^2}$$

Here $f(x) = e^x - 1 - x$, $g(x) = x^2$. Since $\lim_{x\to 0} f(x) = 0 = \lim_{x\to 0} g(x)$, we look at

$$\frac{f''(x)}{g''(x)} = \frac{e^x - 1}{2x}.$$

Here again, $\lim_{x\to 0} f'(x) = 0 = \lim_{x\to 0} g'(x)$, so we look at

$$\frac{f''(x)}{g''(x)} = e^x/2.$$

Since

$$\lim_{x\to 0} \frac{f''(x)}{g''(x)} = \tfrac{1}{2},$$

we know that

$$\lim_{x\to 0} \frac{f'(x)}{g'(x)} = \tfrac{1}{2},$$

and thus

$$\lim_{x\to 0} \frac{f(x)}{g(x)} = \tfrac{1}{2}.$$

That is,

$$\lim_{x\to 0} \frac{e^x - 1 - x}{x^2} = \tfrac{1}{2}. \qquad \triangle$$

## A.5 Taylor's Theorem

Taylor's theorem permits accurate approximation of a function $f$ by means of a polynomial $P$ that has the same value and first $n$ derivatives as $f$ does at some point $a$, and whose $(n + 1)$st derivative is not too large over the interval being considered. The condition that $|f^{(n+1)}|$ is not too large guarantees that $f$ behaves like an $n$th degree polynomial. (If $|f^{(n+1)}| = 0$, then $f$ would be an $n$th degree polynomial.) The precise result is the following theorem.

***5.1 Theorem: Taylor Approximation***

Let $f$ be a real valued function with $n + 1$ derivatives in the real interval $[a, a + x]$, with $f^{(n+1)}$ continuous. Then

$$f(a + x) + f(a) + f'(a)x + f''(a)\frac{x^2}{2!} + \cdots + f^{(n)}(a)\frac{x^n}{n!} + f^{(n+1)}(c)\frac{x^{n+1}}{(n+1)!},$$

where $c$ lies between $a$ and $a + x$. The polynomial

$$P(x) = f(a) + f'(a)x + f''(a)\frac{x^2}{2!} + \cdots + f^{(n)}(a)\frac{x^n}{n!}$$

is called the *Taylor Polynomial* of degree $n$ about $a$ of the function $f$. Note that $P^{(j)}(0) = f^{(j)}(a)$ for $j = 0, \ldots, n$. Thus $P(x)$ and $f(a + x)$ have the same value, same slope, same rate of change of slope, and so forth at $x = 0$.

### 5.2 *Example:* Sin $x$

Let $a = 0$. Then

$$\sin x = \sin 0 + (\sin' 0)x + (\sin'' 0)\frac{x^2}{2!} + (\sin''' c)\frac{x^3}{3!}$$

$$= x - (\cos c)\frac{x^3}{6}.$$

If we use the approximation $\sin x \cong x$, the preceding result shows that we are in error by at most $|x^3/6|$, since $|\cos c| \leq 1$. △

## A.6 Improper Integrals

If for each $a \geq \alpha$, $f$ is Riemann integrable on $[\alpha, a]$ and $\lim_{a\to\infty} \int_\alpha^a f(x)\,dx$ exists, then this limit is denoted by $\int_\alpha^\infty f(x)\,dx$.

Similarly, $\lim_{b\to\infty} \int_{-b}^\beta f(x)\,dx$ is denoted by $\int_{-\infty}^\beta f(x)\,dx$. If both $\int_{-\infty}^0 f(x)\,dx$ and $\int_0^\infty f(x)\,dx$ exist, then $\int_{-\infty}^0 f(x)\,dx + \int_0^\infty f(x)\,dx$ is denoted by $\int_{-\infty}^\infty f(x)\,dx$. If $f$ is unbounded on $[a, b]$ and is Riemann integrable on $[a + \varepsilon, b]$ for each sufficiently small positive $\varepsilon$ (hence, $f$ is bounded on $[a + \varepsilon, b]$), and if

$$\lim_{\substack{\varepsilon\to 0\\ \varepsilon>0}} \int_{a+\varepsilon}^b f(x)\,dx$$

exists, then this limit is denoted by $\int_a^b f(x)\,dx$.

The same notation is used for

$$\lim_{\substack{\varepsilon\to 0\\ \varepsilon>0}} \int_a^{b-\varepsilon} f(x)\,dx.$$

All of the symbols so defined are called improper integrals.

### *6.1 Example:*

$$\int_0^1 \frac{dx}{\sqrt{x}}, \int_0^\infty e^{-x}\, dx, \int_{-\infty}^\infty e^{-|x|}\, dx$$

Since

$$\int_\varepsilon^1 \frac{dx}{\sqrt{x}} = 2\sqrt{x}\Big|_\varepsilon^{x=1} = 2 - 2\sqrt{\varepsilon},$$

we see that $\int_0^1 dx/\sqrt{x} = 2$. Since $\int_0^a e^{-x}\, dx = -e^{-x}\Big|_0^{x=a} = -e^{-a} + 1$, we see that $\int_0^\infty e^{-x}\, dx = 1$. Finally, since $\int_{-b}^0 e^{-|x|}\, dx = 1 - e^{-|-b|}$ and $\int_0^a e^{-|x|}\, dx = -e^{-|a|} + 1$, we see that $\int_{-\infty}^\infty e^{-|x|}\, dx = 2$. △

## A.7 Infinite Series and Convergence Tests

Recall that we say that

$$\lim_{n\to\infty} \sum_{k=k_0}^{n} a_k = L,$$

provided that for large $n$, the sum $\sum_{k=k_0}^{n} a_k$ is close to $L$. If $\lim_{n\to\infty} \sum_{k=k_0}^{n} a_k$ exists, we say that the series $\sum_{k=k_0}^{\infty} a_k$ converges, and also denote its limit, $\lim_{n\to\infty} \sum_{k=k_0}^{n} a_k$, by $\sum_{k=k_0}^{\infty} a_k$.

### *7.1 Example: Geometric Series*

By cross multiplication, we can verify that for $x \neq 1$,

$$\sum_{k=0}^{n} x^k = \frac{1 - x^{n+1}}{1 - x}.$$

Note that if $|x| < 1$, then $\lim_{n\to\infty} x^{n+1} = 0$. Thus, for $|x| < 1$, we have

$$\sum_{k=0}^{\infty} x^k = \frac{1}{1 - x}.$$ △

There are several convenient criteria for determining whether or not a series converges.

### *7.2 Theorem: Comparison Test*

If $|a_i| \leq b_i$ for all $i$ sufficiently large, and $\sum_{k=k}^{\infty} b_k$ converges, then so does $\sum_{k=k_0}^{\infty} a_k$.

### 7.3 *Example:*

$$\sum_{k=1}^{\infty} \frac{1}{k \cdot 2^k}$$

We know that

$$\sum_{k=1}^{\infty} \frac{1}{2^k}$$

converges (essentially Example 7.1 with $x = 1/2$), and note that

$$\left|\frac{1}{k2^k}\right| \leq \frac{1}{2^k}. \qquad \triangle$$

Comparison with a geometric series leads to the following theorem.

### 7.4 *Theorem: Ratio Test*

If the sequence

$$r_1 = \left|\frac{a_2}{a_1}\right|, \qquad r_2 = \left|\frac{a_3}{a}\right|, \ldots, r_k = \left|\frac{a_{k+1}}{a_k}\right|, \ldots$$

has a limit $R < 1$, then $\sum_{k=1}^{\infty} a_k$ converges. If $R > 1$, then $\sum_{k=1}^{\infty} a_k$ does not converge. Here we are essentially comparing $\sum_{k=1}^{n} a_k$ with a multiple of $\sum_{k=0}^{n-1} R^k$.

### 7.5 *Example: Convergence of* $\sum_{k=1}^{\infty} k^2(1/4)^k$

Here

$$r_k = \left|\frac{a_{k+1}}{a_k}\right| = \frac{(k+1)^2(1/4)^{k+1}}{k^2(1/4)^k} = \frac{1}{4}\left(\frac{k+1}{k}\right)^2 = \frac{1}{4}\left(1 + \frac{1}{k}\right)^2 \to \frac{1}{4}.$$

We see that the sequence $r_1, r_2, \ldots$ has limit $1/4$; hence $\sum_{k=1}^{n} k^2(1/4)^k$ has a limit as $n \to \infty$. $\triangle$

### 7.6 *Example:* $e^x$

The Taylor series

$$\sum_{k=0}^{\infty} \frac{x^k}{k!}$$

for $e^x$ converges for each $x$, since

$$r_k = \frac{\left|\dfrac{x^{k+1}}{(k+1)!}\right|}{\left|\dfrac{x^k}{k!}\right|} = \frac{|x|}{k+1}$$

has limit 0 as $k \to \infty$. △

Another useful comparison is the integral comparison test.

### 7.7 *Theorem: Integral Comparison Test*

If $f$ is a positive monotone decreasing function, then $\sum_{k=a}^{\infty} f(k)$ converges if and only if the improper integral $\int_a^\infty f(x)\, dx$ exists. In fact, in the convergent case, $\sum_{k=k_0+1}^{\infty} f(k)$, the error made in approximating $\sum_{k=a}^{\infty} f(k)$ by the finite sum $\sum_{k=a}^{k_0} f(k)$ satisfies the inequality

$$\int_{k_0+1}^{\infty} f(x)\, dx \leq \left|\sum_{k=k_0+1}^{\infty} f(k)\right| \leq \int_{k_0}^{\infty} f(x)\, dx,$$

as can be seen from the picture.

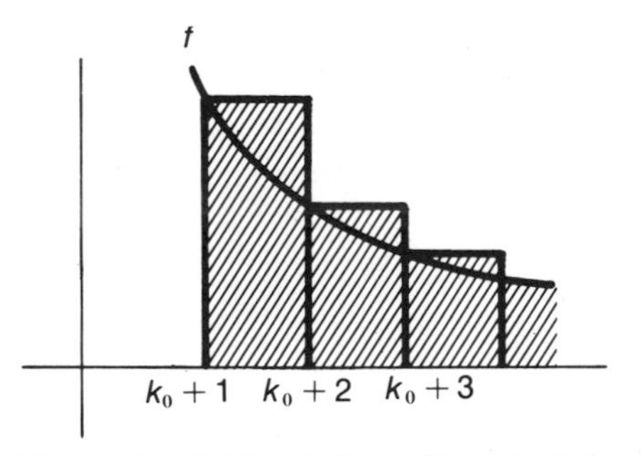


Picture for left-hand inequality—shaded area is the sum.

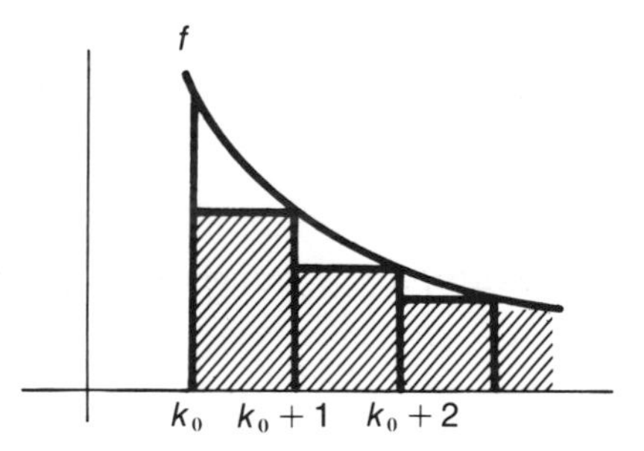


Picture for right-hand inequality—shaded area is the sum.

### 7.8 *Example:* $\sum_{k=1}^{\infty} 1/k^\alpha$

Since

$$\int_1^a \frac{dx}{x^\alpha} = \begin{cases} \left.\dfrac{1}{(1-\alpha)x^{\alpha-1}}\right|_1^{x=a} & \text{for } \alpha \neq 1 \\ \log a & \text{for } \alpha = 1, \end{cases}$$

we see that $\int_1^\infty dx/x^\alpha$ exists only if $\alpha > 1$. Hence $\sum_{k=1}^{\infty} 1/k^\alpha$ converges only for $\alpha > 1$. △

## A.8 Partial Derivatives, Maxima and Minima in Several Variables

Let $F$ be a function of $n$ variables. That is, for each sequence $(x_1, \ldots, x_n)$ of real numbers, $F(x_1, \ldots, x_n)$ is a real number. We say that $F$ has a partial derivative with respect to the $j$th coordinate at $(x_1, \ldots, x_n)$ if the function $G$ given by

$$G(x_j) = F(x_1, \ldots, x_{j-1}, x, x_{+j1j}, \ldots, x_n),$$

with $x_1, \ldots, x_{j-1}, x_{j+1}, \ldots, x_n$ held fixed, is differentiable. The derivative $G'(x_j)$ is called the partial derivative of $F$ with respect to the $j$th coordinate at $(x_1, \ldots, x_n)$ and is denoted by any of the following:

$$F_j(x_1, \ldots, x_n), \qquad D_jF(x_1, \ldots, x_n), \qquad \frac{\partial F}{\partial x_j}(x_1, \ldots, x_n), \quad \text{or} \quad \frac{\partial}{\partial x_j}F(x_1, \ldots, x_n).$$

The partial derivative is obtained by holding all coordinate values except $x_j$ fixed, acting as if $x_1, \ldots, x_{j-1}, x_{j+1}, \ldots, x_n$ are constants, in carrying out the differentiation. For example, if

$$F(x_1, x_2, x_3) = \frac{x_1 \cos x_2}{x_3}$$

for $x_3 \neq 0$, then

$$\frac{\partial F}{\partial x_1}(x_1, x_2, x_3) = \frac{\cos x_2}{x_3}$$

$$\frac{\partial F}{\partial x_2}(x_1, x_2, x_3) = \frac{-x_1 \sin x_2}{x_3}$$

$$\frac{\partial F}{\partial x_3}(x_1, x_2, x_3) = \frac{-x_1 \cos x_2}{x_3^2}.$$

We may take second partial derivatives of the partial derivative as follows: By $F_{ij}$, we mean the partial derivative with respect to the $j$th coordinate of the partial derivative $F_i$. That is,

$$F_{ij} = D_j[D_iF] = \frac{\partial^2 F}{\partial x_j \, \partial x_i},$$

with similar notation for further partial derivatives.

In order to find "the" maximum or minimum of a given function $F$, we proceed along lines analogous to those in the case of functions of one variable. The following is the precise result.

### *8.1 Theorem: Places to Hunt for Maxima and Minima*

The only places at which a function $F$ can have a maximum or minimum are:

(a) places where *all* partial derivatives are 0.

(b) places where some partial derivatives fail to exist.
(c) boundary points of the domain of the function.

In most situations of interest, case (a) will apply. The geometric idea behind case (a) is that at an interior maximum or minimum where the derivatives exist, the slope in every direction must be 0.

This information is of very little use unless we have established that the sought-for maximum or minimum exists. In this regard, the following theorem is of great help.

### 8.2 *Theorem: Existence of Maxima and Minima*

Suppose that $F$ is continuous at each point of the closed, bounded, $n$-dimensional interval $I$ of points $(x_1, \ldots, x_n)$ satisfying $a_i \leq x_i \leq b_i$ for $i = 1, \ldots, n$, where for each $i$, the symbols $a_i$ and $b_i$ denote fixed real numbers. Then there are points $(x_{1m}, \ldots, x_{nm})$ and $(x_{1M}, \ldots, x_{nM})$ in $I$ such that for all $(x_1, \ldots, x_n)$ in $I$,

$$F(x_{1m}, \ldots, x_{nm}) \leq F(x_1, \ldots, x_n) \leq F(x_{1M}, \ldots, x_{nM}).$$

We call $F(x_{1m}, \ldots, x_{nm})$ the *minimum* and $F(x_{1M}, \ldots, x_{nM})$ the *maximum of $F$ on $I$.*

### 8.3 *Example: Minimum of*

$$f(x, y) = x^2 + y^2 - 1.5xy - 2x - 5y + 17$$

We can see by inspection that $f$ has no maximum for real $x, y$, since for $y$ fixed,

$$f(x, y) = x^2\left[1 + \frac{y^2}{x^2} - \frac{1.5y}{x} - \frac{2}{x} - \frac{5y}{x^2} + \frac{17}{x^2}\right],$$

which becomes unboundedly large for large $x$.

To find the minimum of $f$, we set

$$\frac{\partial f}{\partial x} = 0 \qquad \text{and} \qquad \frac{\partial f}{\partial y} = 0,$$

yielding

$$2x - 1.5y - 2 = 0$$
$$2y - 1.5x - 5 = 0$$

or

$$2x - 1.5y = 2$$
$$1.5x - 2y = -5.$$

Solving these equations, we find $x = 64/25$, $y = 52/25$.

If we really want to be sure that $f(64/25, 52/25)$ is the desired minimum, we first note that $f$ is continuous. Also, it is easy to see that

**8.4** $$f(64/25, 52/25) < 9 + 9 + 0 + 0 + 0 + 17 = 35.$$

In addition,

$$f(x, y) = \tfrac{3}{4}(x - y)^2 + \frac{x^2}{4} + \frac{y^2}{4} - 2x - 5y + 17$$

$$= \tfrac{3}{4}(x - y)^2 + \tfrac{1}{4}(x^2 - 8x + 16) + \tfrac{1}{4}(y^2 - 20y + 100) - 12$$

$$= \tfrac{3}{4}(x - y)^2 + \tfrac{1}{4}(x - 4)^2 + \tfrac{1}{4}(y - 10)^2 - 12.$$

So, we see that if $|x| \geq 24$ or $|y| \geq 30$, then $f(x, y) \geq 88$, because $(x - y)^2 \geq 0$, $(x - 4)^2 \geq 0$, $(y - 10)^2 \geq 10$, and we must have at least one of the inequalities $(x - 4)^2 \geq 400$, $(y - 10)^2 \geq 400$. From Theorem 8.2, $f$ achieves a minimum in the region $R$ for which $|x| \leq 24$ and $|y| \leq 30$, and this minimum cannot exceed 35.

*To summarize*, $f$ achieves a minimum in $R$ not exceeding 35, and on the boundary and outside of $R$, $f(x, y) \geq 88$. *Hence, the minimum in $R$ is the overall minimum.* Since, by 8.4, this minimum is not achieved on the boundary, it must be achieved at (64/25, 52/25). △

## A.9 Multiple Integrals

$$\iint_A f(x, y)\, dx\, dy$$

represents the volume lying between the surface $f$ and the set $A$ in the plane, as indicated in Figure A.3. To approximate the value of this integral, we subdivide the set

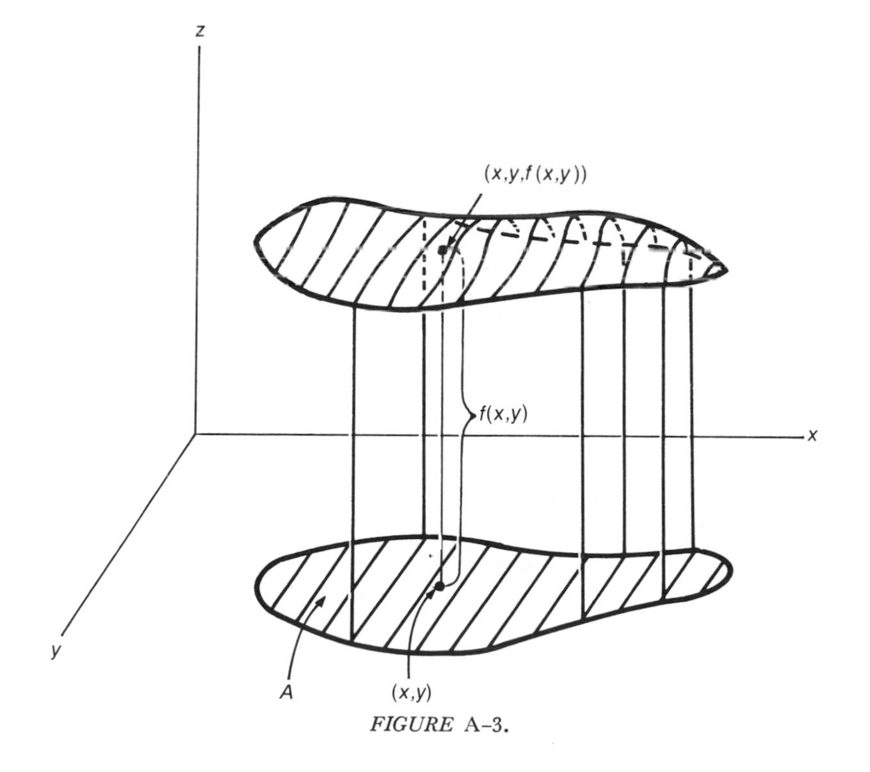


*FIGURE* A-3.

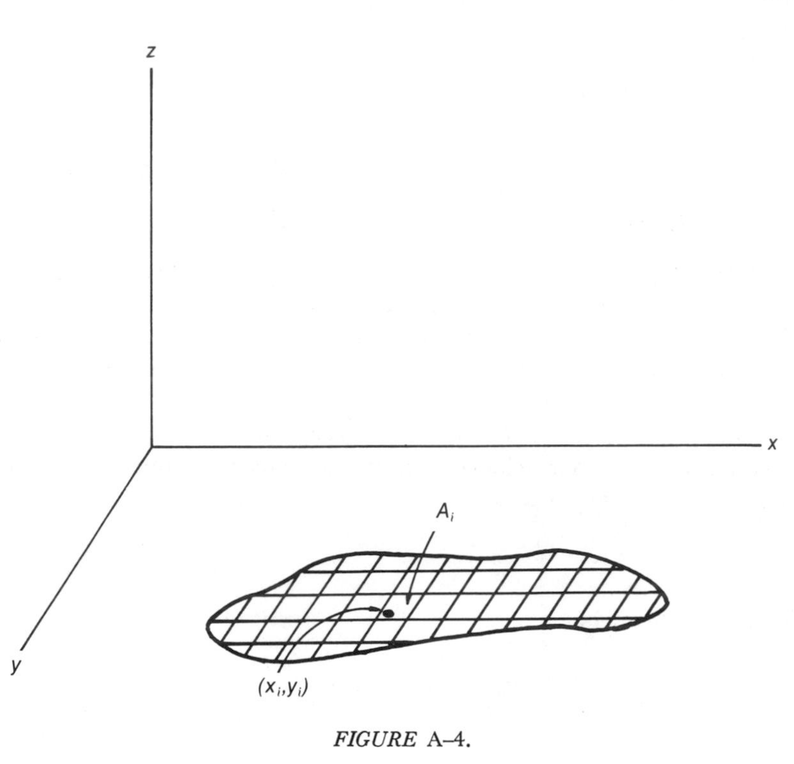


*FIGURE* A–4.

$A$ into rectangles $A_i$ of small diameter, and choose $(x_i, y_i) \in A_i$, as indicated in Figure A.4.

As an approximation to $\iint_A f(x, y)\,dx\,dy$, we take

$$\sum_{\text{all i}} \underbrace{f(x_i, y_i)}_{\text{height}} \underbrace{\mathscr{A}(A_i)}_{\substack{\text{area of} \\ \text{base}}}$$

where $\mathscr{A}(A_i)$ is the area of the set $A_i$. (We ignore those rectangles that cross the border of $A$.) The integral $\iint_A f(x, y)\,dx\,dy$ is actually the limit of such approximations as the diameter of the largest of the $A_i$ tends to 0. Thus, this integral is essentially a sum. We define

$$\iiint_A f(x, y, z)\,dx\,dy\,dz$$

essentially the same way, the modification being that the "small" rectangles become "small" boxes. In higher dimensions than 3, we use an analogous limit of sums where a 4-dimensional box or interval is defined as

$$\{(x_1, x_2, x_3, x_4)\colon\ a_i \leq x_i \leq b_i \qquad \text{for} \qquad i = 1, 2, 3, 4\}$$

having 4-dimensional volume

$$(b_1 - a_1)(b_2 - a_2)(b_3 - a_3)(b_4 - a_4).$$

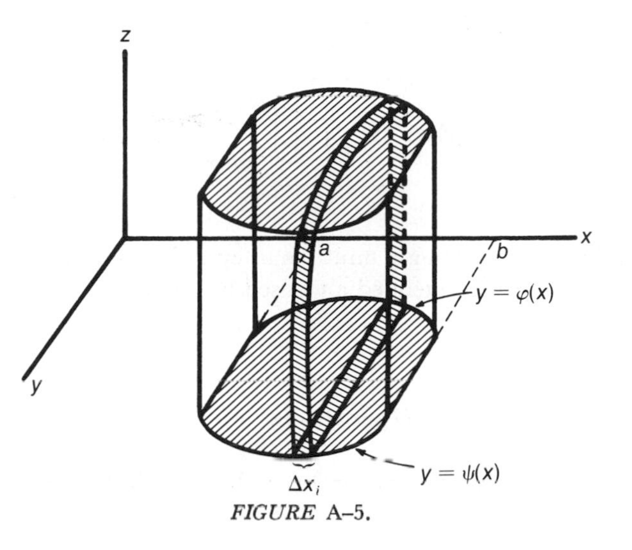


*FIGURE* A–5.

We can sometimes evaluate a multiple integral using a sequence of one-variable integrations. We illustrate how this may be done in Figure A.5. The volume of the thin slice is approximately

$$\underbrace{\int_{\varphi(x_i)}^{\psi(x_i)} f(x_i, y)\, dy}_{\text{area of slice}} \quad \underbrace{\Delta x_i}_{\substack{\text{depth}\\ \text{of slice}}} = g(x_i)\, \Delta x_i.$$

Thus, the total volume is approximately (sum of slice volumes)

$$\sum_i g(x_i)\, \Delta x_i \simeq \int_a^b \left[ \int_{\varphi(x)}^{\psi(x)} f(x, y)\, dy \right] dx.$$

***9.1 Example:***

$$\iint_A f(x, y)\, dy\, dx \qquad \text{where} \qquad A = \{(x, y) : x^2 + y^2 \leq 1\}$$

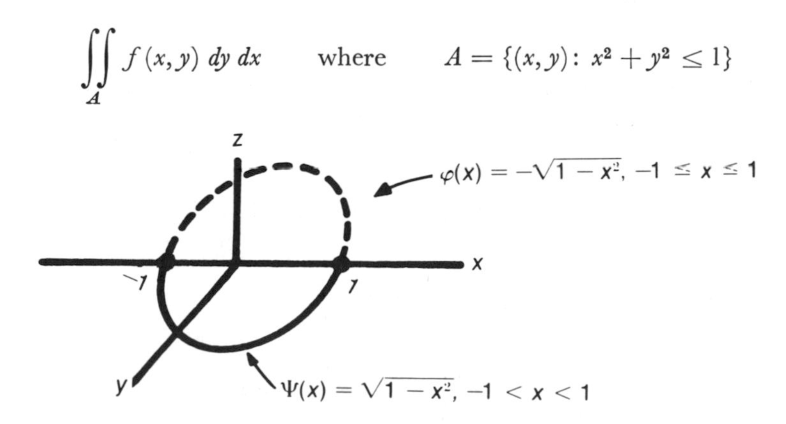

Provided that $\iint_A f(x, y)\, dy\, dx$ exists, we obtain

$$\iint_A f(x, y)\, dy\, dx = \int_{-1}^{1} \left[ \int_{-\sqrt{1-x^2}}^{\sqrt{1-x^2}} f(x, y)\, dy \right] dx.$$

△

This slicing method of evaluating a multiple integral as a sequence of single integrals (iterated integral) is further extended and used in the text.

## A.10 Interchange of Limit Operations

Quite frequently we are faced with the problem of differentiating some function $g$ given by

**10.1**
$$g(x) = \int_a^b f(t, x)\, dt$$

where $f$ is some known function. Since an integral is essentially a sum, and we know that the derivative of a finite sum is the sum of the derivatives, it is reasonable to expect that

**10.2**
$$g'(x) = \int_a^b \frac{\partial f}{\partial x}(t, x)\, dt.$$

There are cases in which 10.2 does not follow from 10.1, because the integral is a *limit* of sums rather than simply a sum. For this reason, it is useful to have a criterion available that will indicate conditions under which it is permissible to differentiate under the integral sign.

### *10.3 Theorem: Differentiation under Integral Sign*

Let $f(t, x)$ be given and suppose that $g(x) = \int_a^b f(t, x)\, dt$ is defined for all $x$ close to $x_0$. Here $a$ may be finite or $-\infty$, and $b$ may be finite or $+\infty$. Suppose further that

**10.4**
$$\int_a^b \frac{\partial f}{\partial x}(t, x)\, dt \qquad \text{is defined}$$

and that there exists a function $q(t)$ such that

**10.5**
$$\left| \frac{f(t, x_0 + h) - f(t, x_0)}{h} \right| \leq q(t)$$

for all $h$ smaller in magnitude than some fixed $h_0$, where $\int_a^b q(t)\,dt$ is finite. Then $g'(x_0)$ exists, with

$$g'(x_0) = \int_a^b \frac{\partial f}{\partial x}(t, x_0)\,dt.$$

The key to checking on whether 10.5 is satisfied is usually the mean value theorem.

### 10.6 *Example: Derivative of* $\int_{-\infty}^{\infty} e^{-xt^2}\,dt$, $x > 0$

For $x > 0$, it is fairly easy to show that the preceding integral exists by comparison with $\int_{-\infty}^{\infty} e^{-x|t|}\,dt$. Similarly,

$$\int_{-\infty}^{\infty} \frac{\partial}{\partial x} e^{-xt^2}\,dt = \int_{-\infty}^{\infty} t^2 e^{-xt^2}\,dt$$

can be shown to exist by comparison with $\int_{-\infty}^{\infty} t^2 e^{-x|t|}\,dt$. Finally, note that $f(t, x_0) = e^{-x_0 t^2}$, and that by the mean value theorem,

$$\frac{f(t, x_0 + h) - f(t, x_0)}{h} = \frac{\partial f}{\partial x}(t, c),$$

where $c$ is between $x_0$ and $x_0 + h$. But

$$\frac{\partial f}{\partial x}(t, c) = -t^2 e^{-ct^2}.$$

If $|h| < x_0/2 = h_0$, we see that

$$\left|\frac{\partial f}{\partial x}(t, c)\right| \le t^2 e^{-(x_0/2)t^2} = q(t)$$

and $\int_{-\infty}^{\infty} q(t)\,dt$ is finite. Hence, if $g(x) = \int_{-\infty}^{\infty} e^{-xt^2}\,dt$ for $x > 0$,

$$g'(x) = \int_{-\infty}^{\infty} -t^2 e^{-xt^2}\,dt. \qquad \triangle$$

In a similar manner, if $g(x) = \sum_{k=k_0}^{\infty} f_k(x)$, we would like to be able to determine conditions that make it legitimate to write $g'(x) = \sum_{k=k_0}^{\infty} f_k'(x)$. The following result, which is similar to Theorem 10.3, is usually adequate for our purposes.

### 10.7 *Derivative of* $\sum_{k=k_0}^{\infty} f_k$

Given functions $f_k$, suppose that

$$g(x) = \sum_{k=k_0}^{\infty} f_k(x)$$

and $\sum_{k=k_0}^{\infty} f_k'(x)$ exist for all $x$ satisfying $x_0 - h \leq x \leq x_0 + h$, where $h > 0$ is fixed. If there is a sequence $M_{k_0}, M_{k_0+1}, \ldots$, with $|f_k'(x)| \leq M_k$ for all $x$ in

$$x_0 - h \leq x \leq x_0 + h$$

such that $\sum_{k=k_0}^{\infty} M_k$ converges, then $g'(x_0)$ exists and is given by

$$g'(x_0) = \sum_{k=k_0}^{\infty} f_k'(x_0).$$

### *10.8 Example: Derivative of $\sum_{k=0}^{\infty} x^k$*

For $|x| < 1$, we know from Example 7.1 that

$$g(x) = \sum_{k=0}^{\infty} x^k = \frac{1}{1-x}.$$

We shall use 10.7 to see that for $0 < x_0 < 1$,

**10.9**
$$g'(x_0) = \sum_{k=0}^{\infty} kx_0^{k-1} = \frac{1}{(1-x_0)^2}.$$

We know from the ratio test (Theorem 7.4) that if

$$M_k = k\left(\frac{1+x_0}{2}\right)^k,$$

then $\sum_{k=0}^{\infty} M_k$ converges, and furthermore, that $|kx^{k-1}| \leq M_k$ for

$$-\left(\frac{1+x_0}{2}\right) \leq x_0 - \frac{1-x_0}{2} \leq x \leq x_0 + \frac{1-x_0}{2} = \frac{1+x_0}{2}.$$

Hence, 10.9 holds. △

## A.11 Density of Functions of Random Variables

In many situations, we are given the joint density $f_{X_1, \ldots, X_n}$ of random variables $X_1, \ldots, X_n$, and want to compute the density $f_{Y_1, \ldots, Y_n}$ where for all $i$ we have $Y_i = g_i(X_1, \ldots, X_n)$, where $g_1, \ldots, g_n$ are given real valued functions. We shall give an intuitive derivation for the case $n = 2$, and state the general result.

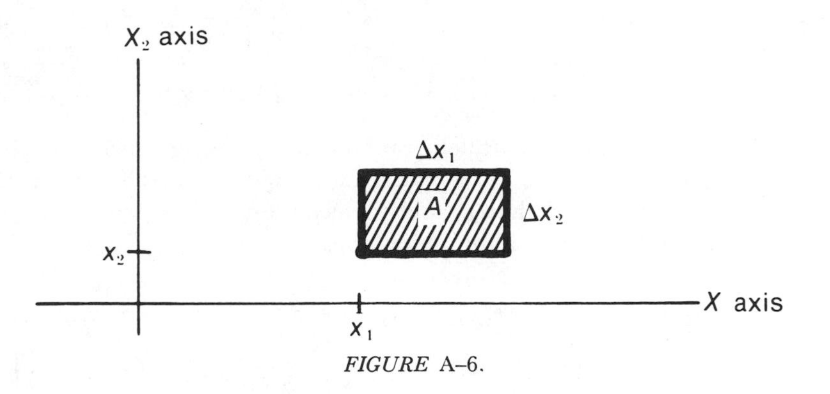


*FIGURE* A–6.

Suppose that we know $f_{X_1,X_2}$ and want to find $f_{Y_1,Y_2}$ where $Y_1 = g_1(X_1, X_2)$ and $Y_2 = g_2(X_1, X_2)$. We know that for the set $A$ in Figure A.6,

$$P((X_1, X_2) \in A) = P\{x_1 \leq X_1 \leq x_1 + \Delta x_1, x_2 \leq X_2 \leq x_2 + \Delta x_2\}$$

$$\cong f_{X_1,X_2}(x_1, x_2)\Delta x_1 \Delta x_2$$

when $\Delta x_1$ and $\Delta x_2$ are small.

Now, if the equations

$$y_1 = g_1(x_1, x_2), \qquad y_2 = g_2(x_1, x_2)$$

can always be solved uniquely to obtain

**11.1** $$x_1 = h_1(y_1, y_2), \qquad x_2 = h_2(y_1, y_2)$$

(that is, if the values $g_1(x_1, x_2)$ and $g_2(x_1, x_2)$ always uniquely determine $x_1$ and $x_2$), then the event $(X_1, X_2) \in A$ corresponds to an event $(Y_1, Y_2) \in \mathscr{A}$. But

$$P\{(Y_1, Y_2) \in \mathscr{A}\} \simeq f_{Y_1,Y_2}(y_1, y_2) \cdot |\mathscr{A}|,$$

where $|\mathscr{A}|$ is the unsigned area of the region $\mathscr{A}$. Thus,

**11.2** $$f_{Y_1,Y_2}(y_1, y_2) \cdot |\mathscr{A}| = f_{X_1,X_2}(x_1, x_2)\Delta x_1 \Delta x_2,$$

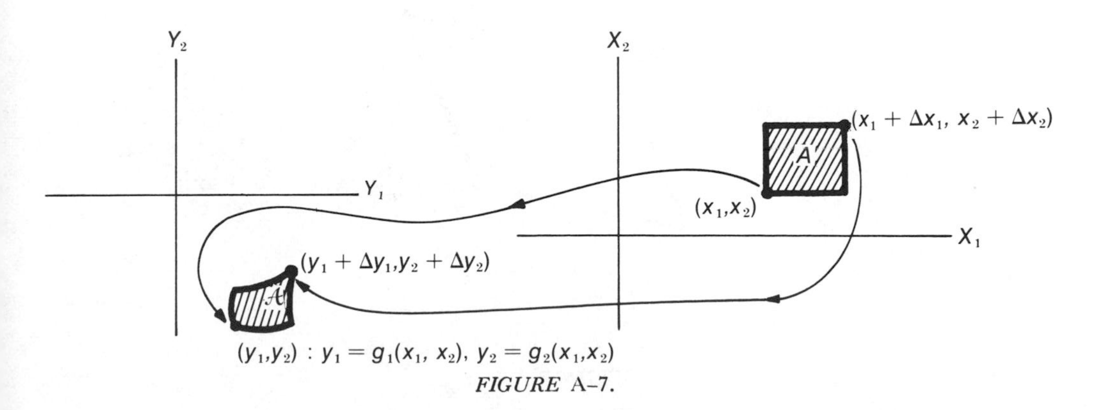


*FIGURE* A–7.

where $y_i = g_i(x_1, x_2)$ and

$$\mathscr{A} = \{(y_1^*, y_2^*) \in \mathscr{R}_2\colon y_1^* = g_1(x_1^*, \ {}_2^*), \ {}_2^* = g_2(x_1^*, x_2^*) \quad \text{for some} \quad (x_1^*, x_2^*) \in A\}.$$

We drew $\mathscr{A}$ as an approximate parallelogram for the following reason: Each point of $\mathscr{A}$ is of the form $(g_1(x_1 + h_1, x_2 + h_2), g_2(x_1 + h_1, x_2 + h_2))$ for some $(h_1, h_2)$ with $0 \leq h_i \leq \Delta x_i$. Provided that $g_1$ and $g_2$ have continuous partial derivatives, then we can write the differential approximations

$$\begin{aligned} g_1(x_1 + h_1, x_2 + h_2) &= g_1(x_1 + h_1, x_2 + h_2) - g_1(x_1 + h_1, x_2) + g_1(x_1 + h_1, x_2) \\ &\cong \left[\frac{\partial g_1}{\partial x_2}(x_1 + h_1, x_2)\right] \cdot h_2 + g_1(x_1, x_2) + \left[\frac{\partial g_1}{\partial x_1}(x_1, x_2)\right] \cdot h_1 \\ &\cong \left[\frac{\partial g_1}{\partial x_2}(x_1, x_2)\right] \cdot h_2 + g_1(x_1, x_2) + \left[\frac{\partial g_1}{\partial x_1}(x_1, x_2)\right] \cdot h_1 \\ &= g_1 + \frac{\partial g_1}{\partial x_1} h_1 + \frac{\partial g_1}{\partial x_2} h_2, \end{aligned}$$

where all functions are assumed evaluated at $(x_1, x_2)$. That is,

$$g_1(x_1 + h_1, x_2 + h_2) \cong g_1 + \frac{\partial g_1}{\partial x_1} h_1 + \frac{\partial g_1}{\partial x_2} h_2.$$

Similarly,

$$g_2(x_1 + h_1, x_2 + h_2) \cong g_2 + \frac{\partial g_2}{\partial x_2} h_1 + \frac{\partial g_2}{\partial x_2} h_2.$$

Thus, we see that if $h_1$ and $h_2$ increase proportionally, tracing out a line in $A$, then

$$g_1(x_1 + h_1, x_2 + h_2) \qquad \text{and} \qquad g_2(x_1 + h_1, x_2 + h_2)$$

increase approximately proportionally, tracing out approximate lines in $\mathscr{A}$. Therefore, it is not difficult to see that $\mathscr{A}$ is approximately a parallelogram with the indicated vertices (all functions involved being evaluated at $(x_1, x_2)$).

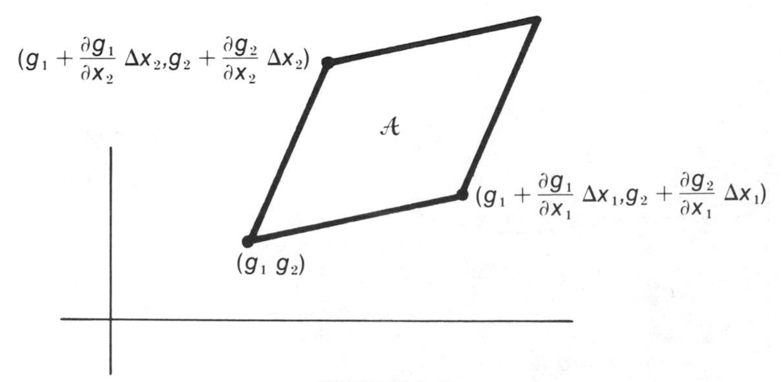


*FIGURE* A-8.

By direct integration, it can be checked that the unsigned area, $|\mathscr{A}|$, of $\mathscr{A}$ is given by

$$|\mathscr{A}| \cong \left| \frac{\partial g_1}{\partial x_1} \frac{\partial g_2}{\partial x_2} - \frac{\partial g_2}{\partial x_1} \frac{\partial g_1}{\partial x_2} \right| \Delta x_1 \, \Delta x_2.$$

We recognize

$$\frac{\partial g_1}{\partial x_1} \frac{\partial g_2}{\partial x_2} - \frac{\partial g_2}{\partial x_1} \frac{\partial g_1}{\partial x_2}$$

as the determinant

$$\begin{Vmatrix} \dfrac{\partial g_1}{\partial x_1} & \dfrac{\partial g_2}{\partial x_1} \\ \dfrac{\partial g_1}{\partial x_2} & \dfrac{\partial g_2}{\partial x_2} \end{Vmatrix}.$$

Thus

$$|\mathscr{A}| \cong \left| \begin{Vmatrix} \dfrac{\partial g_1}{\partial x_1} & \dfrac{\partial g_2}{\partial x_1} \\ \dfrac{\partial g_1}{\partial x_2} & \dfrac{\partial g_2}{\partial x_2} \end{Vmatrix} \right| \Delta x_1 \, \Delta x_2.$$

Substituting this into 11.2 and taking limits as $h_1$ and $h_2$ become small yields the exact relation

$$f_{Y_1, Y_2}(g_1(x_1, x_2), g_2(x_1, x_2)) \left| \begin{Vmatrix} \dfrac{\partial g_1}{\partial x_1} & \dfrac{\partial g_2}{\partial x_1} \\ \dfrac{\partial g_1}{\partial x_2} & \dfrac{\partial g_2}{\partial x_2} \end{Vmatrix} \right| = f_{X_1, X_2}(x_1, x_2).$$

We can in a similar fashion, using 11.1, derive the result

$$f_{Y_1, Y_2}(y_1, y_2) = f_{X_1, X_2}(h_1(y_1, y_2), h_2(y_1, y_2)) \left| \begin{Vmatrix} \dfrac{\partial h_1}{\partial y_1} & \dfrac{\partial h_2}{\partial y_1} \\ \dfrac{\partial h_1}{\partial y_2} & \dfrac{\partial h_2}{\partial y_2} \end{Vmatrix} \right|.$$

Since the main theorem will depend on determinants, we briefly outline their definition.

Given a square array

$$\begin{bmatrix} a_{11} & \cdots & a_{1n} \\ \cdot & & \\ \cdot & & \\ \cdot & & \\ a_{n1} & \cdots & a_{nn} \end{bmatrix}$$

of numbers, the determinant of such an array, written

$$\begin{Vmatrix} a_{11} & \cdots & a_{1n} \\ \cdot & & \\ \cdot & & \\ \cdot & & \\ a_{n1} & \cdots & a_{nn} \end{Vmatrix},$$

is defined for each $n$ as follows:

$$\|a_{11}\| = a_{11}$$

$$\begin{Vmatrix} a_{11} & a_{12} \\ a_{21} & a_{22} \end{Vmatrix} = a_{11}a_{22} - a_{12}a_{21}$$

$$\cdot$$
$$\cdot$$
$$\cdot$$

$$\begin{Vmatrix} a_{11} & \cdots & a_{1n} \\ \cdot & & \\ \cdot & & \\ \cdot & & \\ a_{n1} & \cdots & a_{nn} \end{Vmatrix} = a_{11} \cdot \begin{Vmatrix} a_{22} & \cdots & a_{2n} \\ \cdot & & \\ \cdot & & \\ \cdot & & \\ a_{n2} & \cdots & a_{nn} \end{Vmatrix} - a_{21} \cdot \begin{Vmatrix} a_{12}\, a_{13} & \cdots & a_{1n} \\ a_{32} & & \\ \cdot & & \\ \cdot & & \\ \cdot & & \\ a_{n2} & & a_{nn} \end{Vmatrix}$$

$$+ a_{31} \cdot \| \quad \cdots \quad \| - \cdots + \cdots.$$

Here, the factor of $a_{j1}$ is the determinant of the array formed by omitting the $j$th row and 1st column from the original array. Thus,

$$\begin{Vmatrix} a & b & c \\ d & e & f \\ g & h & i \end{Vmatrix} = a \begin{Vmatrix} e & f \\ h & i \end{Vmatrix} - d \begin{Vmatrix} b & c \\ h & i \end{Vmatrix} + g \begin{Vmatrix} bc \\ ef \end{Vmatrix}$$

$$= a(ei - hf) - d(bi - hc) + g(bf - ce).$$

## *11.3 Theorem: Joint Density of Functions of Random Variables*

Suppose that for all but at most a finite number of values of any of $y_1, \ldots, y_n$, we can always solve the equations

$$y_1 = g_1(x_1, \ldots, x_n), \ldots, y_n = g_n(x_1, \ldots, x_n)$$

for the $x_i$ in the form

$$x_1 = h_1(y_1, \ldots, y_n), \ldots, x_n = h_n(y_1, \ldots, y_n).$$

Then

$$f_{Y_1,\ldots,Y_n}(\lambda_1, \ldots, \lambda_n)$$
$$= f_{X_1,\ldots,X_n}(h_1(\lambda_1, \ldots, \lambda_n), \ldots, h_n(\lambda_1, \ldots, \lambda_n))|J(\lambda_1, \ldots, \lambda_n)|,$$

where $J(\lambda_1, \ldots, \lambda_n)$ is the Jacobian determinant defined by

$$J(\lambda_1, \ldots, \lambda_n) = \begin{Vmatrix} \dfrac{\partial h_1}{\partial y_1} & \cdots & \dfrac{\partial h_n}{\partial y_1} \\ \cdot & & \\ \cdot & & \\ \cdot & & \\ \dfrac{\partial h_1}{\partial y_n} & \cdots & \dfrac{\partial h_n}{\partial y_n} \end{Vmatrix}.$$

Here $\partial h_i/\partial y_j$ stands for the partial derivative of $h_i$ with respect to its $j$th coordinate, evaluated at $(\lambda_1, \ldots, \lambda_n)$. A proof may be found in Bartle [2], under the heading, "Change of Variable in Integration."

We give some examples to illustrate what is involved in using this theorem.

### *11.4 Example: Solution of $x = r\cos\theta$, $y = r\sin\theta$ for $r, \theta$*

If $x = g_1(r, \theta) = r\cos\theta$, $y = g_2(r, \theta) = r\sin\theta$ where $r \geq 0$, $0 \leq \theta < 2\pi$, then $r = h_1(x, y) = \sqrt{x^2 + y^2}$ and

$$\theta = h_2(x, y) = \begin{cases} \arctan y/x & \text{for } x > 0 \\ \arctan y/x + \pi & \text{for } x < 0. \end{cases}$$

Note that $h_2(x, y)$ is defined for all values of $x, y$ except $x = 0$. △

### *11.5 Example: Polar Coordinates*

Let $X$, $Y$ have density $f_{X,Y}$ and let

$$R = \sqrt{X^2 + Y^2}, \quad \Theta = \begin{cases} \arctan Y/X & \text{for } x > 0 \\ \arctan Y/X + \pi & \text{for } x < 0. \end{cases}$$

Then

$$f_{R,\Theta}(r, \theta) = f_{X,Y}(r\cos\theta, r\sin\theta)|J(r, \theta)|.$$

Here, since

$$h_1(r, \theta) = r\cos\theta$$

$$h_2(r, \theta) = r\sin\theta,$$

we have

$$J(r, \theta) = \begin{Vmatrix} \dfrac{\partial h_1}{\partial r} & \dfrac{\partial h_2}{\partial r} \\ \dfrac{\partial h_1}{\partial \theta} & \dfrac{\partial h_2}{\partial \theta} \end{Vmatrix}$$

(note that here $\partial h_1/\partial_r$ stands for the partial derivative of $h_1$ with respect to its first coordinate, evaluated at $(r, \theta)$)

$$= \begin{Vmatrix} \cos\theta & \sin\theta \\ -r\cos\theta & r\sin\theta \end{Vmatrix}$$

$$= r\cos^2\theta + r\sin^2\theta = r.$$

Thus,

$$f_{R,\Theta}(r, \theta) = f_{X,Y}(r\cos\theta, r\sin\theta)r$$

for $r \geq 0$, $0 \leq \theta < 2\pi$. △

### *11.6 Example: Linear Transformation*

Let $X$, $Y$ have density $f_{X,Y}$. Let

$$U = X + Y = g_1(X, Y), \qquad V = X - Y = g_2(X, Y).$$

Then

$$X = \frac{U+V}{2} = h_1(U, V), \qquad Y = \frac{U-V}{2} = h_2(U, V)$$

$$f_{U,V}(u, v) = f_{X,Y}(h_1(u, v), h_2(u, v))\,|J(u, v)|$$

$$= f_{X,Y}\left(\frac{u+v}{2}, \frac{u-v}{2}\right)|J(u, v)|.$$

But

$$J(u, v) = \begin{Vmatrix} \dfrac{\partial h_1}{\partial u} & \dfrac{\partial h_2}{\partial u} \\ \dfrac{\partial h_1}{\partial v} & \dfrac{\partial h_2}{\partial v} \end{Vmatrix} = \begin{Vmatrix} \tfrac{1}{2} & \tfrac{1}{2} \\ \tfrac{1}{2} & -\tfrac{1}{2} \end{Vmatrix} = -\tfrac{1}{2}.$$

Thus

$$f_{U,V}(u, v) = \tfrac{1}{2} f_{X,Y}\left(\frac{u+v}{2}, \frac{u-v}{2}\right).$$ △

## A.12 Evaluation of

$$\int_{-\infty}^{\infty} \frac{1}{\sqrt{2\pi}} e^{-t^2/2}\,dt$$

The theory of the previous section applies to functions of several variables that are not necessarily probability densities. That is, suppose that we try to evaluate

$\iint_A f(x, y)\, dy\, dx$. This may sometimes be accomplished by a change of variable. Suppose that

$$r = g_1(x, y) = \sqrt{x^2 + y^2}$$

**12.1**
$$\theta = g_2(x, y) = \begin{cases} \text{arc tan } y/x & \text{for } \quad x > 0 \\ \text{arc tan } y/x + \pi & \text{for } \quad x < 0. \end{cases}$$

Let

$$\mathscr{A} = \{(r, \theta)\colon\ r = g_1(x, y),\ \theta = g_2(x, y) \text{ for } (x, y) \in A\}.$$

If

$$x = h_1(r, \theta) = r \cos \theta$$

$$y = h_2(r, \theta) = r \sin \theta$$

is the solution of 12.1, then

$$\iint_A f(x, y)\, dy\, dx = \iint_{\mathscr{A}} f(r \cos \theta, r \sin \theta)\, |J(r, \theta)|\, dr\, d\theta$$

$$= \iint_{\mathscr{A}} f(r \cos \theta, r \sin \theta) r\, dr\, d\theta,$$

as shown in the previous section. Now, to evaluate

$$\frac{1}{\sqrt{2\pi}} \int_{-\infty}^{\infty} e^{-t^2/2}\, dt,$$

let

$$N = \frac{1}{\sqrt{2\pi}} \int_{-\infty}^{\infty} e^{-t^2/2}\, dt > 0.$$

Then

$$N = \sqrt{N^2} = \sqrt{\left(\frac{1}{\sqrt{2\pi}} \int_{-\infty}^{\infty} e^{-x^2/2}\, dx\right)\left(\frac{1}{\sqrt{2\pi}} \int_{-\infty}^{\infty} e^{-y^2/2}\, dy\right)}$$

$$= \sqrt{\frac{1}{2\pi} \iint_A e^{-(x^2+y^2)/2}\, dy\, dx}.$$

Here we made use of the basic theorem that a double integral can be evaluated as an iterated integral, and conversely, where the set $A$ is the whole $xy$ plane. Now we make the change of variable to $r$, $\theta$ noting that

$$\mathscr{A} = \{(r, \theta)\colon\ r \geq 0, 0 \leq \theta < 2\pi\},$$

to obtain

$$N = \sqrt{\frac{1}{2\pi}\iint_{\mathscr{A}} e^{-r^2/2} r\, dr\, d\theta} = \sqrt{\frac{1}{2\pi}\int_0^\infty \left[\int_0^{2\pi} re^{-r^2/2}\, d\theta\right] dr}$$

(again writing a double integral as an iterated integral)

$$= \sqrt{\int_0^\infty re^{-r^2/2}\, dr} = \sqrt{-e^{-r^2/2}\Big|_0^{r=\infty}} = \sqrt{1} = 1.$$

That is,

$$N = \frac{1}{\sqrt{2\pi}}\int_{-\infty}^{\infty} e^{-t^2/2}\, dt = 1,$$

as asserted. △

# References


1. Barlow, R.; Proschan, F.: Mathematical Theory of Reliability. New York, Wiley, 1965.
2. Bartle, R.: The Elements of Real Analysis. New York, Wiley, 1964. (In-depth introduction to the background mathematics we use.)
3. Bell, S., et al.: Modern University Calculus. San Francisco, Holden-Day, 1966. (Also good background material.)
4. Brunk, D.: An Introduction to Mathematical Statistics. Waltham, Mass., Blaisdell, 1965. (Covers essentially the same material we do.)
5. Courant, R.: Differential and Integral Calculus. New York, Interscience, 1936. (Background material.)
6. Cramer, H.: Mathematical Methods of Statistics. Princeton, N.J., Princeton University Press, 1951.
7. DeGroot, M.: Optimal Statistical Decisions. New York, McGraw-Hill, 1970. (A more advanced approach to our material.)
8. Feller, W.: An Introduction to Probability Theory and its Applications. New York, Wiley, 1966 (Vol. 2), 1968 (Vol. 1). (Volume 1 covers basically the same material we do; Volume 2 covers it at a more advanced level.)
9. Graybill, F.: An Introduction to Linear Statistical Models. New York, McGraw-Hill, 1961. (Deeper coverage of regression.)
10. Hogg, R. V., and Craig, A. T.: Introduction to Mathematical Statistics, 2nd ed. New York, Macmillan, 1965. (More advanced.)
11. Karlin, S.: A First Course in Stochastic Processes. New York, Academic Press, 1966. (More advanced.)
12. Lehmann, E.: Testing Statistical Hypotheses. New York, Wiley, 1959. (More advanced.)
13. Lindgren, B.: Statistical Theory. New York, Macmillan, 1900. (Covers pretty much what we do.)
14. Neuts, M.: Probability. Boston, Allyn & Bacon, 1973. (More advanced.)
15. Owen, D.: Handbook of Statistical Tables. Reading, Mass., Addison-Wesley, 1962.
16. Parzen, E.: Modern Probability Theory and its Applications. New York, Wiley, 1960. (Covers the same material we do.)
17. Renyi, A.: Probability Theory. New York, American Elsevier, 1970. (More advanced.)
18. Royden, H.: Real Analysis. New York, Macmillan, 1968. (Covers basic mathematics needed for the study of more advanced subjects.)
19. Scheffé, H.: The Analysis of Variance. New York, Wiley, 1959. (Deeper coverage of regression.)
20. Uspensky, J.: Introduction to Mathematical Probability. New York, McGraw-Hill, 1937. (Covers essentially what we do.)

## *Values of the Standard Normal Distribution Function*

$$\Phi(z) = \int_{-\infty}^{z} (1/\sqrt{2\pi})\, exp\, (-u^2/2)\, du = P(Z \leq z)$$

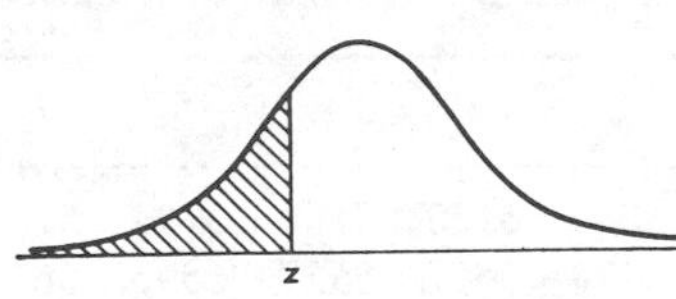


| z | 0 | 1 | 2 | 3 | 4 | 5 | 6 | 7 | 8 | 9 |
|---|---|---|---|---|---|---|---|---|---|---|
| −3.0 | .0013 | .0010 | .0007 | .0005 | .0003 | .0002 | .0002 | .0001 | .0001 | .0000 |
| −2.9 | .0019 | .0018 | .0017 | .0017 | .0016 | .0016 | .0015 | .0015 | .0014 | .0014 |
| −2.8 | .0026 | .0025 | .0024 | .0023 | .0023 | .0022 | .0021 | .0021 | .0020 | .0019 |
| −2.7 | .0035 | .0034 | .0033 | .0032 | .0031 | .0030 | .0029 | .0028 | .0027 | .0026 |
| −2.6 | .0047 | .0045 | .0044 | .0043 | .0041 | .0040 | .0039 | .0038 | .0037 | .0036 |
| −2.5 | .0062 | .0060 | .0059 | .0057 | .0055 | .0054 | .0052 | .0051 | .0049 | .0048 |
| −2.4 | .0082 | .0080 | .0078 | .0075 | .0073 | .0071 | .0069 | .0068 | .0066 | .0064 |
| −2.3 | .0107 | .0104 | .0102 | .0099 | .0096 | .0094 | .0091 | .0089 | .0087 | .0084 |
| −2.2 | .0139 | .0136 | .0132 | .0129 | .0126 | .0122 | .0119 | .0116 | .0113 | .0110 |
| −2.1 | .0179 | .0174 | .0170 | .0166 | .0162 | .0158 | .0154 | .0150 | .0146 | .0143 |
| −2.0 | .0228 | .0222 | .0217 | .0212 | .0207 | .0202 | .0197 | .0192 | .0188 | .0183 |
| −1.9 | .0287 | .0281 | .0274 | .0268 | .0262 | .0256 | .0250 | .0244 | .0238 | .0233 |
| −1.8 | .0359 | .0352 | .0344 | .0336 | .0329 | .0322 | .0314 | .0307 | .0300 | .0294 |
| −1.7 | .0446 | .0436 | .0427 | .0418 | .0409 | .0401 | .0392 | .0384 | .0375 | .0367 |
| −1.6 | .0548 | .0537 | .0526 | .0516 | .0505 | .0495 | .0485 | .0475 | .0465 | .0455 |
| −1.5 | .0668 | .0655 | .0643 | .0630 | .0618 | .0606 | .0594 | .0582 | .0570 | .0559 |
| −1.4 | .0808 | .0793 | .0778 | .0764 | .0749 | .0735 | .0722 | .0708 | .0694 | .0681 |
| −1.3 | .0968 | .0951 | .0934 | .0918 | .0901 | .0885 | .0869 | .0853 | .0838 | .0823 |
| −1.2 | .1151 | .1131 | .1112 | .1093 | .1075 | .1056 | .1038 | .1020 | .1003 | .0985 |
| −1.1 | .1357 | .1335 | .1314 | .1292 | .1271 | .1251 | .1230 | .1210 | .1190 | .1170 |
| −1.0 | .1587 | .1562 | .1539 | .1515 | .1492 | .1469 | .1446 | .1423 | .1401 | .1379 |
| − .9 | .1841 | .1814 | .1788 | .1762 | .1736 | .1711 | .1685 | .1660 | .1635 | .1611 |
| − .8 | .2119 | .2090 | .2061 | .2033 | .2005 | .1977 | .1949 | .1922 | .1894 | .1867 |
| − .7 | .2420 | .2389 | .2358 | .2327 | .2297 | .2266 | .2236 | .2206 | .2177 | .2148 |
| − .6 | .2743 | .2709 | .2676 | .2643 | .2611 | .2578 | .2546 | .2514 | .2483 | .2451 |
| − .5 | .3085 | .3050 | .3015 | .2981 | .2946 | .2912 | .2877 | .2843 | .2810 | .2776 |
| − .4 | .3446 | .3409 | .3372 | .3336 | .3300 | .3264 | .3228 | .3192 | .3156 | .3121 |
| − .3 | .3821 | .3783 | .3745 | .3707 | .3669 | .3632 | .3594 | .3557 | .3520 | .3483 |
| − .2 | .4207 | .4168 | .4129 | .4090 | .4052 | .4013 | .3974 | .3936 | .3897 | .3859 |
| − .1 | .4602 | .4562 | .4522 | .4483 | .4443 | .4404 | .4364 | .4325 | .4286 | .4247 |
| − 0 | .5000 | .4960 | .4920 | .4880 | .4840 | .4801 | .4761 | .4721 | .4681 | .4641 |

***(Continued)***

*Values of the Standard Normal Distribution Function (Continued)*

| z | 0 | 1 | 2 | 3 | 4 | 5 | 6 | 7 | 8 | 9 |
|---|---|---|---|---|---|---|---|---|---|---|
| .0 | .5000 | .5040 | .5080 | .5120 | .5160 | .5199 | .5239 | .5279 | .5319 | .5359 |
| .1 | .5398 | .5438 | .5478 | .5517 | .5557 | .5596 | .5636 | .5675 | .5714 | .5753 |
| .2 | .5793 | .5832 | .5871 | .5910 | .5948 | .5987 | .6026 | .6064 | .6103 | .6141 |
| .3 | .6179 | .6217 | .6255 | .6293 | .6331 | .6368 | .6406 | .6443 | .6480 | .6517 |
| .4 | .6554 | .6591 | .6628 | .6664 | .6700 | .6736 | .6772 | .6808 | .6844 | .6879 |
| .5 | .6915 | .6950 | .6985 | .7019 | .7054 | .7088 | .7123 | .7157 | .7190 | .7224 |
| .6 | .7257 | .7291 | .7324 | .7357 | .7389 | .7422 | .7454 | .7486 | .7517 | .7549 |
| .7 | .7580 | .7611 | .7642 | .7673 | .7703 | .7734 | .7764 | .7794 | .7823 | .7852 |
| .8 | .7881 | .7910 | .7939 | .7967 | .7995 | .8023 | .8051 | .8078 | .8106 | .8133 |
| .9 | .8159 | .8186 | .8212 | .8238 | .8264 | .8289 | .8315 | .8340 | .8365 | .8389 |
| 1.0 | .8413 | .8438 | .8461 | .8485 | .8508 | .8531 | .8554 | .8577 | .8599 | .8621 |
| 1.1 | .8643 | .8665 | .8686 | .8708 | .8729 | .8749 | .8770 | .8790 | .8810 | .8830 |
| 1.2 | .8849 | .8869 | .8888 | .8907 | .8925 | .8944 | .8962 | .8980 | .8997 | .9015 |
| 1.3 | .9032 | .9049 | .9066 | .9082 | .9099 | .9115 | .9131 | .9147 | .9162 | .9177 |
| 1.4 | .9192 | .9207 | .9222 | .9236 | .9251 | .9265 | .9278 | .9292 | .9306 | .9319 |
| 1.5 | .9332 | .9345 | .9357 | .9370 | .9382 | .9394 | .9406 | .9418 | .9430 | .9441 |
| 1.6 | .9452 | .9463 | .9474 | .9484 | .9495 | .9505 | .9515 | .9525 | .9535 | .9545 |
| 1.7 | .9554 | .9564 | .9573 | .9582 | .9591 | .9599 | .9608 | .9616 | .9625 | .9633 |
| 1.8 | .9641 | .9648 | .9656 | .9664 | .9671 | .9678 | .9686 | .9693 | .9700 | .9706 |
| 1.9 | .9713 | .9719 | .9726 | .9732 | .9738 | .9744 | .9750 | .9756 | .9762 | .9767 |
| 2.0 | .9772 | .9778 | .9783 | .9788 | .9793 | .9798 | .9803 | .9808 | .9812 | .9817 |
| 2.1 | .9821 | .9826 | .9830 | .9834 | .9838 | .9842 | .9846 | .9850 | .9854 | .9857 |
| 2.2 | .9861 | .9864 | .9868 | .9871 | .9874 | .9878 | .9881 | .9884 | .9887 | .9890 |
| 2.3 | .9893 | .9896 | .9898 | .9901 | .9904 | .9906 | .9909 | .9911 | .9913 | .9916 |
| 2.4 | .9918 | .9920 | .9922 | .9925 | .9927 | .9929 | .9931 | .9932 | .9934 | .9936 |
| 2.5 | .9938 | .9940 | .9941 | .9943 | .9945 | .9946 | .9948 | .9949 | .9951 | .9952 |
| 2.6 | .9953 | .9955 | .9956 | .9957 | .9959 | .9960 | .9961 | .9962 | .9963 | .9964 |
| 2.7 | .9965 | .9966 | .9967 | .9968 | .9969 | .9970 | .9971 | .9972 | .9973 | .9974 |
| 2.8 | .9974 | .9975 | .9976 | .9977 | .9977 | .9978 | .9979 | .9979 | .9980 | .9981 |
| 2.9 | .9981 | .9982 | .9982 | .9983 | .9984 | .9984 | .9985 | .9985 | .9986 | .9986 |
| 3. | .9987 | .9990 | .9993 | .9995 | .9997 | .9998 | .9998 | .9999 | .9999 | 1.0000 |

Note: Entries opposite 3 and − 3 are for 3.0, 3.1, 3.2, etc., and − 3.0, − 3.1, etc., respectively.

*Percentiles of the Chi-Square Distribution*

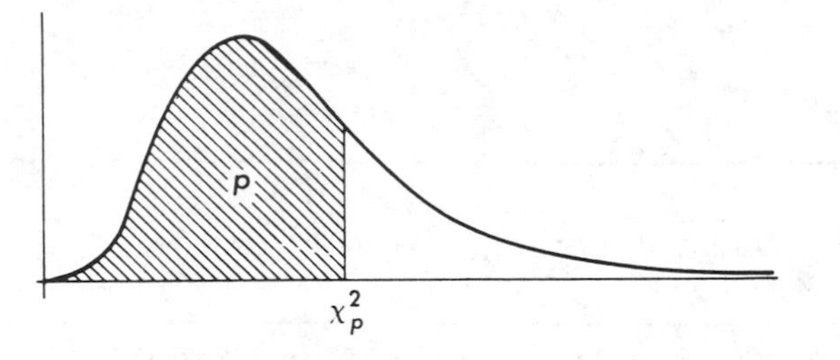


| Degrees of freedom | $\chi^2_{.005}$ | $\chi^2_{.01}$ | $\chi^2_{.025}$ | $\chi^2_{.05}$ | $\chi^2_{.10}$ | $\chi^2_{.20}$ | $\chi^2_{.30}$ | $\chi^2_{.50}$ | $\chi^2_{.70}$ | $\chi^2_{.80}$ | $\chi^2_{.90}$ | $\chi^2_{.95}$ | $\chi^2_{.975}$ | $\chi^2_{.99}$ | $\chi^2_{.995}$ |
|---|---|---|---|---|---|---|---|---|---|---|---|---|---|---|---|
| 1 | .000 | .000 | .001 | .004 | .016 | .064 | .148 | .455 | 1.07 | 1.64 | 2.71 | 3.84 | 5.02 | 6.63 | 7.88 |
| 2 | .010 | .020 | .051 | .103 | .211 | .446 | .713 | 1.39 | 2.41 | 3.22 | 4.61 | 5.99 | 7.38 | 9.21 | 10.6 |
| 3 | .072 | .115 | .216 | .352 | .584 | 1.00 | 1.42 | 2.37 | 3.66 | 4.64 | 6.25 | 7.81 | 9.35. | 11.3 | 12.8 |
| 4 | .207 | .297 | .484 | .711 | 1.06 | 1.65 | 2.20 | 3.36 | 4.88 | 5.99 | 7.78 | 9.49 | 11.1 | 13.3 | 14.9 |
| 5 | .412 | .554 | .831 | 1.15 | 1.61 | 2.34 | 3.00 | 4.35 | 6.06 | 7.29 | 9.24 | 11.1 | 12.8 | 15.1 | 16.7 |
| 6 | .676 | .872 | 1.24 | 1.64 | 2.20 | 3.07 | 3.83 | 5.35 | 7.23 | 8.56 | 10.6 | 12.6 | 14.4 | 16.8 | 18.5 |
| 7 | .989 | 1.24 | 1.69 | 2.17 | 2.83 | 3.82 | 4.67 | 6.35 | 8.38 | 9.80 | 12.0 | 14.1 | 16.0 | 18.5 | 20.3 |
| 8 | 1.34 | 1.65 | 2.18 | 2.73 | 3.49 | 4.59 | 5.53 | 7.34 | 9.52 | 11.0 | 13.4 | 15.5 | 17.5 | 20.1 | 22.0 |
| 9 | 1.73 | 2.09 | 2.70 | 3.33 | 4.17 | 5.38 | 6.39 | 8.34 | 10.7 | 12.2 | 14.7 | 16.9 | 19.0 | 21.7 | 23.6 |
| 10 | 2.16 | 2.56 | 3.25 | 3.94 | 4.87 | 6.18 | 7.27 | 9.34 | 11.8 | 13.4 | 16.0 | 18.3 | 20.5 | 23.2 | 25.2 |
| 11 | 2.60 | 3.05 | 3.82 | 4.57 | 5.58 | 6.99 | 8.15 | 10.3 | 12.9 | 14.6 | 17.3 | 19.7 | 21.9 | 24.7 | 26.8 |
| 12 | 3.07 | 3.57 | 4.40 | 5.23 | 6.30 | 7.81 | 9.03 | 11.3 | 14.0 | 15.8 | 18.5 | 21.0 | 23.3 | 26.2 | 28.3 |
| 13 | 3.57 | 4.11 | 5.01 | 5.89 | 7.04 | 8.63 | 9.93 | 12.3 | 15.1 | 17.0 | 19.8 | 22.4 | 24.7 | 27.7 | 29.8 |
| 14 | 4.07 | 4.66 | 5.63 | 6.57 | 7.79 | 9.47 | 10.8 | 13.3 | 16.2 | 18.2 | 21.1 | 23.7 | 26.1 | 29.1 | 31.3 |
| 15 | 4.60 | 5.23 | 6.26 | 7.26 | 8.55 | 10.3 | 11.7 | 14.3 | 17.3 | 19.3 | 22.3 | 25.0 | 27.5 | 30.6 | 32.8 |
| 16 | 5.14 | 5.81 | 6.91 | 7.96 | 9.31 | 11.2 | 12.6 | 15.3 | 18.4 | 20.5 | 23.5 | 26.3 | 28.8 | 32.0 | 34.3 |
| 17 | 5.70 | 6.41 | 7.56 | 8.67 | 10.1 | 12.0 | 13.5 | 16.3 | 19.5 | 21.6 | 24.8 | 27.6 | 30.2 | 33.4 | 35.7 |
| 18 | 6.26 | 7.01 | 8.23 | 9.39 | 10.9 | 12.9 | 14.4 | 17.3 | 20.6 | 22.8 | 26.0 | 28.9 | 31.5 | 34.8 | 37.2 |
| 19 | 6.83 | 7.63 | 8.91 | 10.1 | 11.7 | 13.7 | 15.4 | 18.3 | 21.7 | 23.9 | 27.2 | 30.1 | 32.9 | 36.2 | 38.6 |
| 20 | 7.43 | 8.26 | 9.59 | 10.9 | 12.4 | 14.6 | 16.3 | 19.3 | 22.8 | 25.0 | 28.4 | 31.4 | 34.2 | 37.6 | 40.0 |
| 21 | 8.03 | 8.90 | 10.3 | 11.6 | 13.2 | 15.4 | 17.2 | 20.3 | 23.9 | 26.2 | 29.6 | 32.7 | 35.5 | 38.9 | 41.4 |
| 22 | 8.64 | 9.54 | 11.0 | 12.3 | 14.0 | 16.3 | 18.1 | 21.3 | 24.9 | 27.3 | 30.8 | 33.9 | 36.8 | 40.3 | 42.8 |
| 23 | 9.26 | 10.2 | 11.7 | 13.1 | 14.8 | 17.2 | 19.0 | 22.3 | 26.0 | 28.4 | 32.0 | 35.2 | 38.1 | 41.6 | 44.2 |
| 24 | 9.89 | 10.9 | 12.4 | 13.8 | 15.7 | 18.1 | 19.9 | 23.3 | 27.1 | 29.6 | 33.2 | 36.4 | 39.4 | 43.0 | 45.6 |
| 25 | 10.5 | 11.5 | 13.1 | 14.6 | 16.5 | 18.9 | 20.9 | 24.3 | 28.2 | 30.7 | 34.4 | 37.7 | 40.6 | 44.3 | 46.9 |
| 26 | 11.2 | 12.2 | 13.8 | 15.4 | 17.3 | 19.8 | 21.8 | 25.3 | 29.2 | 31.8 | 35.6 | 38.9 | 41.9 | 45.6 | 48.3 |
| 27 | 11.8 | 12.9 | 14.6 | 16.2 | 18.1 | 20.7 | 22.7 | 26.3 | 30.3 | 32.9 | 36.7 | 40.1 | 43.2 | 47.0 | 49.6 |
| 28 | 12.5 | 13.6 | 15.3 | 16.9 | 18.9 | 21.6 | 23.6 | 27.3 | 31.4 | 34.0 | 37.9 | 41.3 | 44.5 | 48.3 | 51.0 |
| 29 | 13.1 | 14.3 | 16.0 | 17.7 | 19.8 | 22.5 | 24.6 | 28.3 | 32.5 | 35.1 | 39.1 | 42.6 | 45.7 | 49.6 | 52.3 |
| 30 | 13.8 | 15.0 | 16.8 | 18.5 | 20.6 | 23.4 | 25.5 | 29.3 | 33.5 | 36.2 | 40.3 | 43.8 | 47.0 | 50.9 | 53.7 |
| 40 | 20.7 | 22.1 | 24.4 | 26.5 | 29.0 | 32.3 | 34.9 | 39.3 | 44.2 | 47.3 | 51.8 | 55.8 | 59.3 | 63.7 | 66.8 |
| 50 | 28.0 | 29.7 | 32.3 | 34.8 | 37.7 | 41.3 | 44.3 | 49.3 | 54.7 | 58.2 | 63.2 | 67.5 | 71.4 | 76.2 | 79.5 |
| 60 | 35.5 | 37.5 | 40.5 | 43.2 | 46.5 | 50.6 | 53.8 | 59.3 | 65.2 | 69.0 | 74.4 | 79.1 | 83.3 | 88.4 | 92.0 |

## *Percentiles of the t Distribution*

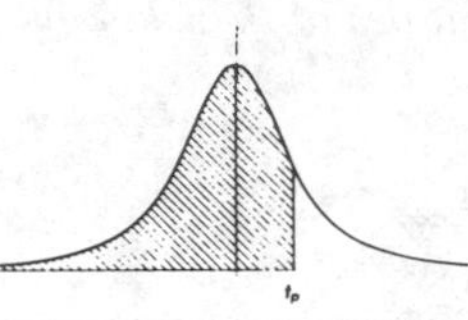


| Degrees of Freedom | $t_{.55}$ | $t_{.60}$ | $t_{.65}$ | $t_{.70}$ | $t_{.75}$ | $t_{.80}$ | $t_{.85}$ | $t_{.90}$ | $t_{.95}$ | $t_{.975}$ | $t_{.99}$ | $t_{.995}$ | $t_{.9995}$ |
|---|---|---|---|---|---|---|---|---|---|---|---|---|---|
| 1 | .158 | .325 | .510 | .727 | 1.00 | 1.38 | 1.96 | 3.08 | 6.31 | 12.7 | 31.8 | 63.7 | 637 |
| 2 | .142 | .289 | .445 | .617 | .816 | 1.06 | 1.39 | 1.89 | 2.92 | 4.30 | 6.96 | 9.92 | 31.6 |
| 3 | .137 | .277 | .424 | .584 | .765 | .978 | 1.25 | 1.64 | 2.35 | 3.18 | 4.54 | 5.84 | 12.9 |
| 4 | .134 | .271 | .414 | .569 | .741 | .941 | 1.19 | 1.53 | 2.13 | 2.78 | 3.75 | 4.60 | 8.61 |
| 5 | .132 | .267 | .408 | .559 | .727 | .920 | 1.16 | 1.48 | 2.01 | 2.57 | 3.36 | 4.03 | 6.86 |
| 6 | .131 | .265 | .404 | .553 | .718 | .906 | 1.13 | 1.44 | 1.94 | 2.45 | 3.14 | 3.71 | 5.96 |
| 7 | .130 | .263 | .402 | .549 | .711 | .896 | 1.12 | 1.42 | 1.90 | 2.36 | 3.00 | 3.50 | 5.40 |
| 8 | .130 | .262 | .399 | .546 | .706 | .889 | 1.11 | 1.40 | 1.86 | 2.31 | 2.90 | 3.36 | 5.04 |
| 9 | .129 | .261 | .398 | .543 | .703 | .883 | 1.10 | 1.38 | 1.83 | 2.26 | 2.82 | 3.25 | 4.78 |
| 10 | .129 | .260 | .397 | .542 | .700 | .879 | 1.09 | 1.37 | 1.81 | 2.23 | 2.76 | 3.17 | 4.59 |
| 11 | .129 | .260 | .396 | .540 | .697 | .876 | 1.09 | 1.36 | 1.80 | 2.20 | 2.72 | 3.11 | 4.44 |
| 12 | .128 | .259 | .395 | .539 | .695 | .873 | 1.08 | 1.36 | 1.78 | 2.18 | 2.68 | 3.06 | 4.32 |
| 13 | .128 | .259 | .394 | .538 | .694 | .870 | 1.08 | 1.35 | 1.77 | 2.16 | 2.65 | 3.01 | 4.22 |
| 14 | .128 | .258 | .393 | .537 | .692 | .868 | 1.08 | 1.34 | 1.76 | 2.14 | 2.62 | 2.98 | 4.14 |
| 15 | .128 | .258 | .393 | .536 | .691 | .866 | 1.07 | 1.34 | 1.75 | 2.13 | 2.60 | 2.95 | 4.07 |
| 16 | .128 | .258 | .392 | .535 | .690 | .865 | 1.07 | 1.34 | 1.75 | 2.12 | 2.58 | 2.92 | 4.02 |
| 17 | .128 | .257 | .392 | .534 | .689 | .863 | 1.07 | 1.33 | 1.74 | 2.11 | 2.57 | 2.90 | 3.96 |
| 18 | .127 | .257 | .392 | .534 | .688 | .862 | 1.07 | 1.33 | 1.73 | 2.10 | 2.55 | 2.88 | 3.92 |
| 19 | .127 | .257 | .391 | .533 | .688 | .861 | 1.07 | 1.33 | 1.73 | 2.09 | 2.54 | 2.86 | 3.88 |
| 20 | .127 | .257 | .391 | .533 | .687 | .860 | 1.06 | 1.32 | 1.72 | 2.09 | 2.53 | 2.84 | 3.85 |
| 21 | .127 | .257 | .391 | .532 | .686 | .859 | 1.06 | 1.32 | 1.72 | 2.08 | 2.52 | 2.83 | 3.82 |
| 22 | .127 | .256 | .390 | .532 | .686 | .858 | 1.06 | 1.32 | 1.72 | 2.07 | 2.51 | 2.82 | 3.79 |
| 23 | .127 | .256 | .390 | .532 | .685 | .858 | 1.06 | 1.32 | 1.71 | 2.07 | 2.50 | 2.81 | 3.77 |
| 24 | .127 | .256 | .390 | .531 | .685 | .857 | 1.06 | 1.32 | 1.71 | 2.06 | 2.49 | 2.80 | 3.74 |
| 25 | .127 | .256 | .390 | .531 | .684 | .856 | 1.06 | 1.32 | 1.71 | 2.06 | 2.48 | 2.79 | 3.72 |
| 26 | .127 | .256 | .390 | .531 | .684 | .856 | 1.06 | 1.32 | 1.70 | 2.06 | 2.48 | 2.78 | 3.71 |
| 27 | .127 | .256 | .389 | .531 | .684 | .855 | 1.06 | 1.31 | 1.70 | 2.05 | 2.47 | 2.77 | 3.69 |
| 28 | .127 | .256 | .389 | .530 | .683 | .855 | 1.06 | 1.31 | 1.70 | 2.05 | 2.47 | 2.76 | 3.67 |
| 29 | .127 | .256 | .389 | .530 | .683 | .854 | 1.05 | 1.31 | 1.70 | 2.04 | 2.46 | 2.76 | 3.66 |
| 30 | .127 | .256 | .389 | .530 | .683 | .854 | 1.05 | 1.31 | 1.70 | 2.04 | 2.46 | 2.75 | 3.65 |
| ∞ | .126 | .253 | .385 | .524 | .674 | .842 | 1.04 | 1.28 | 1.64 | 1.96 | 2.33 | 2.58 | 3.29 |

## Confidence Belts for Proportions
(Confidence coefficient .80)

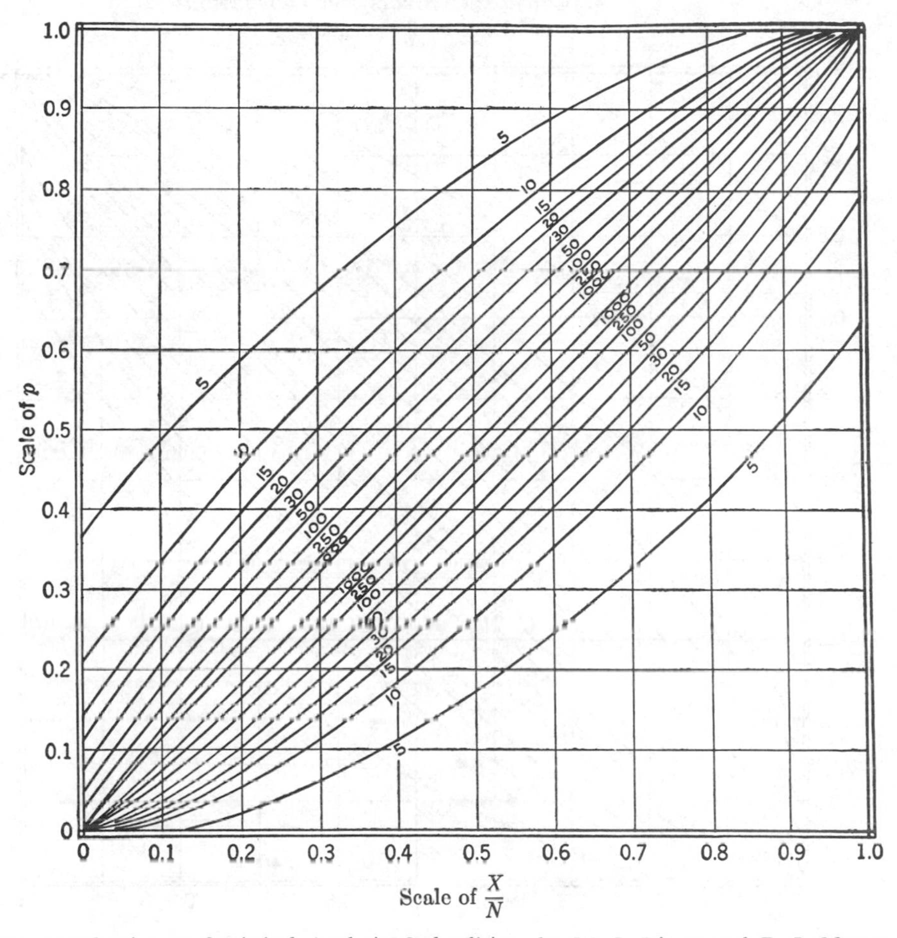


From Introduction to Statistical Analysis, 2nd edition, by W. J. Dixon and F. J. Massey, Jr.

Confidence Belts for Proportions
(Confidence coefficient .90)

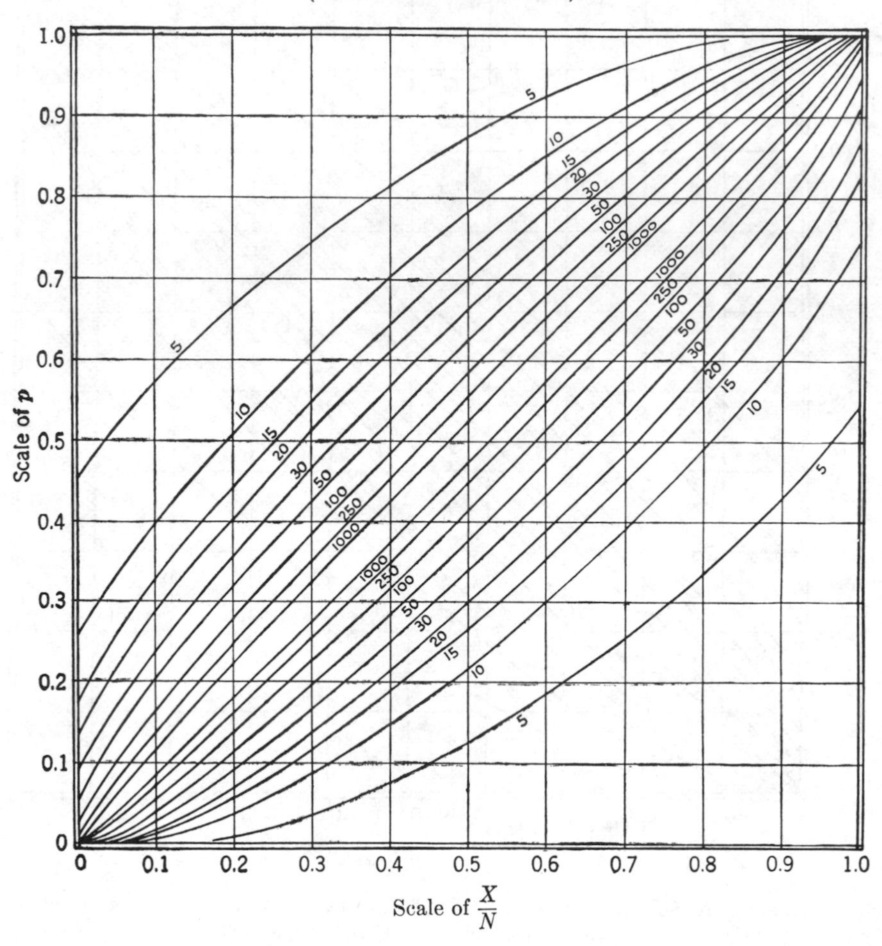

CONFIDENCE BELTS FOR PROPORTIONS*

(Confidence coefficient .95)

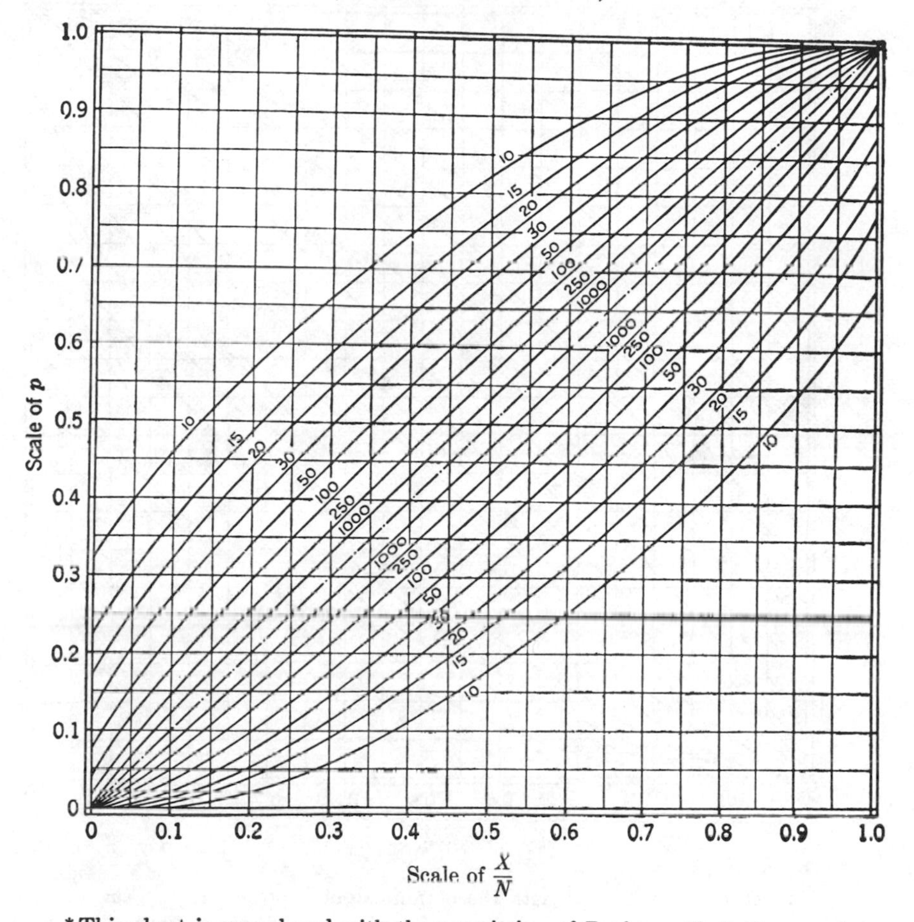


*This chart is reproduced with the permission of Professor E. S. Pearson from C. J. Clopper, E. S. Pearson, "The use of confidence or fiducial limits illustrated in the case of the binomial," *Biometrika*, vol. 26 (1934), p. 404.

Confidence Belts for Proportions*
(Confidence coefficient .99)

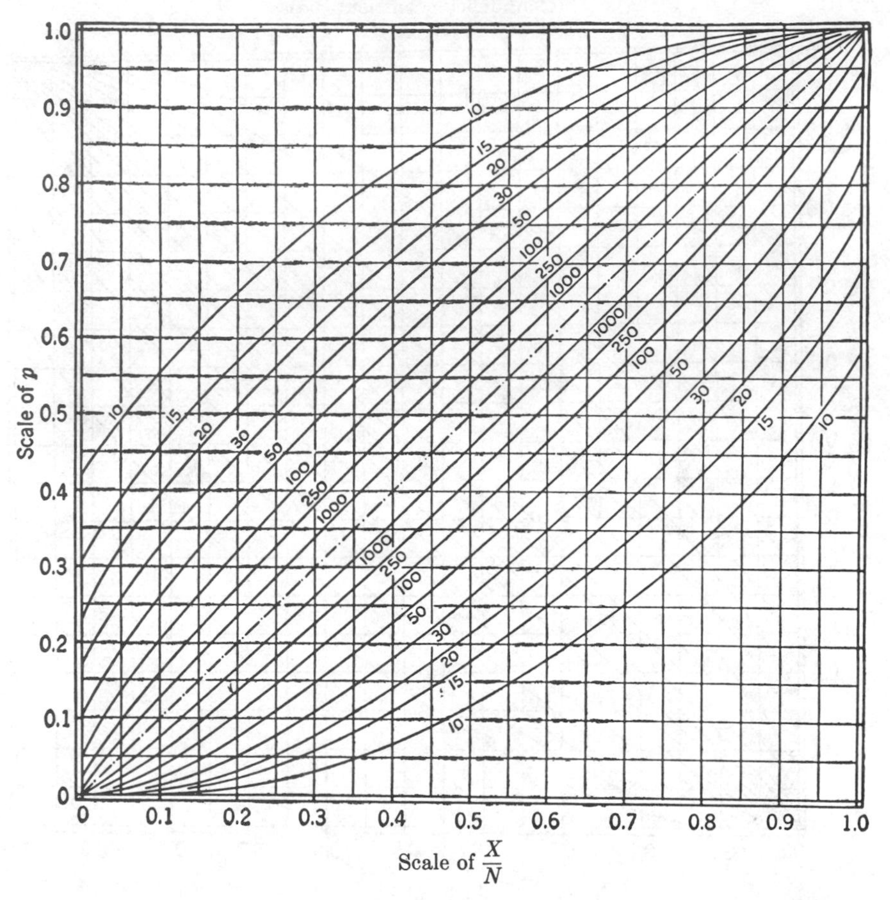


*This chart is reproduced with the permission of Professor E. S. Pearson from C. J. Clopper, E. S. Pearson, "The use of confidence or fiducial limits illustrated in the case of the binomial," *Biometrika,* vol. 26 (1934), p. 404.

## CONFIDENCE BELTS FOR THE CORRELATION COEFFICIENT*
(Confidence coefficient .95)

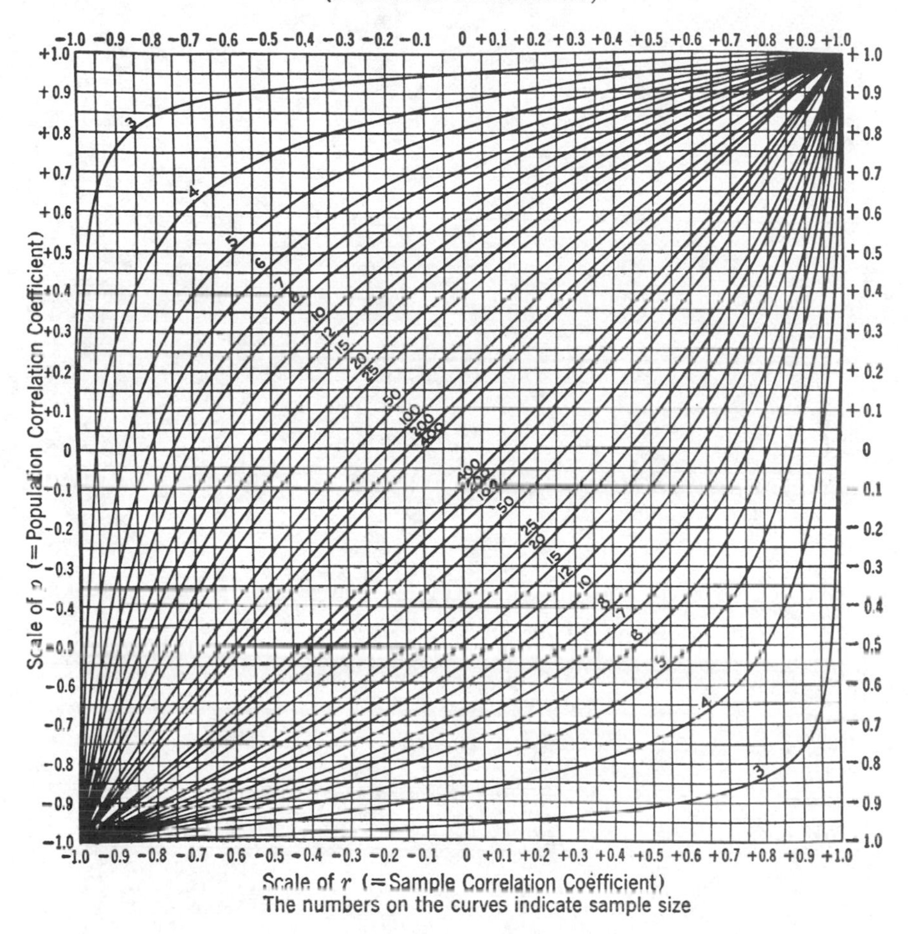


* This chart is reproduced with the permission of Professor E. S. Pearson from F. N. David, *Tables of the Ordinates and Probability Integral of the Distribution of the Correlation Coefficient in Small Samples.* The Biometrika Office, London.

# Index of Special Topics

The topics are numbered according to the Chapter and Examples or Exercises in which they occur. Thus IX.4.2 indicates Example 4.2 (or Theorem 4.2) in Chapter 9, while XII.3.4.1 would indicate Exercise 1 of Exercise or Problem set 3.4 in Chapter 12.

See also the general index, which begins on page 545.

# Subject Index

*Note:* See also Index of Special Topics on page 543 for specific examples and models of importance.